Electronic Circuit Analysis and Design

Electronic Circuit Analysis and Design

Second Edition

William H. Hayt, Jr.
Gerold W. Neudeck

Purdue University

Houghton Mifflin Company · Boston

Atlanta Dallas Geneva, Illinois
Hopewell, New Jersey Palo Alto

Printed in the U.S.A.

Library of Congress Catalog Card Number 83-80943

ISBN: 0–395–32616–8

ABCDEFGHIJ-H-89876543

The following data sheets are reprinted by permission: FD600, used with permission of Fairchild Camera and Instrument Corporation; AD547, courtesy of Analog Devices; LM308, courtesy of National Semiconductor Corporation; HA-2107/2207/2307, courtesy of Harris Intertype Corporation; 2N3823, courtesy of Texas Instruments Incorporated; HP5082-4487, 5082-4488, courtesy of Hewlett Packard; 2N5088, 2N5089, courtesy of Motorola Incorporated; 2N5376, 2N5377, courtesy of Sprague Electric Company.

To our families:

Marly, Peg, and David
Marielle n, Philip, and Alexander

Contents

Preface

One of the subjects in which beginning electrical engineering students profess a strong interest is electronics. This book serves as a basis for a course in electronics that a sophomore or junior will find appealing, motivational, and useful. The course should be preceded by an introductory course in circuit analysis. Students majoring in other disciplines should also find this introduction to electronics and electronic design interesting, helpful, and well within the range of their academic abilities.

We hope that students will find this text useful for individual study, whether at home or in a more formal self-paced course. New topics are introduced with a minimum of theory and derivation. Whenever possible, the subjects are related to familiar material. The examples far outnumber the proofs. As we introduce each new circuit, we analyze it, then give a numerical example to illustrate the procedure and give a feeling for the range of practical values of the parameters. This usually leads to a design example, sometimes to an analysis of the design, and occasionally to a redesign.

Most sections end with one or two drill problems that stress the topics just discussed and provide immediate reinforcement of the material. Answers to all parts of each drill problem are given, in order, immediately after the problem. Problems of a more extensive and challenging nature appear at the end of each chapter. They are presented in the same general order as the text material, and include many design problems. Several problems extend the theory or involve circuits not covered explicitly in the chapter.

Although design problems, unlike analysis problems, usually do not have unique answers, their practical nature has shown them to be highly motivational. Most design problems do not, of course, have a single "best" solution, so our answers are not necessarily any better or worse than yours. For a few design problems, we give two possible solutions.

Answers to odd-numbered problems are given at the back of the book.

The best way to learn electronic circuit design is by designing circuits—over and over again. It's almost as hard to become adept at designing circuits by merely reading a book about the subject as it is to become a good swimmer or tennis player by reading books about swimming or tennis. Working problems is the recommended way to practice circuit design.

Engineers earn their salaries by solving problems. Thus we offer a large selection of problems as the sure road to success.

In this second edition, our basic philosophy of presenting a student-oriented text remains unchanged. Rather than covering a vast number of topics in an encyclopedic fashion, we have chosen to emphasize those basic concepts underlying the analysis and design of all discrete and integrated circuits. We have retained those features that users of the previous edition have found desirable. However, we have incorporated a number of changes that should enhance the book's usefulness. One of the more pertinent changes is the inclusion of integrated-circuit components and configurations throughout the text. Many of the examples and all of the problems are new.

Chapter 1 has been almost completely rewritten. We have expanded the treatment of semiconductor fundamentals slightly to enhance the understanding of the junction diode (and, in a later chapter, the understanding of the differences in the various regions of operation of a transistor, and the current and voltage polarities involved). The concept of the load line is also introduced here. We have added new material on power supplies and Schottky barrier diodes. Useful models for diodes in avalanche breakdown and cutoff also appear for the first time.

Chapter 2 has been revised to emphasize IEEE standard notation for transistor currents and voltages. A simple mathematical treatment is offered for the FET operating in both the ohmic region and in saturation. We have also expanded and modernized the section on integrated-circuit resistors.

Chapter 3 presents dc models for various types of transistors and then uses them to predict circuit operation. We have included a more accurate linear model for the bipolar transistor, and have added load lines as a conceptual tool, although we have not used them as a technique for graphical analysis.

In Chapter 4 we have extended biasing techniques to include integrated-circuit methods, particularly the current mirror. Linear models for FETs have been deleted.

Chapters 5 and 6 have not been substantially changed, although some material has been rewritten to reflect both improved models and the integrated circuits introduced earlier. The concept of the Early voltage is also included for the first time.

Chapter 7 has been rewritten completely to give an improved treatment of frequency response. The high-frequency response of the FET appears first, and the approach is made without specific application of Bode diagrams or complex frequencies. We emphasize the use of open-circuit time constants and establish them as a concept that is used in all calculations for ω_H. We also present a more complete treatment of common-base and common-gate frequency response.

Except for the inclusion of IC circuits, Chapter 8 is basically unchanged. More emphasis is placed on the determination of the high-frequency response.

Chapter 9 is a new chapter, devoted to the operational amplifier and its specifications. We use the concept of the ideal op-amp to introduce the basic circuits, then consider a real op-amp from the viewpoint of its data sheets. The internal circuits of the op-amp are not emphasized. We present basic inverting and noninverting amplifiers, as well as voltage followers.

Chapter 10 is also new. It illustrates the use of the op-amp, with a number of its important applications, including voltage references, comparators, differential amplifiers, bridge amplifiers, differentiators, and integrators. It also covers frequency response and system stability.

We would like to thank the many students, instructors, and engineers who have suggested changes that might be made in this edition. We also wish to thank the reviewers of the second edition: Michael E. McKaughan, U.S. Coast Guard Academy; Paul D. Smith, Rose-Hulman Institute of Technology; and Jacek M. Zurada, University of Louisville. All their comments have been extremely valuable.

To new students, we wish you enjoyment as you design your first electronic circuits. And we assure you that your confidence and competence will grow rapidly together.

William H. Hayt, Jr.
Gerold W. Neudeck

List of Symbols

1

Diodes, diode models, and applications

This chapter has two important objectives. The first follows directly from the chapter title, the introduction of the semiconductor diode as an electronic device.

The second objective is more subtle, as it introduces several analysis and design techniques that we shall apply later to every electronic device that we consider. If we regard the diode as our first example of a general electronic device, then we can view this chapter as a sneak preview of the entire book.

Our study of an electronic device will generally begin with a qualitative description of that semiconductor behavior on which the operation of the element is based. This is followed by a discussion of a voltage–current (*V–I*) characteristic (or characteristics) illustrating the unit's response to dc voltages and currents applied at the external terminals. The *V–I* characteristics are then used to develop a nonlinear dc model. This in turn guides us in establishing models based on a piecewise-linear approximation. These latter models are usually valid only within certain operating ranges of voltage and current. We use both the nonlinear and piecewise-linear dc models to analyze circuits, whereby we find values of the currents and voltages, or determine the operating point, as well as to design the circuits that will establish the desired values of current and voltage.

We then consider the ac or signal model; we shall develop models that are applicable to small signal amplitudes at either low, medium, or high frequencies.

After we have considered both ac and dc models, our final task is to join them in a number of practical examples of signal processing. When we are finished with this chapter, we should be able to set up a dc model and find the dc operating point, determine an appropriate small-signal model at that operating point, and calculate the signal response.

Some skills in the design process should also be present. Given a desired signal response, we should be able to select a suitable circuit configuration, establish values for the circuit elements that will yield an acceptable operating point, and finally check the design by analysis to make certain that it satisfies the required specifications.

We begin in this chapter by taking a look at the circuit aspects of one of the simplest electronic devices: the two-terminal semiconductor diode. We devote only a few paragraphs to the solid-state physics of the diode operation, although we shall make free use of the results of the theory whenever it enables us to analyze or design circuits more intelligently. In neglecting the solid-state physics now, we certainly do not want to minimize its importance. We want only to focus our attention on *circuits* first. We will begin with a rather qualitative description of semiconductors, followed by a few words describing the major phenomena occurring in diode operation. We shall then use these relatively few concepts to describe the circuit models for the diode and, in later chapters, for other semiconductor devices.

There are a very large number of different types of semiconductor devices that are useful in electronic circuitry. We shall have to focus our attention on the most common and most representative types—*diodes* and *transistors*.

1.1 The semiconductor diode

Diode manufacture begins with the growth of a very pure crystal of semiconductor material. Today's diodes are made almost exclusively from silicon (Si) for reasons that involve improved or more convenient methods of fabrication, leading to superior diode characteristics. Pure silicon in itself is neither a good insulator nor a good conductor, as the prefix *semi-* implies. At room temperature its conductivity is about 10^{-10} times that of a metallic conductor and about 10^{14} times that of a good insulator.

However, it is quite easy to increase the conductivity of a semiconductor such as silicon. For example, it may be heated to provide a greater number of charge carriers, causing the conductivity to increase roughly by a factor of ten for every 25°C increase in temperature. The number of free charges within the crystal may also be increased by shining light on the semiconductor material so that the radiant energy can create more of the charge carriers. Although it is evident that these mechanisms may be useful in measuring or controlling temperature and illumination, sensitivity to heat and light are not usually desirable characteristics in, say, a good hi-fi system.

At room temperature, a pure (intrinsic) crystal of silicon has about 5×10^{16} atoms per cubic meter, where about 10^4 electrons per cubic meter are available for conduction, or one electron for every 5×10^{12} silicon atoms. Compared to a metallic conductor in which there is at least one conduction electron for each atom, it is apparent that silicon is not a very good conductor. To increase its conductivity, the number of conduction electrons must be increased.

The most effective and convenient method of providing the desired semiconductor conductivity is through the carefully controlled introduction of impurities, a process known as *doping*. The addition of an n-type impurity, such as antimony or phosphorus, increases the number of *negative* charge carriers, or electrons, in the semiconductor. We find that when one atom in ten million is replaced by an n-type impurity atom, the conductivity of silicon at room temperature increases by a factor of about 10^5. Conductivity increases with the level of the doping, being closely proportional to the number of impurity atoms per unit volume at typical doping levels.

A semiconductor is unusual in that electrons are not the only charge carriers present. Even in intrinsic silicon there are positively charged carriers, called *holes*, that contribute to the conductivity. For pure material, the number of electrons is equal to the number of holes. At room temperature, intrinsic silicon has about 10^4 holes per cubic meter that are available for electrical conduction.

The number of holes can be increased by adding a p-type impurity, such as boron or aluminum, to the silicon crystal. For example, if we add 10^9 boron atoms per cubic meter to the 5×10^{16} silicon atoms per cubic meter, the concentration of holes increases to about 10^9 holes per cubic meter, an increase of five orders of magnitude from the intrinsic value of 10^4 holes per cubic meter. Doping a semiconductor with a p-type impurity has almost the same effect on the conductivity as does the addition of an equal number of atoms of n-type material. As a matter of fact, if equal amounts of n- and p-type impurities are added to a pure semiconductor, there is no appreciable change in conductivity, for the two added types of charge carriers recombine with each other.

A silicon semiconductor junction diode may be produced in a single crystal by introducing p-type impurities in one half and n-type in the other, as suggested by Fig. 1.1*a*. So-called ohmic contacts are provided at the external ends of the two regions to permit low-resistance connections to the external circuitry. Figure 1.1*b* shows the circuit symbol for a semiconductor diode as it is connected to a dc source V through a current-limiting resistor R. Either R or V may be varied to compile data for a V_D-I_D characteristic, such as that shown in Fig. 1.2. The parts of this illustration show several different aspects of the diode characteristic. In Fig. 1.2*a* we can see that large currents may be obtained by connecting the positive terminal of the source to the p-type material, while the negative terminal goes to the n-type material. This large current is called the *forward* current. It results from the application of *forward bias*. This current is said to flow in the *easy* direction.

The forward current is composed of two components, the movements of holes from the p side to the n side of the junction, and of electrons moving in the opposite direction, as illustrated in Fig. 1.1*c*. Their effects are addi-

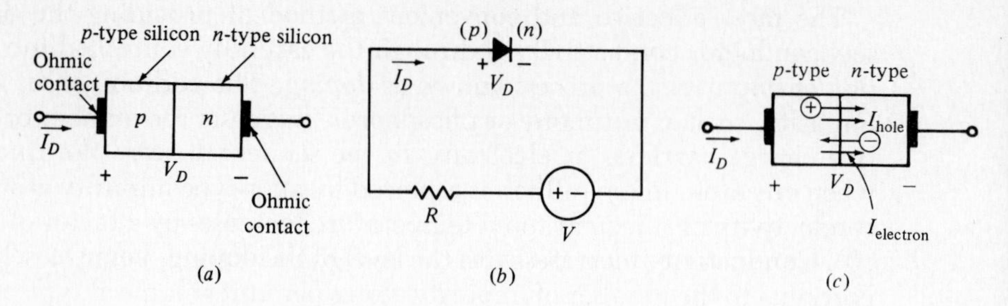

Fig. 1.1 (*a*) The semiconductor diode contains a planar junction between *p*- and *n*-type semiconductor materials. (*b*) When the positive terminal of the source V is connected to the *p*-type material, the diode is forward-biased and I_D is relatively large. The arrow or triangle on the circuit symbol for the semiconductor diode indicates the direction of forward current. (*c*) The diode current I_D has both hole and electron components.

tive, for each is equivalent to a conventional current entering the positive terminal of the diode. If the doping levels are the same on the two sides of the junction, the two components of the current are essentially equal. Many diodes are produced by methods that lead to unequal doping levels, and therefore the current is primarily provided by the movement of the more abundant carrier.

Figure 1.2*b* shows the diode characteristic in the neighborhood of the origin. Note that the diode-current scale is now labeled in nanoamperes (10^{-9} A).

When the diode is *reverse-biased*—or V_D is negative so that the positive-source voltage is connected to the *n*-type material—then a very small *reverse* current flows. When the reverse bias is greater than a few hundredths of a volt, the reverse current remains quite constant at a level known as the *reverse saturation current*, $I_D = -I_0$, as shown in Fig. 1.2*b*. I_0 is taken as a positive quantity.

As the reverse bias becomes much larger, the reverse current begins to increase somewhat, as shown in the region between -50 and -100 V in Fig. 1.2*c*. This is a result of surface leakage around the *p-n* junction, and it follows Ohm's law to a first approximation. At a sufficiently large voltage, the reverse current increases abruptly with only a small change in reverse voltage. Here the electric field in the junction region is strong enough to ionize the semiconductor atoms and produce an avalanche of additional charge carriers. The phenomenon is called *Zener*, or *avalanche, breakdown*, and the magnitude of the voltage at which it occurs is called the *breakdown voltage* V_{BR}. This value is a function of the doping levels in the *p*- and *n*-type regions.

Except in the high-forward-current region, where ohmic voltage drops in the semiconductor material and the contacts become appreciable, and

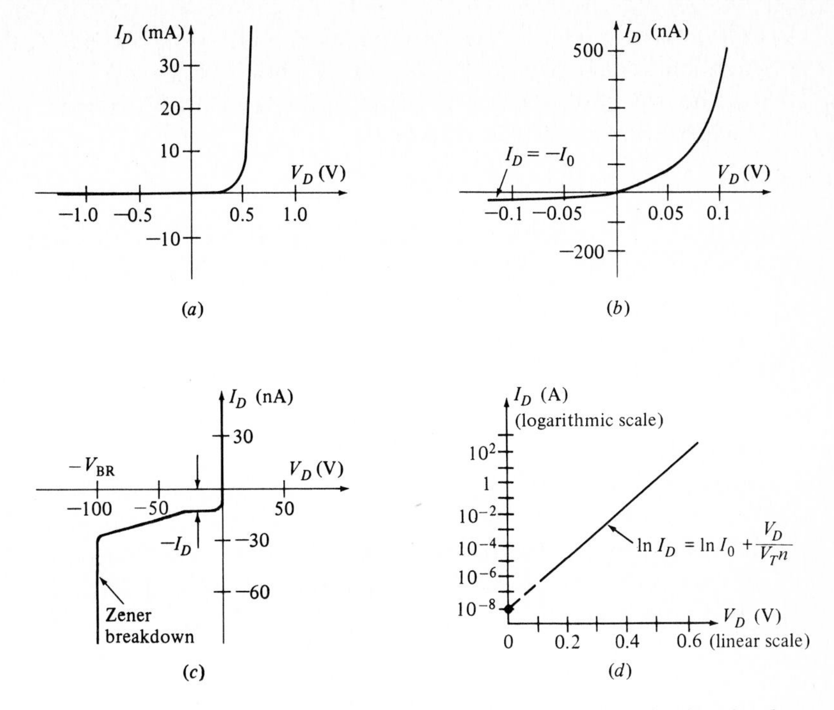

Fig. 1.2 (*a*) The diode $V_D - I_D$ characteristic with scales that display the sharp increase in current when the forward bias is a few tenths of a volt. (*b*) The $V_D - I_D$ characteristic near the origin shows an exponential increase in forward current and the asymptotic approach to the reverse saturation current. (*c*) For larger reverse voltages, we see an increasing surface leakage current and then Zener breakdown at the breakdown voltage V_{BR}. (*d*) A semilog plot of I_D for $V_D > 0.1$ V.

in the Zener breakdown region, the volt-ampere characteristics of Fig. 1.2 obey the Shockley diode equation,

$$I_D = I_0(e^{qV_D/nkT} - 1) \tag{1}$$

Here q is the electronic charge, 1.6022×10^{-19} C; k is Boltzmann's constant, 1.3806×10^{-23} J/K; I_0 is the reverse saturation current, typically between 10^{-16} A and 1 μA; T is the temperature in kelvins (K); and n is a dimensionless factor that is theoretically unity, but ranges from one to two in real diodes. We shall usually let $n = 1$, except on several occasions when we wish to model a particular diode accurately.

At standard temperature (25°C, 77°F, 298.16 K), $q/kT = 38.92\ \text{V}^{-1}$, and for this temperature we have

$$I_D = I_0(e^{38.92V_D/n} - 1) \qquad \text{(at 25°C)} \tag{2}$$

The curves of Fig. 1.2 are plotted for $I_0 = 10$ nA, $T = 298$ K, and $n = 1$; they follow Eq. (2) for $V_D > -30$ V and $I_D < 10$ mA.

The exponent in Eq. (1) or Eq. (2) is often simplified by recognizing the reciprocal of q/kT as a voltage,

$$V_T = \frac{kT}{q} \tag{3}$$

This voltage is a function only of temperature; it is called the *thermal voltage*. Thus we may write

$$I_D = I_0(e^{V_D/V_T n} - 1) \tag{4}$$

At 25°C, $V_T = 25.69$ mV, and we have

$$I_D = I_0(e^{V_D/0.02569n} - 1) \tag{5}$$

It is often possible to neglect the unity term in the parentheses of Eqs. (1), (2), (4), or (5) when the diode is sufficiently forward-biased. Thus, if we let $n = 1$, then a diode voltage $V_D = 0.1$ V leads to an exponential term of $e^{3.892}$, or 49.01. Therefore, if we can tolerate about a 2% maximum error, we may simplify the diode equation by neglecting -1:

$$I_D \doteq I_0 e^{V_D/V_T n} \qquad (\text{for } V_D > 0.1 \text{ V}) \tag{6}$$

It should also be apparent that for reverse bias with $V_D < -0.1$ V, the exponential term can be neglected compared to unity, and therefore

$$I_D \doteq -I_0 \qquad (\text{for } V_D < -0.1 \text{ V}) \tag{7}$$

Figure 1.2*d* illustrates the forward diode characteristic as it is usually plotted on manufacturers' data sheets. Since I_D may vary over a range from picoamperes to amperes, while V_D varies less than one volt, the semilog plot is the most convenient. Taking the natural logarithm of Eq. (6) yields

$$\ln I_D = \ln I_0 + \frac{V_D}{V_T n} \qquad (V_D > 0.1 \text{ V}) \tag{8}$$

This is a straight line on semilog paper. The intercept at $V_D = 0$ (where our approximate relationship is not valid) is $\ln I_D = \ln I_0$, or $I_D = I_0$, whereas the correct value is $I_D = 0$. The plot, however, is useful in that it enables us to see the value of the reverse saturation current readily.

Finally, the slope of the straight line is $1/V_T n$. This value may be found from two representative points on the straight line,

$$\frac{1}{V_T n} = \frac{\ln I_{D1} - \ln I_{D2}}{V_{D1} - V_{D2}} = \frac{\ln (I_{D1}/I_{D2})}{V_{D1} - V_{D2}} \tag{9}$$

D1.1 Let a diode operate at 25°C with a reverse saturation current of 1 fA (1 femtoampere $= 10^{-15}$ A) and $n = 1.1$. Find (a) I_D when $V_D = -12$ V, (b) I_D when $V_D = -0.1$ V, (c) V_D when $I_D = 10$ mA.

Answers. −1.000 fA; −0.971 fA; 0.846 V

D1.2 Experimental data at 25°C indicate that $I_D = 1\ \mu$A at $V_D = 0.59$ V and $I_D = 1$ mA at $V_D = 0.82$ V. Determine (a) n, (b) I_0, (c) I_D at $V_D = 0.6$ V.

Answers. 1.296; 20.2 fA; 1.350 μA

1.2 Diode circuit models: dc

Before we develop our first model for a semiconductor diode, it may be profitable to philosophize for a moment about models in general. What are they used for? What forms do they take? Which one should we select? The answers to these questions are important, for the accuracy of a model, the conditions under which it is applicable, and the frequency or signal-amplitude restrictions on its use are crucial considerations. In designing or analyzing an electronic circuit, we must choose the proper model for our particular problem. This does not mean that there is exactly one correct model, but rather that there is no model that is absolutely accurate. We have to select from a range of models, some of which are more appropriate than others. The choice is governed by the amount of information available on which to base our design or analysis, on the accuracy we require, and on the time we are able to devote to the solution. For a student, of course, time may be unlimited, but when employers are paying a few thousand dollars a month, they may place some severe time restrictions on the problem. In general, the most accurate models are detailed and complicated; solutions are apt to be time-consuming and difficult.

Figure 1.3 shows a broad classification of models for electronic devices. The basic subdivision is between the dc model, which is useful not only in operating at direct current but also in determining the operating point in a circuit where a signal is also present, and the ac model, which applies only to signal analysis. The ac model is used to determine such signal-processing characteristics as voltage gain, current gain, power gain, input impedance, output impedance, voltage isolation, and so forth. Further subdivision is based on the use of linear models and linear mathematical analysis, or nonlinear models and nonlinear techniques, including graphical analysis. Models are also developed for various frequency and signal-amplitude ranges. Although we could make use of a single model that would be applicable under all these conditions, it would probably involve

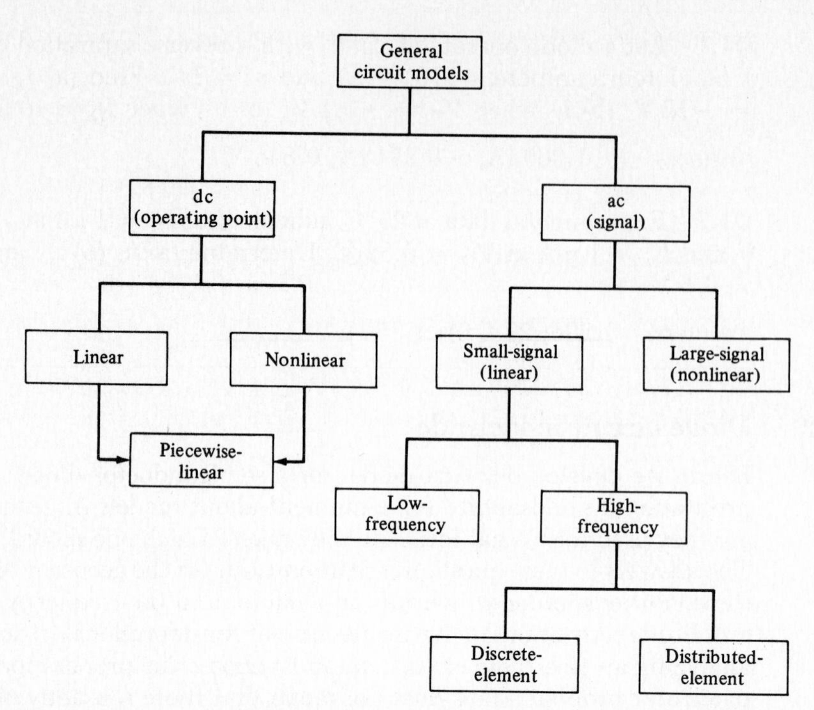

Fig. 1.3 Circuit models for electronic devices are separated into dc models, used to determine operating points, and ac models, which involve only the signal. Other subdivisions are linear and nonlinear, small-signal and large-signal, low-frequency and high-frequency, and discrete- and distributed-element.

a large number of linear elements, a few *v-i* characteristics, an equation or two, several idealized nonlinear elements, and a bottle of aspirin. It is possible to approximate such a general approach by using the storage capacity and the computational abilities of a digital computer, as we shall see later, but there are many problems for which it is not worthwhile bringing all this power into play.

Now let's talk about models for the semiconductor diode. We already have two: the V_D-I_D characteristic of Fig. 1.2; and Eq. (1) of Section 1.1. Both are valid at dc, although Eq. (1) is restricted to the smaller voltage and current amplitudes. These two models may be thought of as dc nonlinear models for the semiconductor diode. In using them we must use either the generally difficult techniques for solving nonlinear equations (such as trial-and-error and numerical iteration, for example) or graphical methods that are often inconvenient or inaccurate or both.

Linear equations are obtained from linear circuits, which are much easier to solve. As a result, we may prefer to use a somewhat inaccurate linear model of a device because it leads to equations that are much more tractable. Beware, however, of the tendency to obtain four significant figures from a set of linear equations applying to a linear model that represents a circuit to no better accuracy than one significant figure.

The choice of model is important, and it is a matter of engineering judgment.

There are several useful linear models for the semiconductor diode. The dc models are *piecewise-linear* models. That is, they approximate the v-i characteristic of Eq. (1) by a series of *straight-line segments.* Figure 1.4 shows four piecewise-linear models. The simplest is the *ideal-diode* model, shown in Fig. 1.4*a*. Here the forward voltage is zero for all values of forward current, and the reverse current is zero for all values of negative voltage. The diode is a short circuit for forward current and an open circuit for reverse current. This very simple model appears quite often, both as a model of the semiconductor diode per se and with added linear elements that provide improved models. The circuit symbol that we shall use for the ideal diode is the same as that for the semiconductor diode, except that the word *ideal* appears adjacent to the symbol.

A somewhat more accurate model is shown in Fig. 1.4*b*. The resistor R_0 enables the forward voltage to take on nonzero values, and the value selected for R_0 may well depend on the general level of the currents and voltages expected in the circuit. In Fig. 1.4*c*, a battery V_0 is added. In using this model we now have the ideal (nonlinear) diode plus two linear elements. The circuit is becoming complicated, and we are also faced with the problem of specifying appropriate values for the resistor and voltage

Fig. 1.4 Four piecewise-linear models. (*a*) The ideal diode. Note that $V_D I_D = 0$. (*b*) The ideal diode plus a forward resistance R_0. (*c*) The ideal diode with resistor R_0 and breakpoint at $V_D = V_0$. (*d*) A model useful in the region of avalanche (Zener) breakdown, which occurs at the breakpoint where $V_D = -V_{BR}$.

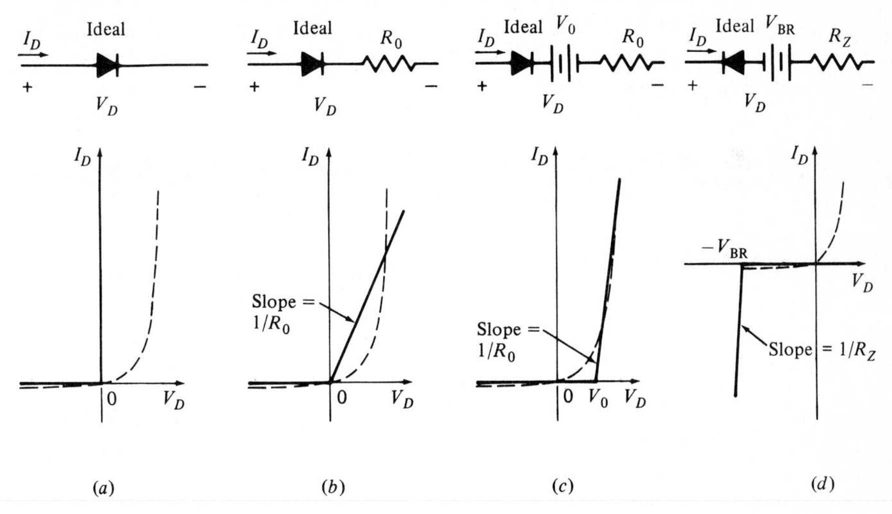

source. Typically, R_0 ranges from 1 to 50 Ω, and V_0 is between 0.4 and 0.7 V for diode currents greater than several milliamperes. Note that the circuit of Fig. 1.4*c* leads to the linear equation

$$V_D = V_0 + I_D R_0 \qquad (\text{for } V_D > V_0) \tag{10}$$

whenever V_D is greater than V_0. This equation is much easier to solve or use in a system of equations than the nonlinear relationship of Eq. (1). Even so, we must always make certain that $V_D > V_0$, usually as a check after the solution is completed.

We first saw the reverse-bias breakdown region in which the current increases very rapidly in Fig. 1.2*c*. This region can be modeled by the circuit shown in Fig. 1.4*d*. Note that the battery has a voltage that is equal to V_{BR} for the diode, and $1/R_Z$ represents the slope of the curve in breakdown. Both the diode and the battery have the opposite polarity in the breakdown model to that in the forward models.

As an example of the use of the piecewise-linear models in finding the values of V_D and I_D at the operating point in a diode circuit, consider Fig. 1.5*a*. Since this is a simple series circuit, the current in each element is I_D, and Kirchhoff's voltage law leads to

$$2 = 100 I_D + V_D \tag{11}$$

Our goal is to determine V_D and I_D, and we shall use four different methods.

Solution 1

We first replace the diode by the simple ideal-diode model of Fig. 1.4*a*; the result appears in Fig. 1.5*b*. Since the polarity of the 2-V source suggests forward bias, then $V_D = 0$, or a short circuit. From Eq. (11), we therefore have $I_D = 20$ mA. Since $I_D > 0$, our assumption of forward bias was correct, and we have an operating point at (0.0 V, 20 mA). If our

Fig. 1.5 (*a*) A simple series circuit containing a semiconductor diode. (*b*) The diode is modeled by an ideal diode. (*c*) A more accurate model for the diode includes $V_0 = 0.7$ V and $R_0 = 5$ Ω.

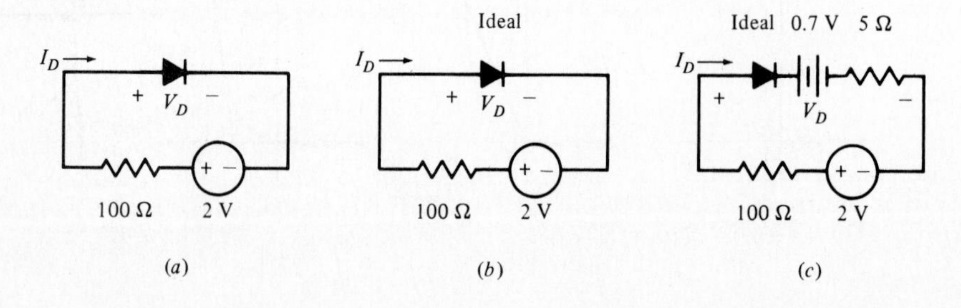

solution for I_D had led to a *negative* value, then it would have been necessary to start over with an assumption of reverse bias.

Solution 2

Let us assume that we need a more accurate solution. It is therefore necessary to use a more complex model, so we first turn to the data sheets for the diode to establish suitable values for V_0 and R_0 in the model of Fig. 1.4*c*. This procedure will be discussed in the following sections, but let us assume for now that $V_0 = 0.7$ V and $R_0 = 5\ \Omega$ are selected. Using these values to model the diode, as shown in Fig. 1.5*c*, we again assume current flow in the easy direction,

$$I_D = \frac{2 - 0.7}{100 + 5} = 12.38 \text{ mA}$$

Thus the diode voltage is

$$V_D = 0.7 + (12.38 \times 10^{-3})\,5 = 0.762 \text{ V}$$

Since both V_D and I_D are positive and $V_D > V_0$, our assumption of forward bias was correct, and we have our new estimate of the operating point (0.762 V, 12.38 mA).

Solution 3

Being desperate for accuracy and fond of algebraic obfuscation, let us try the nonlinear model represented by Eq. (1). The values $n = 1.1$ and $I_0 = 10^{-14}$ A are compatible with our earlier selection of $V_0 = 0.7$ V and $R_0 = 5\ \Omega$. Thus, at 25°C,

$$I_D = 10^{-14}(e^{38.92V_D/1.1} - 1)$$

and

$$2 = 100 \times 10^{-14}(e^{38.92V_D/1.1} - 1) + V_D$$

This is a transcendental equation, and it must be solved by numerical methods. The simplest is probably trial and error, particularly if a programmable calculator is handy. The SOLVE routine also leads to a quick solution. The solution is $V_D = 0.786$ V, and this leads to $I_D = 12.14$ mA. This is the most accurate solution, but it also requires the most work.

Solution 4

Instead of using the nonlinear diode equation, we may also adopt a nonlinear approach that utilizes a graph of I_D vs. V_D, such as that shown in Fig. 1.6. Equation (11) can be plotted on the graph and the intersection of

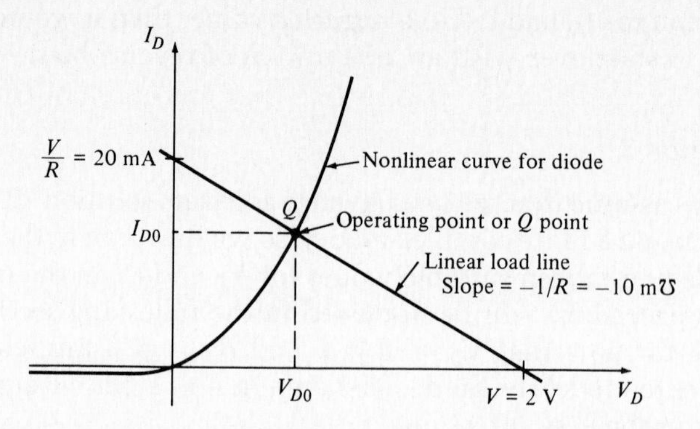

Fig. 1.6 The operating point is the intersection obtained by plotting the linear load line, $2 = 100I_D + V_D$, and the diode V_D vs. I_D curve on the same axes.

this straight line with the V_D-I_D curve for the diode yields the *operating point* or *quiescent point Q*, often called simply the *Q point*. A simple way of plotting the load line is by use of the intercepts:

$$V_D = V = 2\text{ V} \qquad \text{when } I_D = 0$$

and

$$I_D = \frac{V}{R} = 20\text{ mA} \qquad \text{when } V_D = 0$$

Since this is a graphical method, it is rarely used in analyzing circuits. However, it is a valuable conceptual tool. For example, if R were doubled, then the I_D intercept would halve, and both V_D and I_D at the operating point would decrease.

D1.3 (a) Calculate values for R_0 and V_0 in the model of Fig. 1.4*c* so that $I_D = 0.5$ mA for $V_D = 0.6$ V and $I_D = 12$ mA for $V_D = 0.7$ V. (b) How much current will flow if this diode is connected in series with 50 Ω and 1 V, assuming forward bias?

Answers. 8.70 Ω, 0.596 V; 6.89 mA

1.3 Diode circuit models: low-frequency small-signal

The dc models we developed in the previous section serve to establish a dc operating point (V_D, I_D) for the semiconductor diode. Once this has been determined, we may begin to think about the next part of the problem, the application of an ac signal at some point in the circuit. At this time we will consider small or *incremental* signals whose amplitudes are small compared with the dc voltages and currents at the operating point. When the signals are small, a linear signal model is usually sufficient. When the frequency components in the signal are low, the capacitive effects in the diode may be neglected. Therefore we may use a low-frequency small-signal linear model. For the *p-n* junction diode, this model is simply a resistor, which we shall call the *dynamic resistance*.

We begin with the Shockley diode equation, written in Section 1.1 as

$$I_D = I_0(e^{qV_D/nkT} - 1) = I_0(e^{V_D/V_Tn} - 1) \tag{12}$$

The diode current I_D and voltage V_D were written as capital letters with capital-letter subscripts. We shall use this nomenclature consistently to represent dc quantities. Now, however, we are beginning to talk about *signals*, that is, currents and voltages that vary with time. We use i_D and v_D as variables, where the lower-case letter with the capital-letter subscript is reserved exclusively for the *total instantaneous* quantity, the dc plus the signal component. Finally, the *instantaneous signal alone* is represented by a lower-case symbol with a lower-case subscript, such as v_d and i_d. It follows therefore that the following equations may be written:

$$v_D = V_D + v_d \tag{13}$$

$$i_D = I_D + i_d \tag{14}$$

One additional form of nomenclature will be introduced in Section 5.4.

Rewriting Eq. (12) in terms of i_D and v_D, we have

$$i_D = I_0(e^{v_D/V_Tn} - 1) \tag{15}$$

In order to investigate the effects of small variations from the dc values, we first find the slope of the characteristic at the operating point (V_D, I_D),

$$\left.\frac{di_D}{dv_D}\right|_Q = I_0 \frac{1}{V_Tn} e^{v_D/V_Tn}\bigg|_{v_D = V_D} = I_0 \frac{1}{V_Tn} e^{V_D/V_Tn}$$

By solving Eq. (12) for the exponential factor and substituting the result into the equation above, we have the simpler result

$$\left.\frac{di_D}{dv_D}\right|_Q = \frac{1}{V_Tn}(I_D + I_0)$$

The ratio of the current differential to the voltage differential at the operating point is called the *dynamic conductance* g:

$$g = \left.\frac{di_D}{dv_D}\right|_Q = \frac{1}{V_Tn}(I_D + I_0) \tag{16}$$

The *dynamic resistance* r is the inverse ratio, $r = 1/g$:

$$r = \left.\frac{dv_D}{di_D}\right|_Q = \frac{V_Tn}{I_D + I_0} \tag{17}$$

Note that the dynamic resistance and conductance depend on I_D, the dc value of i_D at the operating point. They also depend on V_T and hence on temperature. The fact that the values of the dynamic signal model depend on the dc operating point and on temperature is very important, for it is a concept that affects every signal model for diodes and for transistors.

Figure 1.7 shows an enlarged view of the diode characteristic in the neighborhood of the operating point ($v_D = V_D$, $i_D = I_D$). The slope of the tangent here is g, or $1/r$. When small signals are present, we may use the notation of Eqs. (13) and (14), $v_D = V_D + v_d$ and $i_D = I_D + i_d$, to see that the signals act as small increments in v_D and i_D. Thus the slope may also be expressed as the ratio of these small signals,

$$g = \frac{i_d}{v_d} \tag{18}$$

$$r = \frac{v_d}{i_d} \tag{19}$$

Either of these relationships between the ac components of the voltage and current serves as the complete low-frequency small-signal ac diode model, as shown in Fig. 1.8. Remember that the value of r, or g, depends on the dc operating point selected and on temperature.

Now that we have considered the theoretical problems of the dc and ac models, let us take a slightly more practical approach by trying to determine these models for a diode on which we have the manufacturer's specifications. Figure 1.9 shows the forward characteristic for the FD600 diode. (The complete data sheet for this device appears in Appendix A through the courtesy of the Fairchild Semiconductor Corporation.) Note

Fig. 1.7 The slope of the diode characteristic at the operating point is g or $1/r$. The small signal voltage v_d and small signal current i_d are interpreted as incremental increases in v_D and i_D respectively. Thus $g = 1/r = i_d/v_d$.

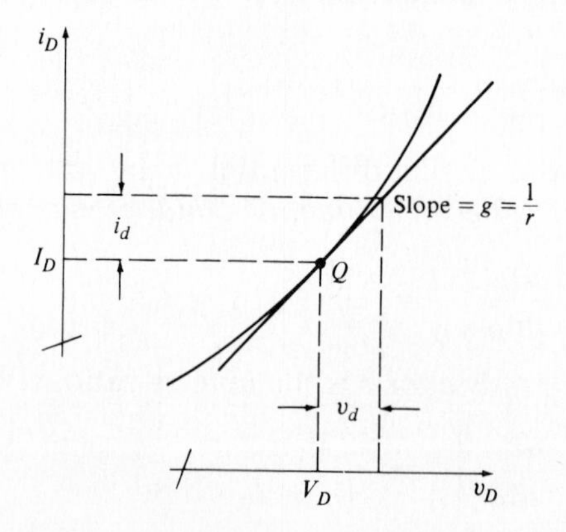

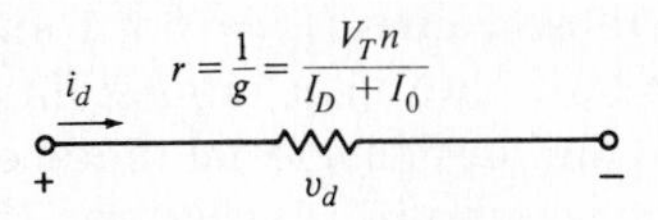

Fig. 1.8 The low-frequency small-signal ac model of the semiconductor diode serves to relate the incremental signal voltage v_d and incremental signal current i_d. The value of r depends on the operating point and on temperature.

that the forward-bias curves are presented on semilog scales similar to those of Fig. 1.2*d*.

We first use the curve for the "typical" diode in Fig. 1.9 to establish values for I_0 and n. With these available, we have a nonlinear model as represented by the diode equation. Let us assume that our model is to be used at an operating point below 10 mA. We therefore select two points on the straight-line portion of the curve, make use of Eq. (9) to determine n, and then find I_0 from Eq. (1), the diode equation. At $V_D = 0.4$ V, we estimate $I_D = 22\ \mu$A, and for $V_D = 0.6$ V, $I_D = 1.3$ mA. Thus,

$$\frac{1}{nV_T} = \frac{\ln (I_{D1}/I_{D2})}{V_{D1} - V_{D2}} = \frac{\ln (1.3/0.022)}{0.6 - 0.4} = 20.4$$

At 25°C, then, $n = 38.92/20.4 = 1.908$, which we may take as $n = 1.9$. To find I_0, we may use either point of the curve. Choosing the larger current, we have

$$1.3 \times 10^{-3} = I_0(e^{38.92 \times 0.6/1.9} - 1)$$

and I_0 is 5.97 nA, which we round off to $I_0 = 6$ nA.

Fig. 1.9 The forward ($V_D - I_D$) characteristics of the FD600 diode as given on the manufacturer's data sheets.

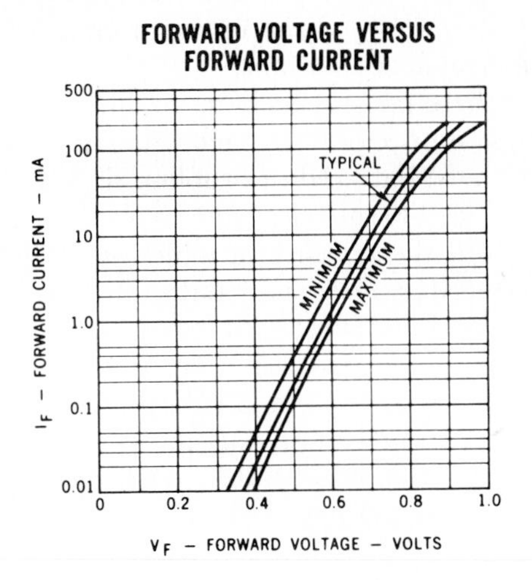

We shall use these values ($I_0 = 6$ nA and $n = 1.9$) in the next section when we obtain both dc and ac models for the FD600 diode, and then use them to analyze and design special diode circuits.

If a more accurate model is necessary, then we might select five or ten pairs of values in the straight-line region of Fig. 1.9 (say $I_D < 10$ mA) and use the linear-regression feature available on many calculators. This would give us the straight line that represents the best least-squares fit to the data.

We should also note that the curves on Fig. 1.9 are not straight lines for $I_D > 10$ mA, or even slightly below that value. This effect may be included by adding a small resistor in series with every diode model to account for ohmic resistances in the diode contacts as well as in the bulk semiconductor material. For the FD600, a series resistance $R_s = 2\ \Omega$ works well, but the improved accuracy this leads to may not be worth the added complexity.

D1.4 Let $I_0 = 2$ pA and $n = 1.2$ for a certain semiconductor diode. (a) Find r at $T = 25°C$ and $I_D = 1$ mA. In order to increase r by 10%, to what value must: (b) T be changed if $I_D = 1$ mA? (c) I_D be changed if $T = 25°C$?

Answers. 30.8 Ω; 54.8°C; 0.909 mA

1.4 Applications of the diode models

In this section we shall consider three examples that involve the use of diode models. In the first example, we must determine an ac model for a given circuit. This model is then used in the second example to find the value of a signal-output voltage. Both these problems have to do with analysis. The third problem has to do with design: We must consider both dc and ac models to design an attenuator circuit that will meet a set of prescribed electrical specifications.

As we look at these progressively more difficult problems, note that we follow a logical sequence of steps each time. The first step is that of *problem identification*. What kind of problem do we have? What kind of information do we need to solve it? Sometimes the type of data we are given or can find governs the manner in which we identify the problem. For example, the complete absence of data on a device would identify our problem as an experimental one, at least at the beginning.

The second step is the establishment of a *plan of attack*. How do we solve the problem? Which method is best? What kind of accuracy is required? We were probably thinking about this step as we identified the problem, and now we should be able to see the sequence of steps that we must take to arrive at the required answer. If several methods are available, we should try to decide which is preferable. Perhaps we will need to start several different methods before we can decide which will be best. And what does "best" mean for our problem? It may mean easiest or quickest if accuracy is not too important, or it may mean the most accurate if we are not concerned with the time it takes.

The last step is the solution itself. If we have done our thinking effectively in the first two steps, we should not find later that we are lost halfway through the problem, or that we are wasting our time in solving some problem that no one is interested in.

Example 1

With these philosophical considerations completed, let's look at the first example, shown in Fig. 1.10*a*. We are asked to find the dynamic resistance of the FD600 diode as it is operated in this circuit at room temperature.

We identify this problem as one in which we have to obtain a value for r, a value that depends on both operating point and temperature. Temperature is fixed at 25°C, so that is no worry here. But we do have to find a dc operating point. Therefore the problem is one of first locating an operating point and then finding r at that point.

As a plan of attack on the operating point, we consider several possibilities. We have data on the FD600, and we have the complete specifications on the circuit in which it is installed. We could write an equation for the linear portion of the circuit and then represent it as a linear load line on the diode characteristic. The intersection would establish the operating point. We might choose to solve the nonlinear Shockley diode equation numerically. Or we could represent the diode by a piecewise-linear dc model and then solve the resulting linear circuit equation. Since we are emphasizing models at this time, let's select this last plan of attack and

Fig. 1.10 (*a*) A circuit containing an FD600 diode whose dynamic resistance is to be determined at 25°C. (*b*) The diode is replaced by a dc model in order to find the operating point. We use $R_0 = 9.3\ \Omega$ and $V_0 = 0.62$ V as the piecewise-linear model.

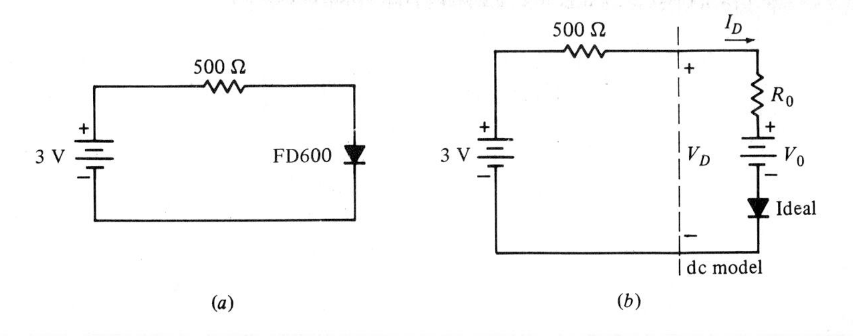

leave nonlinear models for later attention. After we have the operating point located, what is left? A value for the dynamic resistance can be obtained by using an appropriate value for I_D in the simple equation for r, Eq. (17).

To get a linear dc model for the FD600 that will locate an operating point accurately, we select the three-element model, as shown in Fig. 1.10*b*. Values for R_0 and V_0 must be chosen that are valid in the neighborhood of the (as yet unknown) operating point. With the 3-V battery and 500-Ω resistance connected to the diode, it is obvious that the diode current cannot be any larger than 3/500, or 6 mA, the value that would be obtained if the diode were considered to be ideal. With 6 mA through the real diode, Fig. 1.9 indicates that the diode voltage will be about 0.68 V. A better guess at I_D is obtained by assuming 3 − 0.68 = 2.32 V across the 500-Ω resistor, leading to a loop current of 2.32/500, or 4.64 mA. Therefore we are looking for a piecewise-linear model that is useful from perhaps $I_D = 4.5$ to 6 mA.

In Section 1.3, we established $n = 1.9$ and $I_0 = 6$ nA for this diode. We may use the diode equation to find the values of V_D corresponding to $I_D = 4.5$ and 6 mA, obtaining the operating points (0.660 V, 4.5 mA) and (0.674 V, 6 mA). Passing a straight line through these points, we find

$$R_0 = \frac{0.674 - 0.660}{(6 - 4.5)\,10^{-3}} = 9.33\ \Omega$$

$$V_0 = 0.674 - 9.33\,(6 \times 10^{-3}) = 0.618\ \text{V}$$

Let us use $R_0 = 9.3\ \Omega$ and $V_0 = 0.62$ V in the circuit of Fig. 1.10*b*, leading to the operating point

$$I_D = \frac{3 - 0.62}{500 + 9.3} = 4.67\ \text{mA}$$

and

$$V_D = 0.62 + 9.3\,(4.67 \times 10^{-3}) = 0.663\ \text{V}$$

Since the point (0.663 V, 4.67 mA) seems to lie on the FD600 characteristic, this shows that our model was correctly chosen for the problem and that the solution for the Q point is valid. Figure 1.11 shows the diode characteristic and the piecewise-linear approximation on linear axes. If the Q point lies too far from the true curve of I_D vs. V_D, then a new model must be obtained and the analysis repeated.

We complete this example by calculating the dynamic resistance at room temperature with $I_D = 4.67$ mA, $I_0 = 6$ nA, and $n = 1.9$:

$$r = \frac{1.9}{38.92\,(0.00467 + 6 \times 10^{-9})} = 10.45\ \Omega$$

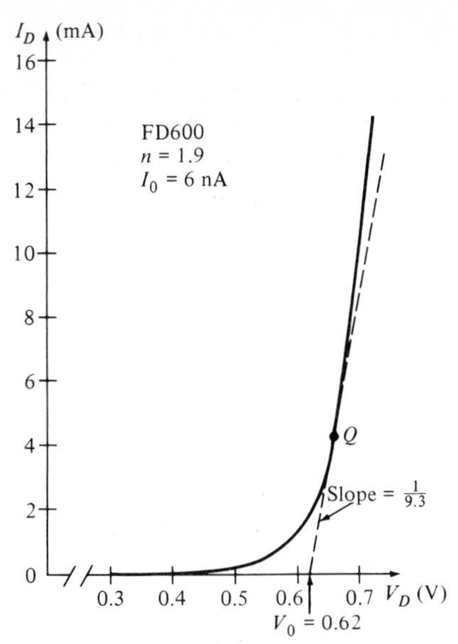

Fig. 1.11 The solid curve is a linear plot of I_D vs. V_D for an FD600 diode with $n = 1.9$ and $I_0 = 6$ nA. The broken line is a piecewise-linear model that is applicable in the neighborhood of the Q point shown. The breakpoint is at $V_D = 0.62$ V and the slope is 1/9.3 Ω.

Example 2

The second example is illustrated by the circuit shown in Fig. 1.12*a*. A signal source, $v_s = 10^{-3} \cos 10t$ volts, has been added to the last example, and we are asked to find v_d, the signal component of v_D. Room temperature is still assumed.

Problem identification comes first, and we see that we have both dc and ac quantities present in the circuit. Although we are asked only for the ac output voltage, both ac and dc models are needed, since the ac model depends on the dc operating point. The circuit contains an FD600 diode, so we will need to make use of its *v-i* characteristics or of models we have already developed.

This is an analysis-type problem. The circuit is given and the element values are specified, and we are asked to compute a specific response, here the signal component of $v_D(t)$.

Next comes the plan of attack. To obtain the output signal voltage as our result, we must have an ac model. Since 10^{-3} V looks like a pretty small voltage compared with 0.5 V or so and 10 rad/s is an exceedingly low radian frequency, we may choose the low-frequency small-signal ac model developed in Section 1.3 (which is just the dynamic resistance r). To

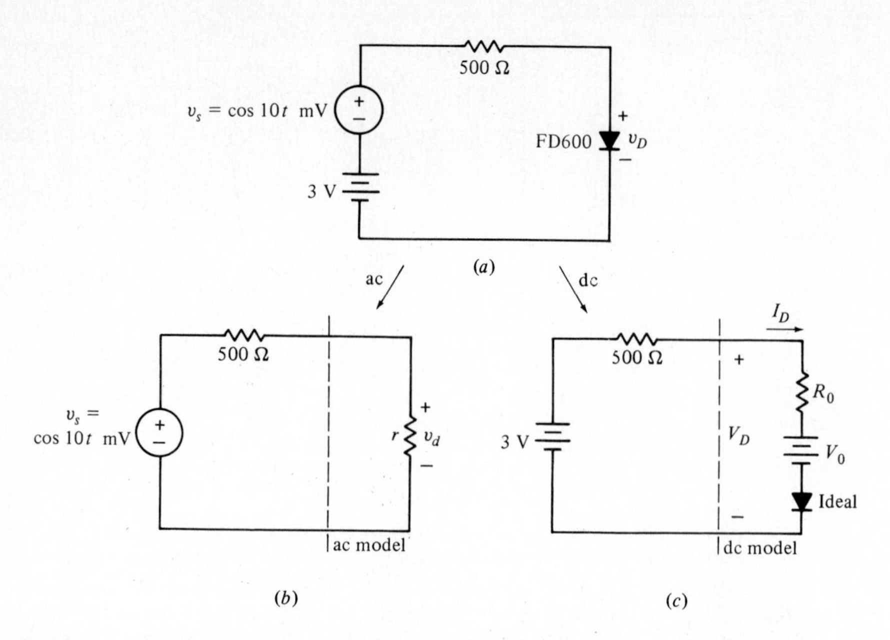

Fig. 1.12 (*a*) A diode circuit containing a signal voltage source v_s. The signal component of v_D is desired. (*b*)The ac-equivalent circuit contains the diode ac model. The signal voltage across the diode is v_d, and the total voltage is $v_D = V_D + v_d$. (*c*) The dc-equivalent circuit contains a dc diode model. It is identical to the circuit of Fig. 1.10*b*.

continue planning our attack, we can find a value for r once we know the dc diode current. This we obtain by working with the dc-equivalent circuit, which contains only the dc diode model and the dc source. The (small) *signal* source is set equal to zero. We shall see that we then have the exact circuit of the first example, and we need not repeat that work in analyzing this circuit. Once is enough.

With these thoughts in mind, we may now carry out the solution. The ac-equivalent circuit with which we shall determine the signal component of the diode voltage is shown in Fig. 1.12*b*. Note that the dc voltage source has been set equal to zero. Once it has been used to determine I_D, we are no longer concerned with it. The signal voltage is identified by the symbol v_d, the lower-case v indicating an instantaneous or time-varying voltage, and the lower-case subscript d signifying the signal component of the diode voltage. Note that a lower-case subscript is also used on the signal input voltage v_s. That is, if we wished, we could let $v_S = v_s + 3$.

From Fig. 1.12*b* and the rules for voltage division, the signal output voltage is found:

$$v_d = 10^{-3}(\cos 10t)\frac{r}{r + 500}$$

Once we obtain a suitable value for the dynamic resistance r, the problem is completed. To do this, we need to determine the dc component of the diode current. We do this with the help of a dc-equivalent circuit, shown in Fig. 1.12*c*. This circuit is the same as the one that we analyzed in the first example. Hence $I_D = 4.67$ mA and $V_D = 0.663$ V again.

Using the results of that earlier example, we have $r = 10.45\ \Omega$, which leads to

$$v_d = 1000 (\cos 10t) \frac{10.45}{10.45 + 500} = 20.5 \cos 10t\ \mu\text{V}$$

Note that the signal has been attenuated from a 1-mV amplitude to about 20 μV; this is a large reduction. The ratio of the signal output voltage v_d to the signal input voltage v_s is called the *attenuation ratio*. In this case it is 0.0205.

Problem 22 at the end of the chapter requests the value of v_d if the 3-V battery voltage is increased to 6 V. Without doing any detailed work, try to visualize the change in operating point and in r, and determine qualitatively whether v_d will increase or decrease. Conversely, what will v_d do as the battery voltage is reduced?

Example 3

As our last example of the use of diode models, let us consider the following problem:

> Design an attenuator for a 1-mV, 100-Hz sine-wave signal source so that the attenuation ratio v_d/v_s can be varied electrically and continuously from 0.1 to 0.5.

No circuit diagram comes with the problem, and there is no unique solution. We are expected to select a circuit and specify suitable element values. When the answer is given and we are asked to determine the circuit and the element values, we have a design problem.

The preceding example showed that a diode and a resistor can be used as an attenuator. It seems possible that the degree of attenuation can be controlled by changing the dynamic resistance of the diode. Since this resistance is a function of temperature, it would be possible to locate a small resistive heating element adjacent to the diode and let the diode temperature change as the resistor current varied. The scheme is possible, but the thermal response time is apt to be rather long and power consumption or heat radiation may cause difficulties. Also, we would need to provide two separate circuits carrying dc currents.

As a better solution, let's plan to vary the value of I_D by changing the value of the dc-source voltage. We propose a circuit of the form of Fig. 1.13*a*.

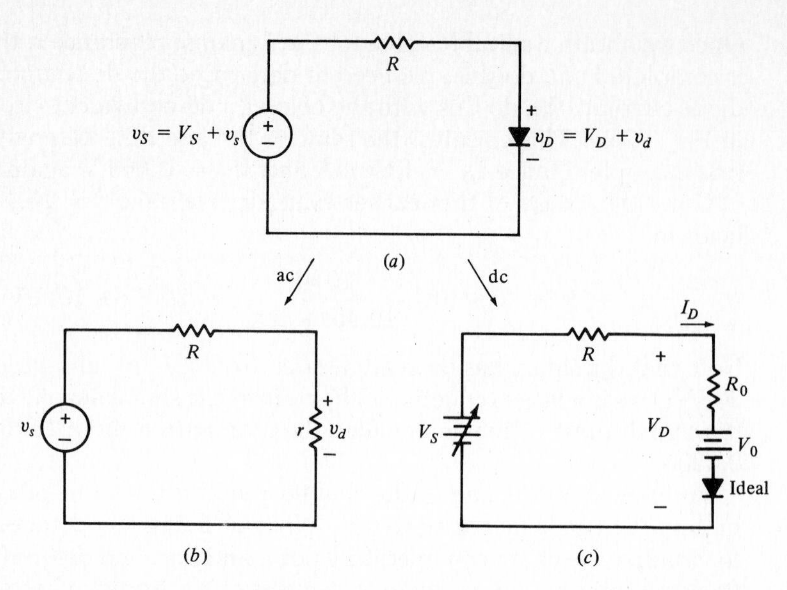

Fig. 1.13 (*a*) The proposed circuit for an electrically controlled attenuator for which $0.1 < v_d/v_s < 0.5$. (*b*) The ac-equivalent circuit for the signal. (*c*) The dc-equivalent circuit that determines the Q point.

For the diode, we should try to use the FD600, since it is the only one we have any data on at this time. We also know that the dynamic resistance is about 10 Ω when $I_D = 4.67$ mA. Because the series resistance R_s is about 2 Ω, the smallest dynamic resistance we can achieve is approximately 3 Ω, although much larger values are available. Notice, however, that we cannot make I_D too small and still satisfy the small-signal approximation $i_d \ll I_D$ and $v_d \ll V_D$. To allow ourselves some leeway, suppose we select the minimum value of r as 10 Ω. If the design does not work out well, we can try another value later. With this first step taken, we can now determine r_{max} and the series resistance R by considering the ac-equivalent circuit, Fig. 1.13*b*. That is, r_{min} provides the attenuation ratio of 0.1, while r_{max} leads to 0.5:

$$\frac{10}{10 + R} = 0.1 \qquad \text{and} \qquad \frac{r_{max}}{r_{max} + R} = 0.5$$

With no great difficulty, we find that $r_{max} = 90$ Ω and $R = 90$ Ω also. If we can vary the dynamic resistance between 10 and 90 Ω, we have a workable design. To do this, we specify room-temperature operation, and it follows from Eq. (17) that

$$r = \frac{0.02569n}{I_D} \qquad (I_D \ll I_0)$$

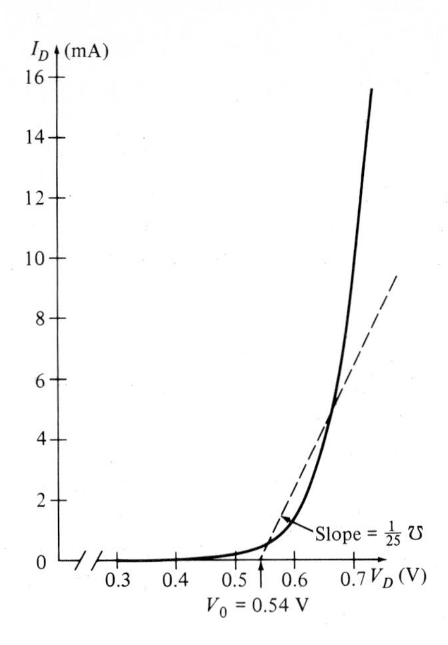

Fig. 1.14 A piecewise-linear approximation to the FD600 characteristic is shown for the range $0.542 \le I_D \le 4.88$ mA.

Therefore

$$I_{D(\min)} = \frac{0.02569(1.9)}{90} = 0.542 \text{ mA}$$

while

$$I_{D(\max)} = \frac{0.02569(1.9)}{10} = 4.88 \text{ mA}$$

To specify the variation of V_S (the dc component of the source voltage) that is needed to provide the required range of I_D, we turn our attention to the dc-equivalent circuit, Fig. 1.13*c*. To use this circuit effectively, we must determine new values for R_0 and V_0 appropriate to the region in which the Q point will be found. The v-i characteristic of the FD600 is shown again as Fig. 1.14. Note the portion of the curve lying between $I_D = 0.542$ and 4.88 mA. A reasonable piecewise-linear approximation here intersects the curve at $I_D = 0.5$ and 5 mA, where V_D is calculated to be 0.553 and 0.666 V, respectively. Therefore

$$R_0 = \frac{0.666 - 0.553}{(5 - 0.5)\,10^{-3}} \doteq 25\ \Omega$$

and

$$V_0 = 0.666 - 25\,(5 \times 10^{-3}) \doteq 0.54 \text{ V}$$

Using these values, we can now find the maximum and minimum values of V_S:

$$\begin{aligned} V_{S(\max)} &= V_0 + I_{D(\max)}\,(R + R_0) \\ &= 0.54 + 0.00488\,(90 + 25) \\ &= 1.10 \text{ V} \end{aligned}$$

and

$$\begin{aligned} V_{S(\min)} &= 0.54 + 0.000542\,(90 + 25) \\ &= 0.602 \text{ V} \end{aligned}$$

Our design may be summarized with reference to Fig. 1.13*a*: $R = 90\ \Omega$, the diode is an FD600, and a variation of V_S from approximately 0.6 V to approximately 1.1 V will cause the ratio of v_d to v_s to vary from 0.5 to 0.1. We also specify that $v_s \ll V_S$. For example, $v_s < 0.6$ V, for a maximum amplitude. Operation is at room temperature.

D1.5 A 2-V dc source and a 100-Ω resistor are in series with an FD600 diode. Calculate the diode voltage and current if the battery is connected to supply (a) forward current, (b) reverse current.

Answers. 0.712 V, 12.88 mA; −2 V, −6 nA

D1.6 Refer to the circuit of Fig. 1.15 and determine (a) I_D, (b) i_d, and (c) v_D.

Answers. 4.02 mA; 0.476 cos 1000*t* mA; 0.597 + 0.00487 cos 1000*t* V

1.5 Diode circuit models: high-frequency small-signal

The final diode models that we shall consider include one or more lumped capacitances that have strong effects on diode operation at high frequencies or for fast transients. Although these models have some application in circuits containing only diodes, their greatest interest for us lies in their direct applicability to transistor models that we shall develop in later chapters.

We first consider the semiconductor diode operating under reverse bias, where only the small reverse saturation current flows. The large number of electrons available on the *n* side and the large number of holes available on the *p* side are not forced across the junction, because the positive terminal of the voltage source is connected to the *n*-type material, and vice versa; see Fig. 1.16. Instead, the electrons in the *n*-type material are pulled away from the junction toward the positive source terminal. A cor-

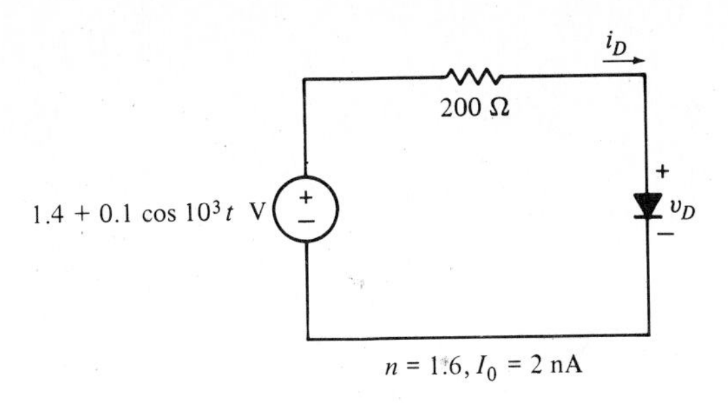

Fig. 1.15 See Problem D1.6.

responding action in the p-type material opens up a region at the junction that is depleted of almost all mobile charges, both electrons and holes. This region at the junction is called the *depletion layer*.

Without free charge carriers, the transition region has the characteristics of an insulator.

Farther away from the junction, both the p- and n-type materials are rich in carriers, and their resistance is therefore low in comparison with that of the transition region. These regions may be compared to the metallic conducting plates on both sides of the dielectric in a capacitor, and the depletion layer correspondingly acts like a capacitor. This effect is termed the *depletion*, or *junction*, *capacitance* C_j.

The depletion capacitance is generally quite small, on the order of several picofarads. Its effect is important in high-frequency diode models. The value of the junction capacitance depends on the dc operating point of the diode; it is represented quite well by the relationship

$$C_j = \frac{C_{j0}}{(1 - V_D/V_{bi})^N} \qquad (V_D \leq 0) \tag{20}$$

Strictly speaking, the expression is valid only under conditions of reverse bias, $V_D \leq 0$, but it holds fairly well for $V_D < 0.2$ V. It is evident that C_{j0} is the depletion capacitance when the external voltage across the diode is zero. The positive voltage V_{bi} is the *built-in potential*. It is a function of the type of semiconductor material, the degree of doping, and the temper-

Fig. 1.16 Under reverse-bias conditions all the mobile charge carriers are forced away from the junction, leaving a depletion layer having no free charge.

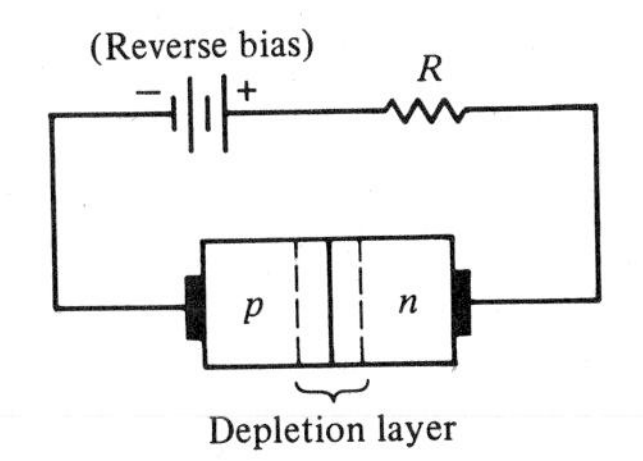

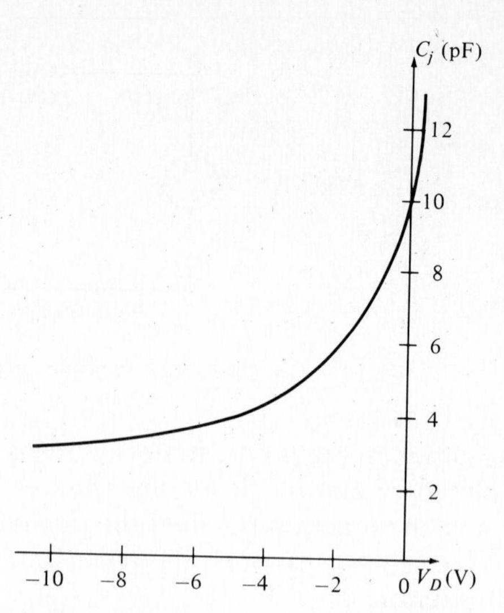

Fig. 1.17 A plot of junction capacitance vs. reverse-bias voltage for a diode with $C_{j0} = 10$ pF, $N = 1/2$, and $V_{bi} = 0.9$ V.

ature. For silicon, typical values range from 0.8 to 0.9 V. The value of the exponent N depends on the characteristics of the junction itself, particularly the manner in which the material varies from n-type to p-type. Typical values of N range between 1/2 for an abrupt change and 1/3 for a linear change. Figure 1.17 illustrates typical behavior of C_j on the operating-point voltage V_D. Note that as V_D becomes slightly positive, the capacitance increases rapidly.

When the junction capacitance is combined with the dynamic resistance, we obtain a first-order model for the reverse-biased semiconductor junction diode at high frequencies for small signals; see Fig. 1.18. The dashed arrows in the diagram are a reminder that the values of the elements vary with the operating point of the diode. The model is not satisfactory for forward bias.

When the junction is forward-biased, there is an additional component of capacitance that is much larger than the junction capacitance. It is

Fig. 1.18 With reverse bias, the small-signal high-frequency diode model is simply the parallel combination of the (very large) dynamic resistance and the junction capacitance, both calculated at the pertinent dc operating point.

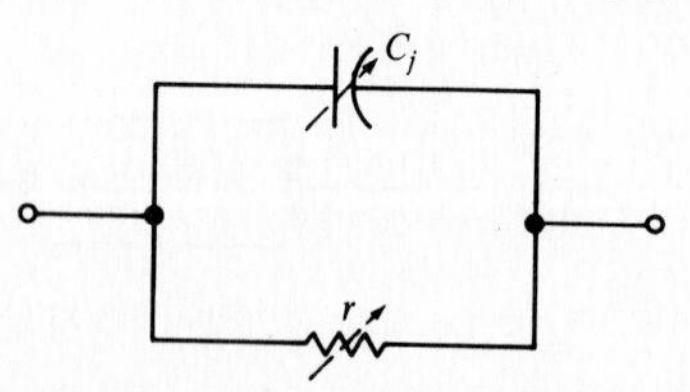

called the *diffusion capacitance*. With forward current, holes are injected from the p side across the junction into the n-type material. They are momentarily stored there before recombining with the large number of free electrons present in the n-type material. In a similar manner, electrons are injected into and stored in the p-type material. Changing the forward current or voltage requires a change in the value of this stored charge, which is again a capacitive effect. It turns out that the diffusion capacitance C_D is proportional to the (forward) current I_D:

$$C_D = BI_D \tag{21}$$

where B is a constant at any given temperature. At a forward current of 1 mA, C_D is several hundred times as large as C_{j0}, typically ranging from 100 to 1000 pF.

The small-signal high-frequency model for forward bias is shown as Fig. 1.19. The series resistance R_s represents contact resistance and bulk resistance. It has a value of several ohms and has the same value as for the dc model at large forward currents.

D1.7 Let $V_{bi} = 0.43$ V for a diode. If $C_j = 4$ pF for $V_D = -6$ V and 7 pF for $V_D = -1$ V, find (a) N, (b) C_{j0}, (c) C_j for $V_D = -10$ V.

Answers. 0.372; 10.96 pF; 3.34 pF

D1.8 Let $C_j = 10$ pF and $R_s = 3\ \Omega$ for a certain forward-biased diode. At $I_D = 1$ mA, $r = 20\ \Omega$, and $C_D = 90$ pF. Find the input impedance of the high-frequency equivalent circuit for (a) $\omega = 5 \times 10^8$ rad/s and $I_D = 1$ mA; (b) $\omega = 2 \times 10^9$ rad/s and $I_D = 1$ mA; (c) $\omega = 5 \times 10^8$ rad/s and $I_D = 2$ mA.

Answers. $16.40\underline{/-37.6°}\ \Omega$; $6.29\underline{/-48.4°}\ \Omega$; $9.65\underline{/-31.2°}\ \Omega$

1.6 Applications of diodes

There are very few electronic circuits that do not contain several semiconductor diodes. These diodes may be used in many different ways. We shall discuss just a few that are of particular importance. In general, we will assume that the diode may be replaced by its ideal model, in order to keep

Fig. 1.19 With forward bias, the diffusion capacitance is present in the small-signal high-frequency model. The series resistance R_s represents contact and bulk resistance of the semiconductor diode.

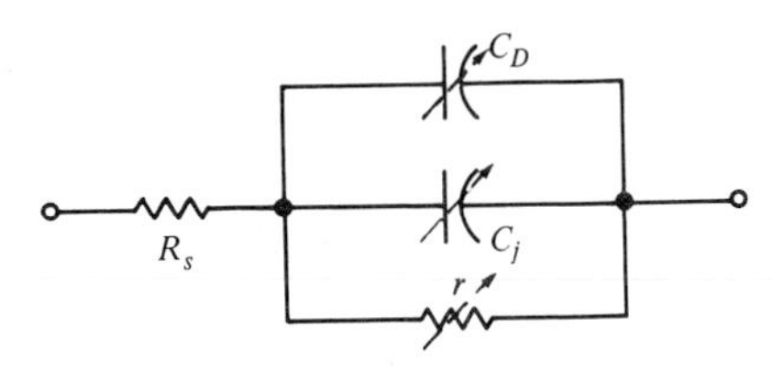

our discussion as simple and brief as possible. More accurate models can always be used if necessary.

Rectifiers

Except for devices that are purely battery-operated, most electronic equipment contains some type of rectifier circuit to convert the sinusoidal voltage present at the electrical power outlet to one or more required dc levels. Figure 1.20*a* shows a simple half-wave rectifier circuit. The transformer serves to isolate the rectifier circuit from the line voltage. It may also provide a voltage v_1 whose peak amplitude is greater or less than $120\sqrt{2}$ V. When the voltage v_1 is positive, the diode is forward-biased, and v_L is equal to v_1. When v_1 is negative, during the succeeding half-cycle, the diode is reverse-biased, no current flows through it, and $v_L = 0$, at least for the case of the ideal diode. The output voltage of Fig. 1.20*b* is thus a replica of the positive half of the input waveform; hence the name half-wave rectifier. A Fourier analysis of the output voltage (or a simple calculation of its average value by integration) shows that its average, or dc, value is $1/\pi$ times the maximum value of v_1. The analysis also shows that the lowest-frequency component is at 60 Hz. Filters must often be sup-

Fig. 1.20 (*a*) A transformer and a half-wave rectifier are used to supply a resistive load R_L with (*b*) a pulsating voltage v_L having a dc average value equal to $1/\pi$ times the crest value of v_1.

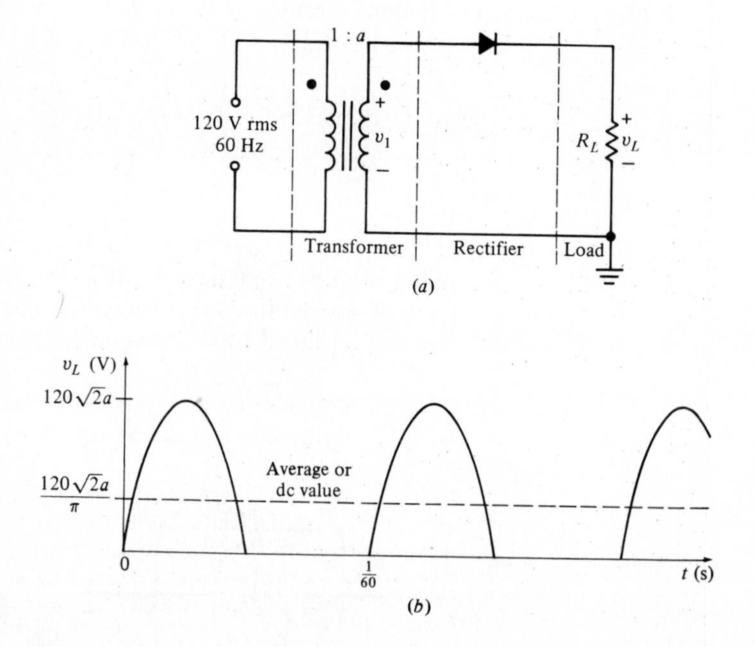

plied to minimize the frequency content of the output, so that it is almost entirely dc. We shall consider such a filter later in this section.

The dc average value of the half-wave rectifier output would be twice as great if the negative half-cycles of the source voltage could somehow be flipped over and used to fill the empty spaces in Fig. 1.20*b*. A circuit that accomplishes this is the full-wave rectifier shown in Fig. 1.21*a*. If we assume that the positive half-cycle of v_1 begins at $t = 0$, then diode D_1 is forward-biased, and current flows through it into the load. Meanwhile v_2 is also positive and diode D_2 is reverse-biased, so that it behaves as an open circuit. In the succeeding half-cycle, v_1 is negative, diode D_1 is nonconducting, v_2 is also negative, diode D_2 is forward-biased, and current therefore flows through it into the load. The output voltage v_L is once again positive. Figure 1.21*b* indicates which diode is conducting each half-cycle, assuming that v_1 is positive in the initial half-cycle shown. It is apparent that the dc, or average, value of the load voltage is twice that for a comparable half-wave rectifier. Moreover, the lowest frequency component of the load voltage is now at 120 Hz and it has a smaller amplitude. It is therefore easier to filter out the high frequencies.

A popular bridge-type full-wave rectifier circuit is shown in Fig. 1.22*a*. This arrangement does not require a center tap on the transformer secondary winding. It also provides a greater dc output voltage than does the

Fig. 1.21 (*a*) A transformer with a center-tapped secondary and a full-wave rectifier are used with a resistive load. (*b*) The average value of the load voltage is twice that of the comparable half-wave rectifier.

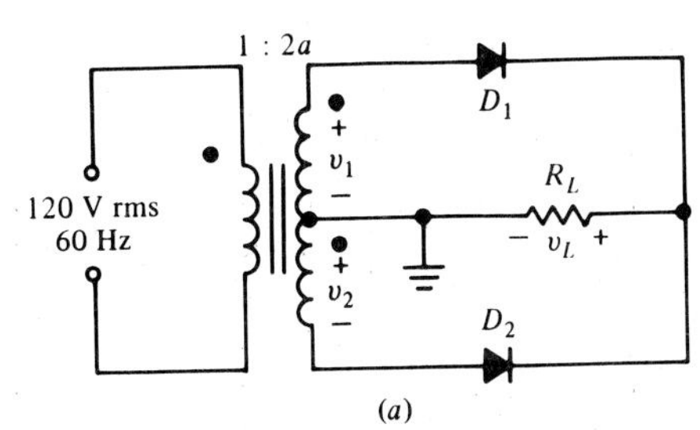

(*a*)

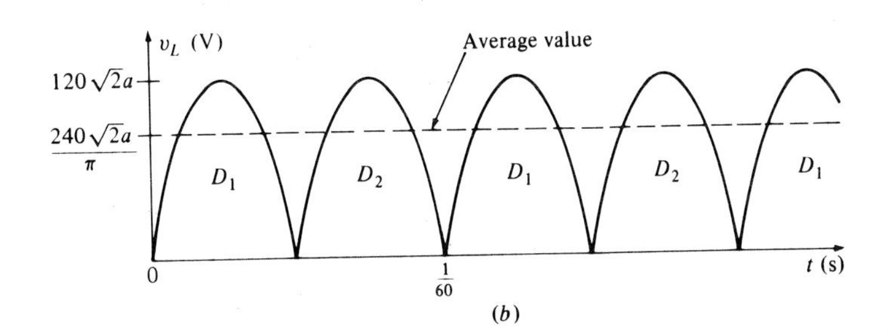

(*b*)

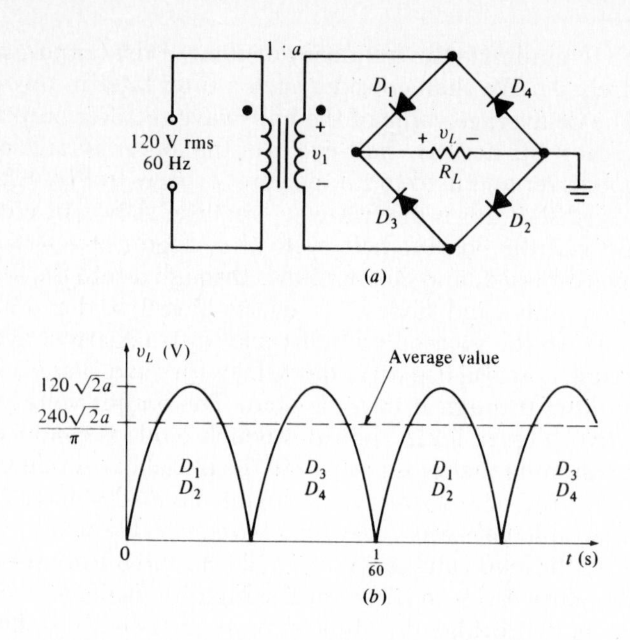

Fig. 1.22 (*a*) The full-wave bridge rectifier with the same transformer used in the half-wave circuit of Fig. 1.19*a* provides an output waveform (*b*) that has twice the dc voltage of the half-wave circuit and a ripple frequency twice as high.

plain full-wave circuit for a transformer with the same turns ratio. In operation, two of the four diodes conduct during each half-cycle. When v_1 is positive, D_1 and D_2 are forward-biased, and current flows through them into the load. During this half-cycle, D_3 and D_4 are reverse-biased and behave as open circuits. In the succeeding half-cycle, D_3 and D_4 conduct while D_1 and D_2 are cut off. Positive current enters the left terminal of R_L in both cases. The output voltage waveform is shown in Fig. 1.22*b*. If the turns ratio of the transformer is $1{:}a$, just as it was in the case of the half-wave rectifier, the average value of the output voltage is $240\sqrt{2}a/\pi$, and the lowest frequency of the output waveform is 120 Hz. This is the best of the three rectifier circuits that we have considered, but it also requires four diodes.

Power supplies

Many applications of rectifiers require that the output of the circuit be much smoother than the waveforms shown in Figs. 1.20 to 1.22. For example, the hum present in a poorly designed (or ailing) audio system is usually intolerable and is often due to the power supply not having a smooth output.

A simple filter that is often satisfactory in smoothing the output consists of a large capacitor placed in parallel with the load resistor R_L. Figure 1.23*a* shows a half-wave rectifier with a capacitor filter. This is called a *low-pass filter*, since it passes lower frequencies (dc) while attenuating higher frequencies (60 Hz and higher). The capacitor charges up during a part of the positive half cycle, as current flows through the diode into C and R_L, and discharges into R_L during the remainder of the cycle when the diode is cut off. Thus the voltage across C and R_L is held more constant. As indicated in Fig. 1.23*b*, the output voltage shows a much smaller variation in amplitude, and its average value is larger than it would be for the rectifier without the filter capacitor. Filter capacitors are also added to the full-wave rectifiers of Figs. 1.21 and 1.22, with similar results.

The peak-to-peak *ripple voltage* Δv is defined in Fig. 1.24*a* as the difference between the peak output (or input) voltage V_m and the minimum value of the output voltage. Note that the diode in Fig. 1.23*a* conducts only when v_1 is greater than v_L. This occurs in the interval $t_1 < t < t_2$, as shown in Fig. 1.24*b*. During this time, current flows through the diode and charges the capacitor, as well as furnishing load current. In the interval $t_2 < t < t_3$, the diode is off and charge flows out of the capacitor and through the load R_L. The time constant R_LC is chosen to be much larger

Fig. 1.23 (*a*) A half-wave rectifier with a capacitor filter. (*b*) The output voltage waveform is shown by the solid curve.

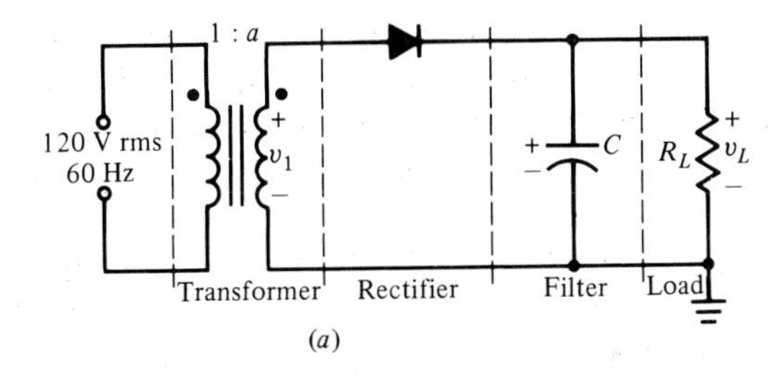

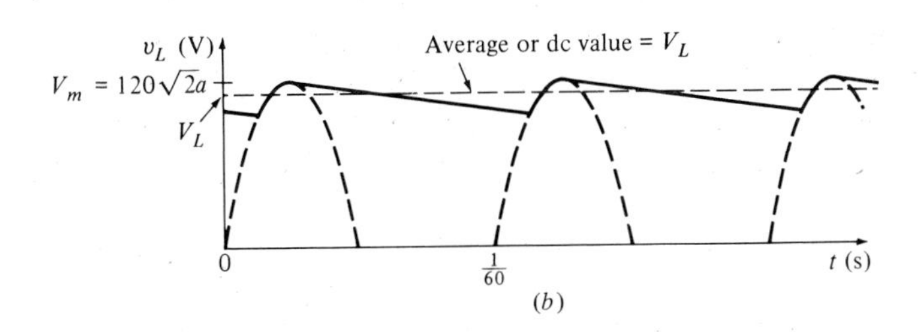

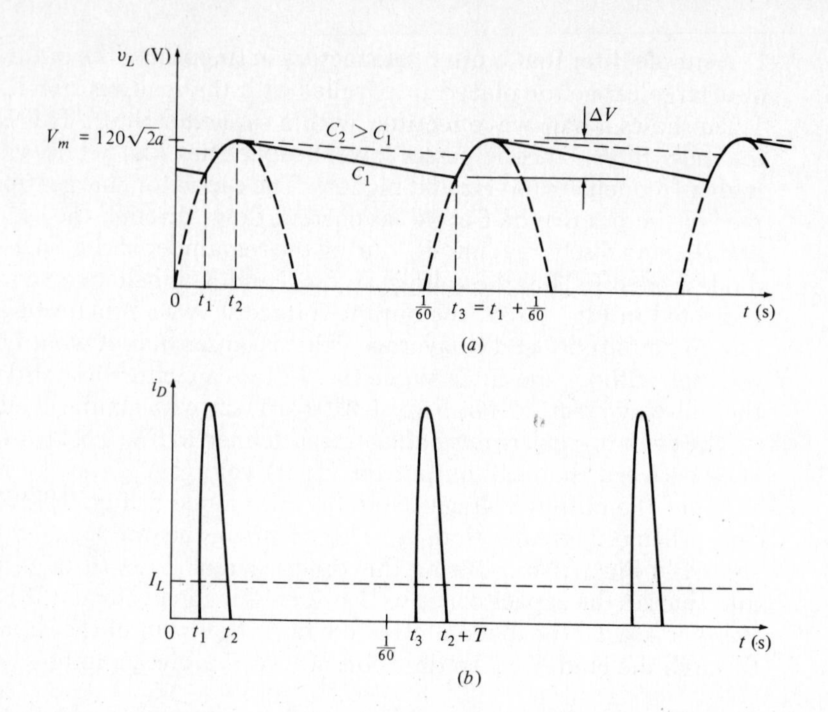

Fig. 1.24 (*a*) The output voltage of a half-wave rectifier with a capacitor filter is shown for two values of capacitance; $C_2 > C_1$. The ripple voltage Δv is indicated for the output when C_1 is used. For this waveform, diode conduction begins at t_1, stops at t_2, and begins again at $t_3 = t_1 + T$. (*b*) The pulsating diode current is shown for the smaller capacitor, C_1. A larger capacitor produces a greater maximum value, a slightly greater average value, and a shorter conduction time.

than the period of the input sine wave in order to reduce Δv, as indicated in Fig. 1.24*a*. During discharge, the capacitor or output voltage v_L is given by the familiar negative exponential,

$$v_L = V_m e^{-(t-t_2)/R_L C} \qquad (t_2 < t < t_3) \tag{22}$$

Since the minimum value of v_L occurs at t_3, while the maximum value is V_m for our ideal diode, the ripple voltage is

$$\Delta v = V_m[1 - e^{-(t_3-t_2)/R_L C}]$$

Well-designed power supplies use large capacitors, and therefore t_2 occurs very near the peak of v_L. Thus $t_3 - t_2 \doteq T$, the period of the input, and this leads to a conservative (slightly too large) estimate of the peak-to-peak ripple voltage:

$$\Delta v \doteq V_m(1 - e^{-T/R_L C}) \tag{23}$$

This result may be simplified if we use the power-series expansion for the exponential,

$$e^x = 1 + x + \frac{x^2}{2!} + \cdots$$

In Eq. (23), $T \ll R_L C$, and we may approximate the series by its first two terms,

$$e^{-T/R_L C} \doteq 1 - \frac{T}{R_L C}$$

Therefore

$$\Delta v \doteq V_m \left[1 - \left(1 - \frac{T}{R_L C}\right)\right]$$

or

$$\Delta v \doteq \frac{V_m T}{R_L C} \tag{24}$$

Note that the larger we make C, the smaller Δv becomes—a very desirable result.

To calculate the average or dc output voltage, shown in Fig. 1.23*b* as $V_L = v_{L\,(\mathrm{av})}$, we assume that the exponential portion of the curve may be approximated by a straight line. Thus $V_L = V_m - \frac{1}{2}\Delta v$, and

$$V_L \doteq V_m \left(1 - \frac{T}{2R_L C}\right) \tag{25}$$

Power supplies may be operated with load resistances that are not constant. For example, increasing the volume level of a hi-fi system is equivalent to a reduction of R_L. Equation (24) shows that the ripple voltage increases as R_L decreases, and Eq. (25) indicates that a small decrease in V_L also accompanies a reduction in R_L. Since V_L remains fairly constant, the dc power varies inversely as R_L.

We define the dc load current as I_L, where $I_L = V_L/R_L$, and it therefore may also vary over a wide range. *No-load* conditions exist when $I_L = 0$, $R_L = \infty$, and $V_L = V_{L(\mathrm{max})} = V_m$. At maximum loading, $I_L = I_{L(\mathrm{max})}$, $R_L = R_{L\,(\mathrm{min})}$, and $V_L = V_{L\,(\mathrm{min})}$. The decrease of V_L is expressed in terms of *regulation*, defined as a percentage:

$$\text{Percentage regulation} = \frac{V_{L(\text{no load})} - V_{L(\text{min})}}{V_{L(\text{no load})}} \times 100\% \tag{26}$$

Diode current flows only in the interval $t_1 < t < t_2$, as illustrated in Fig. 1.24*b*. Because the average value of the current is I_L, it is obvious that very large spikes of diode current must flow in order to average out to I_L. As C is made larger, $t_2 - t_1$ becomes smaller, and the peak current through the diode is even larger. Since we are using ideal diodes, there is no danger of burning them out. But in a real circuit, care must be taken not to exceed the peak-current specifications of the diode.

If we model a real diode with R_0 and V_0 in a piecewise-linear circuit, or even include R_s for very large currents, the effect of R_0 and R_s is a reduction of the peak output voltage caused by the diode voltage drop. Only in low-voltage rectifiers is the effect of V_0 significant. Usually R_0 and R_s are small compared to R_L, and their effects are negligible. But in high-current power supplies, V_L is reduced and the voltage regulation shows even more loading effects.

For a full-wave rectifier, T is replaced by $T/2$ in Eqs. (22) through (25), and thus Δv is reduced by a factor of two, easing the conditions for choosing C. Again we see that there are good reasons for using a full-wave rectifier instead of the half-wave circuit.

Voltage regulators

In order to make a major improvement in the percentage regulation and the magnitude of the ripple voltage, the designer must incorporate an additional circuit in the power supply. Such a *voltage regulator circuit* is shown with a full-wave bridge rectifier in Fig. 1.25.

Many different circuits find use as voltage regulators. Some of these are available as integrated circuits and thus they are quite inexpensive, even though their internal circuitry may be quite involved.

As an example of one type of regulator circuit, we consider the very simple arrangement of Fig. 1.26*a*, consisting of one resistor and a *Zener diode*. This diode application also gives us a chance to model a diode in breakdown. The Zener diode is a diode that is designed to be operated in the avalanche-breakdown region. It is represented by the special diode symbol shown. Note the suggestion of the breakdown characteristic built into the symbol.

Fig. 1.25 A voltage regulator is used between the load and the capacitor filter in a power supply driven by a full-wave bridge rectifier.

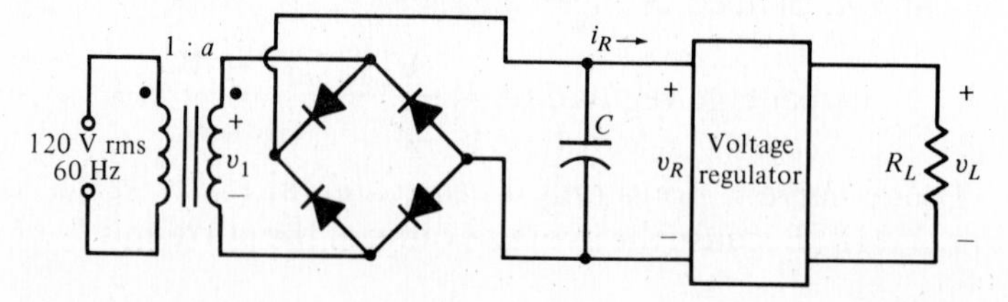

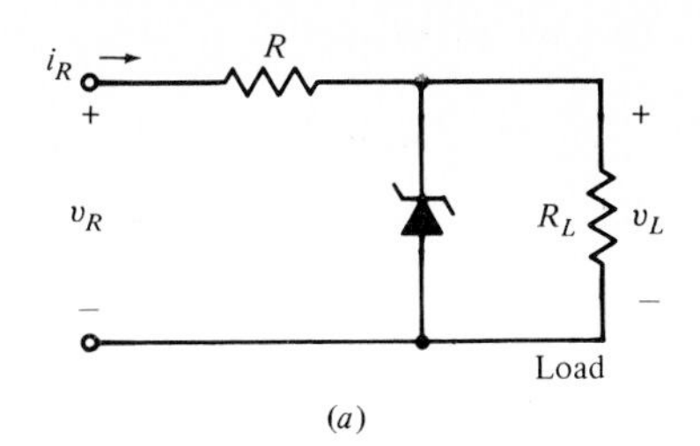

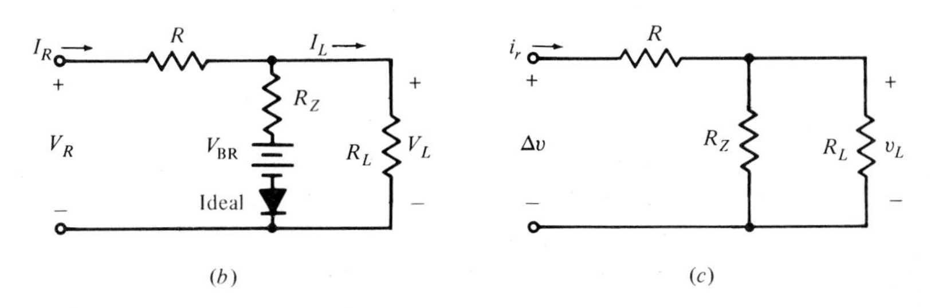

Fig. 1.26 (*a*) A simple shunt regulator circuit includes an added series resistance R and a diode operating in avalanche breakdown. The special symbol represents a Zener diode, one designed to operate in the breakdown region. (*b*) The dc equivalent circuit includes the dc model of a Zener diode, first shown in Fig. 1.4*d*. (*c*) The ac or signal equivalent circuit also uses the resistance R_Z.

Under normal operation, the Zener diode is in avalanche breakdown and the load voltage is closely equal to V_{BR} over a wide range of load currents. Figure 1.26*b* illustrates the dc equivalent circuit for the regulator with the diode replaced by its piecewise-linear equivalent circuit for avalanche breakdown, Fig. 1.4*d*. Normal operation occurs only for $V_R \geq V_{BR}$ and for a diode current having a magnitude greater than the minimum value required to get beyond the knee on the breakdown characteristic and less than some maximum value that will not damage the diode. We note that an ideal Zener diode would have $R_Z = 0$, and thus the dc load voltage V_L remains equal to V_{BR} as R_L is varied. Moreover, I_R would remain constant, since the voltage across R is constant. In other words, when R_L decreases and I_L increases, the diode current must decrease to maintain I_R constant. Thus the regulator current is shared between the diode and the load. The range of possible load currents is therefore equal to the range of values of the Zener diode current. This range is greater than the minimum required to operate in avalanche breakdown, but less than the maximum that the diode can tolerate.

As an example of a regulator in which R_Z is not zero, let us analyze a circuit in which

$$R = 500\ \Omega,$$
$$V_{BR} = 11.8\ \text{V},$$
$$R_Z = 8\ \Omega,$$
$$V_R = 25.5\ \text{V}$$

We first let $R_L = 6\ \text{k}\Omega$.

In the dc equivalent circuit of Fig. 1.26*b*, the ideal diode is a short circuit, and we may replace the network to the left of R_L by its Thévenin equivalent. We find

$$R_{Th} = R \| R_Z = 500 \| 8 = 7.87\ \Omega$$

and an open-circuit voltage,

$$V_{Th} = V_{BR} + \left(\frac{V_R - V_{BR}}{R_Z + R}\right) R_Z$$

$$= 11.8 + \frac{25.5 - 11.8}{8 + 500} 8 = 12.02\ \text{V}$$

Figure 1.27 shows this equivalent circuit. With $R_L = 6\ \text{k}\Omega$, we find that $I_L = 12.02/(7.87 + 6000) = 2.00$ mA, and $V_L = 12$ V.

If R_L now decreases by a factor of 10 to 600 Ω, then $I_L = 12.02/(7.87 + 600) = 19.77$ mA, and $V_L = 11.86$ V. Although the load current has increased almost ten times, the load voltage has decreased by only about 1%. Note that the answer to Drill Problem D1.10 indicates that the output voltage would have changed by more than 16% if the regulator had not been present.

Now let us consider how a voltage regulator can reduce the ripple voltage. To do so, we interpret the ripple as a signal and construct the signal

Fig. 1.27 A load R_L is supplied by the Thévenin equivalent of a Zener diode, shunt regulator circuit.

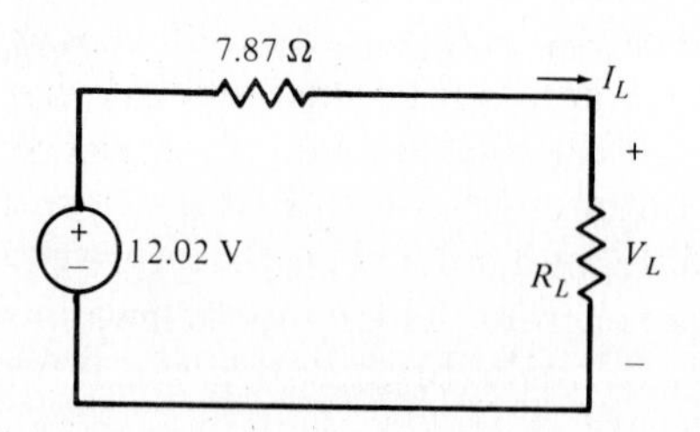

equivalent circuit of the regulator, shown as Fig. 1.26*c*. The ratio of the output ripple voltage v_ℓ to the input ripple voltage Δv is easily found by voltage division,

$$\frac{v_\ell}{\Delta v} = \frac{R_Z \parallel R_L}{R_Z \parallel R_L + R} = \frac{8 \parallel 6000}{8 \parallel 6000 + 500} = 0.0157$$

for $R_L = 6000\ \Omega$. A similar calculation when $R_L = 600\ \Omega$ shows a similar ratio of 0.0155. Thus the ripple voltage is reduced by almost two orders of magnitude. Excessively large values of C are therefore not required.

An integrated circuit (IC) voltage regulator may be obtained with either a fixed or an adjustable output voltage. The former holds the output voltage to less than a few percent of a specified value. The adjustable unit allows an external resistor to be varied to establish V_L. The permissible range of load current is also specified.

Light-emitting diodes

A light-emitting diode, or LED, is a special two-terminal semiconductor diode that emits light when it is forward-biased. The intensity is roughly proportional to the forward current. Appendix A presents the data sheet for the Hewlett-Packard 5082-4487 or 5082-4488 LED. Light-emitting diodes are used to provide the numerical displays in hand-held calculators and other small instruments that use optical read-outs.

Schottky barrier diodes

If a thin layer of aluminum is placed on lightly doped n-type silicon, a *rectifying junction* will form between the metal and the semiconductor. The ohmic or nonrectifying contacts that we have seen on the junction diodes are made when the semiconductor material is heavily doped. The rectifying junction is called a *Schottky barrier*, and the resultant diode is a *Schottky barrier diode*. This device also obeys the Shockley diode equation. Figure 1.28 illustrates its $V_D - I_D$ relationship. Larger values of I_0 and n cause the curve to be shifted to the left, compared to the junction diode.

The important difference between the two types of diode occurs with the diffusion capacitance C_D, which is negligible in the Schottky barrier diode. Only the depletion capacitance C_j appears in the high-frequency model. Therefore the diode operates at higher frequencies than the junction diode, and is several orders of magnitude faster in switching applications.

There are several important applications in which Schottky barrier diodes are used in conjunction with transistors to obtain short switching times.

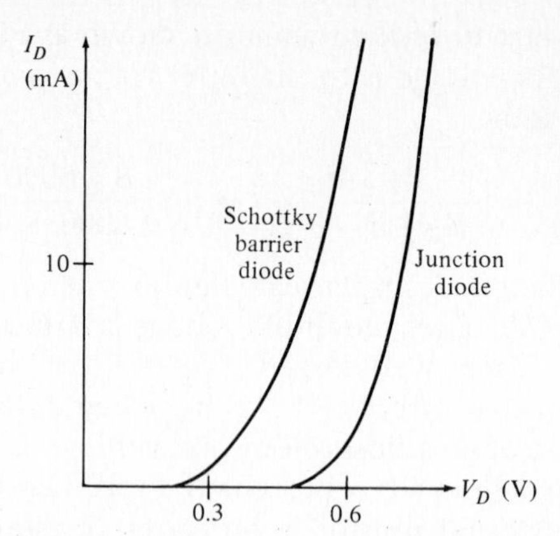

Fig. 1.28 The V_D - I_D characteristic of a Schottky barrier diode is similar to that of a junction diode, except that the breakpoint occurs in the neighborhood of $0.3V$.

Varactor diode

As a final example of the applications of diodes, we note that the depletion capacitance of the diode may be used as an electronic tuning means for FM radio, television circuits, certain microwave oscillators, and any other circuits in which a small variation in capacitance can effect a significant change in frequency. This may be accomplished by placing the diode in parallel with the capacitor in a parallel *RLC* circuit so that the resonant frequency is given by

$$\omega_0 = \frac{1}{\sqrt{L(C + C_j)}}$$

where C_j may be varied electrically. Schemes based on this effect are used to implement automatic frequency control (AFC) circuits.

We have discussed only a few of the more important applications of the semiconductor diode in electronic circuits. Others are described in more advanced treatments, while several appear in problems at the end of this chapter.

From our point of view, perhaps the most important lesson to be learned from this brief study of diodes is the insight into several techniques of analysis and design that we will use in succeeding chapters.

D1.9 In Fig. 1.23*a*, let $f = 60$ Hz and $C = 100\ \mu$F. It is desired that $V_L = 12$ V for $I_L = 2$ mA. Calculate (a) the value of V_m required, (b) Δv, (c) the percentage regulation.

Answers. 12.17 V; 0.338 V; 1.39%

D1.10 In Fig. 1.23*a*, let $f = 60$ Hz, $C = 100\ \mu$F, and $V_m = 12.17$ V. If $I_L = 20$ mA, calculate (a) V_L, (b) Δv, (c) the percentage regulation.

Answers. 10.18 V; 3.99 V; 16.4%

D1.11 Let $V_R = 25.5$ V, $R = 500\ \Omega$, and $V_{BR} = 12$ V for the dc equivalent circuit of Fig. 1.26*b*. Calculate V_L for (a) $R_Z = 4\ \Omega$, $R_L = 6$ kΩ; (b) $R_Z = 4\ \Omega$, $R_L = 600\ \Omega$; (c) $R_Z = 8\ \Omega$, $R_L = 6$ kΩ.

Answers. 12.10; 12.03; 12.20 V

Problems

1. A diode at 25°C has $I_D = 3.373 \times 10^{-10}$ A when $V_D = 0.3$ V, and $I_D = 22.75\ \mu$A at $V_D = 0.6$ V. Calculate (a) n, (b) I_0, (c) I_D for $V_D = 0.7$ V.
2. Let $I_0 = 4$ nA and $n = 1.1$ for a semiconductor diode operating at 25°C. (a) Find I_D when $V_D = 0.4$ V. (b) What value of V_D will cause a current 1 mA greater than the value of (a)?
3. Two data points for a diode are $I_D = 1$ mA at $V_D = 0.35$ V and $I_D = 5$ mA at $V_D = 0.39$ V. Assuming that the Shockley diode equation applies, let $n = 1.05$ and find I_0, V_T, and the operating temperature.
4. A diode is modeled by the equation, $I_D = 20\ (e^{35V_D} - 1)$ nA at 25°C. Sketch the *V-I* characteristic of the combination if the diode is operated with forward bias (a) in series with 25 Ω, (b) in parallel with 25 Ω.
5. The value of I_0 for a junction diode is typically given by $I_0 = KT^3 \times e^{-E_G/kT}$, where the width of the energy gap is $E_G = 1.12$ eV (1 electron-volt $= 1.6022 \times 10^{-19}$ J) for silicon. (a) Determine K if $I_0 = 10^{-14}$ A at 25°C. (b) Let $n = 1.07$ and $I_D = 1$ mA, and calculate V_D at 10°C and 50°C. (c) Derive an equation for the rate of change of V_D with respect to T if I_D is constant and much greater than I_0. (d) Evaluate dV_D/dT for $T = 25$°C, $I_D = 1$ mA.
6. A diode operating at $T = 300$ K has 0.4 V across it when the current through it is 10 mA, and 0.42 V when the current is twice as large. What values of I_0 and n allow this diode to be modeled by the diode equation?
7. A diode for which $I_D = e^{33V_D} - 1$ nA is in series with a dc source of 0.5 V. Sketch a curve of I_D vs. V_{in}, the voltage across the series combination, for the two possible battery polarities. Include both positive and negative values of V_{in}.
8. Two identical diodes are connected in series witn 100 Ω and a 2-V battery such that each diode is forward-biased. Let $I_0 = 10$ fA, $n = 1$, and $T = 25$°C. Calculate the loop current if (a) ideal diodes

are assumed; (b) $V_D = 0.6$ V for each diode; (c) each diode is represented by a piecewise-linear model that is exact at $I_D = 1$ mA and 10 mA; (d) each diode is modeled by the nonlinear diode equation.

9. Two identical diodes, each having $I_0 = 1$ fA and $n = 1.1$, are connected in parallel. The combination is connected in series with 150 Ω and a 1.5-V battery so that the diodes are forward-biased. Calculate the source current at 25°C if (a) each diode is represented by a piecewise-linear model that is exact at 0.5 and 5 mA; (b) each diode is modeled by the nonlinear diode equation.
10. At 25°C, a semiconductor diode has a forward current of 2 mA when a voltage of 0.3 V is applied across it. Assume that $n = 1$. (a) Find I_D when $V_D = 0.4$ V. (b) Determine I_0. (c) Use the diode equation to find I_D if the diode is placed in the circuit of Fig. 1.1*b* with $R = 100$ Ω and $V = 1.5$ V.
11. A semiconductor diode is modeled at 25°C by $n = 1.2$. Let $I_D = 2.5$ mA when $V_D = 0.25$ V. (a) Find V_D when $I_D = 15$ mA. (b) Determine I_0. (c) The diode is now connected in series with 2 V dc and 200 Ω. Find I_D if the diode is operating with forward bias.
12. A diode has the parameters $I_0 = 1$ nA and $n = 1$ and operates at 25°C. (a) Select values of V_0 and R_0 for a dc model so that the model and the diode equation give the same results when $I_D = 5$ mA and 10 mA. The diode is connected in series with 100 Ω and a 1.5-V battery so that it is forward-biased. (b) Represent the diode by its piecewise-linear model and determine I_D. (c) Represent the diode by the nonlinear diode equation and determine I_D.
13. Determine i_x in the circuit of Fig. 1.29 if (a) both diodes are considered to be ideal; (b) both diodes are modeled by $R_0 = 10$ Ω and $V_0 = 0.5$ V. (c) What dc voltage source should be placed in series with the 20-Ω resistor to just reduce i_x to zero? Assume that the piecewise-linear model applies.

Fig. 1.29 See Problem 13.

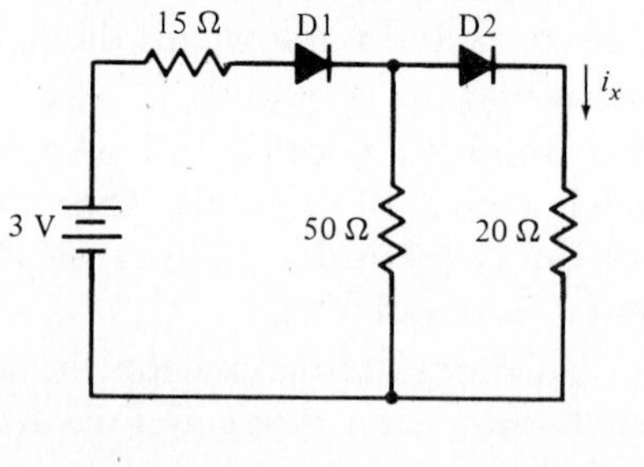

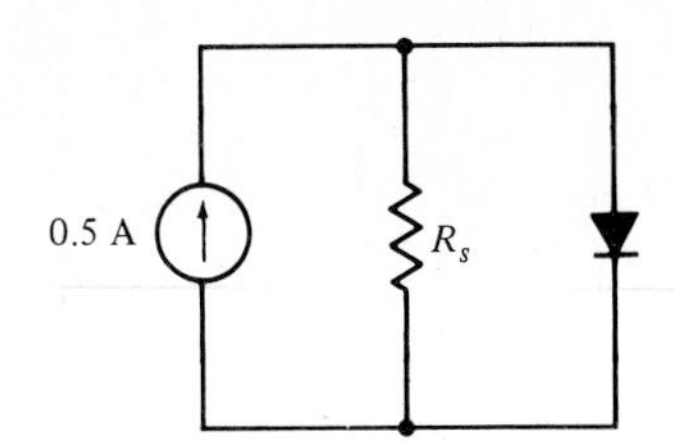

Fig. 1.30 See Problem 14.

14. The diode in the circuit of Fig. 1.30 is characterized by $I_0 = 8$ nA and $n = 1.3$ and is operating at 25°C. Determine the power being absorbed by each of the three circuit elements if (a) $V_D = 0.4$ V; (b) $I_D = 250$ mA; (c) $R_s = 5\ \Omega$.
15. The diode of Fig. 1.31 may be represented by $V_0 = 0.45$ V and $R_0 = 12\ \Omega$ for forward bias, and by $V_{BR} = 10$ V and $R_Z = 8\ \Omega$ for Zener breakdown. Determine V_x if V_s equals (a) 5 V, (b) 2.5 V, (c) −10 V, (d) −30 V, (e) −50 V.
16. Let $V_{BR1} = 20$ V and $R_{Z1} = 20\ \Omega$ for D1, while $V_{BR2} = 25$ V and $R_{Z2} = 35\ \Omega$ for D2. Find I and V for the circuit shown in (a) Fig. 1.32*a*, (b) Fig. 1.32*b*.
17. Let $I_0 = 15$ nA and $q/nkT = 30$ for a semiconductor diode. If $I_D = 7.5$ mA, find (a) V_D, (b) V_D/I_D, (c) r. If I_D varies over the range $7.4 \leq I_D \leq 7.6$ mA, what is the range of (d) V_D? (e) of r?
18. A certain diode is described by the equation $I_D = 16\,(e^{32V_D} - 1)$ nA. Let $V_D = 0.42$ V and find (a) I_D, (b) r, (c) $R_x = V_D/I_D$. (d) Let V_D increase by $\Delta V_D = 1$ mV. Is the change in I_D given more accurately by $\Delta I_D = \Delta V_D/R_x$ or $\Delta V_D/r$?
19. Determine a value for the dynamic resistance of the diode labeled "maximum" in Fig. 1.9 for an operating point at $I_D = 1$ mA.
20. Let $n = 1.2$ and $I_0 = 1.5$ nA for a diode operating at 40°C. Determine V_D if r equals (a) 2 Ω, (b) 200 kΩ.

Fig. 1.31 See Problem 15.

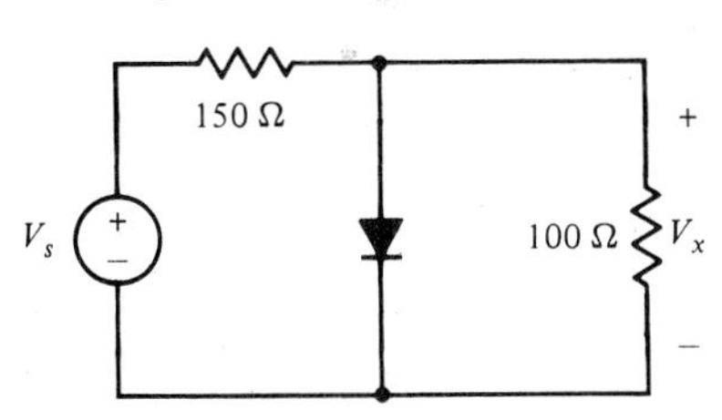

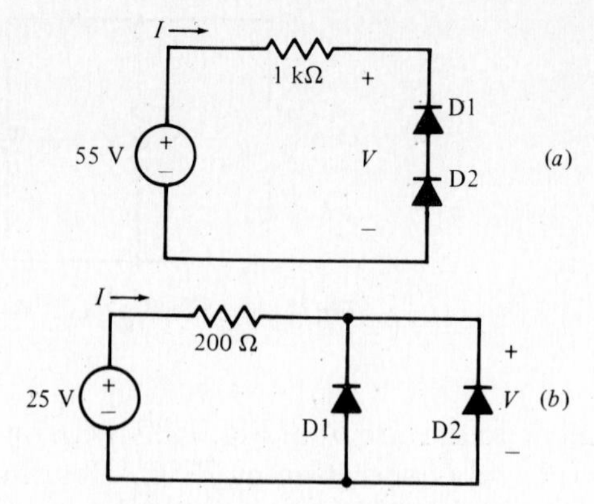

Fig. 1.32 See Problem 16.

21. A diode is modeled by $n = 1.15$ and $I_0 = 0.5$ nA at 25°C. The voltage $v_D = 0.4 + 0.002 \cos 120\pi t$ V is applied across the diode. Find i_d and i_D.
22. In the example illustrated by Fig. 1.12*a*, increase the battery voltage to 6 V and determine the new operating point and the dynamic resistance.
23. The forward characteristic of the diode in the circuit of Fig. 1.33 is given by $i_D = 10^{-8}e^{40v_D}$ A, and v_s is a small signal voltage. Determine a linear relationship between i_D and v_s, such as $i_D = a + bv_s$, where a and b are expressed as numerical values.
24. A source v_S, a 30-Ω resistor, and a diode governed by $i_D = 20(e^{35v_D} - 1)$ nA are in series. (a) If $v_S = 0.5$ V (forward bias), find i_D. (b) If $v_S = 0.5 + v_s$, where v_s is a small signal voltage, find i_D as a function of v_s, such as $i_D = a + bv_s$.
25. (a) Assume that the diode in Fig. 1.34 is ideal, and prepare plots of v_D vs. V_S and v_R vs. V_S, $-2 < V_S < 2$ V. (b) Plot new curves for a dc

Fig. 1.33 See Problem 23.

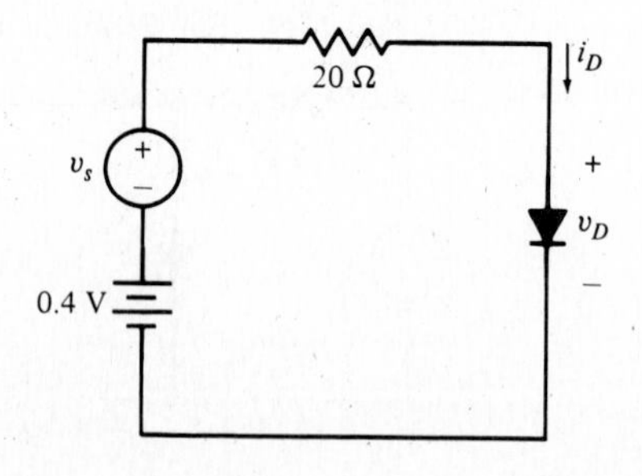

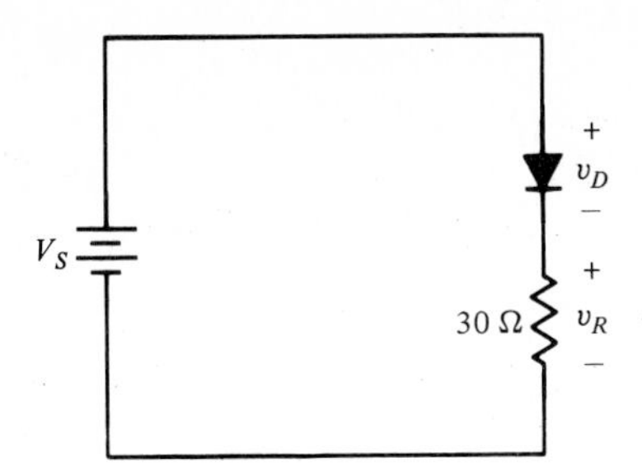

Fig. 1.34 See Problem 25.

diode model having $V_0 = 0.5$ V and $R_0 = 20\ \Omega$. (c) Using the model of (b) and $V_S = 0.8$ V, what ac resistance is offered to a small signal source in series with V_S? Assume that $T = 25°$C and $n = 1$.

26. A series loop contains a diode with $n = 1.5$ and $I_0 = 2$ nA, a 50-Ω resistor, an 0.8-V battery providing forward bias, and a signal voltage source, 6 cos $10^4 t$ mV. Assuming that $T = 25°$C, find (a) I_D, (b) i_d, (c) i_D, (d) v_D.
27. Two diodes are described by $I_{01} = 1$ nA, $n_1 = 1.1$, $I_{02} = 5$ nA, and $n_2 = 1.2$. Calculate the small-signal resistance of the combination at 25°C if the diodes are connected (a) in parallel, with $V_D = 0.45$ V; (b) in parallel, with $V_D = 0.55$ V; (c) in series, with $I_D = 0.5$ mA.
28. Two identical diodes at 25°C have $I_0 = 10$ nA, $n = 1.1$, $V_{BR} = 18$ V, and $R_Z = 25\ \Omega$. They are connected in series, back to back. A source $v_S = v_i + 20$ V and a 1-kΩ resistor are connected in series with the combination. Find the ratio of the signal voltage across the two diodes to v_i.
29. (a) Assume that the ideal diode in Fig. 1.35 is forward-biased and find I_D. (b) Assume that the diode is reverse-biased and find V_D. (c) Which one of these possible conditions is correct? (d) Find I_x.

Fig. 1.35 See Problem 29.

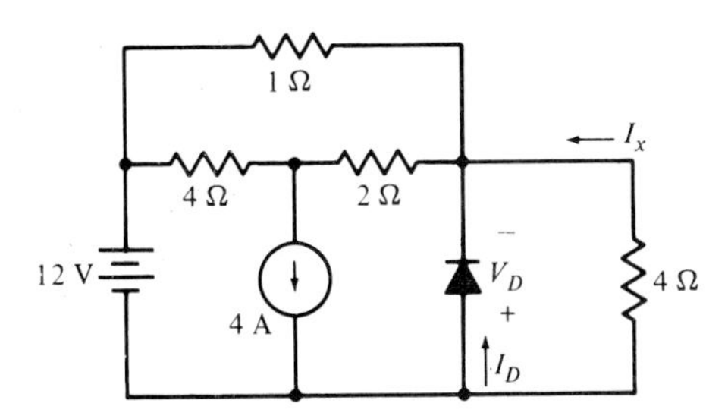

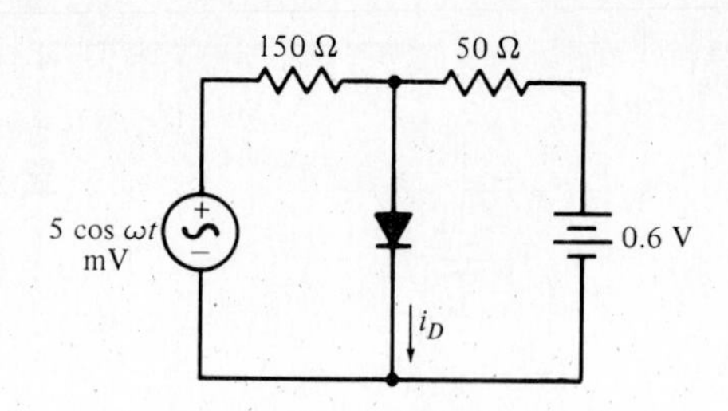

Fig. 1.36 See Problem 30.

30. Two points on the nonlinear characteristic of the diode shown in the circuit of Fig. 1.36 are (0.4 V, 1 mA) and (0.42 V, 2 mA). Assume operation at 320 K and find i_D.
31. Design a low-frequency current attenuator of the form shown in Fig. 1.37 so that the ratio i_d/i_s may be varied electrically and continuously from 0.2 to 0.8. Use a typical FD600 diode at 25°C.
32. Let $I_0 = 10^{-10}$ A, $n = 1$, and $T = 25$°C for both diodes in the circuit of Fig. 1.38. If v_i is a small, low-frequency signal voltage, calculate v_{o1}/v_i and v_{o2}/v_i if V equals (a) 1.6 V, (b) -1.6 V. (c) Which name better describes this circuit: an electronic switch or a diffusion pump?
33. Values suitable for the small-signal, high-frequency model of a certain diode under forward bias conditions are $R_s = 2\ \Omega$, $r = 5\ \Omega$, $C_j = 2$ pF, and $C_D = 48$ pF. Plot curves of the magnitude and angle of the high-frequency diode impedance vs. ω on a linear scale, $0 < \omega < 5 \times 10^{10}$ rad/s.
34. (a) Plot a curve of C_j vs. V_D for a junction diode having $C_{j0} = 1.3$ pF, $V_{bi} = 0.8$ V, and $N = 0.4$. (b) Calculate the value of V_D that would produce parallel resonance at 100 MHz if the diode were in parallel with 2 pF and 1 μH.
35. Let $C_{j0} = 4$ pF, $V_{bi} = 0.75$ V, and $N = 0.48$ for a reverse-biased diode with $V_D = -2$ V. (a) Find C_j. (b) If the diode is in parallel with a 10-pF capacitor and a 1-μH inductor, what change in the reso-

Fig. 1.37 See Problem 31.

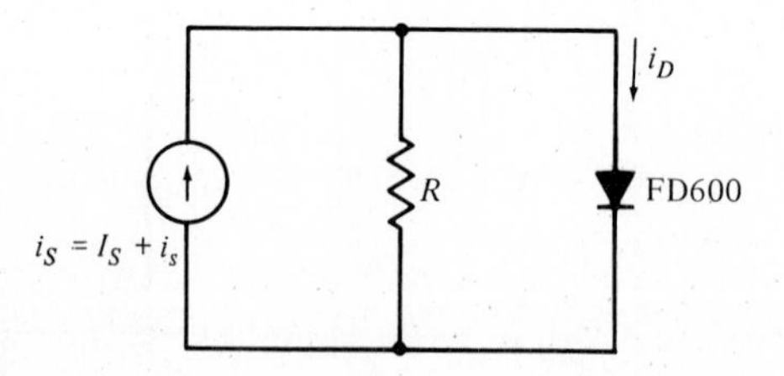

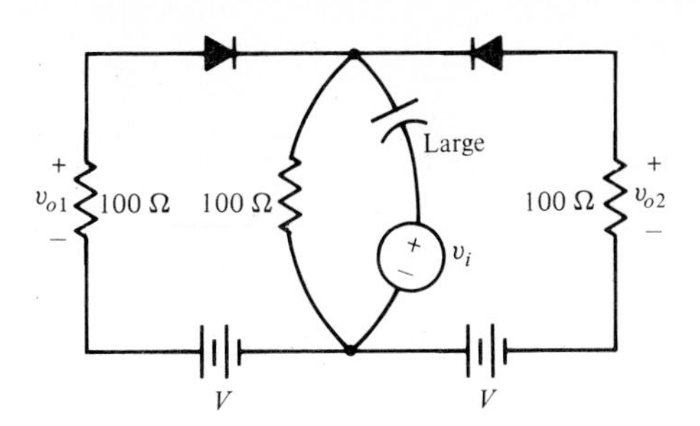

Fig. 1.38 See Problem 32.

nant frequency (in Hz) occurs for an 0.1-V increase in the magnitude of V_D?

36. A curve of the depletion capacitance as a function of reverse voltage includes these three points: (V_D = 0, C_j = 3.5 pF), (−2 V, 2.15 pF), and (−5 V, 1.6 pF). Determine values for C_{j0}, V_{bi}, and N.

37. The dependence of the diffusion capacitance on temperature is expressed by

$$C_D = \frac{q\tau}{2kT} I_D$$

where τ is a constant that is essentially independent of temperature. With $\tau = 10^{-8}$ s and I_D held constant at 1 mA, calculate C_D for T equals (a) 25°C, (b) 0°C, (c) 100°C. (d) Find the rate of change of C_D with T at 25°C.

38. A junction diode operating with a forward current of 1 mA shows an impedance of $7\angle 0°$ Ω at very low frequencies, $1\angle 0°$ Ω at very high frequencies, and an impedance with a phase angle of −45° at $\omega = 10^9$ rad/s. Find $C_j + C_D$.

39. A junction diode is found to have C_{j0} = 1.9 pF, V_{bi} = 0.85 V, and N = 0.5. What fixed capacitance C_0 and fixed inductance L should be placed in parallel with it to provide resonance at $\omega_0 = 1.2 \times 10^9$ rad/s with V_D = −1 V and at $\omega_0 = 1.25 \times 10^9$ rad/s with V_D = −2 V?

40. In Fig. 1.39, let v_s = 100 V for $0 < t < 1$ ms, and v_s = −100 V for $1 < t < 2$ ms. The waveform is periodic with a period of 2 ms. (a) Assume that the diode is ideal and calculate the average voltage across, and current through, the 250-Ω resistor. (b) Repeat for a diode that is modeled by V_0 = 0.5 V and R_0 = 20 Ω.

41. The supply voltage v_s in the power supply of Fig. 1.40*a* is composed of semicircles if the waveform is plotted with a scale of 100 V = 2.5 ms, as shown in Fig. 1.40*b*. (a) Sketch v_o. (b) Determine the average value of v_o and compare the result with that for a sinusoid of equal maximum amplitude and frequency. (c) Sketch v_o if a 50-V battery is

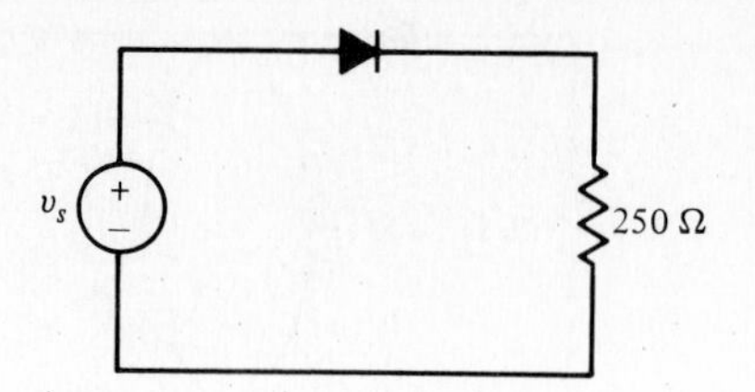

Fig. 1.39 See Problem 40.

placed between the diode and R_L with its positive terminal connected to the diode.

42. Install a 50-μF capacitor across the load resistor in Fig. 1.39 and let v_s = 100 V for $0 < t < 1$ ms, and − 100 V for $1 < t < 2$ ms. The period is 2 ms. Assume an ideal diode. (a) Calculate $v_o(t)$, $0 < t < 2$ ms. (b) Determine V_L and I_L. (c) Find Δv. (d) Calculate the regulation. (e) What is the regulation if R_L is decreased to 125 Ω?
43. The full-wave bridge rectifier shown in Fig. 1.22 has a turns ratio $a = 5$. The diodes are ideal, $R_L = 5$ kΩ, and there is a 25-μF capacitor in parallel with R_L. Determine (a) Δv, (b) V_L, (c) the regulation.
44. A half-wave power supply uses an ideal diode with $V_m = 100$ V, $R_L = 2$ kΩ, and $C = 50$ μF. The sinusoidal source has a period of 1/60 s. Refer to Fig. 1.24 and calculate exact values for t_1, t_2, t_3, and Δv.
45. In the full-wave rectifier of Fig. 1.21, $v_1 = v_2 = 20 \cos \omega t$ V and $R_L = 1200$ Ω. Assume that the diodes are ideal. If $f = 60$ Hz, specify the size of a capacitor to be placed across R_L so that (a) $\Delta v = 1$ V, (b) $V_L = 18$ V, (c) the regulation is 4%.
46. Design a half-wave unregulated power supply to give $V_L = 15$ V at $I_L = 10$ mA with $\Delta v \leq 0.1$ V. Assume a 120-V, 60-Hz supply. What is the regulation of your power supply?

Fig. 1.40 See Problem 41.

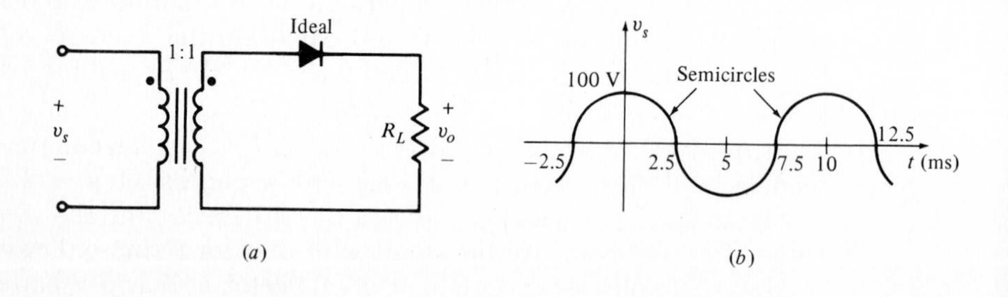

47. Design a full-wave, unregulated, bridge rectifier to give $V_L = -15$ V at $|I_L| = 10$ mA with $\Delta v \leq 0.1$ V. Assume a 120-V, 60-Hz supply. What is the regulation of your power supply?
48. Let $R = 200\ \Omega$, $V_{BR} = 15$ V, and $R_Z = 25\ \Omega$ in the simple voltage-regulator circuit of Fig. 1.26*a*. Given a supply $V_R = 30$ V, calculate V_L when I_L equals (a) 0, (b) 1 mA, (c) 10 mA, (d) 70 mA. (e) Plot V_L vs. I_L. (f) If the ripple voltage in V_R is 0.1 V, what is the ripple voltage in v_L when $I_L = 70$ mA?
49. Values applicable to the regulator circuit and Zener diode of Fig. 1.26*a* are $V_R = 60$ V, $V_{BR} = 30$ V, $R_Z = 25\ \Omega$, and $5000 < R_L < 30{,}000\ \Omega$. The diode current should remain within a range from 2 to 10 mA. Specify a suitable value for R and then calculate the minimum and maximum values of V_L, I_L, and $|I_D|$.
50. There are numerous applications, such as Schottky-clamped TTL (transistor-transistor-logic) integrated circuits, in which Schottky barrier diodes and junction diodes are placed in parallel. Assume that $I_0 = 1$ nA and $n = 1.9$ for the SBD, and $I_0 = 1$ pA and $n = 1.02$ for the junction diode. (a) What forward voltage across the parallel combination will cause the diodes to carry equal currents? (b) At $V_D = 0.5$ V, what is the ratio of the current in the junction diode to that in the SBD?
51. (a) Design a circuit in which a light-emitting diode will supply at least 1.5 mcd of luminous intensity on its central axis when a 25-V source is applied. (See Appendix A for an LED.) (b) Modify your circuit so that the 25-V source may have either polarity.

2

Bipolar and field-effect transistors

The functions involved in amplifying small signals and processing digital information are generally accomplished through the use of solid-state devices, most often transistors. We shall devote the next several chapters to such devices. We begin in this chapter by looking at the way the bipolar and field-effect transistors operate, the nature of their voltage-current characteristics, and the circuit symbols and conventions that are used. In the next chapter we proceed with the circuit use of transistors by working with dc models and the problems associated with establishing an operating point. Chapter 4 describes methods of circuit design to provide a stable operating point, and Chapter 5 deals with the small-signal transistor models. Chapter 6 combines the results of Chapters 4 and 5 to develop methods of analyzing and designing practical amplifier circuits.

With the exception of the elementary concepts presented in the next few sections, we shall not become involved in the solid-state theory on which transistor operation is based. It's easier to accept the transistor as a three-terminal device, even with an insufficient background in solid-state physics, than it is to absorb the physics without any idea of how it is applied.

2.1 *npn* bipolar transistors

Bipolar transistors appear in several different forms, each appropriate for a particular application. They are used at high frequencies, for switching circuits, in high-power applications, and under extreme environmental stress. In this section we will discuss characteristics that all these transistors have in common.

The bipolar transistor may appear in discrete form as an individually encapsulated component, in monolithic form (made in and from a common material) in integrated circuits, or as a "chip" in a hybrid integrated circuit. Regardless of its use or the manufacturing process, each form contains an emitter and a collector region separated by a very thin base region. This is illustrated symbolically in Fig. 2.1*a*, which shows an *npn* bipolar transistor. These transistors appear in two configurations, *npn* or *pnp*. The *npn* is much more widely used than the *pnp*, and we therefore look at it first and in much more detail.

We note first that the bipolar transistor possesses two junctions, the emitter-base junction and the collector-base junction. It is possible to describe a large part of transistor operation by interpreting the device somewhat like a pair of back-to-back diodes. The emitter-base junction is usually forward-biased, or $V_{BE} > 0$. This voltage is small, but the forward current across the junction is relatively large, just as in the forward-biased diode. To see the direction of this current, we note that the majority carriers in the emitter region (*n*-type) are electrons. These are injected across the junction into the base region, which is quite thin, being on the order of 0.5 to 2 μm. Now the base-collector junction is reverse-biased ($V_{CB} > 0$), and thus the holes that are present in the *p*-type base material do not cross it. However, the electrons that have just been injected into the very thin base region from the emitter diffuse across the base-collector junction and are then collected under the influence of the positive collector potential, as illustrated by Fig. 2.1*b*.

This electron current across the base-collector junction is almost as large as the electron current crossing from the emitter into the base region. In general, it runs from 99 to 99.99% of the emitter current, the small decrease being accounted for by electrons that are lost in the narrow base region. In terms of electron currents, we might find that for every hundred electrons flowing into the emitter from the external contact and

Fig. 2.1 (*a*) A symbolic picture of an *npn* bipolar transistor. Ohmic contacts to the semiconductor material are provided at the base, emitter, and collector. (*b*) The movement of electrons and holes across the two junctions is indicated.

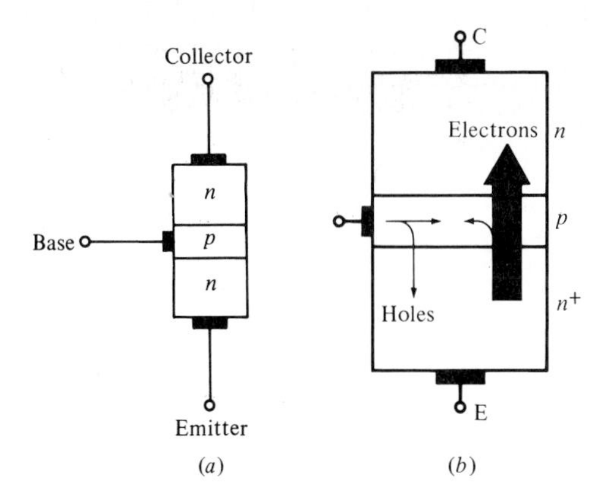

crossing into the base region, perhaps one flows out the base lead into the external circuit, whereas 99 cross over into the collector and flow out into the external collector circuit.

To find the total current across any junction, however, we have to consider both hole and electron motion. In the semiconductor diode operating with forward bias, electrons are injected from the *n* region into the *p* region, while holes move in the opposite direction. The sum of these two currents crossing the junction determines the total diode current. This is also true for the emitter-base junction in the *npn* transistor; there is a hole-current component produced by holes in the *p*-type base material crossing over to the *n*-type emitter (Fig. 2.1*b*). However, this current is usually several orders of magnitude less than the electron current, a consequence of much heavier doping in the emitter region. This heavily doped *n*-type region is often signified by the term n^+, as indicated in Fig. 2.1*b*. The heavier doping means that the supply of electrons is simply much greater than the supply of holes. The small hole current contributes to the current entering the base lead.

The emitter and collector current magnitudes are thus about equal, and the base current is very much smaller than either.

The circuit symbol for an *npn* bipolar transistor is shown in Fig. 2.2*a*. The direction of the arrowhead on the emitter lead indicates whether the transistor is *npn* or *pnp* by showing the sense of the current as if it were caused by the motion of *positive* charges. Since we actually have electrons flowing in from the emitter to the base, the sense of the emitter current in the *npn* transistor is the same as if positive charges were leaving that terminal. Hence, the arrowhead points outward.

Figure 2.2*b* shows the *npn* symbol with the IEEE (Institute of Electrical and Electronics Engineers) standard reference directions for currents. Note that I_E, I_B, and I_C all have a reference direction *into* the specified terminal. As a consequence of this definition and Kirchhoff's current law, at least one of the currents must be negative, unless all are zero.

Keeping in mind the electron currents by which we first described the operation of the *npn* transistor, let's consider the polarities of I_E, I_B, and I_C, as defined in Fig. 2.2*b*. Remember that hole currents are very small compared with electron currents, and total currents are thus almost entirely electron currents. The emitter current I_E is actually negative, the collector current I_C is positive and about equal to the magnitude of the emitter current, and the base current I_B is a small positive quantity. As we establish an operating point for this transistor and analyze its operation, we find that the base and collector currents appear much more often in our work than the emitter current; these are both positive values for the *npn* transistor. We might thus find that we are considering a certain *npn* transistor having an emitter current of -10 mA, a base current equal to 0.1 mA, and a collector current of 9.9 mA.

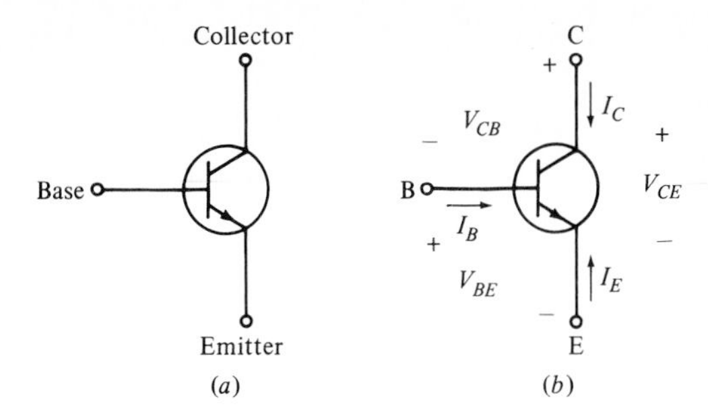

Fig. 2.2 (*a*) The circuit symbol for an *npn* bipolar transistor. The arrow on the emitter indicates that the sense of the emitter current is that of positive charges flowing *out* of the emitter. (*b*) The IEEE standard reference direction for all currents is *into* the device, even though some may have negative values.

The voltages between electrodes are identified in Fig. 2.2*b* by the familiar double-subscript notation, such as V_{CE} or V_{EB}. The first subscript denotes the assumed positive voltage reference.

In describing transistor operation above, we have talked as if the emitter were the input and the collector the output. In practice, by far the greatest use of the transistor is with the base as the input and the collector as the output. Certainly, it does not make any difference which *two* of the three currents we establish in any (linear or nonlinear) three-terminal device; the third is then fixed by Kirchhoff's current law. So let us think of squirting a small current into the base as our input, 0.1 mA in the example above, and obtaining a relatively large current into the collector as the output (9.9 mA). Current gain is obviously present. We shall show a little later that voltage gain and therefore power gain are also available in this configuration. This configuration is called the *common-emitter* (CE) configuration since the emitter is common to both the input and output.

Voltage-current characteristics for a low-power *npn* transistor as it is normally operated are shown in Fig. 2.3. Figure 2.3*a* gives the input characteristics, a family of curves of I_B vs. V_{BE} for various values of V_{CE}. Note that a single curve is sufficient if $V_{CE} \geq 1$ V, a condition that applies almost all the time. This curve is nearly identical to the forward characteristic of a semiconductor diode, even though it shows I_B instead of I_E.

Typical output characteristics appear in Fig. 2.3*b*. These are curves of I_C vs. V_{CE} with the base current I_B as an independent parameter. Without studying these curves for too long, it is easy to see two general characteristics. First, once V_{CE} is greater than 1 or 2 V, I_C is relatively independent of V_{CE}; that is, each curve flattens out. Second, in this region I_C is about 100 times I_B (or the collector current is about 99% of the emitter current).

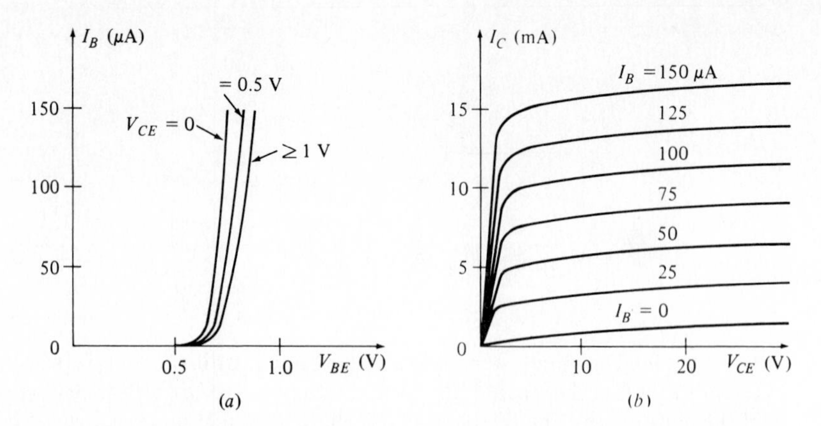

Fig. 2.3 (*a*) The CE input characteristics of a typical low-power *npn* bipolar transistor. When $V_{CE} > 1$ V, the curves coincide. (*b*) The CE output characteristics. As a first approximation, $I_C \doteq 100 I_B$ when V_{CE} is greater than a few volts.

The *v-i* characteristics emphasize the regions where the emitter-base junction is forward-biased ($V_{BE} > 0$) and the collector-base junction is reverse-biased ($V_{CB} > 0$). These bias conditions define the *active region* for a bipolar transistor. A quick glance at the input characteristics shows that $I_B > 0$ under these conditions. Thus, the active region on the output characteristics must lie *above* the $I_B = 0$ curve. Next, we note that V_{CB} is not shown explicitly on the output characteristics (nor on the input characteristics either, for that matter), and it is therefore a little difficult to identify a sharp upper-left edge for the active region. We usually consider it to include everything to the right of the "knees" on the I_B = constant curves, and above the $I_B = 0$ curve. The active region is crosshatched in Fig. 2.4.

When both junctions are reverse-biased, any currents that flow are of the order of magnitude of reverse saturation currents for the emitter-base and collector-base diodes, and thus I_B and I_C are very small (as is I_E). This condition evidently occurs near the V_{CE} axis and is termed the *cutoff region*. This region is also shown in Fig. 2.4. The boundary between the active and cutoff regions occurs at $I_B = 0$.

If both junctions are forward-biased, this defines the *saturation region*. Both V_{BE} and V_{BC} are small positive values, and V_{CE}, their difference, is also small and positive. The boundary between the active and saturation regions occurs at $V_{CB} = 0$.

D2.1 If 10^7 electrons per nanosecond are moving from left to right across a certain junction while 10^5 holes per nanosecond are moving from

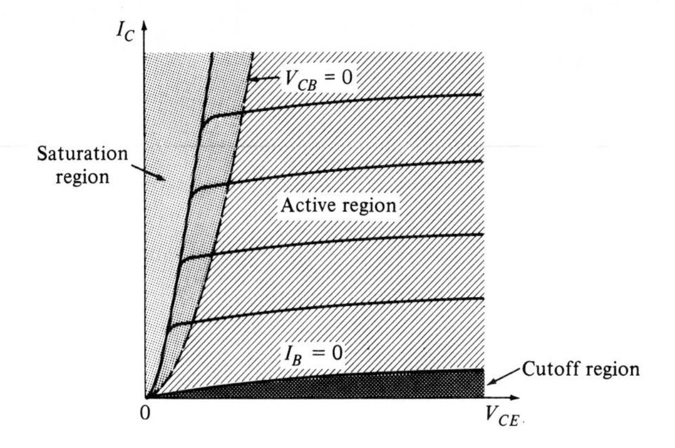

Fig. 2.4 Normal transistor operation is obtained in the active region where the emitter-base junction is forward-biased and the collector-base junction is reverse-biased. For cutoff, both junctions are reverse-biased; and in saturation, both junctions are forward-biased.

right to left, what is (a) the hole current from left to right, (b) the electron current from left to right, (c) the total current from left to right?

Answers. -0.016 mA; -1.602 mA; -1.618 mA

D2.2 Let $I_B = 50\ \mu\text{A}$, $I_C = 15$ mA, $V_{BE} = 0.65$ V, and $V_{CE} = 5$ V for an *npn* transistor operating in the active region. Calculate (a) I_E, (b) V_{CB}, and (c) the total power dissipated by the transistor.

Answers. -15.05 mA; 4.35 V; 75.0325 mW

D2.3 Let $V_{BE} = 0.78$ V, $V_{CE} = 0.21$ V, $I_C = 8$ mA, and $I_B = 1.2$ mA for an *npn* transistor operating in the saturation region. Find (a) V_{CB}, (b) I_E, and (c) the total power dissipated.

Answers. -0.57 V; -9.2 mA; 2.616 mW

2.2 *pnp* bipolar transistors

The *pnp* transistor contains a narrow *n*-type base layer sandwiched between *p*-type emitter and collector regions. The physical structure is suggested by Fig. 2.5*a*. As compared with the *npn* transistor, the *n*- and *p*-type regions are interchanged; we term this a *complementary* arrangement. Forward bias on the emitter-base junction causes holes to be injected into the base region, most of which diffuse across the base and are collected by the reverse-biased base-collector junction. The total emitter

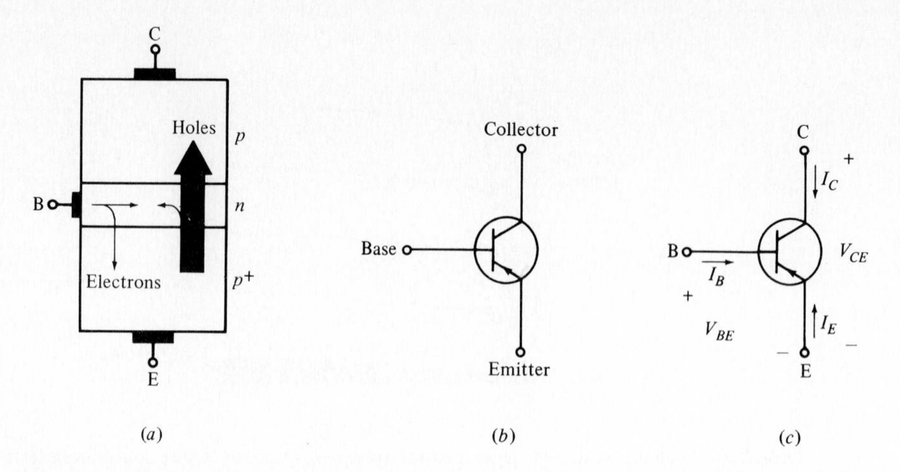

Fig. 2.5 (*a*) A symbolic picture of a *pnp* bipolar transistor. (*b*) The circuit symbol for a *pnp* bipolar transistor. Note that the emitter arrowhead is directed inward. (*c*) The IEEE standard reference direction for all currents is *into* the device, just as for the *npn* transistor. In normal operation, I_B and I_C and V_{BE} and V_{CE} are all negative quantities for the *pnp* transistor.

current I_E is thus composed of the sum of this large hole current plus a much smaller electron current directed from the base toward the emitter. The direction of conventional current is thus into the transistor at the emitter terminal, and this is the basis for the arrowhead on the circuit symbol shown in Fig. 2.5*b*.

The collector current magnitude is just slightly less than that of the emitter current. Referring to Fig. 2.5*c*, we see that the IEEE standard reference directions for all currents are into the device (as they are for the *npn* transistor), and therefore I_C and I_B are both negative quantities for the *pnp* transistor in the active region.

Figures 2.6*a* and *b* show the input characteristics and output characteristics, respectively. The currents and voltages are those defined in Fig. 2.5*c*.

The three regions of operation are also defined in the same manner as they are for the *npn* transistor. In the active region, the emitter-base junction is forward-biased and the collector-base junction is reverse-biased. In cutoff, both junctions are reverse-biased, and saturation occurs when the junctions are both forward-biased.

D2.4 If $|I_E| = 5$ mA, $|I_B| = 0.15$ mA, $|V_{BE}| = 0.7$ V, and $|V_{EC}| = 4.3$ V for a certain *pnp* transistor, find (a) I_C, (b) V_{CB}, and (c) the total power dissipated in the transistor.

Answers. -4.85 mA; -3.6 V; 20.96 mW

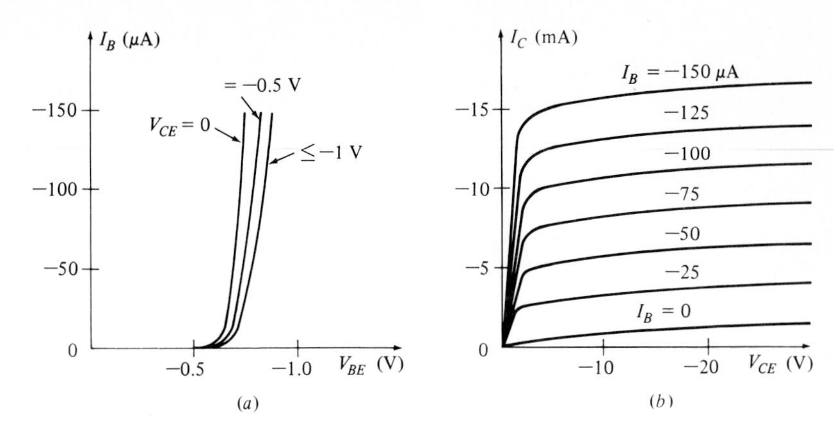

Fig. 2.6 (*a*) The CE input characteristics of a representative low-power *pnp* bipolar transistor. (*b*) The CE output characteristics.

2.3 Junction field-effect transistors: *n*-channel

To complete our discussion of transistors, the next several sections describe field-effect transistors. The name derives from the effect produced on an output current by an electric field inside the body of the device. Field-effect transistors are also called *unipolar transistors* because the charge motion takes place in a single type of material, either *n*- or *p*-type. The operation should be contrasted with that of the bipolar junction transistor, where the carriers move through both *n*- and *p*-type semiconductor material in series.

Transistors may be broadly classified as shown in Fig. 2.7. We have already considered the bipolar transistors at the left side of the drawing; we now turn to the field-effect transistors. Note that they fall into two broad classifications, depending on the method by which they are fabricated and the resultant mode of operation. The two major categories are the *j*unction *f*ield-*e*ffect *t*ransistors (JFET) and the *i*nsulated gate *f*ield-*e*ffect *t*ransistors (IGFET). The most common type of IGFET is termed the *m*etal—silicon di*o*xide—*s*ilicon *f*ield-*e*ffect *t*ransistor (MOSFET).

We shall begin with the junction field-effect transistor, covering first the *n*-channel type and then (very briefly) the *p*-channel type in the following section. We shall look at circuit symbols, current reference directions, and voltage reference polarities and try to gain some preliminary understanding of their physical operation. The *n*-channel JFET forms the basis for most of our discussion of FETs because many of the ideas used in explaining its operation carry over to the other FETs. Later we will

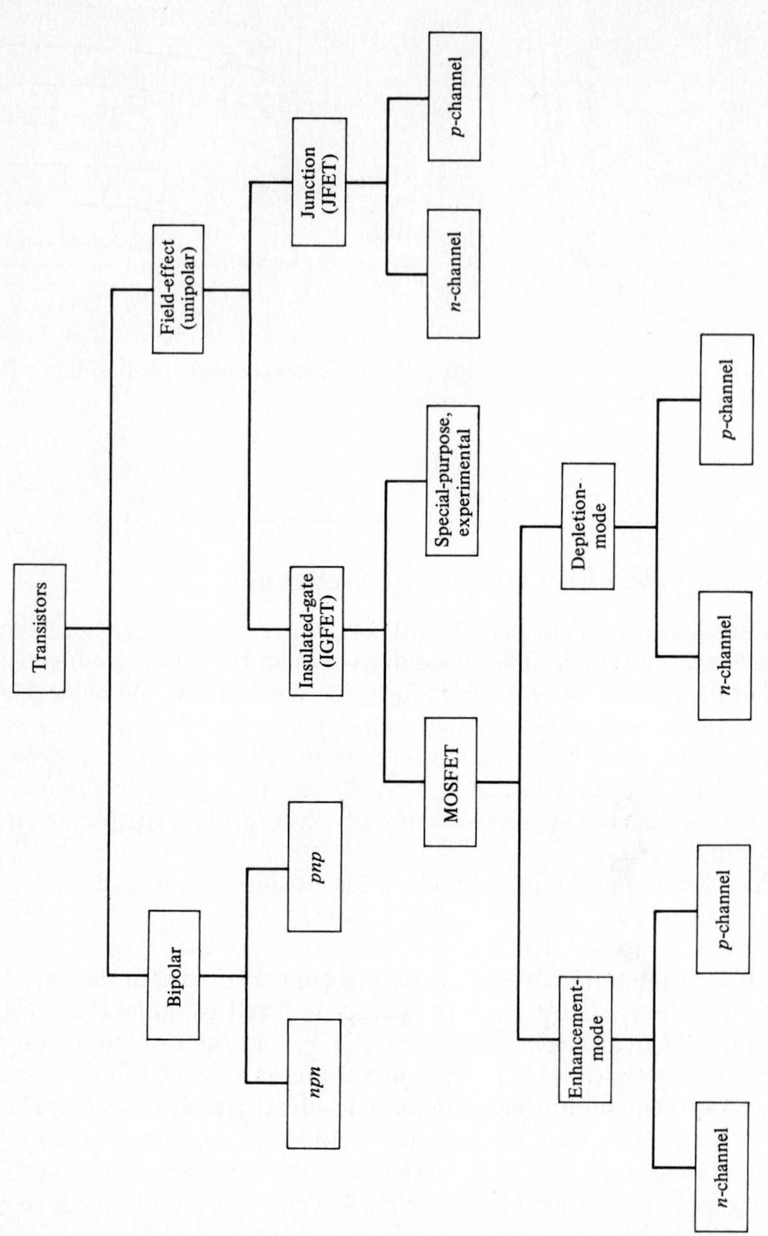

Fig. 2.7 A classification scheme for bipolar and field-effect transistors. We shall emphasize the *npn* bipolar, the *n*-channel JFET, and the *n*-channel MOSFET types.

emphasize the circuit performance of the JFET because it is often used in linear amplifier applications, especially at high frequencies.

The construction of the n-channel JFET is suggested by the sketch shown in Fig. 2.8*a*. The channel is formed from lightly doped (low conductivity) n-type material, usually silicon, with ohmic contacts at the two ends of the channel. The two gate regions are made of heavily doped (high conductivity) p^+-type material, and they are usually tied together electrically. We shall always assume that this is the case when referring to "the gate." We shall also keep the drawings cleaner by showing the connection to only one gate. In reference to Fig. 2.8*a*, if we should connect a low-voltage source of a few tenths of a volt between the two ends (top and bottom) of the channel, the current that would flow could be calculated from Ohm's law by determining the resistance of the channel between source and drain:

$$R_{SD} = \frac{l}{\sigma A}$$

where l is the length (m) of the channel from source to drain, σ is the conductivity (℧/m) of the n-type material, and A is the cross-sectional area (m^2) of the channel.

Fig. 2.8 (*a*) An n-channel JFET contains a channel of n-type material from source to drain. The channel passes between p^+-type gate regions. (*b*) When reverse bias is applied between the gate and channel, the cross-sectional area of the channel is reduced, the channel resistance increases, and I_D decreases.

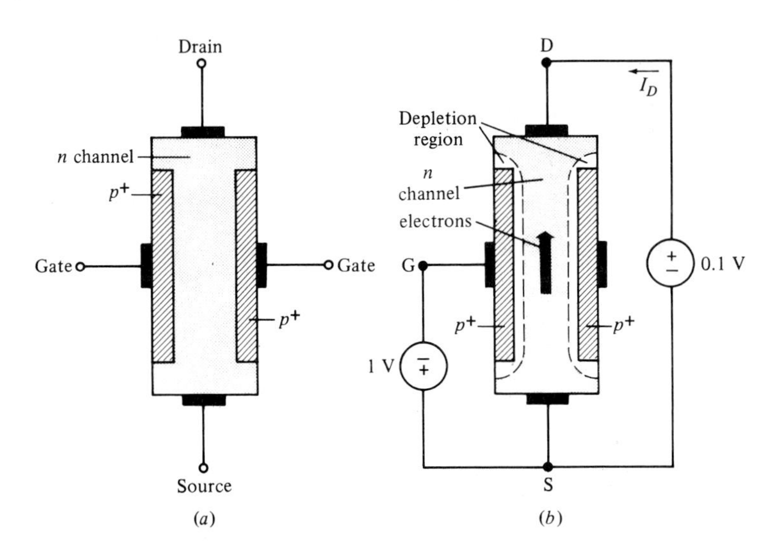

Now let's apply a small reverse-bias voltage to the *p-n* junction formed by the gate and the channel; we establish such a reverse bias in Fig. 2.8*b* by setting $V_{GS} = -1$ V. This reverse bias creates a depletion region that is nearly void of mobile carriers, just as was the case with the reverse-biased diode. The depletion region effectively narrows the width of the channel, and its resistance therefore increases. Note that a positive drain-to-source voltage is applied ($V_{DS} = 0.1$ V), and that we are keeping it small for the present. The majority carriers in the *n*-type channel are electrons. They flow from source to drain, as indicated in Fig. 2.8*b*; we thus have an *n*-channel JFET. The drain current I_D is directed into the transistor, and is a positive current.

As the magnitude of the reverse bias on the gate is increased, the width of the depletion region increases while the channel width decreases. We are effectively increasing the value of the resistance R_{SD}. Curves of I_D vs. V_{DS} for several different reverse-bias voltages are shown in Fig. 2.9. The drain-to-source voltage is still kept small, say $0 < V_{DS} < 0.5$ V. The resistance R_{SD} is the reciprocal of the slope of the straight line corresponding to the gate-to-source voltage applied. Note that we have a voltage-controlled linear resistance in this region of small values for V_{DS}. We also say that V_{GS} modulates, or changes, the drain-to-source resistance.

There are numerous applications of the JFET in the linear resistance region, but larger values of signal gain are achieved for larger values of the drain-to-source voltage, as we shall see shortly.

When V_{GS} becomes sufficiently negative, the width of each depletion layer increases until the channel vanishes, as shown in Fig. 2.10. We say that the channel is "pinched off," and I_D is essentially zero. Alternatively,

Fig. 2.9 For small values of the drain-to-source voltage, the JFET acts as a linear resistor whose resistance is controlled by V_{GS}, the reverse-bias voltage across the gate-to-channel junction.

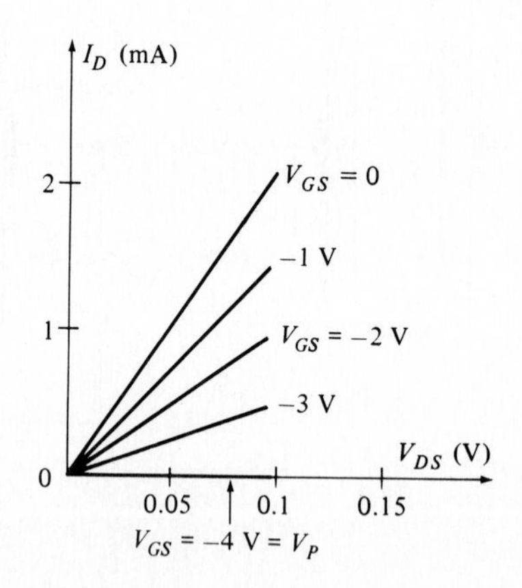

R_{SD} is nearly infinite. The value of V_{GS} for which this occurs is termed the *pinch-off voltage* V_p. It is evaluated for small V_{DS}:

$$V_P = V_{GS}\Big|_{I_D=0,\ V_{DS}\doteq 0}$$

For the n-channel JFET illustrated in Fig. 2.9 or 2.10, we see that $V_P = -4$ V.

JFET characteristics in the ohmic region obey the relationship

$$I_D = D\left[(V_{GS} - V_P)V_{DS} - \frac{V_{DS}^2}{2}\right] \qquad \text{(small } V_{DS}\text{)} \qquad (1)$$

where D is a constant that depends on the material parameters and the geometry of the device. The magnitude of V_{DS} is limited to a few tenths of a volt. The curves shown in Fig. 2.9 are obtained by letting $V_P = -4$ V and $D = 5 \times 10^{-3}$ A/V^2. Their linear nature shows that the $V_{DS}^2/2$ term is negligible when $V_{DS} \le 0.1$ V.

Now let us look at the region of the characteristics used in linear amplification by considering larger values of V_{DS}. We begin by setting $V_{GS} = 0$; this is the maximum value that we normally apply, for satisfactory operation of the JFET is not obtained when the gate-to-channel junction is forward-biased. The depletion region and channel are shown in Fig. 2.11*a*, *b*, *c* for successively larger drain-to-source voltages. In Fig. 2.11*a*, $V_{DS} = 2$ V. Note that the gate-to-channel voltage is essentially the same as V_{GS} at the source end of the channel, but that it is more closely V_{GD}

Fig. 2.10 When V_{GS} becomes sufficiently negative, the channel is pinched off and I_D drops to zero. For small V_{DS}, this value of V_{GS} is called the pinch-off voltage V_P. Here, $V_P = -4$ V.

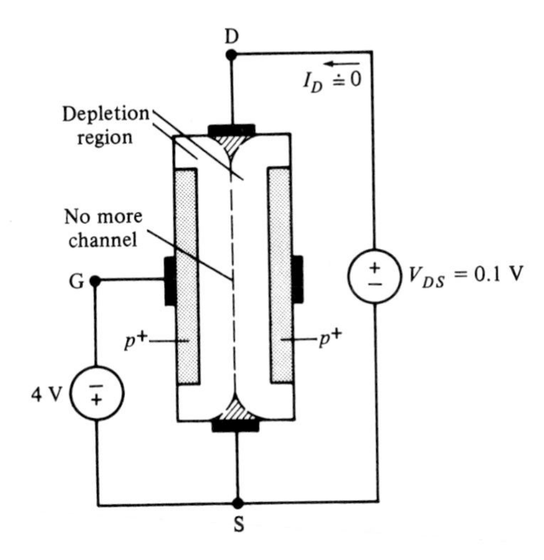

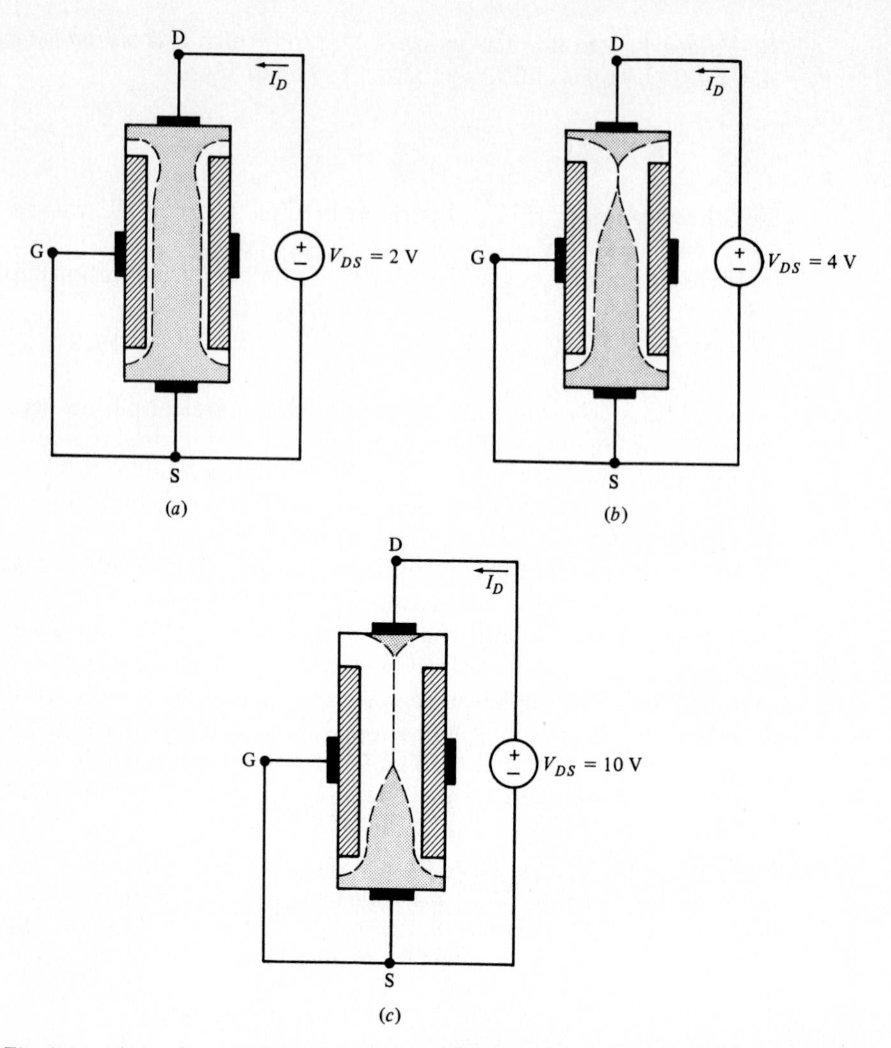

Fig. 2.11 An n-channel JFET is shown with $V_{GS} = 0$ for successively larger values of V_{DS}. The channel narrows at the drain end because the gate-to-channel voltage is approximately V_{GS} ($=0$) at the source end and V_{GD} ($= -V_{DS}$) at the drain end. (*a*) Below pinch-off. (*b*) At pinch-off. (*c*) Beyond pinch-off (in saturation).

$= V_{GS} - V_{DS} = -2$ V at the drain end. The depletion region is thus wider at the top of the channel than it is at the bottom, so the channel itself is narrower at the drain end than it is at the source end. There is an increase in R_{SD}, and the drain current is a little less than a straight-line I_D-vs.-V_{DS} relationship would predict.

In Fig. 2.11*b*, $V_{DS} = 4$ V, and thus the gate-to-channel voltage is about -4 V near the drain ($V_{GD} \doteq -4$ V). This is the pinch-off voltage, and the

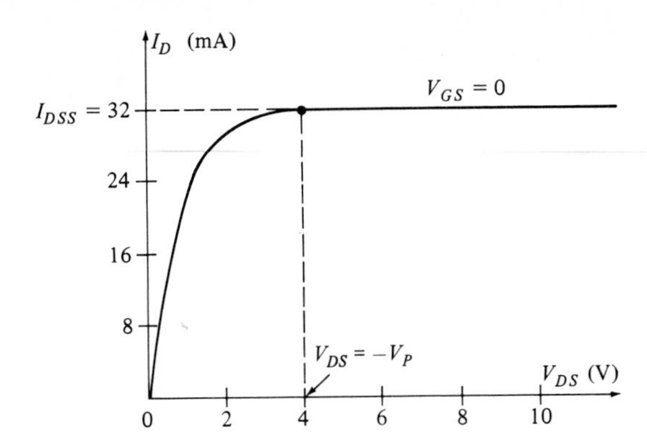

Fig. 2.12 With $V_{GS} = 0$, the relationship between I_D and V_{DS} is seen to be linear for small V_{DS}, nonlinear for intermediate values, and saturated at I_{DSS} for $V_{DS} > -V_P$.

channel is pinched off only at the drain end. Although we describe the channel as pinched off, it might be better to say it is pinched down, for with these larger values of V_{DS}, I_D is not zero. The value of I_D when $V_{DS} = -V_P$ (with $V_{GS} = 0$) is about as large as it can ever be. This is shown in Fig. 2.12, and this maximum value of I_D is termed I_{DSS}, the drain-to-source current saturation value:

$$I_{DSS} = I_D \Big|_{V_{DS} = -V_P,\ V_{GS} = 0}$$

For the transistor shown, $I_{DSS} = 32$ mA.

As V_{DS} increases to 10 V, Fig. 2.11*c*, a gate-to-channel voltage of -4 V or less is found across a longer section of the channel, and a longer section of the channel is pinched down. We call this operating condition *beyond pinch-off*, or *saturation*, and we say the FET is operating in the *beyond-pinch-off region* or the *saturation region*. The drain-to-channel voltage and the length of the constricted section of the channel increase almost exactly in proportion, and the current remains constant (saturated) at the value I_{DSS}. In practice, there is a very slight increase in I_D as V_{DS} increases. We shall take this into account later whenever it becomes necessary.

When the gate-to-source voltage is reduced below zero, pinch-off and saturation occur at smaller values of V_{DS}. Figure 2.13 shows curves for several values of V_{GS}. Note that saturation occurs at a value of V_{DS} such that the gate-to-channel voltage equals V_P at the drain end:

$$V_{GD} = V_{GS} - V_{DS} = V_P$$

or

$$V_{DS} = V_{GS} - V_P \qquad (2)$$

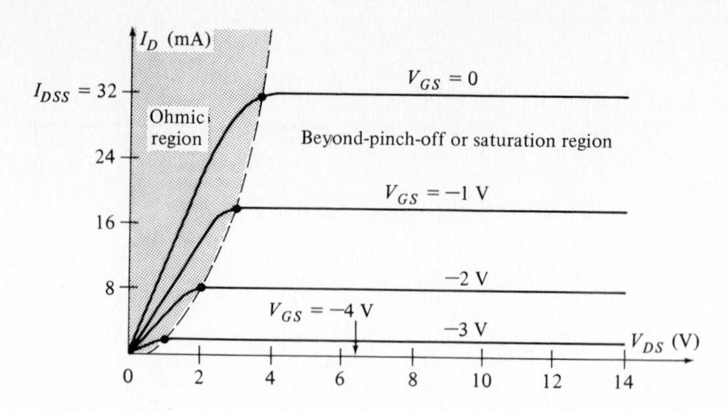

Fig. 2.13 The common-source output characteristics of an *n*-channel JFET show curves of I_D vs. V_{DS} for various values of V_{GS}, $V_P \le V_{GS} \le 0$. We see that $I_{DSS} = 32$ mA and $V_P = -4$ V.

Equation (2) represents the border between the ohmic and saturation regions. Larger values of V_{DS} cause operation to be in the region beyond pinch-off,

$$V_{DS} \ge V_{GS} - V_P \qquad \text{(saturation, or beyond pinch-off)} \tag{3}$$

while smaller values of V_{DS} cause operation in the ohmic region. For the transistor we have been considering, operation is in the beyond-pinch-off region when $V_{DS} \ge V_{GS} + 4$, assuming, of course, that $-4 \le V_{GS} \le 0$.

When $V_{GS} = -2$ V in Fig. 2.13, any V_{DS} greater than 2 V is in the region where the current is nearly constant. Thus to the right of the broken line in Fig. 2.13 the JFET is beyond pinch-off and in current saturation; it acts much like a voltage- (V_{GS}) controlled current source. To the left, operation is in the ohmic region, linear for small V_{DS} and nonlinear for slightly larger values. The complete set of curves represents the common-source output characteristics.

Figure 2.14 illustrates the circuit symbol for the *n*-channel JFET. Note that the arrow on the gate follows the convention used for the diode: It points into the device at a terminal connected to *p*-type material. Also the positive reference direction for all currents is consistently taken *into* the device. Thus I_D is normally positive and I_S is normally negative. Since the gate is reverse-biased, I_G is negative, essentially the reverse saturation current of the gate-to-channel diode, a fraction of a nanoampere for silicon devices. We shall find that the *input* characteristics of a JFET are well represented by a single number, the value of the gate leakage current. Furthermore, many of the dc and ac models we develop will be based largely on just two values, V_P and I_{DSS}.

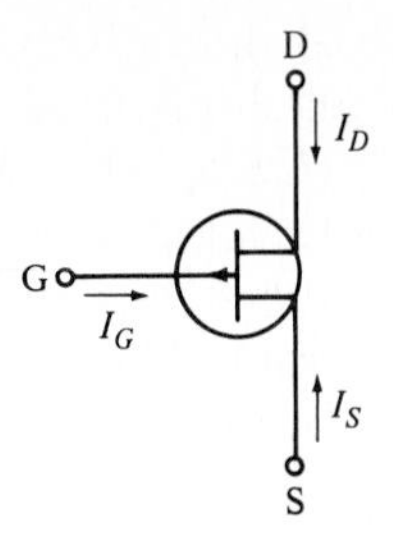

Fig. 2.14 The circuit symbol and current reference directions for the *n*-channel JFET.

D2.5 An *n*-channel JFET has $V_P = -2.8$ V. It is operating in the ohmic region with $V_{GS} = -1$ V, $V_{DS} = 0.05$ V, and $I_D = 0.3$ mA. Find I_D if (a) $V_{GS} = -1$ V, $V_{DS} = 0.08$ V; (b) $V_{GS} = 0$, $V_{DS} = 0.1$ V; (c) $V_{GS} = -3.2$ V, $V_{DS} = 0.06$ V.

Answers. 0.476 mA; 0.930 mA; 0

D2.6 The border between the ohmic and saturation regions for a certain *n*-channel JFET occurs when $V_{GS} = -1$ V and $V_{DS} = 2.2$ V. Find the value of V_{GS} that just produces pinch-off when V_{DS} equals (a) 2 V, (b) 3 V. Determine whether this transistor is operating in the ohmic, saturation, or cutoff region when (c) $V_{GS} = 0$, $V_{DS} = 4$ V; (d) $V_{GS} = -4$ V, $V_{DS} = 8$ V; (e) $V_{GS} = -2$ V, $V_{DS} = 1$ V.

Answers. -1.2 V; -0.2 V; saturation; cutoff; ohmic

2.4 The *p*-channel JFET

The *p*-channel JFET is the exact complement of the *n*-channel JFET because the *n*- and *p*-type regions are interchanged. We again take the positive-current reference directions into the transistor, and all voltages and currents are therefore the negative of those we just finished considering for the *n*-channel case. Figure 2.15*a*, *b* shows the circuit symbol (where only the arrowhead on the gate is reversed in direction) and a representative set of common-source output characteristics. In this case we see that I_{DSS} is -20 mA while V_P is $+3$ V.

Equation (1) of Section 2.3 continues to apply in the ohmic region:

$$I_D = D\left[(V_{GS} - V_P)V_{DS} - \frac{V_{DS}^2}{2}\right] \qquad \text{(small } V_{DS}\text{)}$$

although D is now found to be a negative constant.

In order to establish an operating point for the p-channel JFET that lies in the region beyond pinch-off, it is necessary that $V_{DS} \leq V_{GS} - V_P$, as well as $0 \leq V_{GS} \leq V_P$. Comparing these conditions with those for the n-channel device, we note that all inequality signs are reversed, a consequence of multiplying every term in an inequality by -1.

Either the n- or p-channel JFET may be used as a current-limiting device to protect circuits from current overloads. By connecting the gate to the source ($V_{GS} = 0$), Fig. 2.13 or 2.15*b* indicates that the drain-current magnitude is limited to $|I_{DSS}|$. We shall assume that the drain-to-source voltage is not so large that avalanche breakdown between the gate and drain results, for this could destroy the transistor if the power dissipation ($V_{DS}I_D$) is too great. Since there are effectively only two connections to the external circuit, this configuration is sometimes called a *current-limiting diode*. In integrated circuits, it is called a *pinch resistor*.

D2.7 Identify the region of operation for a JFET having (a) $V_{DS} = -4$ V, $V_{GS} = 2$ V, $V_P = 1$ V; (b) $V_{DS} = -4$ V, $V_{GS} = 1$ V, $V_P = 2$ V; (c) $V_{DS} = 2$ V, $V_{GS} = 0$, $V_P = -3$ V; (d) $V_{DS} = 4$ V, $V_{GS} = -2$ V, $V_P = -4$ V; (e) $V_{DS} = 2$ V, $V_{GS} = -2$ V, $V_P = -1$ V; (f) $V_{DS} = -3$ V, $V_{GS} = 1$ V, $V_P = 4$ V.

Answers. cutoff; saturation; ohmic; saturation; cutoff; ohmic-saturation edge

2.5 Insulated-gate FET: depletion mode

We now leave the junction FET and consider the insulated-gate FET, or IGFET. The most important member of this class is the MOSFET, so named because it has layers of metal (usually aluminum), oxide (usually silicon dioxide), and silicon (either n- or p-type), in that order. A similar type of IGFET has a gate made of polycrystalline silicon (polysilicon) that is doped so that it is conducting. It is called a *silicon-gate* MOS.

The MOSFET can be constructed to operate either in the depletion mode, in the enhancement mode, or in both (but not at the same time). In this section we shall look at the depletion-mode MOSFET because its operation has many similarities to that of the JFET we just considered. We shall limit our attention mainly to the n-channel device, for again the p-channel unit is its complement.

The term *n-channel depletion-mode* suggests the manner of operation. As shown in Fig. 2.16*a*, there is a channel of n-type material, and it connects two very heavily doped n-type regions, customarily designated as n^+-type. The channel itself is only lightly doped and therefore has a low conductivity. The two n^+-type regions act as low-resistance connections to the source and drain ends of the channel. Ohmic contacts are provided to the n^+-regions for connection to the external circuit. The drain, source, and

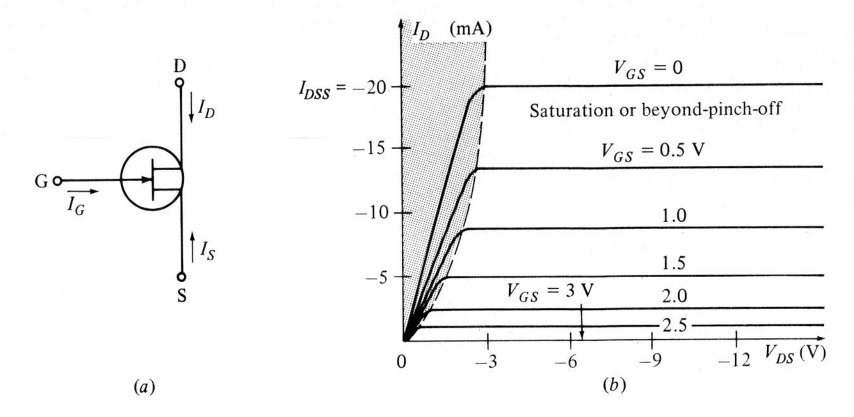

Fig. 2.15 (*a*) The circuit symbol for the *p*-channel JFET differs from that for the *n*-channel JFET in only one way: the arrowhead on the gate is reversed in direction. (*b*) Common-source output characteristics for the *p*-channel JFET: $V_P = +3$ V.

channel are built on a *p*-type substrate. Under normal operating conditions the *p*-*n* junction formed between these three regions and the substrate is reverse-biased, and the resultant depletion region adjacent to the substrate leads to a negligible current through the substrate; the substrate therefore plays a relatively insignificant role in the operation of the device.

The gate electrode is insulated from the channel by a layer of silicon dioxide that provides a resistance from the gate to any other terminal that

Fig. 2.16 (*a*) The physical construction of an *n*-channel depletion-mode MOSFET. The solid black regions are metal, usually aluminum. (*b*) With the substrate connected to the source and with $V_{DS} > 0$, the *p*-*n* junction between substrate and source has zero bias, and the junctions between substrate and channel and substrate and drain are reverse-biased. This results in the depletion regions shown.

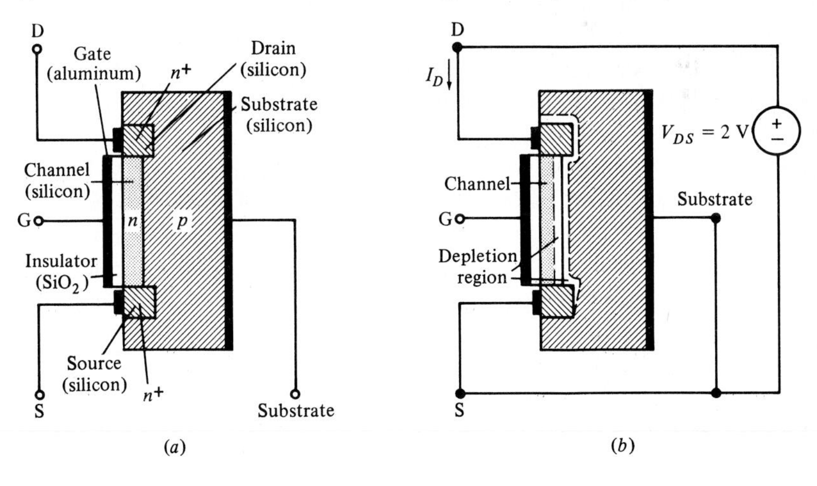

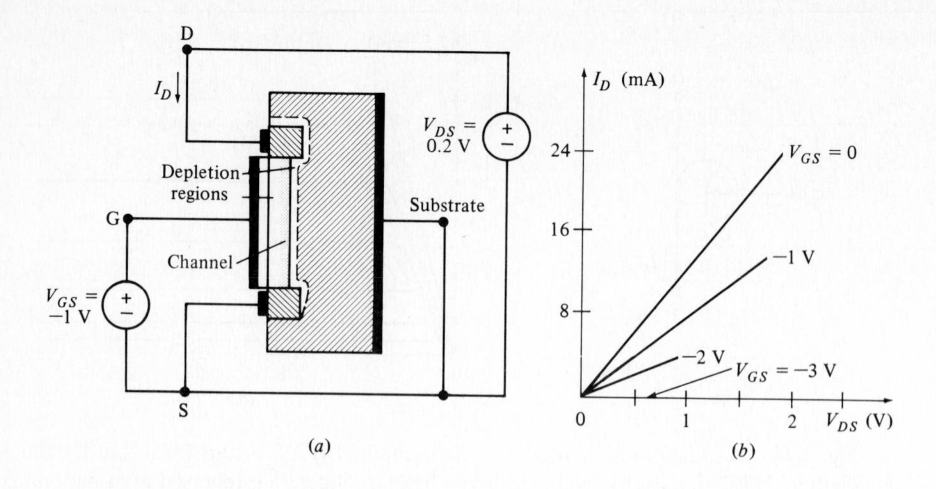

Fig. 2.17 (*a*) With a negative voltage applied between gate and source on an *n*-channel depletion-mode MOSFET, a depletion region is established in the channel; V_{DS} is small. (*b*) The channel acts like a variable linear resistor with its value controlled by V_{GS}.

is typically 10^{12} to 10^{15} Ω. We shall consider the gate current to be zero.[1] Thus the only current we need consider is along the channel.

Let us first leave the gate floating, as shown in Fig. 2.16*b*, and apply a small positive drain-to-source voltage, $V_{DS} = 2$ V. The substrate is often connected to the source, and we therefore would see a voltage across the *p-n* junction varying from zero at the source end of the channel to −2 V at the drain end. This is the reverse bias mentioned above, and the depletion region shown in Fig. 2.16*b* effectively isolates the channel from the substrate. The depletion region lies on both sides of the channel-substrate junction.

The electron current flows from source to drain, and therefore $I_D > 0$. The channel acts as a resistor, the value depending on the dimensions and the conductivity of the *n*-type silicon.

Now let us apply a negative voltage to the gate: $V_{GS} = -1$ V. We let V_{DS} be very small, say $V_{DS} = 0.2$ V, as shown in Fig. 2.17*a*. The negative potential on the gate acts through an electric field in the oxide layer to repel electrons in the *n*-type channel material. A depletion layer adjacent to the insulator thus develops and the channel conductivity decreases. When V_{GS} is reduced to the pinch-off voltage V_P, the channel is pinched off and $I_D = 0$. For these small values of V_{DS}, we obtain a family of

[1] With 0.16 mV applied across 10^{15} Ω, only one electron passes by each second, on the average.

straight lines for I_D vs. V_{DS}, as shown in Fig. 2.17*b*. The result is like that for the JFET, with

$$I_D = D\left[(V_{GS} - V_P)V_{DS} - \frac{V_{DS}^2}{2}\right] \qquad \text{(small } V_{DS}\text{)} \tag{4}$$

and we again have a voltage-controlled linear resistance. From these curves we can see that $V_P = -3$ V for this device.

Once more, our main interest lies in the region beyond pinch-off (saturation), where V_{DS} is larger. In Fig. 2.18*a* we increase V_{DS} to 5 V while keeping $V_{GS} = -1$ V. It is evident now that much of the channel is pinched off, because at one end, $V_{GS} = -1$ V (below pinch-off), whereas

Fig. 2.18 (*a*) Beyond pinch-off we find the channel is pinched down over part of its length. Here I_D stays relatively constant as V_{DS} increases. (*b*) The output characteristics for the *n*-channel depletion-mode MOSFET; $V_P = -3$ V and $I_{DSS} = 32$ mA.

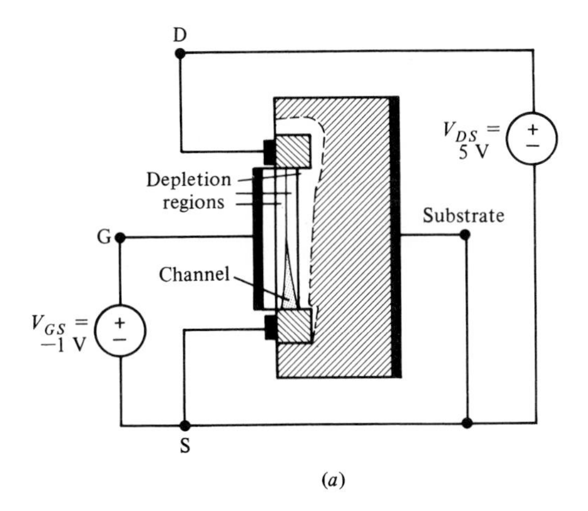

(*a*)

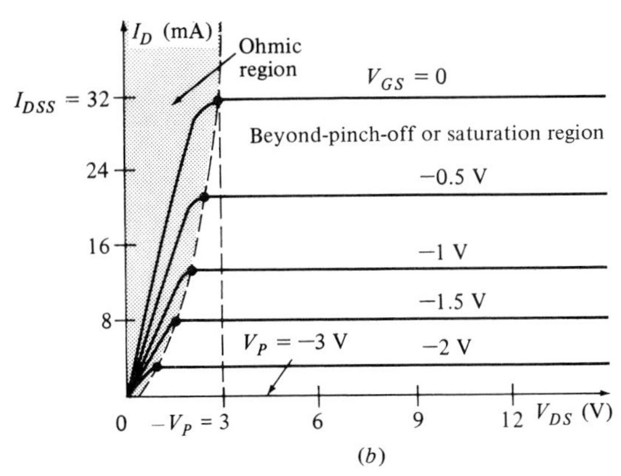

(*b*)

at the other, $V_{GD} = -6$ V (beyond pinch-off). The drain current I_D is essentially constant for all values of V_{DS} that are larger than $V_{GS} - V_P$:

$$V_{DS} \geq V_{GS} - V_P \qquad \text{(beyond pinch-off)}$$

We see that this constant value is about 14 mA in Fig. 2.18*b*, which shows the complete output characteristics for the *n*-channel depletion-mode MOSFET. When $V_{GS} = 0$, this constant value of I_D is again called I_{DSS}. For this transistor, it is 32 mA.

Figure 2.19 shows the circuit symbol for an *n*-channel depletion-mode MOSFET in which the substrate and source are connected internally. The indication that the device is *n*-channel is the arrow on the substrate, showing that the easy direction of current flow is from the *p*-type substrate to the *n*-type channel. The open circuit at the gate signifies the insulated gate. All currents have a positive reference direction that is inward, and therefore I_S is negative for normal operation; I_G is so small that we usually need not bother to place it on the diagram, for it rarely enters into a circuit calculation.

We can explain the operation of the *p*-channel depletion-mode MOSFET in terms of the *n*-channel unit by interchanging all *n*'s and *p*'s and changing the sign of all voltages and currents. Figure 2.20 shows the symbol for the *p*-channel unit; only the direction of the arrow at the substrate is reversed. The pinch-off voltage is now positive, I_{DSS} is negative, and I_D is normally negative. Thus, Eq. (4) continues to apply in the ohmic region, while saturation occurs whenever

$$V_{DS} \leq V_{GS} - V_P \qquad \text{(beyond pinch-off)}$$

D2.8 A *p*-channel depletion-mode MOSFET has a pinch-off voltage of 3.7 V. (a) For what range of V_{GS} is operation beyond pinch-off if $V_{DS} = -3.1$ V? (b) For what range of V_{DS} is operation beyond pinch-off if $V_{GS} = 1.7$ V?

Answers. $0.6 \leq V_{GS} \leq 3.7$ V; $V_{DS} \leq -2$ V

Fig. 2.19 The circuit symbol for an *n*-channel depletion-mode MOSFET having the substrate and source connected.

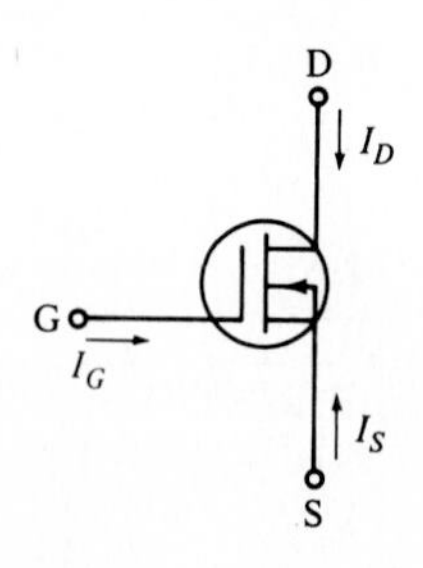

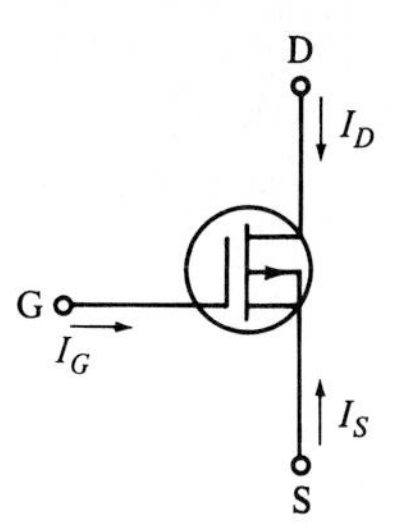

Fig. 2.20 The circuit symbol for a p-channel depletion-mode MOSFET with the source connected to the substrate.

2.6 The enhancement-mode MOSFET

An n-channel enhancement-mode MOSFET differs in its construction from the depletion-mode unit in that it does not have a region of n-type material built into the unit as a channel. Instead, the effect of the electric field between the gate and substrate enhances the conductivity of a portion of the substrate and produces a channel.

Figure 2.21*a* suggests the structure of the device. The substrate is lightly doped p-type material; it has about the same conductivity as the substrate used with the depletion mode. The source and drain regions are again n^+, and an insulating layer separates the metal gate from the substrate. If we

Fig. 2.21 (*a*) The structure of an n-channel enhancement-mode MOSFET shows that the n-type channel is not built into the transistor. (*b*) For a sufficiently large V_{GS}, part of the p-type substrate inverts to n-type; this forms the channel.

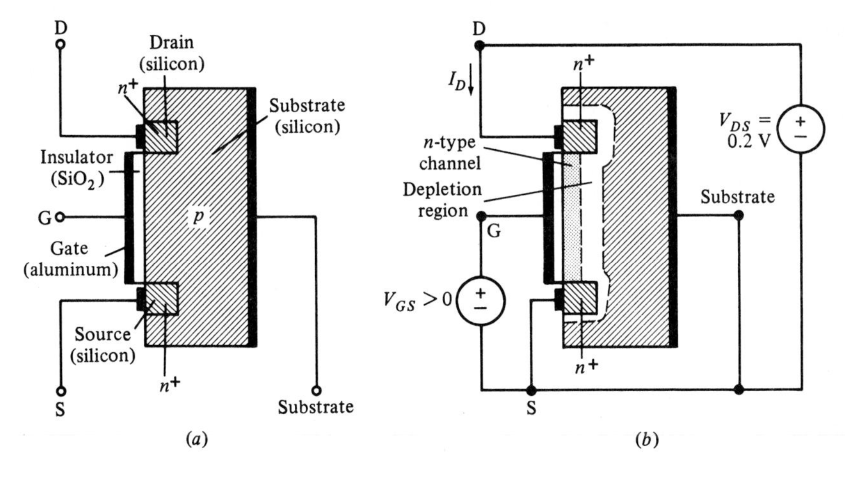

leave the gate open-circuited and apply a voltage (of either polarity) between drain and source, there is only a negligible current. It is a reverse-saturation current because either the substrate-source or substrate-drain *p-n* junction is reverse-biased. There is no channel.

Now let us apply a positive V_{GS} and a small V_{DS}; see Fig. 2.21*b*. An electric field is established across the insulating layer, which acts to repel positive carriers in the substrate and to attract negative carriers. As a result, a layer of substrate near the insulator becomes less *p*-type and its conductivity is reduced. As V_{GS} increases, this surface region of the substrate eventually has more electrons than holes, and it inverts to *n*-type. Figure 2.21*b* shows this *n*-channel. Between the *p*-type substrate and the *n*-type channel is a depletion region that serves to isolate the substrate from the channel.

The smallest value of V_{GS} that will produce a channel and a resultant value of I_D greater than the few nanoamperes of reverse-saturation current is called the *threshold voltage* V_T, typically 0.5 to 3 V. As V_{GS} increases beyond V_T, more electrons are pulled into the channel and its width and electron concentration increase. Thus the channel resistance decreases and I_D increases for the same source-to-drain voltage. In Fig. 2.22*a* we see a family of curves showing I_D vs. V_{DS} for several values of V_{GS}, where V_{DS} is small. We again see a voltage-controlled linear resistance. Note that the threshold voltage is +2 V. These curves[2] are described by

$$I_D \doteq K\left[(V_{GS} - V_T)V_{DS} - \frac{V_{DS}^2}{2}\right] \quad (V_{DS} \text{ small and } V_{GS} \geq V_T) \tag{5}$$

where K is a positive constant for the *n*-channel device.

The complete output characteristics are shown in Fig. 2.22*b*. When V_{DS} becomes sufficiently large, the gate-to-drain voltage is less than V_T, and pinch-off occurs at the drain end of the channel; that is, $V_{GD} = V_{GS} - V_{DS} \leq V_T$. Further increases in V_{DS} do not lead to larger values of I_D, and the transistor is operating in the region beyond pinch-off. This region is identified in the figure. The device behaves as a voltage-controlled current source, just as every FET does in saturation. For very large values of V_{DS}, avalanche breakdown in the channel can lead to catastrophic values of I_D. For very large values of V_{GS}, dielectric breakdown in the oxide layer will also ruin the device.

Figure 2.23 shows the circuit symbol for an *n*-channel enhancement-mode MOSFET with the substrate connected to the source. Note that the

[2] This equation yields slightly larger currents than a real device, but it is often used in circuit design because of its simplicity. See Vol. IV of G. W. Neudeck and R. F. Pierret, *Modular Series on Solid State Devices* (Reading, MA: Addison-Wesley, 1983).

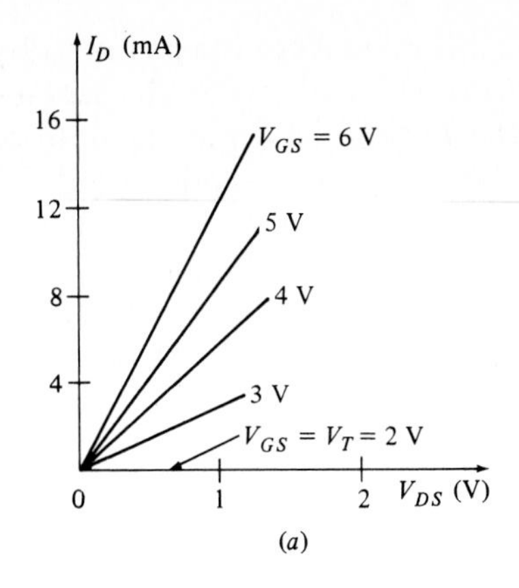

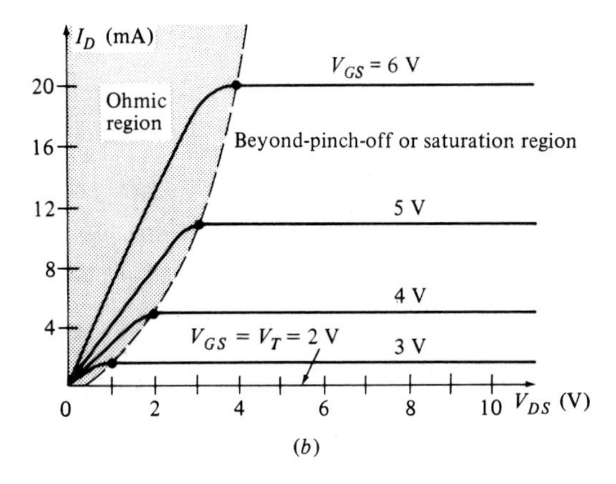

Fig. 2.22 (*a*) The *n*-channel enhancement-mode MOSFET is a voltage-controlled linear resistance for small V_{DS}. (*b*) The complete output characteristics. Operation is beyond pinch-off when $V_{DS} \geq V_{GS} - V_T$.

Fig. 2.23 The circuit symbol for an *n*-channel enhancement-mode MOSFET with an internal connection between substrate and source.

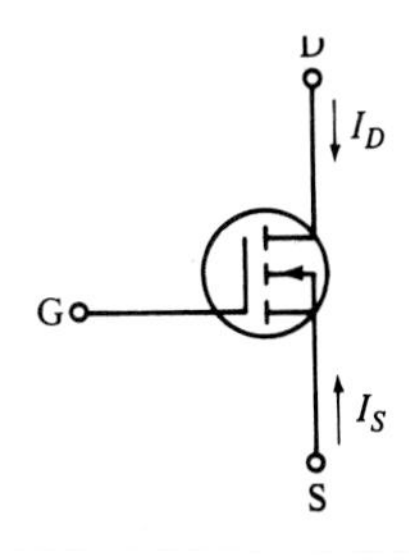

arrow shows the direction from the p side (substrate) to the n side (channel) of the junction, as usual, while the segmented line indicates the enhancement mode; no channel is present until enhancement occurs.

The p-channel enhancement-mode MOSFET is the complement of the n-channel device. It has an n-type silicon substrate in which a p-type channel is induced by making the gate sufficiently negative, $V_{GS} < V_T$. The threshold voltage is a negative number. The family of curves in the ohmic region may be represented by the equation

$$I_D \doteq K\left[(V_{GS} - V_T)V_{DS} - \frac{V_{DS}^2}{2}\right] \quad (V_{DS} \text{ small, and } V_{GS} \leq V_T) \tag{6}$$

where K is a negative constant, and I_D, V_{GS}, V_{DS}, and V_T are negative. Operation is in the region beyond pinch-off when $V_{DS} \leq V_{GS} - V_T$. Figure 2.24 shows the circuit symbol.

Figure 2.25 brings together the eight transistor circuit symbols that we shall be using. All current arrows are directed into the device in accordance with IEEE standards.

Before we conclude this initial look at IGFETs and MOSFETs, one additional point is of interest. The depletion-mode MOSFET studied in Section 2.5 can operate in *either* the enhancement or depletion mode. If, for example, we have a low-conductivity n-type channel, the application of a *positive* voltage between gate and source will draw additional electrons to the channel-insulator surface, thus increasing the conductivity of the channel. Figure 2.26 shows representative output characteristics.

D2.9 An n-channel enhancement-mode MOSFET with $V_T = 1.5$ V has a drain current of 0.6 mA when $V_{GS} = 3$ V and $V_{DS} = 1$ V. Find I_D if (a) $V_{GS} = 3.1$ V and $V_{DS} = 1.1$ V; (b) $V_{GD} = 2$ V and $V_{GS} = 3.2$ V. (c) What

Fig. 2.24 The circuit symbol for a p-channel enhancement-mode MOSFET with an internal connection between substrate and source.

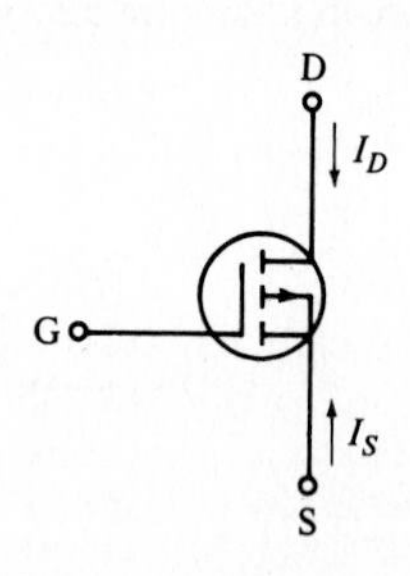

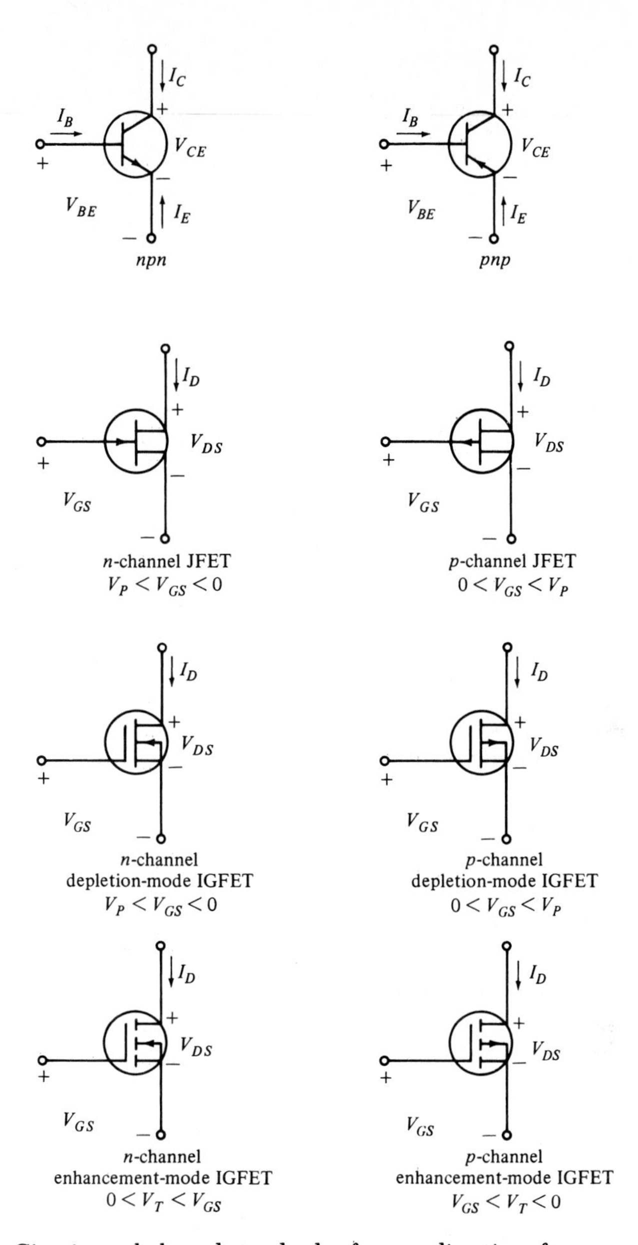

Fig. 2.25 Circuit symbols and standard reference directions for currents and voltages.

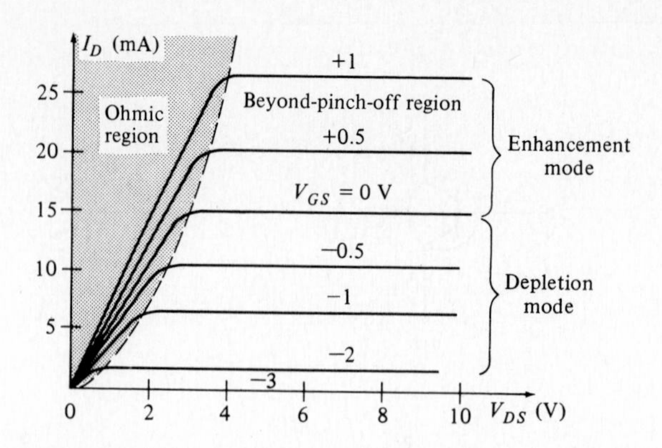

Fig. 2.26 An *n*-channel depletion-mode MOSFET may operate in either the depletion mode or the enhancement mode, as the output characteristics show. Similar *p*-channel devices exist.

is the equivalent resistance of the channel when V_{GS} = 3.8 V, V_{DS} = 0.1 V?

Answers. 0.693 mA; 0.792 mA; 741 Ω

2.7 Monolithic resistors

A monolithic integrated circuit has all its components produced by various processes on or within a single substrate, usually silicon. The components are interconnected along the surface by vaccum-deposited aluminum. If about 20 or fewer circuits are interconnected on one chip, we term it *medium-scale integration* (MSI); *large-scale integration* (LSI) refers to a chip containing several hundred circuits; and *very large-scale integration* (VLSI) is typically reserved for a chip containing an entire system, often more than 1000 circuits or 10,000 elements. Each circuit may contain on the order of two transistors, several diodes, and the associated resistors.

Monolithic integrated circuits are composed of four basic types of circuit elements: resistors, capacitors, diodes, and transistors. All the resistors in the integrated circuit are made at the same time. One common method is by the diffusion of a *p*-type channel into an *n*-type substrate, or vice versa. The thickness or depth of the channel and its conductivity are therefore the same for every resistor; the resistance is found by

$$R = \frac{l}{\sigma A} = \frac{l}{\sigma t w}$$

where l, t, and w are the length, thickness, and width respectively, as shown in Fig. 2.27. If the length and width are equal, the resistor occupies a square portion of the surface and its *sheet resistance* is

$$R_s = \frac{1}{\sigma t} \qquad \text{(ohms per square)} \tag{7}$$

The units are commonly called ohms per square, but the phrase "ohms for a square" would be better because Eq. (7) shows that the dimensions are ohms and not ohms per square meter. With the range of conductivity and thickness that are practical, R_s commonly ranges between one or two and several hundred ohms per square. Other techniques can provide larger values of sheet resistance, ranging up to 25 kΩ per square, as for example a pinch resistor.

For a sheet resistance R_s, the total resistance between the ohmic contacts is

$$R = R_s\left(\frac{l}{w}\right)$$

The ratio l/w (the number of "squares") may vary between perhaps one-quarter and fifty, the latter design requiring a serpentine course along the surface in order to maintain reasonable dimensions.

An important design consideration for resistors is their change of value with temperature. The temperature coefficient of resistance (TCR) α_T enables us to find the resistance R at a temperature T, knowing the resistance R_{ref} at a temperature T_{ref}:

$$R = R_{\text{ref}} + \alpha_T(T - T_{\text{ref}})R_{\text{ref}}$$

or

$$R = R_{\text{ref}}(1 + \alpha_T \Delta T) \tag{8}$$

The temperature difference is commonly measured in degrees Celsius, and α_T is expressed in parts per million per degree Celsius. Thus, a TCR of

Fig. 2.27 The resistance of the p-type channel is $l/\sigma tw$, where σ is the conductivity of the p-type material. We assume that the p-n junction is reverse-biased to isolate the channel from the substrate.

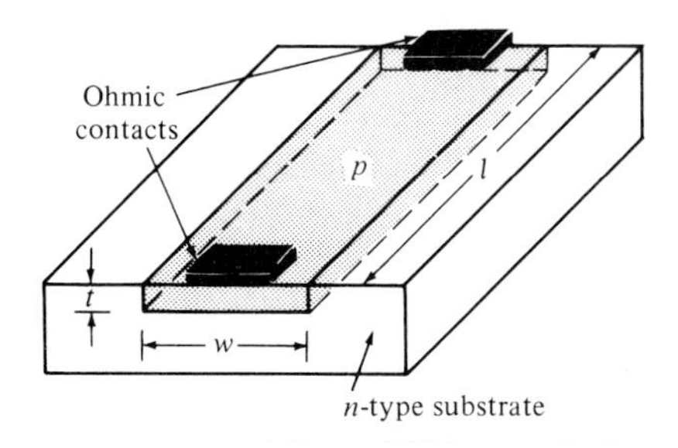

2000 ppm/°C would lead to a 10% increase in resistance for a temperature rise of 50°C, $(1 + 2000 \times 10^{-6} \times 50 = 1.10)$. The value of α_T may range from -500 to $+2000$ ppm/°C for monolithic resistors, depending on the method of manufacture and the materials used.

One of the disadvantages of monolithic resistors is the large tolerance in their manufacture, as high as $\pm 20\%$. However, all the resistors on the substrate are apt to be above or below the nominal value by the same amount. Thus we might find that two resistors that we wished were 100 Ω ended up as 119 and 120 Ω. However, it is possible to design circuits so that their operation is more sensitive to the ratio of two resistances than it is to either absolute value, and the ratios can be maintained within several percent. Since all the resistors in the same substrate have the same α_T, the resistance ratio is constant over a wide range of temperature, and this is a distinct advantage of their design.

A typical sheet resistor, formed at the same time that the diffusion creating the base region takes place, has a sheet resistance of 200 Ω per square, a tolerance of $\pm 20\%$, a matching tolerance with respect to other units of $\pm 1\%$, and a temperature coefficient of resistance of $+2000$ ppm/°C.

D2.10 A monolithic resistor is constructed with a sheet resistance of 200 Ω per square at 25°C, a length $l = 500\ \mu$m, and a width $w = 40\ \mu$m. (a) Find R. If R_s has a tolerance of $\pm 15\%$, l has a tolerance of $\pm 2\%$, w has a tolerance of $\pm 8\%$, $\alpha_T = 2000$ ppm/°C at 25°C, and a temperature range from -55°C to 125°C is possible, find (b) the maximum expected value of R, (c) the minimum expected value of R.

Answers. 2500 Ω; 3825 Ω; 1620 Ω

Problems

1. Active-region operation of an *npn* transistor is present with junction voltages of 0.65 and 8.2 V, and with a collector current–to–base current ratio of 100. If $I_E = -4$ mA, find I_B, I_C, V_{BE}, V_{CE}, V_{CB}, and the total power dissipated by the device.
2. An *npn* transistor operating in the active region has $|I_E| = 8$ mA, $|I_B| = 150\ \mu$A, $|V_{BE}| = 0.6$ V, and $|V_{CE}| = 15$ V. Find (a) I_E, (b) I_B, (c) I_C, (d) V_{BE}, (e) V_{CE}, (f) V_{CB}, (g) the power entering the B-E port P_{BE}, and (h) the power entering the C-E port P_{CE}. If the transistor were connected common-base, what power would enter (i) the E-B port? (j) the C-B port?
3. The active-region operating point for an *npn* transistor is located at $|I_B| = 120\ \mu$A, $|I_C| = 6$ mA, $|V_{EB}| = 0.7$ V, $|V_{CB}| = 8$ V. Find (a)

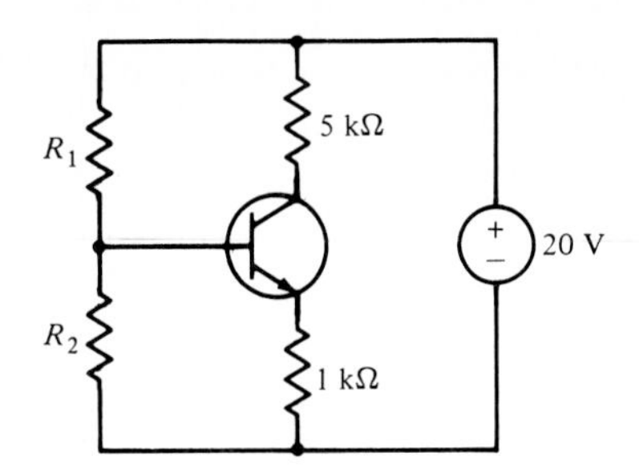

Fig. 2.28 See Problem 4.

I_B, (b) I_C, (c) I_E, (d) V_{BE}, (e) V_{CB}, (f) V_{CE}, (g) the power entering the base-emitter port P_{BE}, assuming the emitter is the common terminal; (h) the power entering the emitter-base port P_{EB}, assuming the base is the common terminal, and (i) the total power being dissipated by the transistor.

4. Determine the region of operation for the transistor shown in Fig. 2.28 if (a) $I_C = 1$ mA, $I_B = 20\ \mu$A, $V_{BE} = 0.7$ V; (b) $I_C = 3.2$ mA, $I_B = 0.3$ mA, $V_{BE} = 0.8$ V; (c) $I_C = 3$ mA, $I_B = 1.5$ mA, $V_{BE} = 0.85$ V.
5. Determine the region of operation for transistors $T1$ and $T2$ in the circuit shown in Fig. 2.29 if (a) $I_{C1} = I_{C2} = 1$ mA, $I_{B1} = I_{B2} = 20\ \mu$A, $V_1 = V_2 = 21$ V; (b) $I_{E1} = -1$ mA, $I_{E2} = -1.5$ mA, $I_{B1} = 20\ \mu$A, $I_{B2} = 25\ \mu$A, $V_{BE1} = 0.5$ V, $V_{BE2} = 0.55$ V. (c) Determine the power dissipated in each of the eight circuit elements in (b).
6. Find I_C, V_{BE}, V_{CE}, and V_{CB} for the transistor shown in Fig. 2.30 if it is known that $I_E = -2$ mA and $I_B = 25\ \mu$A.

Fig. 2.29 See Problem 5.

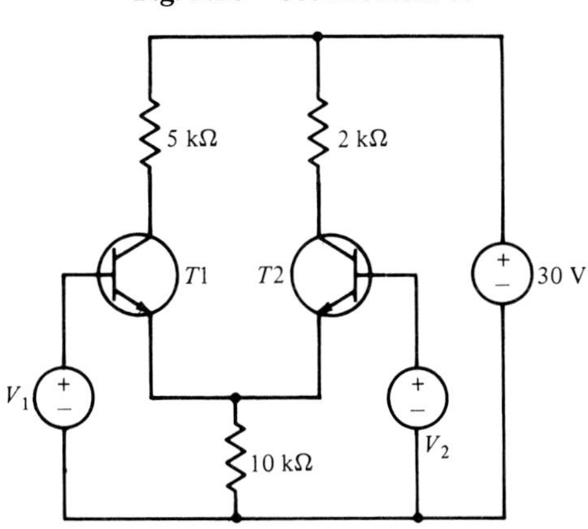

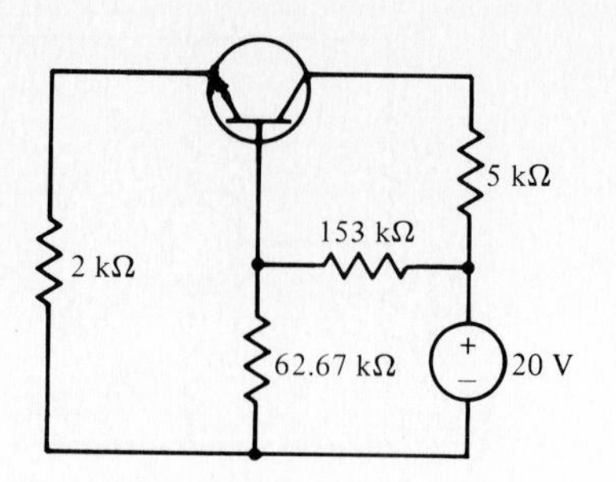

Fig. 2.30 See Problem 6.

7. For each transistor shown in Fig. 2.31, state whether each junction is forward- or reverse-biased, and whether the transistor is operating in the active, cutoff, or saturation region.
8. Determine the region of operation for each of these transistors: (a) *npn*, $V_{BE} = 0.8$ V, $V_{CE} = 0.4$ V; (b) *npn*, $V_{CB} = 1.5$ V, $V_{CE} = 2$ V; (c) *pnp*, $V_{CB} = 0.9$ V, $V_{CE} = 0.4$ V; (d) *npn*, $V_{BE} = -1$ V, $V_{CB} = 0.6$ V; (e) *pnp*, $V_{CB} = 0.7$ V, $V_{CE} = 1.5$ V
9. (a) The potentials of the three terminals of an *npn* transistor are measured with respect to ground as: emitter, 5.5 V; base, 6.3 V; and collector, 6.0 V. Is the transistor cut off, saturated, or operating in

Fig. 2.31 See Problem 7.

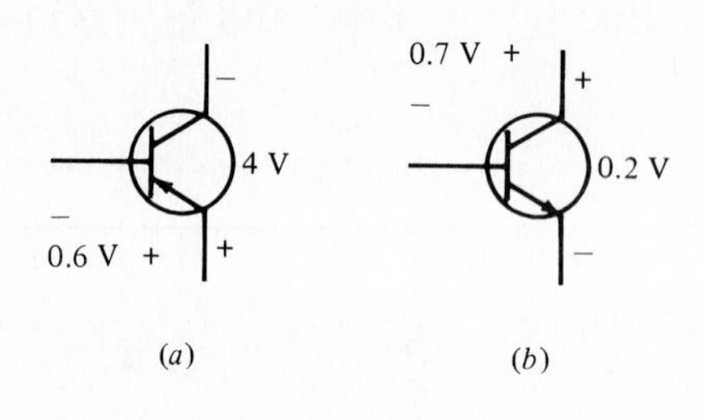

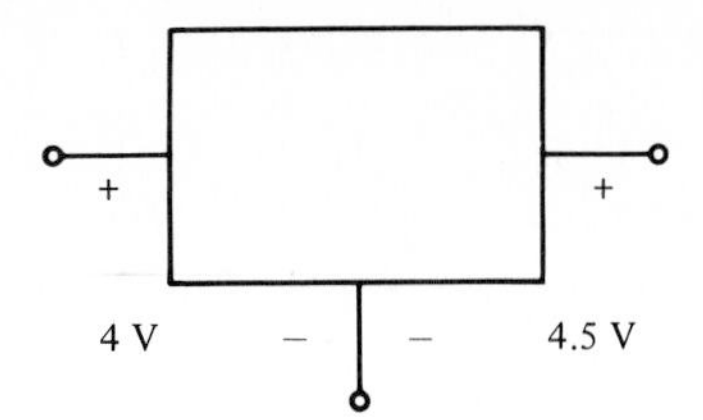

Fig. 2.32 See Problem 10.

the active region? (b) Repeat for a *pnp* transistor with: emitter, −3 V; base, −3.5 V; collector, −5.5 V.

10. (a) An *npn* bipolar transistor is operating in the active region with $|I_C| = 2$ mA, $|I_B| = 0.1$ mA, $|V_{BE}| = 0.8$ V, and $|V_{BC}| = 3$ V. Find I_C, I_B, I_E, V_{BE}, V_{CB}, V_{CE}, and the total power dissipated by the transistor. (b) Repeat if the transistor is *pnp*. (c) The bipolar transistor shown in Fig. 2.32 is operating in the active region. Identify the emitter, collector, and base terminals, and draw the transistor symbol inside the box.
11. The JFET whose output characteristics are shown in Fig. 2.13 is operating with $V_{DS} = 0.1$ V. Determine the source-to-drain resistance for V_{GS} equals (a) 0, (b) −1 V, (c) −2 V.
12. An *n*-channel JFET with $V_P = -5$ V has a drain current of 2.5 mA when $V_{GS} = -3$ V and $V_{DS} = 2$ V. (a) What is the drain-to-source channel resistance when $V_{GS} = -4$ V and V_{DS} is small? (b) Calculate I_D for $V_{GS} = -1$ V, $V_{DS} = 3$ V.
13. The gate leakage current of the JFET whose characteristics are shown in Fig. 2.13 is −40 nA. (a) If $V_{GS} = -2$ V and $V_{DS} = 8$ V, what is the total power being dissipated by the transistor? (b) Repeat for $V_{GS} = -2$ V, $V_{DS} = 1$ V.
14. Install the correct circuit symbol for a JFET in each box of Fig. 2.33, showing proper connections to the external terminals. Identify the source and drain. Assume normal operation in the region beyond pinch-off.
15. In the normal operating region, a JFET has these values: $I_{DSS} = 20$ mA, $V_P = -3$ V, and $I_D = (A - BV_{GS})^2$ (mA, V). (a) Is the JFET *n*-channel or *p*-channel? (b) Find A and B. (c) Find I_D if $V_{GS} = -1$ V and $V_{DS} = 8$ V. (d) Find V_{GS} if $I_D = 10$ mA and $V_{DS} = 6$ V.
16. In the region beyond pinch-off, a certain JFET is described by the relationship $I_D = -(5 - 2V_{GS})^2$ (mA, V). (a) Is the JFET *n*-channel or *p*-channel? (b) Find V_P. (c) Find I_{DSS}. Identify the following operating points as ohmic, cutoff, or in saturation: (d) ($V_{GS} = 2$ V, $V_{DS} = -1$ V); (e) (1, −2); (f) (3, −6); (g) (1.5, −0.5). (h) Give the equa-

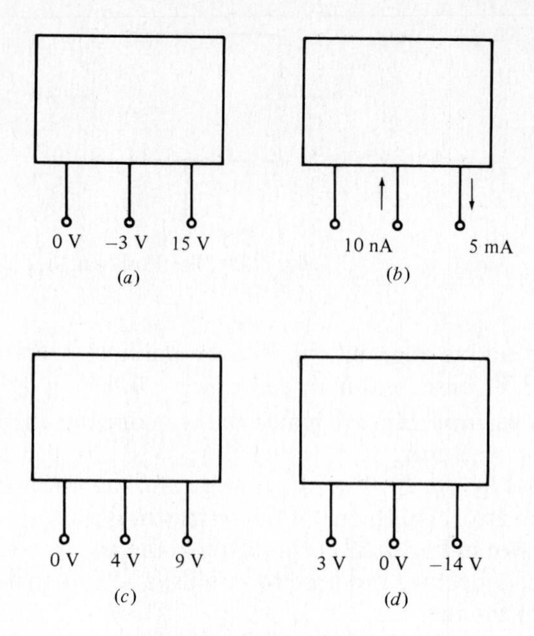

Fig. 2.33 See Problem 14.

tion of the curve (I_D as a function of V_{DS}) that represents the boundary between the ohmic and saturation regions.

17. (a) An n-channel JFET is operating in the region beyond pinch-off with $|I_D| = 2$ mA, $|I_G| = 0.1\ \mu$A, $|V_{DS}| = 5$ V, $|V_P| = 2.5$ V, and $|V_{GS}| = 0.5$ V. Find I_D, I_G, I_S, V_{GS}, V_P, V_{DS}, V_{DG}, and the total power dissipated by the transistor. (b) Repeat if the JFET is p-channel. (c) An unknown JFET having $|V_P| = 3$ V is operating normally in the saturation region. If Terminal A is at 4 V, B is at 0 V, and C is at 4.5 V, state whether the JFET is n-channel or p-channel, and identify the gate, source, and drain.
18. If each of the following facts applies to normal operation of a depletion-mode MOSFET in the saturation region, identify the unit as DN (definitely n-channel), DP (definitely p-channel), or MBE (might be either): (a) $I_D = 4$ mA, (b) $V_P = -3.5$ V, (c) $I_G = 20$ pA, (d) $I_{DSS} = -8$ mA, (e) $V_{GD} = -2$ V, (f) $I_G \doteq 0$, (g) $I_S = 3.5$ mA, (h) $V_{DS} = 10$ V, (i) $V_{DG} = 6$ V, (j) $V_{GS} = 2$ V, (k) $V_{GS} = 0$.
19. (a) Normal operation in the region beyond pinch-off for an n-channel depletion-mode MOSFET occurs where $|I_D| = 5$ mA, $|I_G| = 2$ nA, $|V_{DS}| = 6$ V, $|V_P| = 3$ V, and $|V_{GS}| = 1$ V. Find I_D, I_G, I_S, V_{DS}, V_{GS}, V_P, V_{DG}, and the power being dissipated by the transistor. (b) Repeat if the MOSFET is p-channel. (c) Another depletion-mode MOSFET operating in the saturation region has voltages between ter-

minals of $V_{AB} = 5$ V and $V_{BC} = -6$ V. Identify the gate, source, and drain.

20. Each of the transistors shown in Fig. 2.34 is characterized either by $|V_P| = 4$ V or $|V_T| = 3$ V. Identify each device and state the region in which it is operating.
21. Let $|I_D| = 1$ mA, $I_G \doteq 0$, $|V_{DS}| = 3$ V, and $|V_{GS}| = 2$ V for a MOSFET operating normally in the region beyond pinch-off. Find I_D, I_S, V_{GS}, V_{DS}, and the total power dissipated by the transistor if it is (a) an n-channel depletion-mode unit with $|V_P| = 3$ V; (b) a p-channel depletion-mode unit with $|V_P| = 2.5$ V; (c) an n-channel enhancement-mode unit with $|V_T| = 1$ V; (d) a p-channel enhancement-mode unit with $|V_T| = 1.5$ V.
22. Draw the circuit symbols of appropriate depletion-mode MOSFETs in the circuit diagram of Fig. 2.35.
23. In Fig. 2.36, let $V_T = 1$ V for $T1$ and $V_P = -4$ V for $T2$. For each pair of voltage values $(v_{\text{in}}, v_{\text{out}})$ given below, state the region of operation of each MOSFET and specify whether or not I_D O: (a) (8, 0.2), (b) (4, 1), (c) (2, 3), (d) (1.5, 5), (e) (1, 8).
24. Two resistors, $R_1 = 1$ kΩ and $R_2 = 3$ kΩ, are made by a diffusion process in which $R_s = 200 \pm 10$ Ω per square. Assume that R_s is uniform over the chip. Calculate the maximum and minimum values of (a) $R_1 + R_2$, (b) $R_1/(R_1 + R_2)$. (c) Calculate the percentage errors in (a) and (b).

Fig. 2.34 See Problem 20.

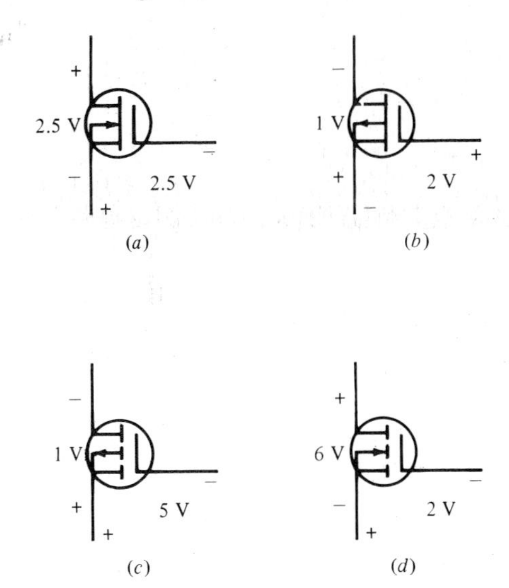

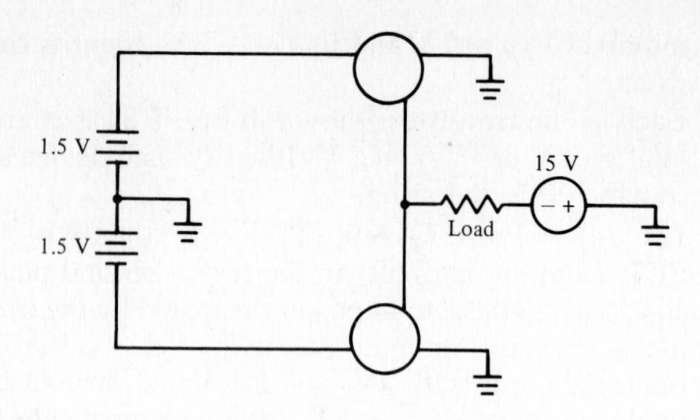

Fig. 2.35 See Problem 22.

25. Two monolithic resistors, $R_A = 500\ \Omega$ and $R_B = 1000\ \Omega$, are made by using $R_s = 200\ \Omega$ per square and a channel width having an uncertainty of $\pm 10\%$. If the channel length is assumed to be exact, determine maximum and minimum values of (a) R_A, (b) R_B, (c) $R_A + R_B$, (d) $R_A \parallel R_B$, (e) $R_A/(R_A + R_B)$.
26. A sheet resistance of 250 Ω per square is used to form the monolithic resistor shown in Fig. 2.37. The width of the path is 100 μm. (a) Find R. (b) What are the worst-case values of R that might be obtained if the dimensional tolerance is $\pm 10\%$ for the length, $\pm 15\%$ for the width, and $\pm 20\%$ for R_s?
27. A conducting material with $\sigma = 250$ ℧/m is printed uniformly as a layer 0.025 mm thick to form a thick-film resistor on an insulating substrate. The region is a rectangle with sides 1.5 and 7.5 mm. Find the total resistance between (a) the two shorter edges, (b) the two longer edges. (c) Find the width and length of a 2000-Ω resistor having the same surface area as the 1.5 × 7.5 mm unit.

Fig. 2.36 See Problem 23.

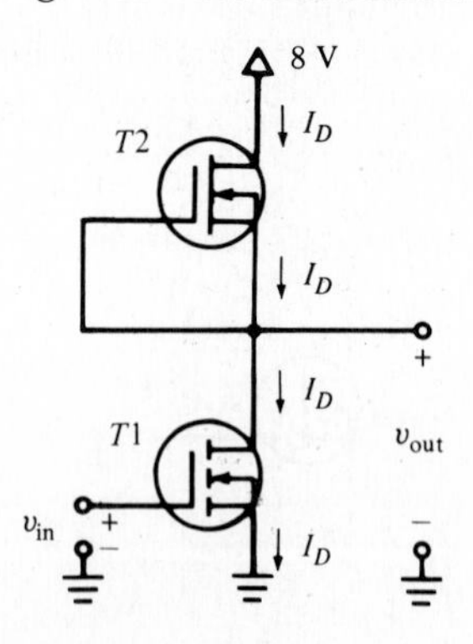

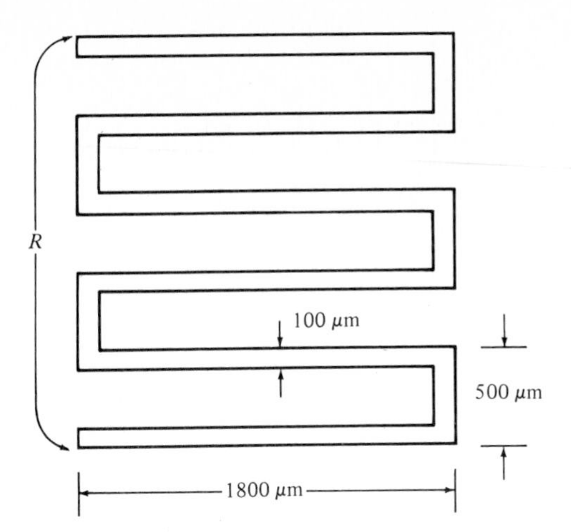

Fig. 2.37 See Problem 26.

28. A resistor has a resistance of 4704 Ω at 25°C and 4818 Ω at 55°C. Assume a linear relationship between R and T in degrees Celsius. (a) Find α_T in ppm/°C, using $T_{\text{ref}} = 25°\text{C}$. (b) Find R at 105°F. (c) Estimate the temperature in degrees Fahrenheit at which $R = 4400\ \Omega$. (d) Assuming that your linear relationship may be extrapolated, at what temperature in kelvins would $R = 0$?
29. Some resistors in integrated circuits show a nonlinear variation of R with T. As an example in which this is represented by a piecewise-linear model, let α_T be 800 ppm/°C for $T < 0°\text{C}$ and 1800 ppm/°C for $T > 0°\text{C}$. Determine R (a) at 30°C if $R = 450\ \Omega$ at $-20°\text{C}$; (b) at $-55°\text{C}$ if $R = 2000\ \Omega$ at 125°C. (c) As another example, consider a resistance having a value R_0 at 150°C, $0.9R_0$ at 25°C, and $0.8R_0$ at $-55°\text{C}$. Assume a piecewise-linear model for R vs. T having a break point at 0°C, and calculate values for α_T for $T < 0°\text{C}$ and $T > 0°\text{C}$.
30. A certain material has a surface resistivity of 700 Ω per square at 0°C and 750 Ω per square at 60°C. Find α_T using a reference temperature of (a) 0°C, (b) 60°C. Calculate the predicted resistivity at 20°C using (c) $\alpha_{T,0}$, (d) $\alpha_{T,60}$, (e) linear interpolation of the original data.

3

Transistor dc models

In analyzing or designing electronic circuits, one of the most difficult questions to answer concerns the selection of the model for the device. As we found in Chapter 1, there is a trade-off between accuracy and complexity. To obtain an extremely accurate model we may have to determine data experimentally, as well as using powerful circuit-analysis techniques to arrive at a solution. Linear models are simple and may be analyzed easily, but they are applicable only to certain problems. Thus the selection of a model often depends on the available data and our own abilities and eagerness. Our guiding philosophy will be to use the simplest model that is consistent with the needed accuracy. We cannot offer a single answer to the model-selection question; it is often a matter of preference or opinion. (In case of a tie, the instructor wins.)

As with the semiconductor diode, we may subdivide the model collection into dc and ac models. In determining or selecting an operating point, we need a dc model; in specifying or obtaining a certain voltage or power gain, we require an ac model. Figure 3.1 suggests how we might classify the bipolar transistor models. This chapter and the following one refer to the dc models; the ac models are discussed in Chapter 5.

3.1 Bipolar transistor linear dc models

As we have done before, we will pay attention first to the *npn* transistor, for the extension to the *pnp* bipolar transistor then follows easily.

There are three regions to consider—active, saturation, and cutoff. Rather than use one complicated model that may serve all three possibilities, we will develop a simpler model for each condition alone. We consider the active region first, since it is the region in which most signal-amplifying devices operate.

Active-region model

Following the discussion of the bipolar transistor and its characteristics in the previous chapter, our first and simplest attempt at a model for the *npn* transistor of Fig. 3.2*a* is shown in Fig. 3.2*b*. We have provided a dependent current source that supplies a collector current precisely β_{dc} times as

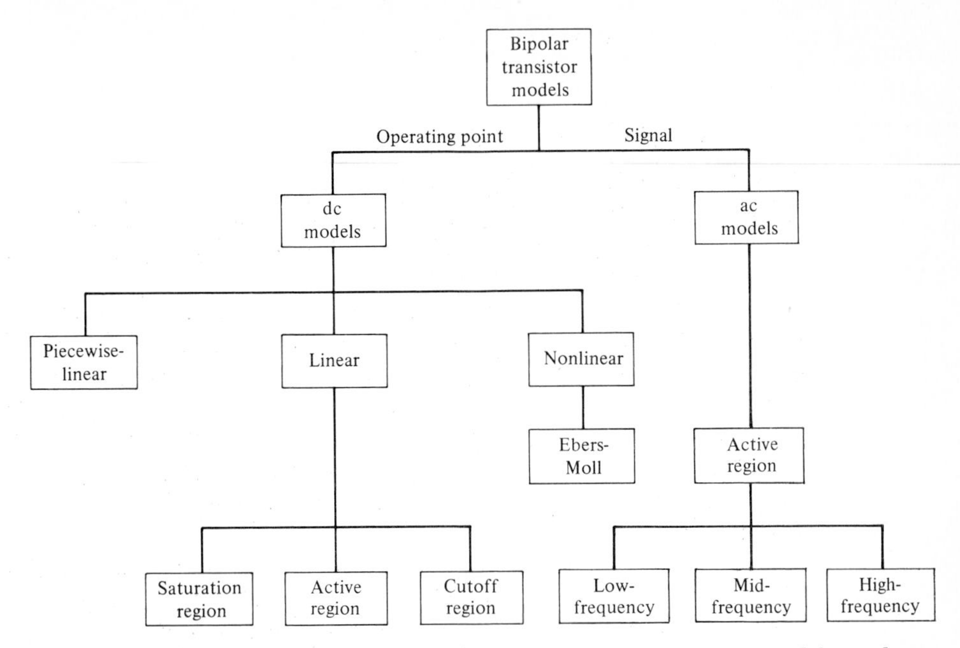

Fig. 3.1 A classification scheme for bipolar transistor models. The ac models are discussed in Chapter 5.

large as the base current, and a voltage V_0 between base and emitter. From the characteristic curves discussed before (Fig. 2.3*b*, for example), we see that β_{dc} may range from about 50 to perhaps 1000. Figures 3.2*c* and *d* compare the characteristics predicted by this model with those of a real device. Note that the model is reasonably accurate in the active region, but that it does not indicate where the active region ends and saturation begins. We will have more to say about this later.

The accuracy of this model is improved by the addition of a base resistor R_{BB} and the independent current source I_{CEO} in Fig. 3.3. The collector current is now the sum of two source currents, $\beta_{dc} I_B$ as a dependent source and I_{CEO} as an independent source. This latter current is defined as

$$I_{CEO} = I_C|_{I_B=0} \qquad (V_{CE} > 0)$$

The current I_{CEO} flows from *c*ollector to *e*mitter with the base *o*pen-circuited, as the three subscripts suggest. It ranges in value from the order of 1 nA to 1 μA for silicon transistors at room temperature, but increases exponentially as the temperature increases.

To improve the input characteristics of the model, we insert a resistor R_{BB} in the base lead. Note the similarity between the base-emitter branch of the model and the diode model of Chapter 1. The value of V_0 is typically between 0.5 and 0.75 V, while R_{BB} ranges from 1 kΩ to 20 kΩ.

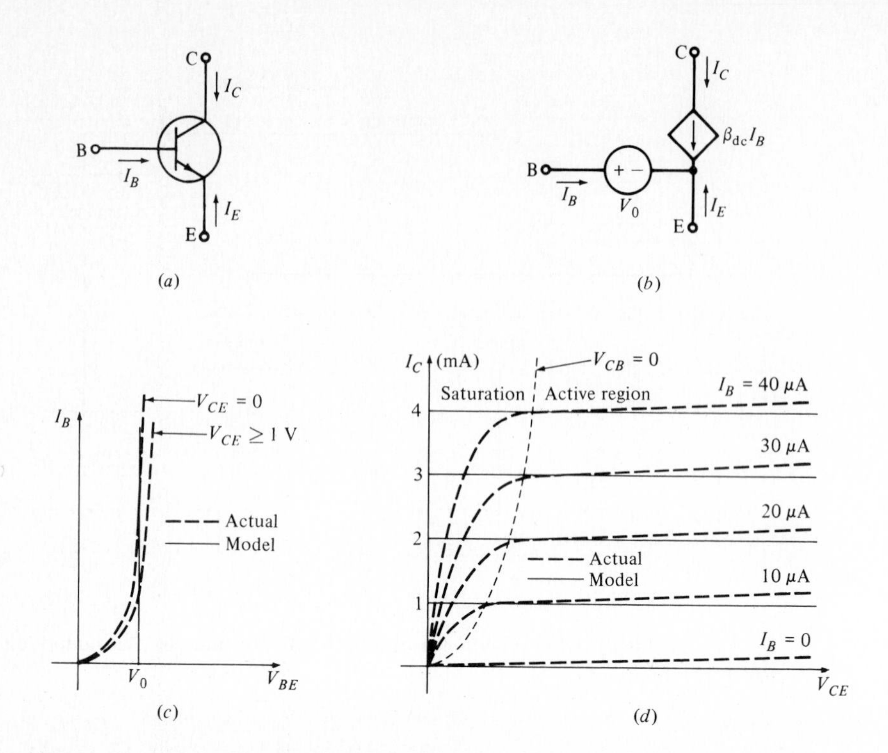

Fig. 3.2 (*a*) The circuit symbol for the *npn* bipolar transistor. (*b*) A simple dc model for the active region. (*c*) Input and (*d*) output characteristics predicted by the model compared with typical curves for a real device.

The effect of R_{BB} on the input characteristics predicted by the model is that the solid line in Fig. 3.2*c* leans slightly to the right, thus agreeing better with the actual characteristics. The addition of the independent current source I_{CEO} raises every one of the horizontal lines in Fig. 3.2*d* slightly. They remain horizontal, however.

Before we make one last addition to our model, let us spend a few algebraic moments relating the several currents and defining two other useful quantities, α_{dc} and I_{CBO}.

The basic relationship among the currents arises from Kirchhoff's current law,

$$I_C + I_B + I_E = 0 \tag{1}$$

where I_C and I_B are positive in the active region for the *npn* transistor. From the linear model of Fig. 3.3, we have

$$I_C = \beta_{dc} I_B + I_{CEO} \tag{2}$$

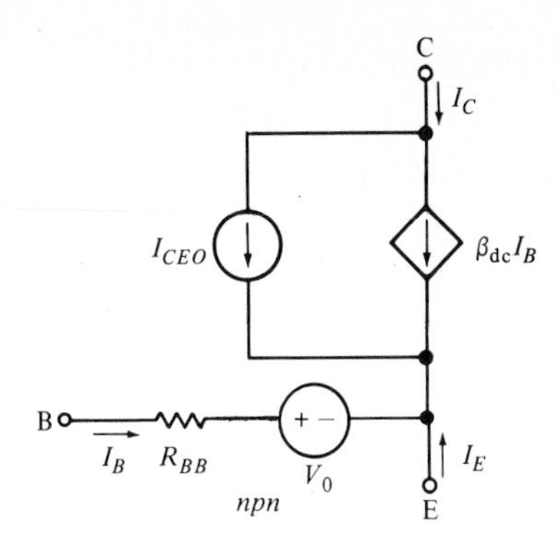

Fig. 3.3 A more accurate dc model for the bipolar transistor in the active region includes an independent current source I_{CEO} and a resistor R_{BB} in the base lead.

Using Eq. (1) to eliminate I_B, we find

$$I_C = \beta_{dc}(-I_E - I_C) + I_{CEO}$$

which may be solved easily for I_C:

$$I_C = \frac{-\beta_{dc} I_E}{1 + \beta_{dc}} + \frac{I_{CEO}}{1 + \beta_{dc}} \tag{3}$$

If we now define

$$\alpha_{dc} = \frac{\beta_{dc}}{1 + \beta_{dc}} \tag{4}$$

then

$$I_C = -\alpha_{dc} I_E + \frac{I_{CEO}}{1 + \beta_{dc}} \tag{5}$$

Since β_{dc} ranges typically from 50 to 1000, we see that α_{dc} is a fraction just slightly less than unity.

Our final new quantity is defined by inspecting Eq. (3) or (5) above. When the emitter is open-circuited, the current from the collector to base (collector to *b*ase with emitter *o*pen-circuited) is

$$I_{CBO} = I_C|_{I_E = 0} = \frac{I_{CEO}}{1 + \beta_{dc}} \qquad (V_{CE} > 0) \tag{6}$$

If I_{CEO} is of the order of 1 μA, then I_{CBO} is of the order of 10 nA.

Equation (3) or (5) may now be written concisely in terms of these new parameters:

$$I_C = -\alpha_{dc} I_E + I_{CBO} \tag{7}$$

A simple relationship between I_B and I_E may also be obtained from Eqs. (3) and (6):

$$I_C = -I_E - I_B = -\frac{\beta_{dc}}{1+\beta_{dc}} I_E + I_{CBO}$$

$$-I_B = I_E\left(1 - \frac{\beta_{dc}}{1+\beta_{dc}}\right) + I_{CBO}$$

or

$$I_B = -\frac{1}{1+\beta_{dc}} I_E - I_{CBO} \tag{8}$$

Therefore, whenever I_{CBO} may be neglected in comparison with I_B, we have

$$I_B \doteq -\frac{1}{1+\beta_{dc}} I_E$$

The two collector-current expressions, Eq. (2) in terms of β_{dc} and I_{CEO}, and Eq. (7) in terms of α_{dc} and I_{CBO}, are well worth memorizing. A third memorable equation is Eq. (4), the relationship between β_{dc} and α_{dc}.

As our final and most accurate model for the bipolar transistor, we provide a slight slope to the output characteristics of the model by including a resistor R_0 between the collector and emitter terminals. The slope of the I_C vs. V_{CE} curves is therefore $1/R_0$. The model is shown in Fig. 3.4;

$$I_C = \beta_{dc} I_B + I_{CEO} + \frac{V_{CE}}{R_0}$$

Typical values of R_0 range from 50 kΩ to 2 MΩ.

Fig. 3.4 This dc model for the bipolar transistor allows for a slight slope to the output characteristics.

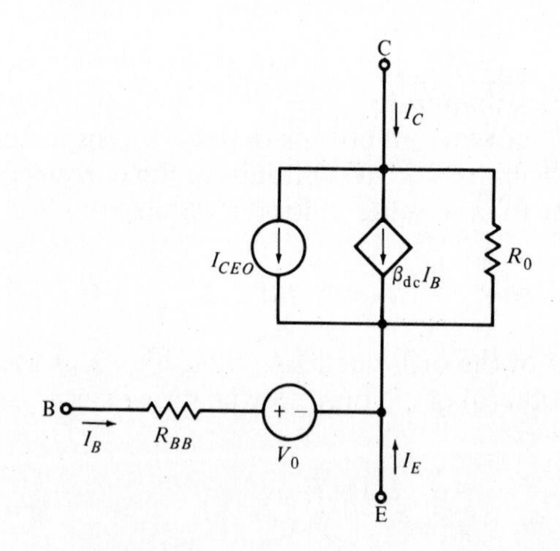

For the *pnp* transistor, Eqs. (1) through (8) apply to the active region, where I_E is positive and I_B, I_C, I_{CEO}, I_{CBO}, and V_0 are all negative. Also, V_{BE} and V_{CE} are negative.

As an example of the differences and similarities of the *npn* and *pnp* devices, consider a set of experimental data for the *npn* transistor, as modeled by Fig. 3.3:

$$I_B = 20\ \mu\text{A}$$
$$\beta_{dc} = 100$$
$$I_{CEO} = 0.1\ \mu\text{A}$$
$$V_0 = 0.6\ \text{V}$$
$$R_{BB} = 5\ \text{k}\Omega$$

Therefore

$$I_C = 2000 + 0.1 = 2000.1\ \mu\text{A}$$
$$I_E = -2000.1 - 20 = -2020.1\ \mu\text{A}$$
$$V_{BE} = 0.6 + 20 \times 10^{-6}(5000) = 0.7\ \text{V}$$

We also find that $\alpha_{dc} = 100/101 = 0.990$, and $I_{CBO} = 10^{-7}/101 = 0.990$ nA.

For a comparable *pnp* transistor:

$$I_B = -20\ \mu\text{A}$$
$$\beta_{dc} = 100$$
$$I_{CEO} = -0.1\ \mu\text{A}$$
$$V_0 = -0.6\ \text{V}$$
$$R_{BB} = 5\ \text{k}\Omega$$

Therefore

$$I_C = -2000 - 0.1 = -2000.1\ \mu\text{A}$$
$$V_{BE} = -0.6 - 20 \times 10^{-6}(5000) = -0.7\ \text{V}$$
$$V_{EB} = 0.7\ \text{V}$$
$$\alpha_{dc} = 0.990$$
$$I_{CBO} = -0.990\ \text{nA}$$

For the *npn* transistor above, if we let $V_{CE} = 10$ V and $R_0 = 50$ kΩ, then I_C increases by $10/(50 \times 10^3)$ A, or 200 μA, to 2200.1 μA, and I_E is − 2220.1 μA.

Saturation model

Saturation is characterized by having both junctions (emitter-base and collector-base) operating with forward bias and very small voltage drops. The voltages at the three transistor terminals rarely differ by more than several tenths of a volt, which leads to the simple model shown in Fig. 3.5*a*. The dc source V_0 in the base lead has a value of 0.6 to 0.8 V, R_{BB} ranges from 1 to 20 kΩ, and the saturation resistance R_{sat} in the collector lead is typically 1 to 200 Ω. We can obtain a reasonable value for R_{sat} by dividing the manufacturer's values for the collector-emitter saturation voltage $V_{CE(sat)}$ (0.2 to 1 V) by the collector current at which it is measured, $I_{C(sat)}$. Figures 3.5*b* and *c* show the input and output characteristics of the model.

Cutoff model

At cutoff, both junctions are reverse-biased, and only small leakage currents flow to the external leads. Figure 3.6 illustrates a model applicable to the boundary between the cutoff and the active region. Note that $I_C = I_{CEO}$, since $I_B = 0$. Smaller collector currents are possible if I_B is made negative, although this is not reflected in the model. When the emitter-base junction is made sufficiently reverse-biased, $I_C \doteq I_{CBO}$.

D3.1 A bipolar transistor is operating with $\beta_{dc} = 80$, $V_0 = 0.65$ V, $R_{BB} = 4$ kΩ, $I_{CEO} = 2$ μA, and $I_E = -1.5$ mA. (a) Is the transistor *npn* or *pnp*? Determine (b) I_B, (c) I_C, (d) V_{BE}, (e) α_{dc}.

Answers. *npn*; 18.49 μA; 1.482 mA; 0.724 V; 0.988

D3.2 A certain transistor in saturation is modeled with $V_0 = -0.62$ V, $R_{BB} = 2$ kΩ, and $R_{sat} = 40$ Ω. If $I_E = 6$ mA and $V_{EB} = 0.78$ V, determine (a) I_B, (b) I_C, (c) V_{CE}.

Answers. − 80 μA; − 5.92 mA; − 0.237 V

D3.3 Let $V_0 = 0.6$ V, $R_{BB} = 5$ kΩ, $I_{CEO} = 1$ μA, $\beta_{dc} = 120$, and $R_{sat} = 40$ Ω for an *npn* transistor. The transistor is operated in the circuit of Fig. 3.7. Let $V_{BB} = 0$ and find (a) V_{BE}, (b) V_{CE}. Now let $V_{BB} = -1$ V and find (c) V_{BE}, (d) V_{CE}.

Answers. 0; 5.99; − 1; 6 V

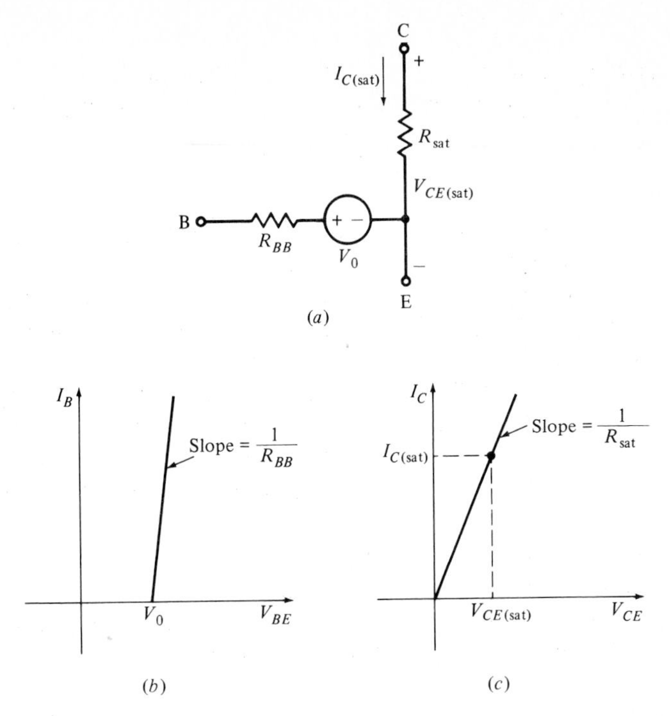

Fig. 3.5 (*a*) A saturation model for a bipolar transistor that is valid for both *npn* and *pnp* units. The input (*b*) and output (*c*) characteristics predicted by the model.

Fig. 3.6 This cutoff dc model for the bipolar transistor is a simple approximation for the boundary between the cutoff and active regions.

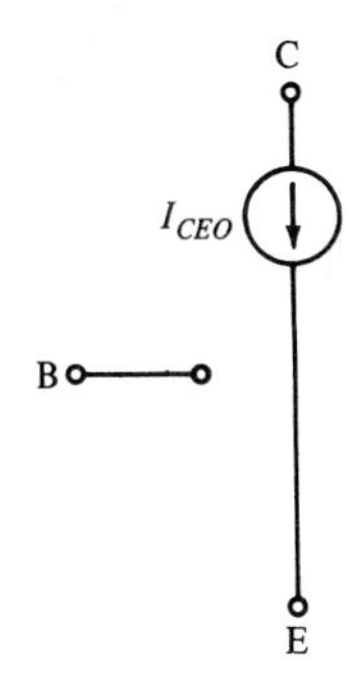

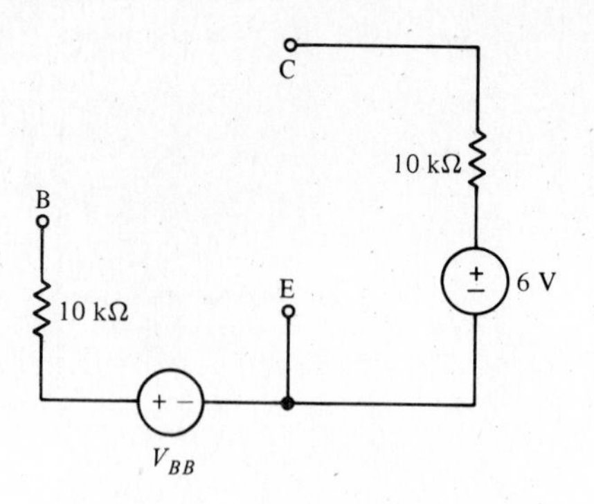

Fig. 3.7 See Problem D3.3

3.2 Examples of the use of dc models

We are now ready to look at several examples that illustrate the use of the various dc models for the bipolar transistor. The examples will progress from analysis to design as we get smarter.

The first example is presented in Fig. 3.8. Note the use of the ground symbol on the emitter lead and the voltage-supply symbols on the upper ends of the base and load resistors to permit neater schematics. They indicate that 15-V independent dc voltage sources are connected between each upper end and ground. The dashed lines in Fig. 3.8 indicate an equivalent arrangement using a single 15-V source. In this problem we are asked to find all currents and voltages, or to specify the operating point completely.

We identify the problem as a dc-analysis problem that requires the use of a transistor dc model. Our plan of attack is to select a dc model using the parameter values given in Fig. 3.8. Which type of model do we choose—active, saturation, or cutoff? We do not in general know in which region the device is operating. As a first attempt, let's try an active-region model; if we're wrong, we'll find that one or more currents or voltages have inappropriate signs, indicating that we're not located in the active region. We can then try a saturation or cutoff model.

In Fig. 3.9 we show the active-region model inserted into the specified circuit. Around the left mesh (through a 15-V source), and using a consistent set of units (V, mA, kΩ), we have

$$15 = 100I_B + 0.6 + 1I_B$$

so that

$$I_B = 0.1426 \text{ mA} = 142.6\ \mu\text{A}$$

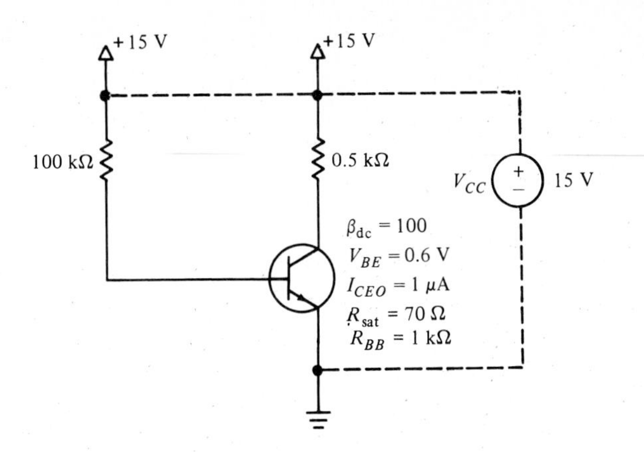

Fig. 3.8 A circuit for which we are asked to find all the transistor voltages and currents.

The collector current is the sum of the two source currents,

$$I_C = 0.001 + 100(0.1426) = 14.26 \text{ mA}$$

Writing the loop equation from the collector, through the 0.5-kΩ load resistor and the power supply, back to the emitter, gives us

$$V_{CE} = 15 - (14.26)(0.5) = 7.87 \text{ V}$$

Also,

$$V_{BE} = 0.6 + 1I_B = 0.743 \text{ V}$$

Fig. 3.9 The active-region dc model for the *npn* bipolar transistor is installed in the circuit of Fig. 3.8.

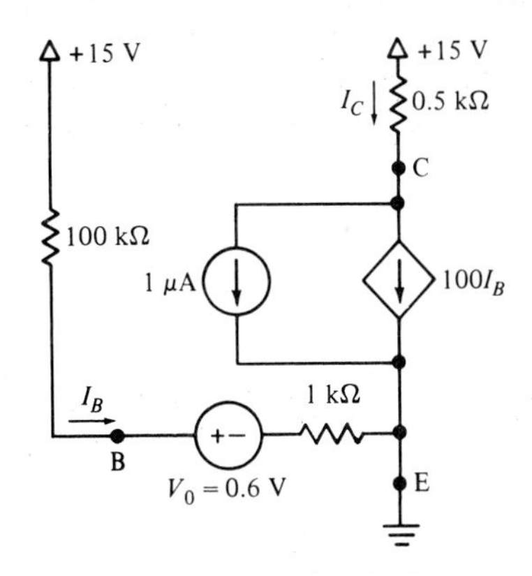

and therefore

$$V_{CB} = 7.87 - 0.74 = 7.13 \text{ V}$$

The last current to be found is

$$I_E = -I_B - I_C = -0.1426 - 14.26 = -14.40 \text{ mA}$$

This completes the solution, provided that our assumption of an active-region model was correct. We now see that it was, for the set of currents and voltages we have obtained locate the operating point in the active region.

Now let us change the problem slightly by increasing the value of the load resistance from 0.5 to 2 kΩ. Selecting the same active-region model, we again obtain $I_B = 142.6\ \mu\text{A}$ and $I_C = 14.26$ mA. But next we see that

$$V_{CE} = 15 - (14.26)(2) = -13.52 \text{ V}$$

This indicates that something has gone astray, for V_{CE} must be positive for an *npn* transistor in the active region. It follows that we are not operating in the active region, so we need a different model. Since the emitter-base junction is still forward-biased, let us assume that the increase in load resistance has caused the collector-base junction to change from reverse- to forward-bias. Thus we try a saturation-region model, as shown in Fig. 3.10. Once again, $I_B = 142.6\ \mu\text{A}$; but we now see that I_C does not depend on that value. It is

$$I_{C(\text{sat})} = \frac{15}{2 + 0.07} = 7.25 \text{ mA}$$

It then follows that

$$V_{CE(\text{sat})} = 7.25(0.07) = 0.507 \text{ V}$$

Fig. 3.10 When the load resistor of Fig. 3.8 is increased to 2 kΩ, operation is in the saturation region.

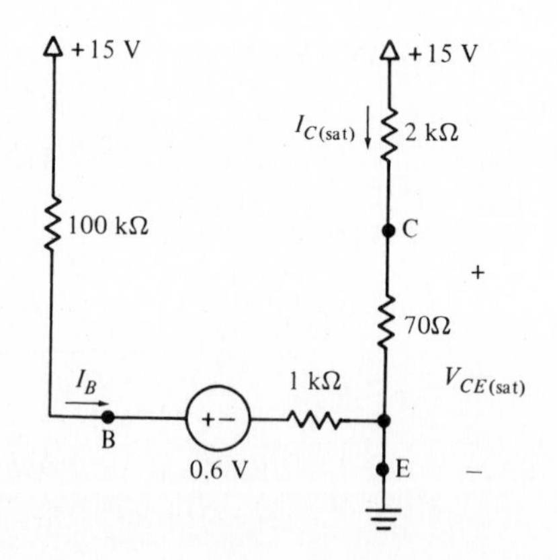

and

$$V_{CB} = 0.507 - 0.6 - (0.1426)(1) = -0.235 \text{ V}$$

This is forward bias for the collector-base junction (collector is *n*-type and base is *p*-type), and we are operating in the saturation region. The model and the solution therefore agree.

If we had assumed saturation operation for the circuit of Fig. 3.8 and used the saturation model for the *npn* transistor, we would have obtained

$$I_{C(\text{sat})} = \frac{15}{0.5 + 0.07} = 26.3 \text{ mA}$$

and

$$V_{CE} = (26.3)(0.07) = 1.842 \text{ V}$$

Thus

$$V_{CB} = 1.842 - 0.743 = 1.099 \text{ V}$$

This indicates that the collector-base junction is reverse-biased, which does not agree with our assumption. Therefore the model and the solution do *not* agree, and the transistor is *not* in saturation.

Some care must be taken when selecting R_{sat} to model a particular device. If the value used is too small for the given values of V_0 and R_{BB}, then an ambiguous situation can result in which the active-region model says "saturation," while the saturation-region model screams "active region."

Let us now try our hands at a design problem, as outlined by Fig. 3.11. The same *npn* transistor we have been working with must operate at $I_C = 2$ mA, $V_{CE} = 5$ V; the form of the external circuit is specified. Note that for a design problem we are given the answer (the operating point) and must find the circuit elements necessary to yield that result; it is often the

Fig. 3.11 Using the circuit and transistor parameter values given, we are to provide an operating point at $I_C = 2$ mA, $V_{CE} = 5$ V.

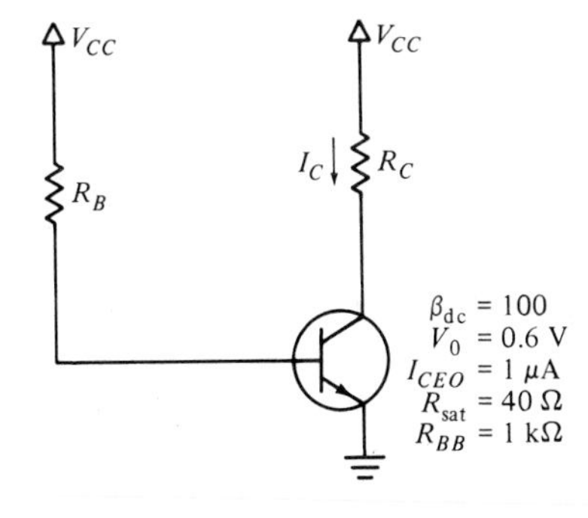

case that the circuit and the circuit elements required are not unique. We identify this as a dc-design problem that involves the use of a dc model for the *npn* transistor. Since the given values of I_C and V_{CE} provide operation in the active region, the active-region model is obviously required. A solution consists of suitable values for the three unspecified quantities on the circuit diagram: V_{CC}, R_B, and R_C. Our plan of attack will be to use the simplest active-region model, to make any arbitrary choices as intelligently as possible, and to check our results by an analysis using our most accurate dc model.

We begin with the simple model, as shown in Fig. 3.12*a*, and select a value for the supply voltage V_{CC}. It certainly must be larger than V_{CE}, which is 5 V. Also, the difference between V_{CC} and V_{CE} must appear across R_C, through which the collector current of 2 mA passes. As V_{CC} is made larger, the power dissipated in R_C increases proportionately. On the other hand, as we shall discover later, greater values of signal gain are associated with larger values of R_C and V_{CC}. Let us select $V_{CC} = 20$ V as a reasonable compromise. The voltage across R_C is then 15 V, and $R_C = 15/2 = 7.5$ kΩ. Since the dependent source shows that

$$I_C = 100I_B = 2 \text{ mA}$$

we find that $I_B = 20\ \mu$A, and thus $V_{BE} = 0.6 + 0.02\,(1) = 0.62$ V, and $R_B = (20 - 0.62)/0.02 = 969$ kΩ.

This is the last of the three circuit values that are required, but the design problem is not quite complete. A circuit analysis must still be performed as a check on the design. In this case we are being a little unfair to

Fig. 3.12 (*a*) The transistor in Fig. 3.11 is replaced by a simple dc model. (*b*) A more detailed dc model is used to check the design.

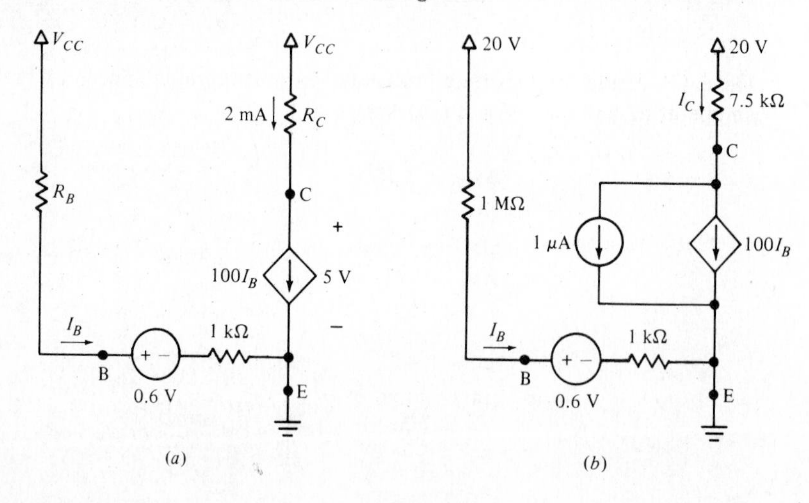

ourselves in analyzing our design with a more accurate dc model than we used to develop it, as shown in Fig. 3.12*b*. But let us see how the operating point comes out. We find that

$$20 = 969I_B + 0.6 + 1I_B \qquad \text{and} \qquad I_B = 20\ \mu\text{A}$$

Also,

$$I_C = 0.001 + 100(0.02) = 2.001\ \text{mA}$$

$$V_{CE} = 20 - 2.001(7.5) = 4.993\ \text{V}$$

which is not a serious deviation from the desired operating point.

We shall take up other applications of the dc models for the bipolar transistor after we have gained some familiarity with the dc models for the FETs.

D3.4 Determine I_C and V_{CE} for the circuit of Fig. 3.11 if $V_{CC} = 12$ V, $R_C = 800\ \Omega$, and R_B equals (a) 150 kΩ, (b) 50 kΩ.

Answers. 7.55 mA, 5.96 V; 14.29 mA, 0.571 V

D3.5 Specify a set of circuit values for the circuit of Fig. 3.11 that will establish an operating point at $I_C = 3$ mA, $V_{CE} = 4.5$ V.

Answers. $V_{CC} = 15$ V; $R_C = 3.5$ kΩ; $R_B = 479$ kΩ

3.3 Nonlinear dc models: JFETs and depletion-mode MOSFETs

If we refer to Fig. 2.7, we see that there are many more varieties of FETs than there are of bipolar transistors. Fortunately, however, the number of nonlinear dc models required is considerably less than the number of varieties of FETs. In the last chapter we found that both the JFET and the depletion-mode MOSFET (or IGFET) are depletion-mode devices. They have circuit models of the same form. The FETs therefore fall into two groups for modeling. The first group includes the JFET and the depletion-mode MOSFET, in either *n*-channel or *p*-channel form; the second contains the enhancement-mode MOSFET, either as *n*-channel or *p*-channel.

Thus, two nonlinear dc models are sufficient to encompass all *n*-channel FETs, and a simple reversal of sign for all voltages and currents is the only change required for the *p*-channel models.

We begin with the nonlinear dc model for the *n*-channel JFET. It is normally biased in the region beyond pinch-off where the signal gain is the greatest; we shall content ourselves with a single model for this region. To use this model, we must thus show that the transistor operates in the region beyond pinch-off.

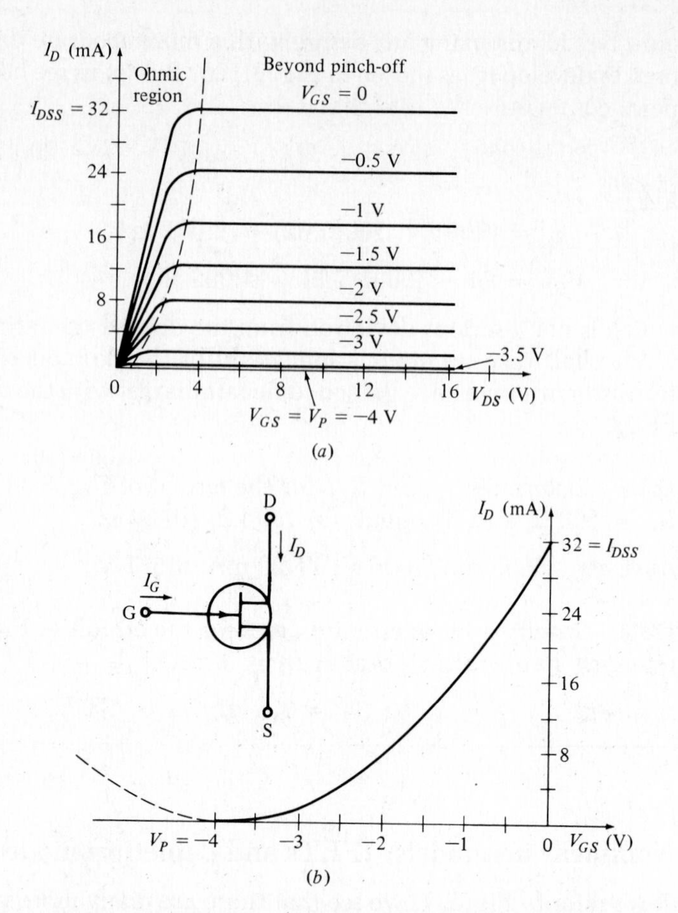

Fig. 3.13 (*a*) The drain characteristics for an *n*-channel JFET. Here $I_{DSS} = 32$ mA and $V_P = -4$ V. (*b*) A plot of I_D vs. V_{GS} is a portion of a parabola in the region beyond pinch-off, and it can be approximated by $I_D = (I_{DSS}/V_P^2)(V_{GS} - V_P)^2$.

The drain current is relatively independent of the drain-to-source voltage in the region beyond pinch-off, as shown in Fig. 3.13*a*. Because these curves have almost zero slope in the region beyond pinch-off, this region can be modeled by a transfer characteristic (I_D vs. V_{GS}) at any value of V_{DS} that provides operation in the region beyond pinch-off. Such a transfer characteristic is shown in Fig. 3.13*b*.

In Chapter 2 we found that for operation at or beyond pinch-off, it was necessary that

$$V_{DS} \geq V_{GS} - V_P \qquad (n\text{-channel})$$

Since the characteristics are flat for $V_{DS} \geq V_{GS} - V_P$, we may simplify matters by selecting the limiting condition in which

$$V_{DS} = V_{GS} - V_P$$

Substituting this equation into Eq. (1) of Chapter 2, we eliminate V_{DS} and have

$$I_D = D\left[(V_{GS} - V_P)(V_{GS} - V_P) - \frac{(V_{GS} - V_P)^2}{2}\right]$$

or

$$I_D = \frac{D}{2}(V_{GS} - V_P)^2 \quad \text{for } V_{DS} \geq V_{GS} - V_P \text{ and } V_P \leq V_{GS} \leq 0 \tag{9}$$

This equation may be simplified by remembering that I_{DSS} is defined as I_D when $V_{GS} = 0$ and $V_{DS} = -V_P$. Thus

$$I_D|_{V_{GS}=0} = \frac{D}{2}(-V_P)^2 = I_{DSS}$$

and

$$D = \frac{2I_{DSS}}{V_P^2}$$

We thus have the important result,

$$I_D = \frac{I_{DSS}}{V_P^2}(V_{GS} - V_P)^2 \quad \text{for } V_{DS} \geq V_{GS} - V_P \text{ and } V_P \leq V_{GS} \leq 0 \tag{10}$$

Note that I_{DSS} is positive, while V_{GS} and V_P are both negative quantities for the n-channel JFET. In the case of the p-channel unit, I_{DSS} is negative and V_{GS} and V_P are positive.

Equation (10) graphs in the form of a parabola, as illustrated in the curve of Fig. 3.13b. Note that the curve is completely specified by two values, the drain-to-source saturation current I_{DSS} and the pinch-off voltage V_P.

Equation (10) provides us with 90% of the desired nonlinear model for the n-channel JFET. We may need to consider the gate leakage current I_{GSS}, but since it is the leakage current of the gate-to-drain diode under reverse bias, it is often small enough to be neglected.

It should be noted that although Eq. (10) defines a complete parabola, the only part of the curve that represents the n-channel JFET is the right half, for which $V_P \leq V_{GS} \leq 0$ and $I_D \leq I_{DSS}$. We may therefore obtain two formal mathematical solutions to a certain problem, corresponding to points on the two halves of the parabola; one makes sense, the other does not. Only the solution for which $V_P \leq V_{GS} \leq 0$ applies to the real device.

Let us use this nonlinear model to calculate the operating point for the FET shown in Fig. 3.14. From the data given near the circuit diagram, Eq. (10) becomes

$$I_D = \frac{10}{(-2)^2}[V_{GS} - (-2)]^2 \quad \text{mA}$$

Since the gate current equals the gate leakage current when $V_P \le V_{GS} \le 0$ and $I_{GSS} = -1$ nA, the voltage across R_B is only 1 mV. Thus for all practical purposes, V_{GS} is V_G. From Eq. (10),

$$I_D = \frac{10}{4}(-1 + 2)^2 = 2.5 \text{ mA}$$

The loop equation through the drain-supply voltage leads to

$$V_{DS} = V_{DD} - R_D I_D = 20 - 2(2.5) = 15 \text{ V}$$

We thus find that we are well beyond pinch-off at an operating point specified by $V_{DS} = 15$ V, $I_D = 2.5$ mA, and $V_{GS} = -1$ V. That is, $15 \ge -1 - (-2) = 1$, and the model agrees with the solution.

Now let us make two changes in our viewpoint. Instead of analysis, let us attempt a design problem; we also use a *p*-channel JFET. The problem is outlined in Fig. 3.15*a*: We want an external circuit (of the form indicated) that will provide an operating point at a drain current of -8 mA and a drain-to-source voltage of -10 V. Operation in the saturation region is assumed. The problem involves finding a suitable set of values for V_G, V_{DD}, R_D, and R_B.

For this *p*-channel JFET, operation beyond pinch-off requires that $V_{DS} = -10 \le V_{GS} - V_P = V_{GS} - 2$, where V_{GS} and V_P are both positive.

Fig. 3.14 An example in which an *n*-channel JFET is analyzed to determine the operating point.

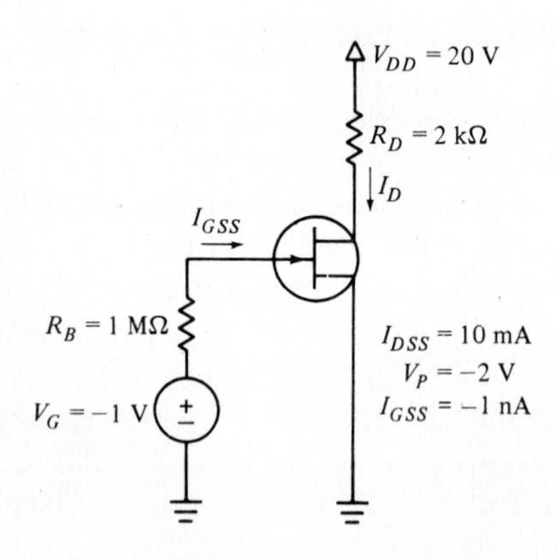

Also,

$$I_D = \frac{I_{DSS}}{V_P^2}(V_{GS} - V_P)^2 \quad \text{for } V_{DS} \le V_{GS} - V_P \text{ and } 0 \le V_{GS} \le V_P \tag{11}$$

so that

$$I_D = \frac{-10}{(2)^2}(V_{GS} - 2)^2 = -8 \text{ mA}$$

Solving for V_{GS}, we find that

$$\frac{-8}{-10}(2^2) = (V_{GS} - 2)^2$$

and

$$V_{GS} = 2 \pm \sqrt{3.2} = 2 \pm 1.788 \text{ V}$$

Since V_{GS} is constrained to be *less* than V_P(2 V), we see that the only root that makes sense is 2 − 1.788, or 0.212 V. This is illustrated by Fig. 3.15*b*, which shows the transfer characteristic for the *p*-channel unit and identifies the two mathematical results. We conclude then that

$$V_{GS} = 2 - 1.788 = 0.212 \text{ V}$$

The gate leakage current is once again negligible, and we therefore set $V_G = 0.212$ V. The value of R_B does not affect the operating point, unless it is

Fig. 3.15 (*a*) Circuit values are to be specified for this *p*-channel JFET that will provide an operating point at $I_D = -8$ mA, $V_{DS} = -10$ V. (*b*) The left half of the parabola forms the transfer characteristic for the *p*-channel unit. On this portion of the curve, $0 \le V_{GS} \le V_P = 2$ V.

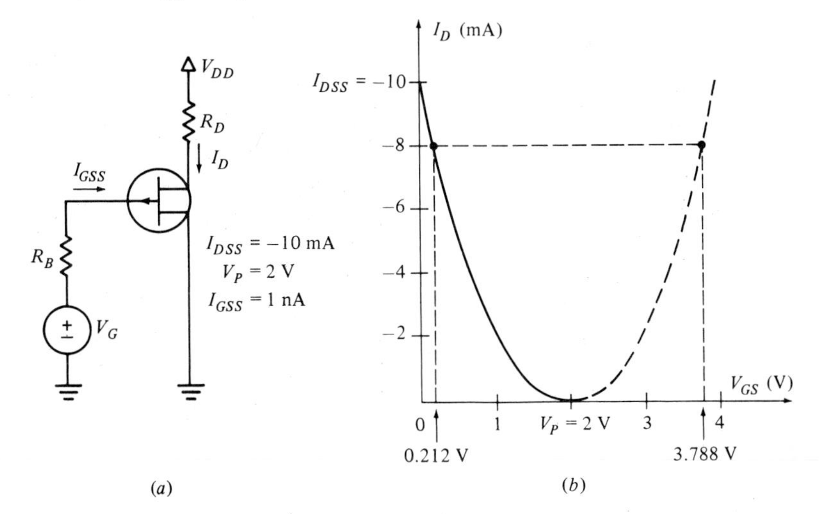

large enough to produce an appreciable voltage drop when I_{GSS} (= 1 nA) flows through it. We arbitrarily select $R_B = 1\ \text{M}\Omega$.

In the drain circuit,

$$V_{DS} + I_D R_D = V_{DD}$$

or

$$-10 - 8R_D = V_{DD}$$

Arbitrarily picking $V_{DD} = -26$ V, a value that is several times greater in magnitude than V_{DS}, we have

$$R_D = \frac{-26 + 10}{-8} = 2\ \text{k}\Omega$$

Thus $V_G = 0.212$ V, $V_{DD} = -26$ V, $R_D = 2\ \text{k}\Omega$, and $R_B = 1\ \text{M}\Omega$ provide a suitable set of values for the external circuit. These values are certainly not unique, and other considerations might very well lead to different selections. For example, with an available voltage supply of -18 V we might select $R_D = 1\ \text{k}\Omega$, assuming other considerations were not seriously affected.

In checking our results by analysis, we can use only the same nonlinear model, Eq. (10), and the result therefore checks perfectly.

Equations (10) and (11) also serve as nonlinear models for the *n*- and *p*-channel depletion-mode MOSFETs respectively. Values for I_{DSS} and V_P are in the same range as those for the JFETs, the only appreciable difference in the dc models being the magnitude of the gate leakage current I_{GSS}. Since the MOSFET (or IGFET) has an insulated gate, I_{GSS} is quite small, usually in the neighborhood of 1 pA (10^{-12} A). For the JFET, it is more often around 1 nA (10^{-9} A).

D3.6 An *n*-channel JFET with $I_{DSS} = 12$ mA, $V_P = -1.8$ V, and $I_G = 100$ pA is used in a circuit similar to that of Fig. 3.15*a*. Specify values for V_G, V_{DD}, and R_D if $R_B = 12\ \text{k}\Omega$, $V_{DS} = 9$ V, R_D dissipates 25 mW, and the JFET dissipates 20 mW.

Answers. -1.025 V; 20.25 V; 5.06 kΩ

3.4 Nonlinear dc models: enhancement-mode MOSFETs

Let us review by comparing the operation of depletion-mode and enhancement-mode MOSFETs. The *n*-channel depletion-mode device we considered in the previous section is normally operated in the region beyond pinch-off with V_{GS} greater than the negative pinch-off voltage V_P but less than zero. The *n*-channel enhancement-mode unit, however, is normally biased with V_{GS} greater than the positive threshold voltage V_T (a symbol that should not be confused with $V_T = kT/q$ for the bipolar transistor). The drain current increases as the difference $(V_{GS} - V_T)$ increases, assuming that the drain-to-source voltage is sufficiently large, $V_{DS} \geq V_{GS} - V_T$. The device is then operating in the region beyond pinch-off.

Figure 3.16*a* shows the output characteristics and Fig. 3.16*b* illustrates the parabolic transfer characteristic relating I_D and V_{GS} for an *n*-channel enhancement-mode MOSFET in the region beyond pinch-off. Using Eq. (5) from Chapter 2 at the boundary between the ohmic and saturation regions, $V_{DS} = V_{GS} - V_T$, we have

$$I_D = K\left[(V_{GS} - V_T)(V_{GS} - V_T) - \frac{(V_{GS} - V_T)^2}{2}\right]$$

Fig. 3.16 (*a*) The output characteristics for an *n*-channel enhancement-mode MOSFET. Here, $V_T = 2$ V. (*b*) A plot of I_D vs. V_{GS} is a portion of a parabola and follows the relationship $I_D = (K/2)(V_{GS} - V_T)^2$ in the region beyond pinch-off.

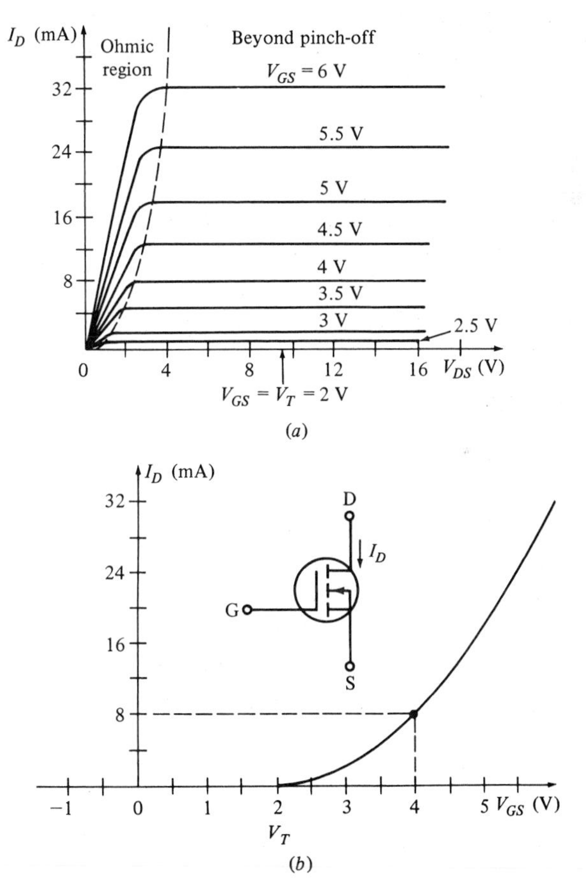

or

$$I_D = \frac{K}{2}(V_{GS} - V_T)^2 \quad \text{for } V_{DS} \geq V_{GS} - V_T \text{ and } V_{GS} \geq V_T \tag{12}$$

where K is a positive constant that must be found from a knowledge of the value of I_D at some specific V_{GS}.

For the device of Fig. 3.16*a* or *b*, we see that $V_T = 2$ V, while I_D is 8 mA when V_{GS} is 4 V. Therefore

$$I_D = 8 = \frac{K}{2}(4 - 2)^2$$

and $K = 4$ mA/V^2. So the nonlinear model for this *n*-channel enhancement-mode IGFET is

$$I_D = 2(V_{GS} - 2)^2 \quad \text{(mA)}$$

To illustrate the use of this nonlinear dc model, let us determine the operating point for the FET shown in Fig. 3.17. From the symbol, we see that we have an *n*-channel enhancement-mode MOSFET. The data given adjacent to the circuit indicate that $V_T = 1$ V, $I_D = 10$ mA when $V_{GS} = 5$ V, and the gate saturation current $I_{GSS} = -1$ pA, a truly negligible value. We decide that the values of I_D, V_{DS}, and V_{GS} must be determined.

Since we are able to neglect I_G, resistors R_1 and R_2 act as a simple voltage divider, and we find the gate-to-source voltage,

$$V_{GS} = 15 \frac{0.1 \times 10^6}{0.4 \times 10^6 + 0.1 \times 10^6} = 3 \text{ V}$$

Fig. 3.17 An *n*-channel enhancement-mode IGFET appears as an example in which the operating point is desired.

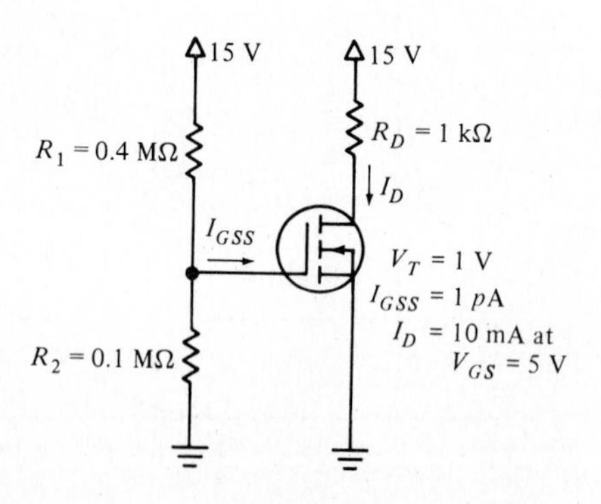

This is greater than the threshold voltage; therefore the transistor is not cut off. We select the nonlinear model, Eq. (12), valid in the region beyond pinch-off, as our first try at this analysis problem:

$$I_D = \frac{K}{2}(V_{GS} - 1)^2$$

Using the data supplied for the MOSFET, we determine K first:

$$10 = \frac{K}{2}(5 - 1)^2$$

or

$$K = 1.25 \text{ mA/V}^2$$

Since $V_{GS} = 3$ V in the circuit given, we find

$$I_D = \frac{1.25}{2}(3 - 1)^2 = 2.5 \text{ mA}$$

Thus the voltage across R_D is 2.5 V and $V_{DS} = 12.5$ V. This indicates that operation is well beyond pinch-off, $12.5 \geq 3 - 1$, and our model is appropriate.

Once more, we have slighted the p-channel unit, but it should suffice to remind ourselves that all currents and voltages simply have the opposite sense; even K becomes negative. Thus the transfer characteristic takes the form shown in Fig. 3.18. Problem D3.8 gives us the opportunity to practice this art of sign reversal.

D3.7 In the circuit and data of Fig. 3.17, let $V_{DD} = 12$ V, $R_1 = 0.4$ MΩ, $V_T = 1.2$ V, $I_{GSS} = 100$ fA, and $I_D = 4$ mA at $V_{GS} = 4.4$ V. Select values

Fig. 3.18 The transfer characteristic for a p-channel enhancement-mode IGFET for which $V_T = -2$ V.

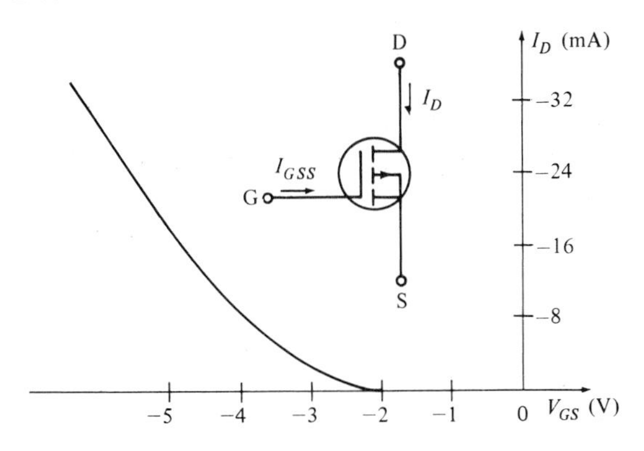

for R_D and R_2 that will establish an operating point at $V_{DS} = 6$ V, $I_D = 2$ mA.

Answers. 3 and 162.2 kΩ

D3.8 A *p*-channel enhancement-mode MOSFET has $|V_T| = 2.4$ V and $|I_D| = 6$ mA at $|V_{GS}| = 4.8$ V. Assume that I_{GSS} is negligible and find (a) I_D when $|V_{GS}| = 4$ V, (b) V_{GS} when $|I_D| = 5$ mA, (c) the value of V_{DS} at the boundary between the ohmic and saturation regions when $|I_D| = 5$ mA.

Answers. − 2.67 mA; − 4.59 V; − 2.19 V

3.5 Load Lines

The concept of the load line was introduced for the diode in Section 1.2 of Chapter 1 as a simple means of analyzing or visualizing the solution of a nonlinear problem. Many times the effect of making a certain change in a circuit can be seen qualitatively by a rough sketch of the load line on the nonlinear characteristics of the device. Let us see how we can use the load line in analyzing circuits containing bipolar and field-effect transistors.

Bipolar load lines

Figure 3.19*a* shows a bipolar transistor in a circuit of the type we analyzed in Section 3.2 using linear, active-region models. Now let us see how the use of a load line enables us to answer the question, "How is the operating point affected by changing R_C?" We write the loop equation for the output,

$$V_{CC} = I_C R_C + V_{CE} \tag{13}$$

This linear equation in the variables I_C and V_{CE} is called the *load-line equation*, and it is plotted as a straight line on the I_C-V_{CE} output characteristics in Figure 3.19*b*. Note the intercepts at $I_C = 0$, $V_{CE} = V_{CC}$, and $V_{CE} = 0$, $I_C = V_{CC}/R_C$. The operating point (O.P.) lies at the intersection of the load line and the output curve for the specified value of I_B.

From Fig. 3.19*b* we see that any increase in I_B results in an increase in I_C and a decrease in V_{CE}. We also observe that if I_B is made sufficiently large, the device will go into saturation.

The slope of the load line is $-1/R_C$. Thus, as R_C decreases, the magnitude of the slope increases, as shown in Fig. 3.19*b*. We now see clearly an important effect of changing R_C: If R_C is made large enough, the operating point will move into the saturation region for a fixed value of I_B.

We may also consider the input characteristics (I_B vs. V_{BE}) of Fig.

3.19*c*. The loop equation around the input circuit is

$$V_{CC} = I_B R_B + V_{BE} \tag{14}$$

This equation is linear in I_B and V_{BE}. It leads to the *input load line*, as plotted on the input characteristics in Fig. 3.19*c*. Note that the intersection of the load line and the nonlinear input characteristic establishes the operating point, (I_B, V_{BE}). Again, it is easy to see the effect of changing R_B or V_{CC} on the operating point. We will use these concepts in Chapters 4 and 5 in designing circuits to provide a desired operating point.

Field-effect load lines

Only an output load line is needed for a JFET or a MOSFET, because I_G is negligibly small. Figure 3.20 illustrates the use of the load line for an *n*-channel JFET. The load-line equation is

$$V_{DD} = I_D R_D + V_{DS}$$

or

$$I_D = -\frac{1}{R_D} V_{DS} + \frac{V_{DD}}{R_D} \tag{15}$$

The intercepts are at $V_{DS} = 0$, $I_D = V_{DD}/R_D$, and $V_{DS} = V_{DD}$, $I_D = 0$. For a fixed value of V_{GS}, note that V_{DS} decreases as R_D increases. In fact, if

Fig. 3.19 (*a*) An *npn* transistor is shown in a simple linear external circuit. (*b*) The output load line is drawn on the output characteristics, and an operating point is identified for $I_B = I_{B2}$. A change in R_C causes the slope of the load line to change as indicated. (*c*) An input load line is shown. Note that $V_{CC} \gg V_{BE}$ at the O.P., and it is necessary to show a break in the abscissa and in the load line.

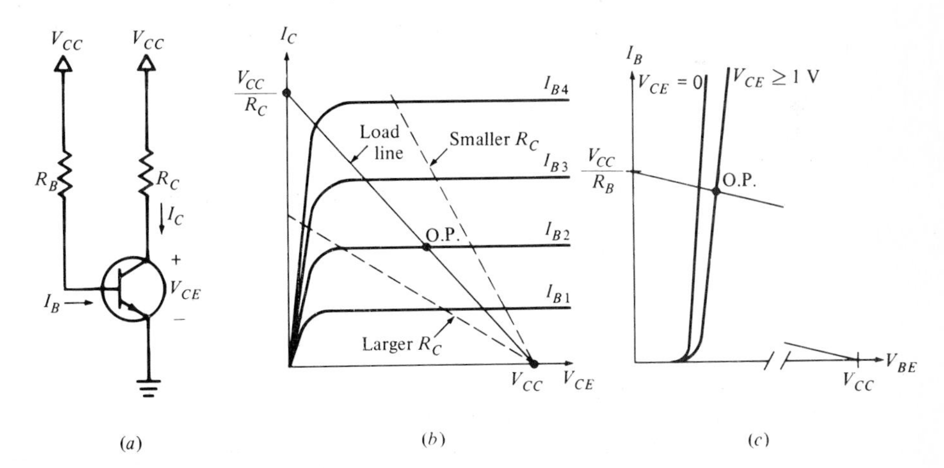

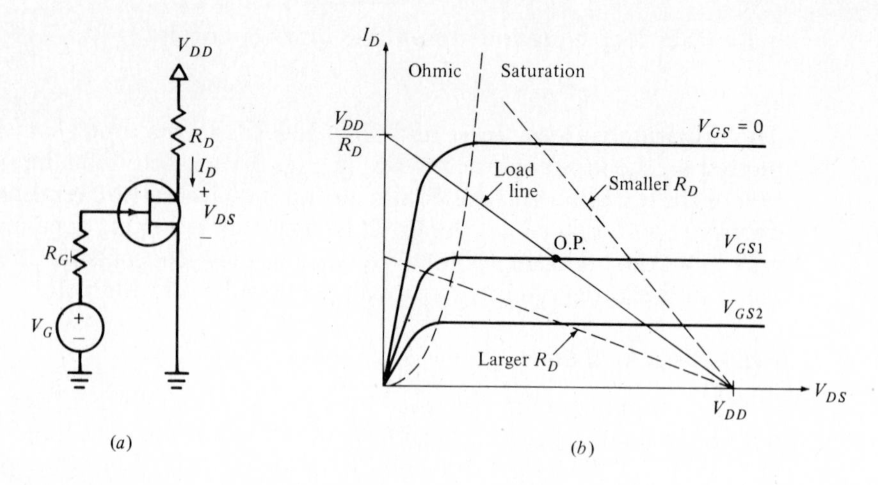

Fig. 3.20 (*a*) An *n*-channel JFET is shown in a linear external circuit. (*b*) An output load line is drawn on the output characteristics. Increasing R_D without changing V_{GS} may cause the operating point to shift into the ohmic region.

R_D is made large enough, the operating point will go from the saturation region into the ohmic region.

Figure 3.21 shows another application of load lines that applies *only* when the FET is operating beyond pinch-off. If we write the equation around the input loop of Fig. 3.21*a*,

$$V_G = I_G R_B + V_{GS} + I_D R_{SS} \doteq V_{GS} + I_D R_{SS}$$

Neglecting I_G,

$$I_D = -\frac{1}{R_{SS}} V_{GS} + \frac{V_G}{R_{SS}} \tag{16}$$

Note that Eq. (16) is a linear equation in the variables I_D and V_{GS}. Therefore this equation appears as a straight line on the transfer characteristic of Fig. 3.21*b*. The intersection of the load line and the transfer characteristic locates the operating point (I_D, V_{GS}). The value of V_{DS} at the operating point is obtained by solving the equation $V_{DD} = I_D R_D + V_{DS} + I_D R_{SS}$ for V_{DS}.

Figure 3.21*b* also illustrates the effect of changing R_{SS} on the operating point. For example, if R_{SS} is made larger, I_D decreases. Similarly, an increase in the value of V_G moves the V_{GS} intercept to the right and results in a larger value of I_D at the operating point.

Load lines similar to these are obtained for *p*-channel JFETs, as well as *n*- and *p*-channel MOSFETs.

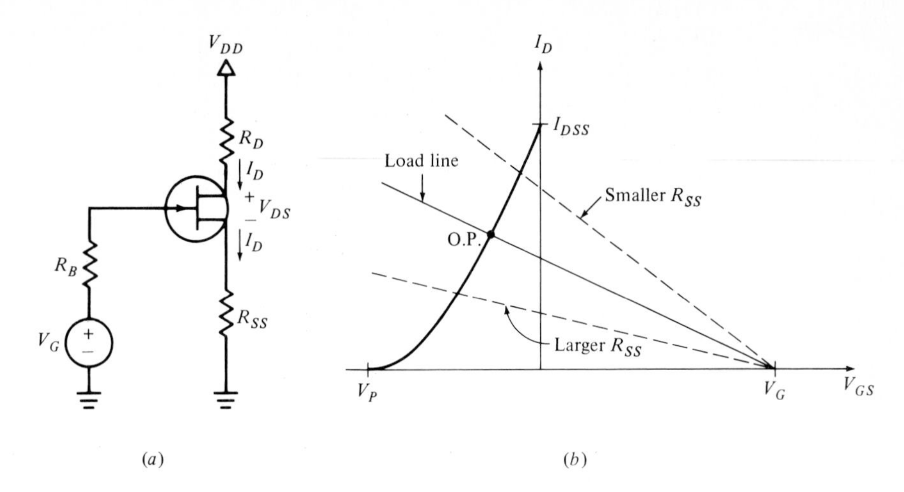

Fig. 3.21 (*a*) An n-channel JFET is shown with a linear external circuit that includes a resistor in the source lead. (*b*) When the FET is operating in the region beyond pinch-off, a load line with slope $-1/R_{SS}$ may be drawn on the transfer characteristic.

D3.9 A p-channel JFET is used in a circuit similar to that of Fig. 3.21*a*, but having $V_{DD} = -25$ V, $V_G = -4$ V, $R_B = 1$ MΩ, and $R_D = 2$ kΩ. The transfer characteristic is given in Fig. 3.15*b*. Use a load line to estimate I_D if R_{SS} equals (a) 0.5 kΩ, (b) 1 kΩ, (c) 4 kΩ.

Answers. -8.3; -4.6; -1.3 mA

D3.10 An n-channel enhancement-mode MOSFET with $V_T = 1$ V and $I_D = 4$ mA when $V_{GS} = 2$ V is operating in a circuit with $V_G = 4$ V, $R_B = 1$ MΩ, $V_{DD} = 20$ V, $R_D = 2$ kΩ, and $R_{SS} = 0.5$ kΩ. Let the value of R_{SS} increase. State whether each of the following quantities will increase or decrease in value: (a) I_D, (b) V_{DS}, (c) V_{GS}.

Answers. Decrease; increase; decrease

Problems

1. Let $R_{BB} = 6$ kΩ, $V_0 = 0.57$ V, $\beta_{dc} = 120$, and $I_{CEO} = 1.6\ \mu$A in the active-region model of an *npn* transistor. Determine I_C if (a) $I_B = 250\ \mu$A, and (b) $I_E = -2.8$ mA. (c) Find V_{BE} when $I_C = 2.5$ mA.
2. For an *npn* transistor, let $\beta_{dc} = 200$, $I_{CEO} = 1\ \mu$A, $R_0 = 50$ kΩ, $V_0 = 0.6$ V, and $R_{BB} = 5$ kΩ. Let $V_{CE} = 5$ V and calculate I_C if (a) $V_{BE} = 0.65$ V, (b) $V_{CB} = 4.3$ V, (c) $V_{BE} = 0.75$ V.

3. In Problem 2(a), calculate the power dissipated at (a) the emitter-base junction, (b) the collector-base junction, (c) the entire transistor.
4. Let the active region model of an *npn* transistor contain $R_{BB} = 1\ \text{k}\Omega$, $V_0 = 0.6$ V, $\beta_{dc} = 49$, $I_{CEO} = 1.5\ \mu$A, and $R_0 = 100\ \text{k}\Omega$. If $I_B = 100\ \mu$A and $V_{CE} = 4$ V, find the power supplied by each of the three ideal sources in the model.
5. Let $I_{CEO} = -1\ \mu$A, $V_0 = -0.6$ V, $R_{BB} = 3\ \text{k}\Omega$, and $\beta_{dc} = 50$ for a bipolar transistor. Calculate (a) I_{CBO}, (b) I_C when $V_{BE} = -0.65$ V, (c) I_E when $V_{BE} = -0.7$ V.
6. Knowing that $I_{CBO} = -24$ nA, $\alpha_{dc} = 0.997$, and $V_0 = -0.66$ V for a *pnp* transistor operating with $I_E = 3$ mA, find I_C and I_B.
7. (a) Establish suitable values for R_{sat}, R_{BB}, and V_0 if measurements show that $V_{CE} = 0.4$ V and $V_{BE} = 0.68$ V when $I_B = 100\ \mu$A and $I_C = 12$ mA, but V_{BE} increases to 0.8 V when $I_B = 200\ \mu$A. (b) Calculate I_C, I_E, and V_{CE} when $I_B = 0.3$ mA and $V_{CB} = -0.3$ V.
8. A saturation model of the type shown in Fig. 3.5*a* includes $R_{sat} = 50\ \Omega$, $V_0 = 0.6$ V, and $R_{BB} = 700\ \Omega$. Find I_C, I_E, and V_{CE} if $I_B = 0.2$ mA and $V_{BC} = 0.28$ V.
9. Modify the cutoff model of Fig. 3.6 by adding R_0 between collector and emitter. (a) Express I_C as a function of V_{CE}. (b) Plot the output characteristics on I_C-V_{CE} axes if $I_{CEO} = 1\ \mu$A and $R_0 = 50\ \text{k}\Omega$.
10. Parameters for an *npn* model are $R_{BB} = 2\ \text{k}\Omega$, $V_0 = 0.68$ V, $\beta_{dc} = 120$, $I_{CEO} = 1\ \mu$A, and $R_{sat} = 60\ \Omega$. (a) If $I_B = 180\ \mu$A, find the values of I_C, V_{BE}, and V_{CE} for the point on the output characteristics lying on the border between the active and saturation regions in the model. (b) If $V_{CE} = 4$ V, determine I_C and I_B at the point separating the active and cutoff regions in the model.
11. The dc model of Fig. 3.12*b* is improved by adding a 20-kΩ resistor across the dependent current source. Find I_C, I_B, and V_{CE}.
12. Parameter values for the transistor of Fig. 3.22 are: $\beta_{dc} = 75$, $V_0 = 0.6$ V, $I_{CEO} = 1.3\ \mu$A, and $R_{sat} = 40\ \Omega$. (a) Find α_{dc} and I_{CBO}. Let $V_{BB} = 6$ V, $R_B = 27\ \text{k}\Omega$, and $V_{CC} = 9$ V. (b) Calculate I_C if $R_C = 400\ \Omega$. (c) Find I_C if $R_C = 800\ \Omega$.

Fig. 3.22 See Problems 12, 16, and 18.

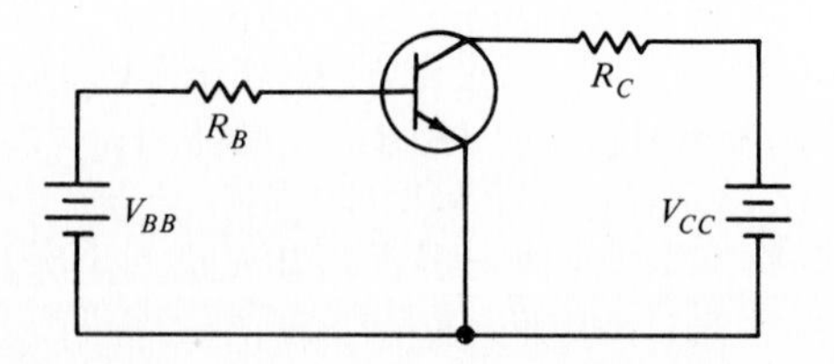

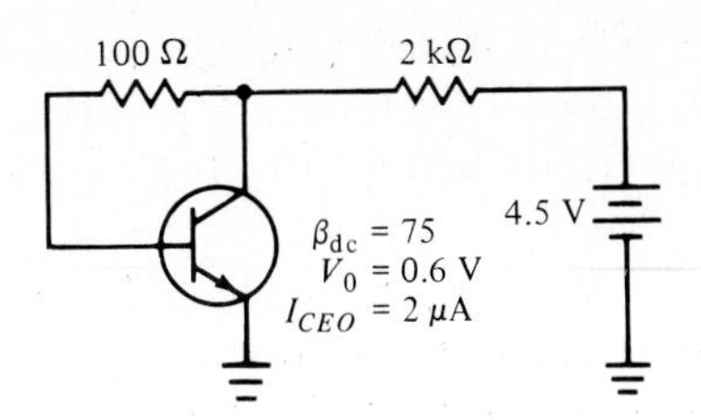

Fig. 3.23 See Problem 13.

13. The transistor in the circuit of Fig. 3.23 is operating essentially as a diode. Find I_C and V_{CE}.
14. The transistor used in Fig. 3.24 has $\beta_{dc} = 75$, $V_0 = 0.6$ V, and $I_{CEO} = 2\ \mu$A. Let $V_{CC} = 15$ V and $R_2 = 50$ kΩ, and select values for R_1 and R_C so that $I_C = 2$ mA and $V_{CE} = 8$ V.
15. Let $\beta_{dc} = 60$, $V_0 = 0.65$ V, $R_{sat} = 50\ \Omega$, and $I_{CEO} \doteq 0$ for the transistor shown in Fig. 3.24. If $R_C = 4$ kΩ and $V_{CC} = 10$ V, select values for R_1 and R_2 so that $I_C = 1.8$ mA while the downward current in R_1 is $30 I_B$.
16. Circuit values for Fig. 3.22 are $V_{BB} = 9$ V, $R_B = 120$ kΩ, $R_C = 3$ kΩ, and $V_{CC} = 13.5$ V. The transistor may be modeled by $V_0 = 2/3$ V, $R_{BB} = 1$ kΩ, $I_{CEO} = 4\ \mu$A, $R_{sat} = 50\ \Omega$, and $\beta_{dc} = 50$. (a) Find I_C and V_{CE}. (b) Determine I_C and V_{CE} if V_{CC} is reduced to 9 V.
17. Data for the RCA 40319, a silicon *pnp* power transistor, are given as: $I_{CBO} = -0.25\ \mu$A, $\beta_{dc(min)} = 35$, $\beta_{dc(max)} = 200$, and $V_{BE} = -1$ V. The maximum collector dissipation ($V_{CE} I_C$) is listed as 1 W at 25°C. (a) If $I_B = -4$ mA, find I_C and I_E for both $\beta_{dc(min)}$ and $\beta_{dc(max)}$. (b) If

Fig. 3.24 See Problems 14 and 15.

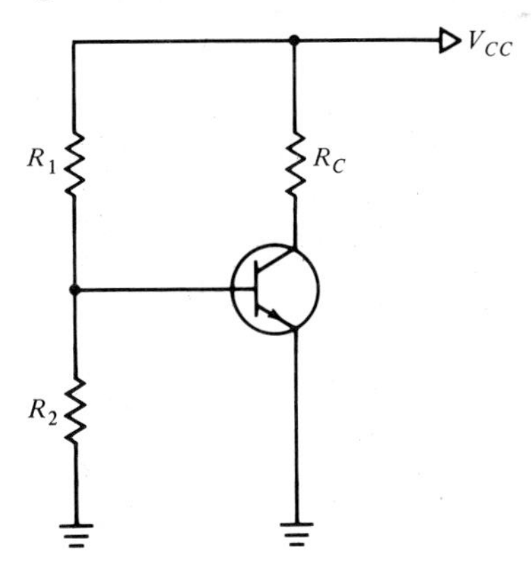

$V_{CE} = -20$ V, find the values of I_B that will lead to maximum collector dissipation for both $\beta_{dc(min)}$ and $\beta_{dc(max)}$.

18. Voltage-source values for Fig. 3.22 are $V_{BB} = 9$ V and $V_{CC} = 6$ V; the transistor has $V_0 = 0.65$ V, $I_{CEO} = 1\ \mu$A, $\beta_{dc} = 90$, and $R_{sat} = 40\ \Omega$. Assume that R_{BB} may be neglected. (a) Select values for R_B and R_C so that the transistor operates with $I_C = 1$ mA, $V_{CE} = 2$ V. (b) Find the power dissipated in each of the five circuit elements of Fig. 3.22. (c) What would the operating point be if β_{dc} decreased to 50? (d) What would the operating point be if β_{dc} increased to 200?
19. The transistor in Fig. 3.25*a* may be modeled by using $\beta_{dc} = 100$, $V_0 = 0.7$ V, $R_{BB} = 2.5$ kΩ, $R_{sat} = 50\ \Omega$, and $I_{CEO} = 0$. (a) Find I_C and V_{CE}. (b) Specify a new value for the 315-kΩ resistor so that $V_{CB} = 0.5$ V. (c) A 500-kΩ resistor is connected between the base terminal and ground. Find I_C.
20. Calculate new values of I_C and V_{CE} if the dc model shown in Fig. 3.12*b* also includes a resistor $R_0 = 50$ kΩ between collector and emitter.
21. With reference to Fig. 3.11, let $V_{CC} = 25$ V and select circuit values to establish an operating point at $I_C = 1$ mA, $V_{CE} = 10$ V. Use the parameter values given, and in addition let $R_0 = 40$ kΩ.
22. Select element values for the circuit of Fig. 3.25*b* that will provide an operating point at $I_C = -1$ mA, $V_{CE} = -3$ V. Use a dc model with $V_0 = -0.65$ V, $R_{BB} = 7$ kΩ, $I_{CEO} = -2\ \mu$A, $R_0 = 25$ kΩ, and $\beta_{dc} = 75$.
23. In Fig. 3.26, let $R_2 = 50$ kΩ, $R_D = 2.5$ kΩ, $R_{SS} = 300\ \Omega$, and $V_{DD} = 22.5$ V. The transistor is characterized by $I_{DSS} = 12$ mA, $I_{GSS} = -2$ nA, and $V_P = -3$ V. Find I_D, V_{GS}, and V_{DS}.

Fig. 3.25 (*a*) See Problem 19. (*b*) See Problem 22.

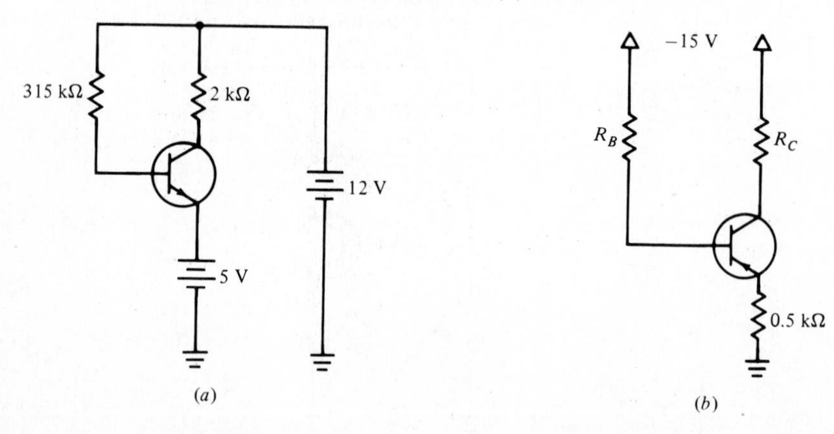

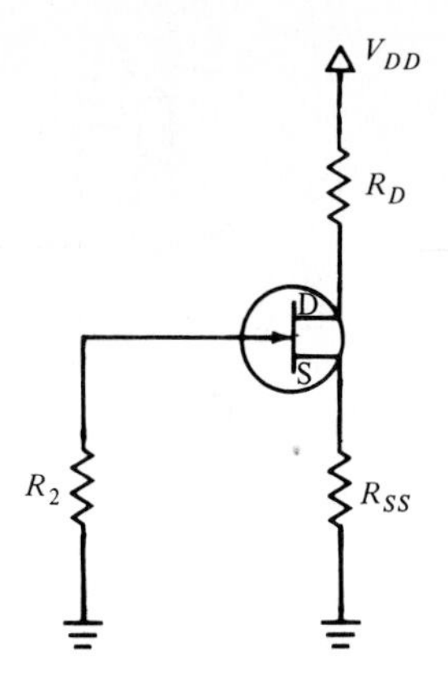

Fig. 3.26 See Problems 23 and 24.

24. An 850-kΩ resistor is connected between the gate and 22.5-V supply of the JFET of Problem 23 and Fig. 3.26. Find I_D, V_{GS}, and V_{DS}.
25. The transistor shown in Fig. 3.27 is operating with $I_D = 4$ mA. If $I_{GSS} \doteq 0$, (a) find V_{DS}. If the 0.3-V source is reduced to 0.2 V, I_D increases to 6 mA. Find (b) V_P, (c) I_{DSS}.
26. The MOSFET shown in Fig. 3.28 is characterized by $I_{DSS} = -15$ mA, $V_P = 2.5$ V, and $I_{GSS} \doteq 0$. Select values for R_D, V_G, and V_{DD} so that $I_D = -12$ mA, $V_{DS} = 0.5V_{DD}$, and the transistor dissipates a total power of 150 mW.
27. Let $I_{DSS} = 9$ mA, $V_P = -2$ V, and $I_{GSS} \doteq 0$ for the JFET of Fig. 3.29. If $R_1 = 500$ kΩ, $R_2 = 400$ kΩ, and $R_D = 1$ kΩ, (a) find I_D and V_{DS}. (b) What is the smallest value that R_2 may have without causing cutoff? (c) What is the largest value that R_2 may have without causing operation in the ohmic region?

Fig. 3.27 See Problem 25.

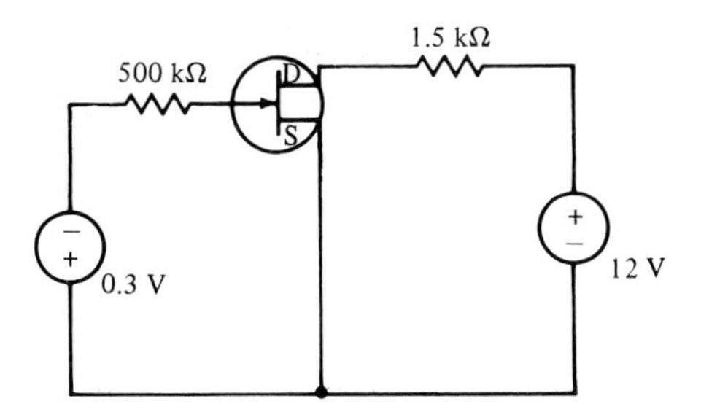

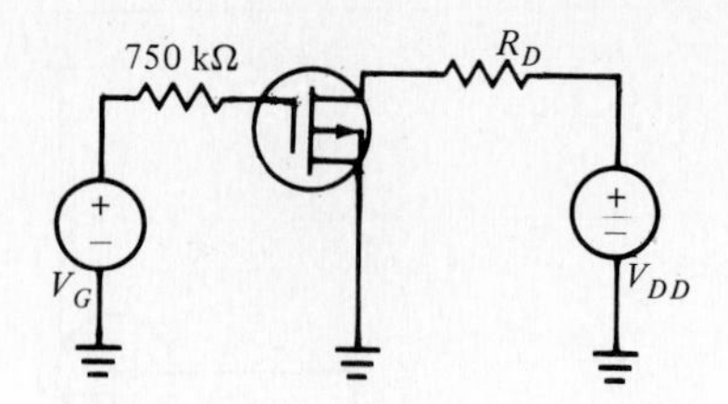

Fig. 3.28 See Problems 26 and 31.

28. (a) If $R_D = 2.5\ \text{k}\Omega$ in the circuit shown in Fig. 3.30, find the power dissipated in the FET. (b) How large may we make R_D before operation slides into the ohmic region?
29. An n-channel JFET has a set of output characteristics that includes the two operating points $I_D = 2$ mA, $V_{GS} = -3$ V, and $I_D = 8$ mA, $V_{GS} = -1.5$ V. (a) Find V_P and I_{DSS}. (b) Find V_{GS} if $I_D = 10$ mA. (c) Find I_D if $V_{GS} = -5$ V.
30. The transistor in Fig. 3.31 is operating with $|I_D| = 12$ mA and $|V_{DS}| = 6$ V. If $|V_P| = 3$ V and $|I_{DSS}| = 20$ mA, specify values for V_G, V_{SS}, R_{SS}, and R_B.
31. The transistor in the circuit of Fig. 3.28 is changed to an n-channel enhancement-mode device having $V_T = 2.5$ V. If $I_D = 10$ mA when $V_{GS} = 4$ V, specify suitable values for V_G, R_D, and V_{DD} to obtain an operating point at $I_D = 6$ mA, $V_{DS} = 3$ V.
32. The n-channel enhancement-mode MOSFET described in Problem 31 is used in a circuit like that shown in Fig. 3.31. Specify values for

Fig. 3.29 See Problem 27.

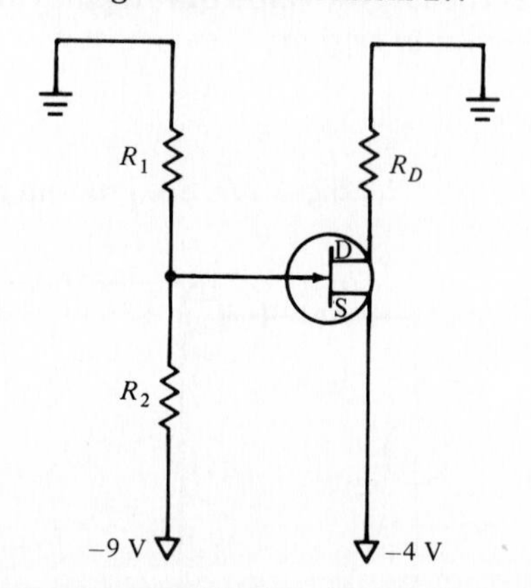

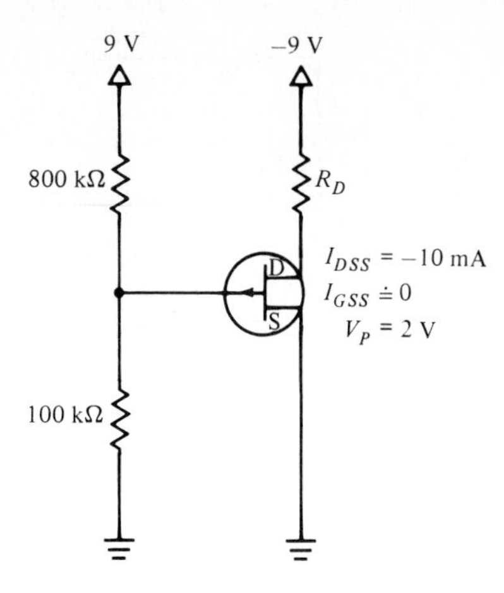

Fig. 3.30 See Problem 28.

V_G, R_B, V_{SS}, and R_{SS} that will provide an operating point at $I_D = 8$ mA, $V_{DS} = 4$ V.

33. Let $R_1 = 500$ kΩ, $R_2 = 100$ kΩ, $R_D = 1.5$ kΩ, and $V_{DD} = 24$ V in the circuit shown in Fig. 3.32. The transistor has a threshold voltage of 1.6 V, and the drain current is 8 mA when $V_{GS} = 3.6$ V. The gate current is negligible. Find I_D and V_{DS} for R_{SS} equals (a) 0, (b) 200 Ω.
34. Element values for Fig. 3.32 are $R_1 = 210$ kΩ, $R_2 = 90$ kΩ, $R_D = 2$ kΩ, $R_{SS} = 0.5$ kΩ, and $V_{DD} = 20$ V. The vertex of the parabola representing the transfer characteristic of the n-channel enhancement-mode MOSFET lies at $V_G = 2.5$ V, $I_D = 0$. Another point on the curve is $V_{GS} = 3$ V, $I_D = 2$ mA. Assume that $I_G = 0$. (a) Find I_D and V_{DS}. (b) What value for R_2 will cause I_D to equal 6 mA?

Fig. 3.31 See Problems 30 and 32.

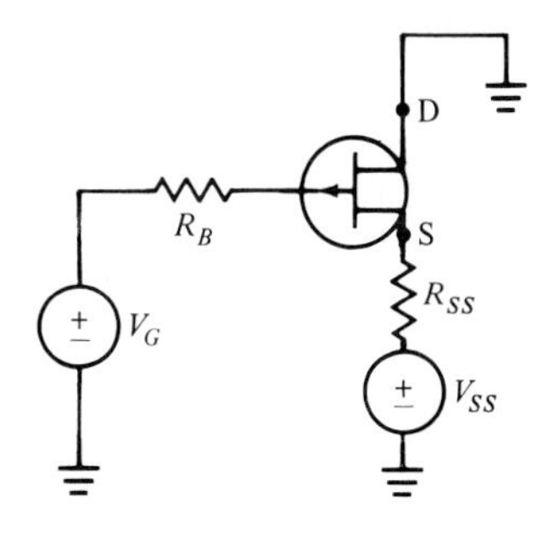

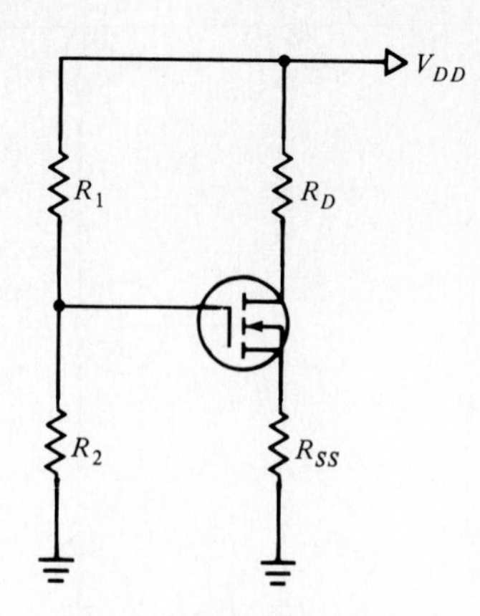

Fig. 3.32 See Problems 33, 34, and 36.

35. The transistor shown in Fig. 3.33 provides 12 mA with $V_{GS} = 6$ V and 4 mA with $V_{GS} = 5$ V. Find (a) V_T, I_D, and V_{DS}, (b) the power absorbed by the transistor and each resistor and the power supplied by each battery. (c) Find the slope (in mA/V) of the transfer characteristic at the operating point.
36. The transistor of Fig. 3.32 is replaced with a *p*-channel enhancement-mode unit having the transfer characteristic $I_D = -4(V_{GS} + 2.5)^2$ (mA). If $R_1 = 300$ kΩ, $R_2 = 200$ kΩ, $R_{SS} = 600$ Ω, $R_D = 1.6$ kΩ, and $V_{DD} = -25$ V, find I_D and V_{DS}.
37. A transfer characteristic for the trilateral transfister of Fig. 3.34*a* is given in Fig. 3.34*b*. Find I_3 and V_{32}.
38. The transistor that is indicated in the circuit of Fig. 3.35 has the characteristics given in Fig. 2.3. If $V_{CC} = 20$ V and $R_B = 200$ kΩ, find

Fig. 3.33 See Problem 35.

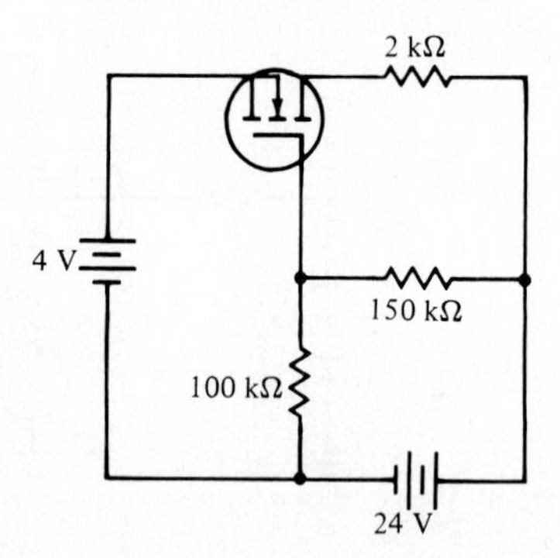

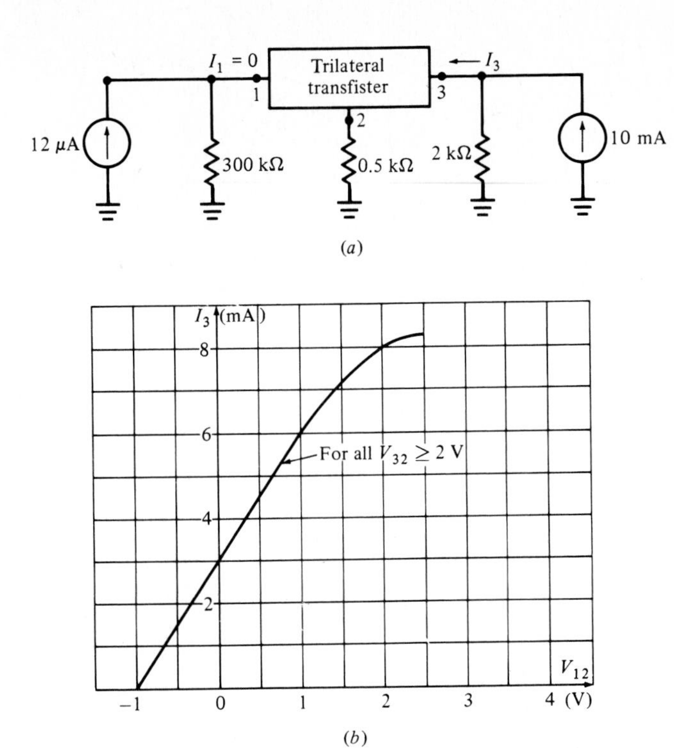

Fig. 3.34 See Problem 37.

Fig. 3.35 See Problems 38 and 39.

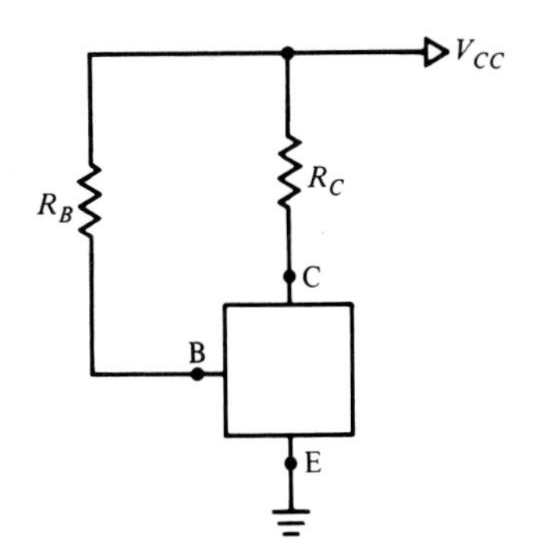

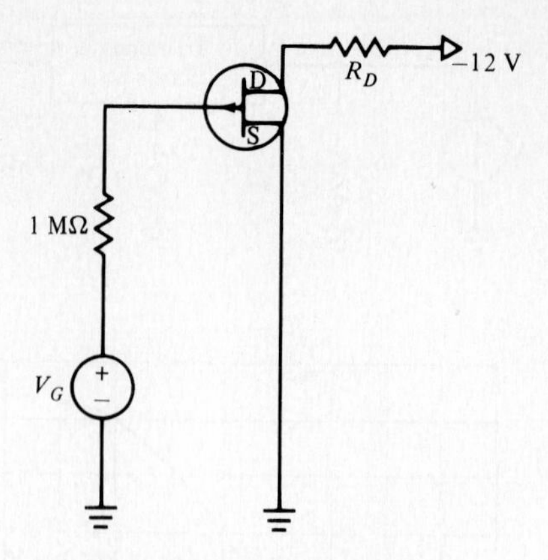

Fig. 3.36 See Problem 40.

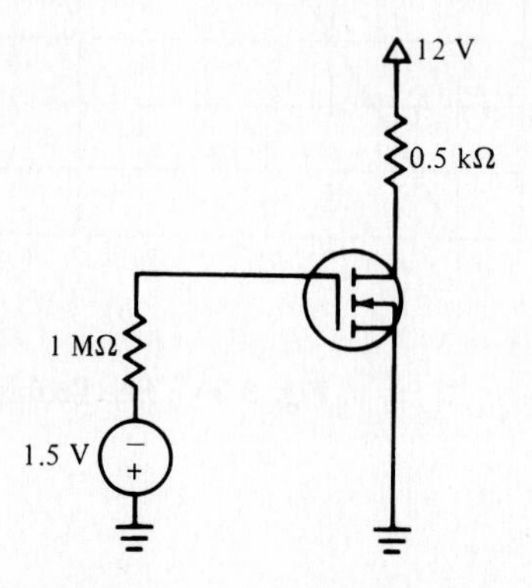

Fig. 3.37 See Problem 41.

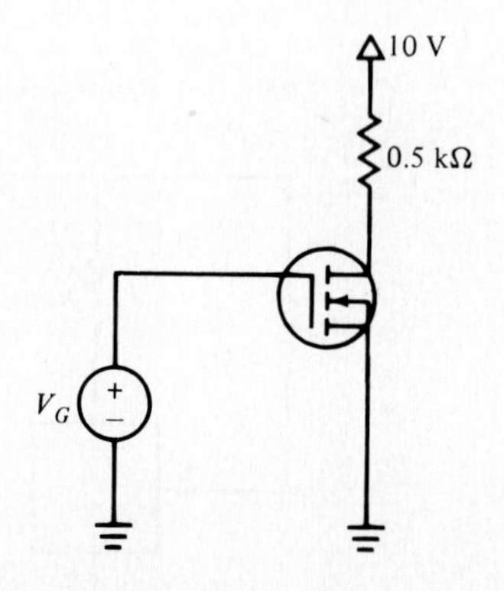

Fig. 3.38 See Problem 42.

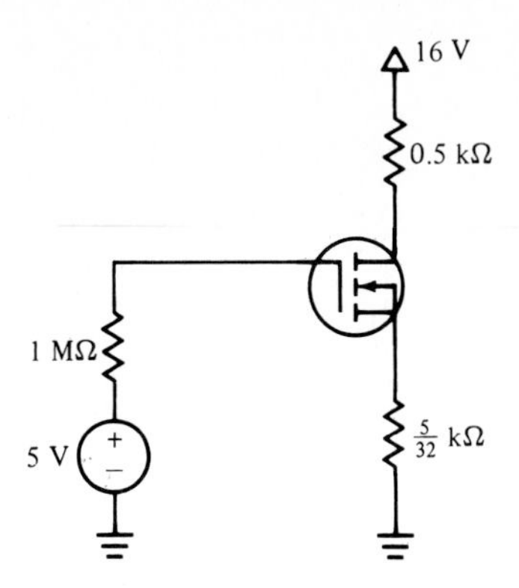

Fig. 3.39 See Problem 43.

approximate values for I_C, I_B, V_{CE}, and V_{BE} if R_C equals (a) 2 kΩ, (b) 1 kΩ, (c) 4 kΩ.

39. The input and output characteristics of Fig. 2.6 apply to the transistor of Fig. 3.35. Estimate values for I_C, I_B, V_{CE}, and V_{BE} if $V_{CC} = -20$ V, $R_B = 200$ kΩ, and R_C equals (a) 2 kΩ, (b) 1 kΩ, (c) 4 kΩ.
40. Let $V_G = 1$ V in Fig. 3.36. Use the characteristics given in Fig. 2.15*b* to determine approximate values for I_D and V_{DS} if R_D equals (a) 600 Ω, (b) 1200 Ω.
41. Give approximate values of I_D and V_{DS} for the transistor in the circuit of Fig. 3.37 if its output characteristics are those of Fig. 2.18*b*.
42. The characteristics of Fig. 2.22 apply to the transistor of Fig. 3.38. Find I_D and V_{DS} if V_G equals (a) 4 V, (b) 5 V, (c) 6 V.
43. Use the output and transfer characteristics shown in Fig. 3.16 to determine I_D and V_{DS} for the circuit of Fig. 3.39.

4

Designing for a stable operating point

We shall now apply our knowledge of dc-equivalent circuits for the various types of transistors to the design of practical circuits. Given transistors (of the same type number) that may have a wide variation of β_{dc} and that may be subject to wide variations in temperature, to nonconstant values of V_{BE} and I_{CBO}, or perhaps even to variable supply voltages, we seek to ensure almost identical operating points. We will accomplish this for n-type and p-type bipolar transistors, for n-channel and p-channel JFETs and MOSFETs in either the depletion or enhancement mode, and for circuits in which any of the three terminals is the signal ground at both input and output.

In doing this, our considerations are practical. Variations in β_{dc} are greater among transistors purchased in large lots at low unit cost. A smaller range in beta could be supplied, but at a greater unit price. In integrated circuits, while all the transistors on one chip might have closely matched values of β_{dc}, there may be wide variations from wafer to wafer. As for temperature variation, it is usually more economical to design a circuit that is insensitive to such changes than it is to provide a constant ambient temperature.

In maintaining a specific active-region operating point, we are able to guarantee that operation is not subject to saturation or cutoff and also will stay within the maximum voltage, current, and power ratings for the transistor.

We shall work with manufacturers' specifications for commercially available transistors in both integrated and discrete form.

4.1 Operating-point design against variation in β_{dc}: common-emitter

All the circuits that are developed to establish a specific operating point for the bipolar transistor have one thing in common—they fix the value of the emitter current. Once this is done, the collector current lies between

0.98 and 0.999 times that value, which fixes its value very closely, and this in turn fixes V_{CE}, assuming that any resistors in the collector or emitter circuits maintain constant values.

Figure 4.1*a* illustrates a very common circuit for discrete transistors that we will show can satisfy the criteria above. We have drawn the circuit so as to suggest common-emitter operation, although the emitter terminal does not appear to be common to the base-ground signal input or to the collector-ground signal output (neither of which are shown). When we begin to apply a signal in the next chapter, however, we shall place a large capacitor C_E across R_E, and this effectively places the emitter terminal at signal ground but has no effect on the dc operating point.

We now replace the transistor with its dc-equivalent circuit, as developed earlier in Fig. 3.3*a*. The result in shown in Fig. 4.1*b*. The portion of the circuit to the left of the base terminal is replaced with its Thévenin equivalent, which yields the simpler circuit of Fig. 4.1*c*. Since I_B and I_C are both positive quantities for this *npn* transistor in the active region, it is evident that the voltage across R_E *opposes* V_{Th}. Thus any tendency for I_C (and hence $|I_E|$) to increase reduces the net voltage acting to establish the value of I_B. This is negative feedback and it provides the mechanism by which I_E and I_C are kept almost constant.

Analysis

Let us analyze the circuit algebraically now to see if we can illustrate these conclusions. Around the base-emitter loop in Fig. 4.1*c* we have

$$V_{\text{Th}} = R_{\text{Th}}I_B + R_{BB}I_B + V_0 + R_E(I_B + I_C) \tag{1}$$

Note that R_{Th} and R_{BB} are in series and they will always appear together as $(R_{\text{Th}} + R_{BB})$ in any equations based on Eq. (1). We shall simplify our expressions by using R_{Th} in place of $(R_{\text{Th}} + R_{BB})$, but it is always easy to include the effect of R_{BB} by simply adding it to R_{Th}.

The collector current is the sum of the currents supplied by the two current sources:

$$I_C = \beta_{dc}I_B + I_{CEO}$$

Solving for I_B,

$$I_B = \frac{1}{\beta_{dc}} I_C - \frac{1}{\beta_{dc}} I_{CEO} \tag{2}$$

Substituting Eq. (2) into Eq. (1), we find

$$V_{\text{Th}} = (R_{\text{Th}} + R_E)\left(\frac{1}{\beta_{dc}}I_C - \frac{1}{\beta_{dc}} I_{CEO}\right) + R_E I_C + V_0$$

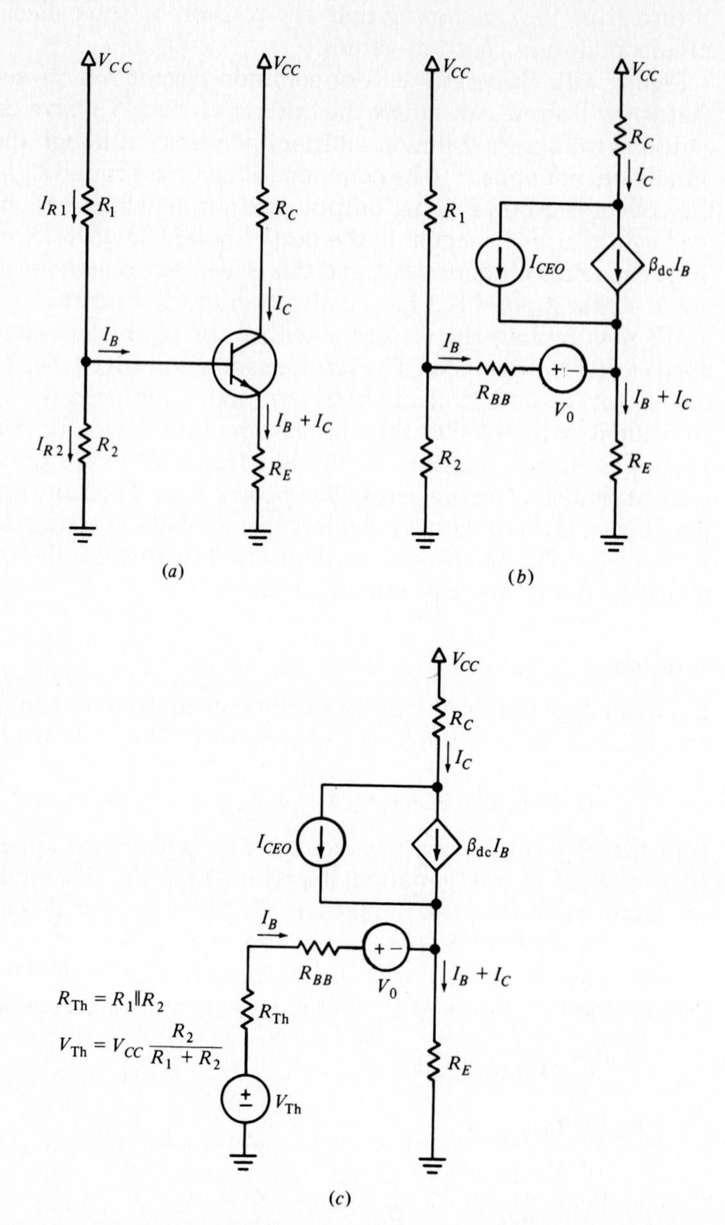

Fig. 4.1 (*a*) A circuit that can provide a stable operating point for an *npn* bipolar transistor. (*b*) The transistor is replaced by its dc-equivalent circuit. (*c*) The input circuit is replaced by its Thévenin equivalent.

When we solve this relationship for I_C, we obtain

$$I_C = \frac{\beta_{dc}(V_{Th} - V_0) + (R_{Th} + R_E)I_{CEO}}{R_{Th} + (\beta_{dc} + 1)R_E} \tag{3}$$

To understand how this expression shows that I_C is relatively independent of β_{dc}, let us consider the simpler case in which I_{CEO} is negligible:

$$I_C = \frac{\beta_{dc}(V_{Th} - V_0)}{R_{Th} + (\beta_{dc} + 1)R_E} \qquad (I_{CEO} \doteq 0) \tag{4}$$

If we select R_E so that $(\beta_{dc} + 1)\ R_E \gg R_{Th}$, we find next that

$$I_C = \frac{\beta_{dc}}{\beta_{dc} + 1}\,\frac{V_{Th} - V_0}{R_E}$$

$$= \alpha_{dc}\frac{V_{Th} - V_0}{R_E} \qquad \text{if } (\beta_{dc} + 1)R_E \gg R_{Th} \tag{5}$$

Recall the relationship between α_{dc} and β_{dc}: $\alpha_{dc} = \beta_{dc}/(\beta_{dc} + 1)$. Thus, I_C is essentially independent of β_{dc}, a quantity that may easily vary over a three-to-one range for some types of transistors, while α_{dc} changes by less than 1%.

The operating point is uniquely determined by specifying values for any two of the four quantities, I_C, V_{CE}, I_B, or V_{BE}. If I_C is constant, then an equation around the collector-emitter loop shows that

$$V_{CE} = V_{CC} - R_C I_C + R_E I_E \tag{6}$$

Thus V_{CE} is fixed, since $|I_E| \doteq I_C$, and I_C is essentially constant.

As we consider ways of designing circuits to maintain a fixed operating point, we will base our preliminary design on the simplest equations, Eqs. (5) and (6), and then check our work using more accurate formulas.

Design

The design process involves selecting values for the four resistors and the supply voltage in the circuit of Fig. 4.1*a*. We assume that desired values of I_C and V_{CE} for the operating point have been given to us, although part of our task in later work will be to select a suitable operating point as well.

The procedure that we shall follow in obtaining a *preliminary design* involves three assumptions, or rules of thumb, that successful designs seem to follow: (1) V_{CC} is three to five times V_{CE}; (2) the voltage across R_E is equal to or slightly less than V_{CE}; and (3) the current through R_1 is 10 to 100 times I_B. Considerable latitude is available for these choices, and the exact values chosen may depend on other criteria. For example, the value of V_{CC} may be affected by supply voltages available elsewhere in the circuit or by commercially available battery voltages; the value of the current in the input voltage divider may depend on the power available from the supply voltage or the degree to which a prior circuit can be loaded. We shall consider some of these points as we attempt a few designs.

Using these three assumptions, we can proceed with the design in a clockwise direction around the circuit of Fig. 4.1*a*, beginning in the upper-right corner. Each element may be found in turn.

Let us try this suggested technique by attempting to design a circuit for a Sprague 2N5377 transistor that will give an operating point at $I_C = 1$ mA, $V_{CE} = 5$ V for room-temperature operation (25°C). A portion of the bulletin giving the details for this transistor is shown in Fig. 4.2. We note that it is an *npn* bipolar silicon transistor, and that the value of β_{dc} for I_C

Fig. 4.2 The first page of the bulletin for the 2N5377 transistor furnished by the Sprague Electric Company, duplicated with permission. For $I_C = 1$ mA and $V_{CE} = 5$ V, we see that β_{dc} (= h_{FE}) may range from 100 to 500.

NPN Silicon Planar

TYPE 2N5376 AND 2N5377 PREMIUM PERFORMANCE ECONOLINE® TRANSISTORS

For Industrial Small-Signal, Low-Noise, Low-Power Audio Frequency Applications

ABSOLUTE MAXIMUM RATINGS at 25 C Free-Air Temperature (unless otherwise noted)

Collector-Base Voltage 60V
Collector-Emitter Voltage (See Note 1) 30V
Emitter-Base Voltage 5V
Total Dissipation at 25 C Free-Air Temperature (See Note 2) 360mW
Collector Current 500mA
Junction Temperature, Operating .... +150 C
Lead Temperature 1/16 Inch from Case for 10 Seconds .... 260 C
Storage Temperature Range .... −55 C to +150 C

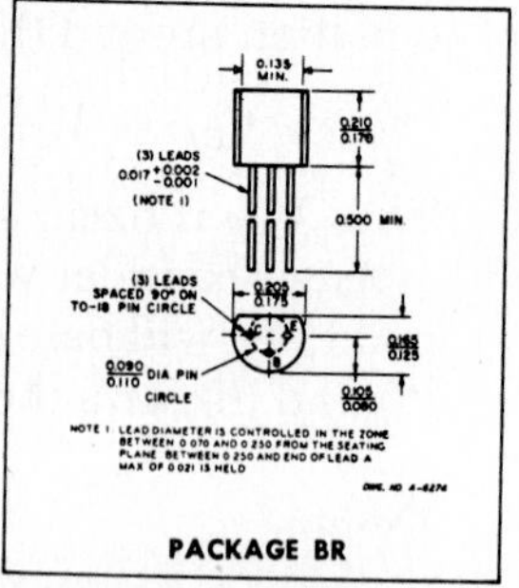

PACKAGE BR

ELECTRICAL CHARACTERISTICS: at T_A = 25 C (unless otherwise noted)

Parameter	Symbol	Test Conditions	2N5376 Min.	2N5376 Max.	2N5377 Min.	2N5377 Max.	Units
Collector-Base Breakdown Voltage	$V_{(BR)CBO}$	I_C = 10μA, I_E = 0	60	—	60	—	V
Collector-Emitter Breakdown Voltage	$V_{(BR)CEO}$	I_C = 10mA, I_B = 0 (See note 3)	30	—	30	—	V
Emitter-Base Breakdown Voltage	$V_{(BR)EBO}$	I_E = 100μA, I_C = 0	5	—	5	—	V
Collector Cut-off Current	I_{CBO}	V_{CB} = 30V, I_E = 0	—	10	—	10	nA
Static Forward Current Transfer Ratio	h_{FE}	I_C = 10μA, V_{CE} = 5V	100	500	40	200	—
		I_C = 1mA, V_{CE} = 5V	120	600	100	500	—
		I_C = 10mA, V_{CE} = 5V (See Note 3)	150	—	120	—	—
Collector-Emitter Saturation Voltage	$V_{CE(SAT)}$	I_C = 10mA, I_B = 1mA	—	0.20	—	0.20	V
Base-Emitter Saturation Voltage	$V_{BE(SAT)}$	I_C = 10mA, I_B = 1mA	0.65	0.80	0.65	0.80	V

Notes:
1. This value applies when the base-emitter diode is open-circuited.
2. Derate linearly to 150 C free air temperature at the rate of 2.88 mW/°C.
3. Pulse test: Pulse width = 300 μsec, duty cycle ≤2%.

= 1 mA is listed as h_{FE}, a term introduced in the next chapter. The value of h_{FE} is given as lying in the range from 100 to 500. We also note that $I_{CBO(max)} = 10$ nA, and therefore I_{CEO} is 101 to 501 times as large, or in the range of 1 to 5 μA. This is negligible compared with either I_E or I_C, so that we need not consider it in our preliminary design. Our main problem lies with the wide range of values for β_{dc}, which we shall try to overcome by designing to hold I_C constant.

Let us first make three selections in accord with the rules of thumb presented above. We want V_{CC} to be three to five times V_{CE}, so we select

$$V_{CC} = 5V_{CE} = 25 \text{ V}$$

The voltage across R_E is to be equal to or slightly less than V_{CE}; therefore, let's choose

$$V_{RE} = V_{CE} = 5 \text{ V}$$

Finally, the current through R_1 should be 10 to 100 times I_B. We select a factor of 50. From Eq. (2), we see that I_B is essentially equal to I_C divided by β_{dc}, and therefore the worst transistor ($\beta_{dc} = 100$) has the maximum base current $I_B = 10^{-3}/100 = 10\ \mu$A, and

$$I_{R1} = 50(10 \times 10^{-6}) = 500\ \mu\text{A}$$

Having made these three decisions, we may now complete the design for the case where $\beta_{dc} = 100$. We begin with R_C and work our clockwise way around the circuit.

$$V_{RC} = V_{CC} - V_{CE} - V_{RE} = 25 - 5 - 5 = 15 \text{ V}$$

and

$$R_C = \frac{V_{RC}}{I_C} = \frac{15}{1} = 15 \text{ k}\Omega$$

while

$$R_E = \frac{V_{RE}}{-I_E} \doteq \frac{V_{RE}}{I_C} = \frac{5}{1} = 5 \text{ k}\Omega \tag{7}$$

We next consider R_2. The current through it is $I_{R1} - I_B = 500 - 10\ \mu$A, while the voltage across it is $V_{BE} + V_{RE} = V_0 + R_{BB}I_B + V_{RE}$. The 2N5377 bulletin gives a range for $V_{BE(sat)}$ of 0.65 to 0.8 V. This value represents the base-emitter voltage when large emitter, base, and collector currents are flowing. Since we are operating at $I_B = 10\ \mu$A, a relatively small value, we select $V_{BE(sat)} = 0.65$ V. This voltage is the sum of V_0 and $R_{BB}I_B$. For this design, let us neglect R_{BB} and use $V_0 = V_{BE(sat)} = 0.65$ V. Thus, $V_{R2} = 0.65 + 5 = 5.65$ V, and

$$R_2 = \frac{V_{RE} + V_0}{I_{R1} - I_B} = \frac{5.65}{0.49} = 11.5 \text{ k}\Omega$$

Finally,

$$R_1 = \frac{V_{CC} - V_{R2}}{I_{R1}} = \frac{25 - 5.65}{0.5} = 38.7 \text{ k}\Omega$$

We now have values for V_{CC} and the four resistors appearing in Fig. 4.1*a*, but we must *check our design* for a transistor whose β_{dc} is 500. We let $V_0 = 0.65$ V and $R_{BB} = 0$ again, since no better information is available on the data sheet, but this time we shall include the effect of I_{CEO}. From Fig. 4.1*c*, we have

$$R_{Th} = \frac{R_1 R_2}{R_1 + R_2} = \frac{(38.7)(11.5)}{38.7 + 11.5} = 8.88 \text{ k}\Omega$$

and

$$V_{Th} = V_{CC}\frac{R_2}{R_1 + R_2} = 25\frac{11.5}{38.7 + 11.5} = 5.74 \text{ V}$$

From Eq. (3), we find

$$I_C = \frac{\beta_{dc}(V_{Th} - V_0) + (R_{Th} + R_E)I_{CEO}}{R_{Th} + (\beta_{dc} + 1)R_E}$$

$$= \frac{500(5.74 - 0.65) + (8.88 + 5)(501 \times 10^{-5})}{8.88 + (501)(5)}$$

$$= \frac{2545 + 0.07}{2514} = 1.012 \text{ mA}$$

This is certainly close enough to our design objective.

The effect of letting $R_{BB} = 5$ kΩ is quite small; it is shown in the answer to Drill Problem D4.2 at the end of this section.

We also check the collector-emitter voltage:

$$V_{CE} = V_{CC} - R_C I_C - R_E(I_C + I_B)$$

$$= 25 - 15(1.012) - 5(1.012 + 0.002)$$

$$= 25 - 15.18 - 5.07 = 4.75 \text{ V}$$

This guarantees operation in the active region.

Without giving all the details, we should point out that a similar careful analysis for $\beta_{dc} = 100$ produces the results $I_C = 0.990$ mA, $V_{CE} = 5.14$ V. Incidentally, this would have come out exactly 1.000 mA if we had not made the assumption in Eq. (7) that $|I_E|$ and I_C were equal. The expected range of β_{dc} thus leads to operation in the active region very close to the desired operating point of 1 mA and 5 V.

Our success in achieving the same operating point for transistors differing so widely in their values of β_{dc} is directly attributable to the use of negative feedback in fixing the emitter-current value. If we assume that the current through R_1 in Fig. 4.1*a* is large compared with the maximum base current, then the voltage divider composed of R_1 and R_2 provides a fixed value of V_{R2}, independent of I_B. Then any tendency for I_C to increase tends to increase V_{RE}, which will reduce the voltage difference between V_{R2} and V_{RE}; this leads to a smaller I_B and a smaller $\beta_{dc} I_B$, and thus a reduced collector current. That is, the circuit tends to *oppose the change* in I_C. Similarly, if I_C tries to decrease, V_{RE} will decrease, $V_{R2} - V_{RE}$ will increase, I_B will increase, and $\beta_{dc} I_B$ will increase; I_C again tends to be restored to its original value.

The algebraic result of this negative feedback shows up in Eqs. (4) and (5), since larger values of R_E (more negative feedback) and smaller values of R_{Th} (larger values of I_{R1} and hence more constant V_{R2}) cause I_C to be less dependent upon the value of β_{dc}.

In the next chapter we shall see that while this selection of values for R_E and R_{Th} may lead to a more stable operating point, it tends to produce a smaller signal gain. The optimum design must involve a typical engineering trade-off between stability and gain.

As a slightly different example, consider a case in which V_{CC} is specified, say at 9 V, and a very small value for I_C is desired, say $I_C = 10\ \mu A$ at $V_{CE} = 5$ V. We again design for the 2N5377 transistor. From the data sheet of Fig. 4.2, $\beta_{dc(min)} = 40$ at this lower collector current. Let's assume that V_{BE} is 0.6 V this time and begin our design with

$$I_B \doteq \frac{10 \times 10^{-6}}{40} = 0.25\ \mu A$$

so that

$$I_E = -10.25\ \mu A$$

With $V_{CE} = 5$ V and $V_{CC} = 9$ V, we only have 4 V to distribute between R_C and R_E. We try 2 V across each (although we could have $V_{RC} = 3$ V and $V_{RE} = 1$ V). Thus

$$R_E = \frac{2}{10.25 \times 10^{-6}} = 195.1\ k\Omega$$

and

$$R_C = \frac{2}{10 \times 10^{-6}} = 200\ k\Omega$$

Since I_B is very small, let us use $I_{R1} = 200 I_B = 50\ \mu A$. Thus $I_{R2} = 49.75\ \mu A$; if we use $V_0 = 0.6$ V, we may complete the design:

$$V_{R2} = V_{RE} + V_0 = 2.6 \text{ V}$$

$$R_2 = \frac{2.6}{49.75 \times 10^{-6}} = 52.3 \text{ k}\Omega$$

$$V_{R1} = 9 - 2.6 = 6.4 \text{ V}$$

$$R_1 = \frac{6.4}{50 \times 10^{-6}} = 128 \text{ k}\Omega$$

To check the design, we calculate $V_{Th} = 2.61$ V and $R_{Th} = 37.1$ kΩ, and then analyze the circuit at $\beta_{dc(min)} = 40$ and at $\beta_{dc(max)} = 200$:

$\beta_{dc(min)} = 40$	$\beta_{dc(max)} = 200$
$I_{CEO} = (40 + 1)\, 10 \times 10^{-9} = 410$ nA	$I_{CEO} = 2.01\ \mu$A
$I_C = 10.02\ \mu$A	$I_C = 10.25\ \mu$A
$I_B = (I_C - I_{CEO})/\beta_{dc} = 0.240\ \mu$A	$I_B = 0.0412\ \mu$A
$I_E = -10.26\ \mu$A	$I_E = -10.29\ \mu$A
$V_{CE} = 4.99$ V	$V_{CE} = 4.94$ V

The operating point stays in the active region for both cases.

D4.1 Suppose that the circuit designed for the 2N5377 transistor with $I_C = 1$ mA is actually used with a 2N5376 transistor. Use $V_{BE} = 0.65$ V and find (a) $I_{C(max)}$, (b) $I_{C(min)}$, (c) $V_{CE(max)}$, (d) $V_{CE(min)}$.

Answers. 1.013 mA; 0.995 mA; 5.06 V; 4.76 V

D4.2 For the 2N5377 example with $I_C = 1$ mA discussed in this section, let $R_{BB} = 5$ kΩ while $V_0 = 0.6$ V, and calculate (a) $I_{C(max)}$, (b) $I_{B(max)}$, (c) $V_{BE(max)}$, (d) $V_{CE(min)}$.

Answers. 1.020 mA; 9.91 μA; 0.650 V; 4.58 V

4.2 Operating-point design against variation in temperature: common-emitter

In addition to the variability of β_{dc} among supposedly identical transistors, the designer must guard the operating point against the effects of temperature change. Many circuits must operate over a temperature range from −55 to 125°C, and such a variation has several effects on transistors and diodes. The value of V_{BE} decreases by 1.8 to 2.5 mV for every 1°C increase in temperature. A typical value is 2.3 mV/°C. The value of I_{CEO} is also very sensitive to temperature.[1] Finally, β_{dc} increases nonlinearly with temperature. There is no simple rule of thumb to estimate the change; it is either measured or lumped into the larger variation expected among different units of the same type.

The greatest effects of temperature on circuits using silicon transistors usually result from changes in V_{BE} and β_{dc}. The reverse-saturation current is usually so small compared to I_C that even a hundredfold increase in its value is negligible. This is not true with some integrated circuit devices, however, and the variation of I_{CEO} can pose a severe problem, since the value of I_{CEO} at 25°C may itself be a noticeable fraction of I_C.

To stabilize the operating point against these effects, our primary goal once again is to maintain a constant emitter current. The use of an emitter resistor to provide negative feedback is helpful against changes in both β_{dc} and V_{BE}. Stability with respect to V_{BE} is also improved by providing a larger value of V_{Th} in the circuit considered previously in Fig. 4.1*a* and its equivalent in Fig. 4.1*c*. Repeating Eq. (4) from Section 4.1,

$$I_C = \frac{\beta_{dc}(V_{Th} - V_0)}{R_{Th} + (\beta_{dc} + 1)\,R_E} \qquad (I_{CEO} \doteq 0) \tag{4}$$

Here, V_0 is the voltage we have selected to model V_{BE}. It is apparent that the change in I_C is proportional to the change in the difference $V_{Th} - V_0$. We can thus minimize the effect of a change in V_{BE} and V_0 by providing a larger value of V_{Th}.

The effect of a change in I_{CEO} can be seen from Eq. (3):

$$I_C = \frac{\beta_{dc}(V_{Th} - V_0)}{R_{Th} + (\beta_{dc} + 1)R_E} + \frac{(R_{Th} + R_E)I_{CEO}}{R_{Th} + (\beta_{dc} + 1)R_E} \tag{8}$$

We note that the ratio of the change in I_C to the change in I_{CEO} is given by the fraction

$$\frac{\Delta I_C}{\Delta I_{CEO}} = \frac{dI_C}{dI_{CEO}} = \frac{(R_{Th} + R_E)}{R_{Th} + (\beta_{dc} + 1)\,R_E} \tag{9}$$

We can minimize this ratio by making $\beta_{dc}R_E$ considerably larger than $R_{Th} + R_E$. Remember, however, that R_{BB} appears as a part of R_{Th} if it is made part of the transistor model.

We should design the circuit by assuming "worst-case" conditions. Thus Eq. (4) shows that the minimum collector current in a silicon transistor will occur when V_0 is at a maximum (minimum temperature) and β_{dc} is at

[1] $I_{CEO} = KT^3e^{-E_G/kT}(\beta_{dc} + 1)$, where the width of the energy gap $E_G = 1.12$ eV (1 electron volt $= 1.6022 \times 10^{-19}$ J) for silicon and K is a constant.

a minimum (also at minimum temperature). We also must consider the unit having the smallest β_{dc}. Therefore,

$$I_{C(\min)} = \frac{\beta_{dc(\min)}(V_{Th} - V_{0(\max)})}{R_{Th} + (\beta_{dc(\min)} + 1)R_E} \tag{10}$$

Correspondingly, $I_{C(\max)}$ occurs at the maximum expected temperature, where V_0 is at a minimum and β_{dc} is at a maximum. We again assume the worst possible case and select the unit having the highest β_{dc}:

$$I_{C(\max)} = \frac{\beta_{dc(\max)}(V_{Th} - V_{0(\min)})}{R_{Th} + (\beta_{dc(\max)} + 1)R_E} \tag{11}$$

To see how such a design is accomplished, let us reconsider the problem we solved in Section 4.1. We have a 2N5377 transistor and must try to provide an operating point of $V_{CE} = 5$ V, $I_C = 1$ mA, both within ±10%, for a temperature range from −55 to 125°C. We shall *assume* that β_{dc} varies from 0.8 of its 25°C value at −55°C to 1.4 times the 25°C value at 125°C. Thus, since we are faced with a variation from 100 to 500 at room temperature, we have a worst-cast minimum,

$$\beta_{dc(\min)} = 0.8(100) = 80$$

and a worst-case maximum,

$$\beta_{dc(\max)} = 1.4(500) = 700$$

The minimum and maximum values of V_0 are calculated by increasing the value at 25°C, 0.65 V, by 2.3 mV for each 1°C decrease in temperature:

$$V_0 = V_{0(25°C)} - 2.3 \times 10^{-3}(T_{°C} - 25) \tag{12}$$

Thus the range of V_0 extends from a minimum value at 125°C,

$$V_{0(\min)} = 0.65 - 2.3 \times 10^{-3}(125 - 25) = 0.42 \text{ V}$$

to a maximum at the lower temperature, −55°C,

$$V_{0(\max)} = 0.65 - 2.3 \times 10^{-3}(-55 - 25) = 0.834 \text{ V}$$

We may now carry out the design using the same procedures described in Section 4.1. The supply voltage is chosen to be three to five times V_{CE}, and we again select 25 V. The voltage across R_E is selected as equal to or slightly less than V_{CE}; we again choose 5 V. Finally, the current through R_1 is 10 to 100 times I_B. Since the center values of I_C and β_{dc} are unchanged, we may again let I_{R1} be $50I_B$, obtaining identical values for the four resistances: $R_1 = 38.7$ kΩ, $R_2 = 11.53$ kΩ, $R_E = 5$ kΩ, and $R_C = 15$ kΩ. We again have $R_{Th} = 8.88$ kΩ, and $V_{Th} = 5.74$ V. Thus the design is the same, and we must now check it for the two worst cases.

The minimum value of I_C is calculated by using Eq. (10):

$$I_{C(\text{min})} = \frac{80(5.74 - 0.834)}{8.88 + (81)5} = 0.948 \text{ mA}$$

The maximum value is

$$I_{C(\text{max})} = \frac{700(5.74 - 0.420)}{8.88 + (701)5} = 1.060 \text{ mA}$$

These values both lie within the ±10% limits and we have a successful design. As a matter of fact, if we wish, we can *increase* the value of R_{Th} (or R_1 and R_2) by choosing to send a smaller current through the input voltage divider. Higher resistances would lead to an increased signal gain, as we indicated earlier, but this is not one of our immediate concerns and we shall let it lie until the following chapter.

Had we not met the ±10% tolerance on I_C, we might have modified our design by increasing I_{R1}, which reduces R_{Th}; by increasing R_E; or by doing both.

D4.3 Decrease V_0 to 0.6 V at 25°C for the 2N5377 transistor described above, but keep all other characteristics the same. Let $V_{CC} = 24$ V, $R_C = 15$ kΩ, $R_E = 4$ kΩ, $R_1 = 120$ kΩ, and $R_2 = 30$ kΩ in the external circuit, and determine $I_{C(\text{min})}$ and $I_{C(\text{max})}$ under the same temperature and tolerance worst-case conditions.

Answers. 0.923; 1.097 mA

4.3 The thermal environment and maximum junction temperature

The average power that any diode or transistor can dissipate safely is determined by the temperature of the semiconductor junction itself. The maximum temperature at which the device itself (not its external case) may operate is referred to as the *maximum junction temperature.* Its value depends on the type of semiconductor material and the doping levels used, the type of encapsulation, the solder or bonding material used, and the degree of reliability desired. A typical value for the maximum junction temperature of a germanium transistor or diode is about 100°C, while values from 135 to 200°C are more typical for silicon devices. The higher values are associated with hermetically sealed metal-encapsulated units. The data sheet for the 2N5377 we considered earlier in this chapter (Fig. 4.2) gives the maximum junction temperature as 150°C in the list headed "Absolute Maximum Ratings" under the entry "Junction Temperature, Operating." For the 2N5088 described in Fig. 4.3, the maximum junction temperature is found to be 135°C.

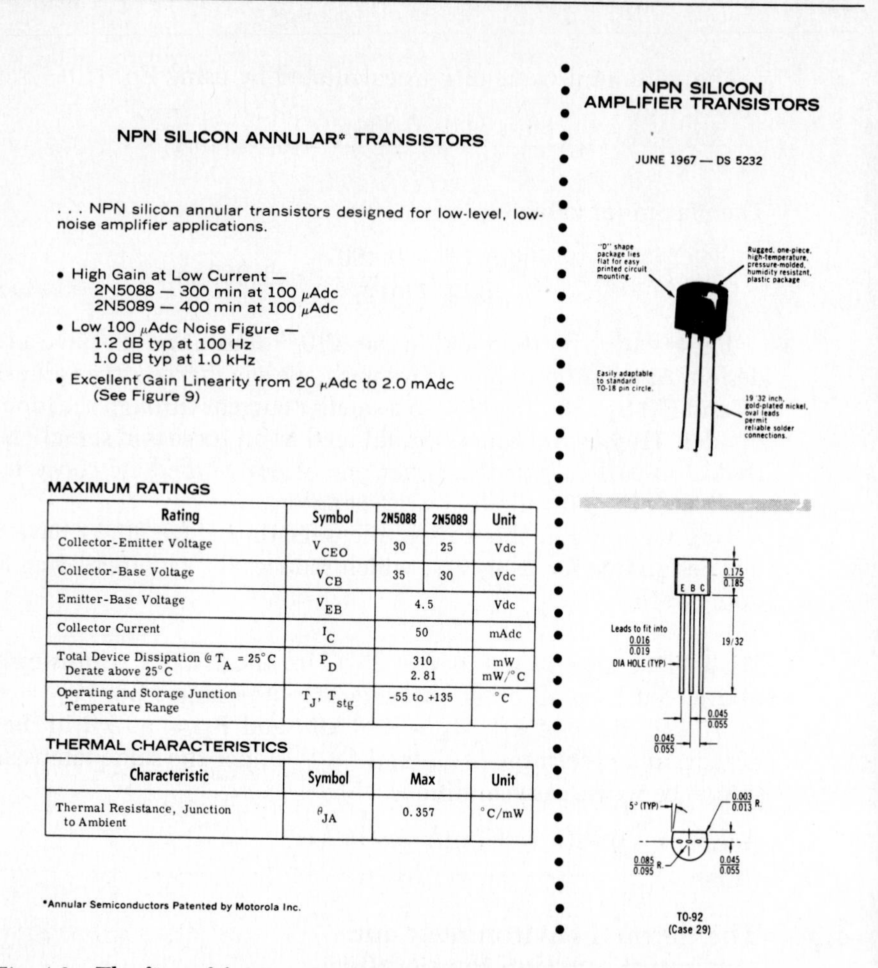

NPN SILICON ANNULAR* TRANSISTORS

. . . NPN silicon annular transistors designed for low-level, low-noise amplifier applications.

- High Gain at Low Current —
  2N5088 — 300 min at 100 μAdc
  2N5089 — 400 min at 100 μAdc
- Low 100 μAdc Noise Figure —
  1.2 dB typ at 100 Hz
  1.0 dB typ at 1.0 kHz
- Excellent Gain Linearity from 20 μAdc to 2.0 mAdc
  (See Figure 9)

NPN SILICON AMPLIFIER TRANSISTORS

JUNE 1967 — DS 5232

MAXIMUM RATINGS

Rating	Symbol	2N5088	2N5089	Unit
Collector-Emitter Voltage	V_{CEO}	30	25	Vdc
Collector-Base Voltage	V_{CB}	35	30	Vdc
Emitter-Base Voltage	V_{EB}	4.5		Vdc
Collector Current	I_C	50		mAdc
Total Device Dissipation @ T_A = 25°C Derate above 25°C	P_D	310 2.81		mW mW/°C
Operating and Storage Junction Temperature Range	T_J, T_{stg}	-55 to +135		°C

THERMAL CHARACTERISTICS

Characteristic	Symbol	Max	Unit
Thermal Resistance, Junction to Ambient	θ_{JA}	0.357	°C/mW

TO-92
(Case 29)

*Annular Semiconductors Patented by Motorola Inc.

Fig. 4.3 The first of four pages of data furnished by Motorola Semiconductors for their 2N5088 bipolar transistor, reprinted by permission.

There are several factors that determine the junction temperature when the transistor is in thermal equilibrium at some specified operating point. These are the ambient temperature T_A of the medium surrounding the circuit, the average power dissipated at the collector junction $I_C V_{CB}$, and the thermal characteristics of the materials and path from the junction to the ambient environment. These thermal characteristics are described quantitatively in terms of a *thermal resistance*. The name is appropriate because of an analogy that can be drawn between a linear resistive circuit and a thermal system in equilibrium. If we think of a thermally conducting object as having two isothermal (of uniform temperature) surfaces with an average thermal dissipation power P_D entering the hotter surface, then the temperature difference ΔT between the two faces is given by the product of the power and the thermal resistance θ, which has the units of degree Celsius per watt:

$$\Delta T = \theta P_D \tag{13}$$

This corresponds to Ohm's law:

$$V = RI$$

where V is a potential difference. Low values of θ signify more effective heat radiation, in that the junction temperature is closer to the temperature of the environment.

In case the thermal circuit consists of several elements through which the thermal power flows in series, such as from the junction at T_J to the case at T_C to the heat sink at T_H and finally to the ambience at T_A, we may model the circuit as shown in Fig. 4.4 and write

$$T_J - T_A = (\theta_{JC} + \theta_{CH} + \theta_{HA})P_D \tag{14}$$

Similar relationships relate the temperature difference across other combinations of elements. If no heat sink is used,

$$T_J - T_A = (\theta_{JC} + \theta_{CA})P_D \tag{15}$$

As an example of the use of such a thermal equivalent circuit, let us use the data supplied for the 2N5088 transistor to find the maximum power that may be dissipated safely at the collector junction for an ambient temperature of 75°C.

From Fig. 4.3 we obtain three needed nuggets of information. First, the total device dissipation at $T_A = 25°C$ is given as 310 mW; second, the thermal resistance θ_{JA} is specified as 0.357°C/mW (or 357°C/W); and third, the maximum operating junction temperature is listed as 135°C.

Fig. 4.4 A thermal circuit in equilibrium is analogous to a linear resistive circuit: $P_D \leftrightarrow I$, $T_J - T_C \leftrightarrow V_J - V_C$, and $\theta_{CH} \leftrightarrow R_{CH}$, or $T_J - T_C = \theta_{JC}P_D$.

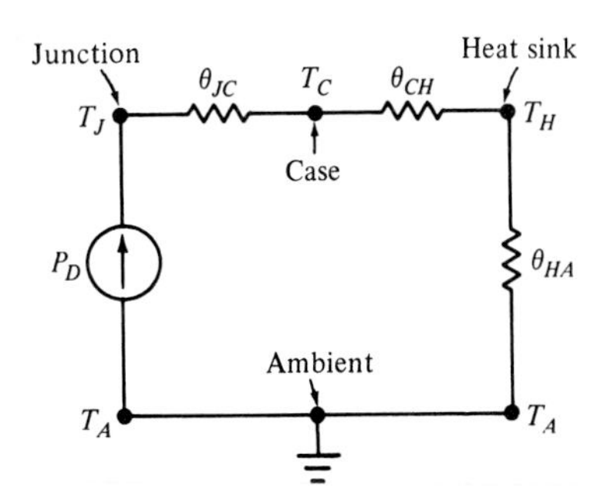

Our thermal equivalent circuit therefore has a temperature difference of

$$T_J - T_A = 135 - 75 = 60°\text{C}$$

Therefore from Eq. (13) we find

$$P_D = \frac{\Delta T}{\theta_{JA}} = \frac{60}{0.357} = 168 \text{ mW}$$

Note that as the ambient temperature increases, the maximum power that can be dissipated decreases.

As another example, suppose we have a 2N5088 transistor operating with $I_C = 2$ mA and $V_{CE} = 20$ V at an ambient temperature of 25°C. What is the junction temperature?

To find the power being dissipated at the collector-base junction, we need the collector-base voltage

$$V_{CB} = V_{CE} - V_{BE} = 20 - 0.7 = 19.3 \text{ V}$$

(where we have used an average value of V_{BE} for a silicon transistor). Therefore

$$P_D = I_C V_{CB} = 2(19.3) = 38.6 \text{ mW}$$

One could also use $P_D \doteq I_C V_{CE}$, since this includes the power dissipation at the emitter-base junction; however, this adds only a few milliwatts.

From the thermal equivalent circuit, we have

$$\Delta T = \theta_{JA} P_D = 0.357(38.6) = 13.8°\text{C}$$

Since $T_A = 25°$C, we find that $T_J = 38.8°$C, well within the 135°C limit.

One other way of specifying thermal information is by using a derating factor. Figure 4.5 shows a typical derating curve that may be drawn from a knowledge of only three numbers: the maximum power dissipation $P_{D(\max)}$, the reference temperature below which it applies, and the maximum junction temperature.

The derating curve for the 2N5088 transistor may be constructed from the information we have already gleaned from the data sheet. We set $P_{D(\max)} = 310$ mW, locate the breakpoint at $T_{A0} = 25°$C, and set $T_{J(\max)} = 135°$C. Then, for $25 \leq T_A \leq 135°$C, we use the properties of a straight line to write

$$135 - T_A = P_D \frac{135 - 25}{310 - 0} = 0.355 P_D$$

Here we see that $\theta_{JA} = 0.355°$C/mW, which checks rather closely with the value of 0.357°C/mW given on the data sheet. This little exercise also in-

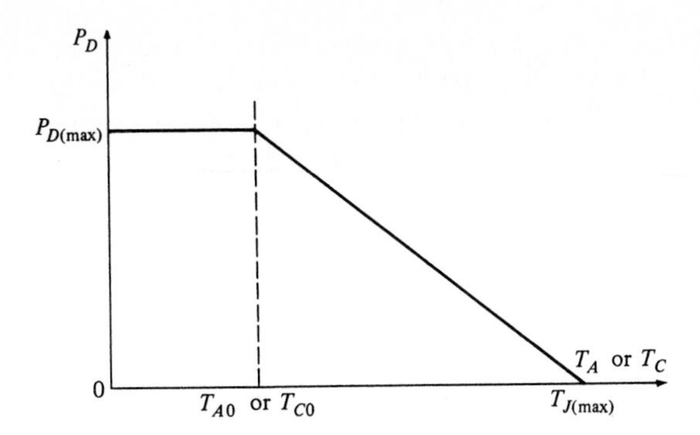

Fig. 4.5 The derating curve for a semiconductor device exhibits a linear decrease from a maximum allowable average power dissipation valid at or below a specified ambient or case temperature to the maximum allowable junction temperature.

dicates that θ_{JA} is the reciprocal of the magnitude of the slope of the linearly decreasing portion of the derating curve; that is, the derating factor is $1/\theta_{JA}$.

If the data were given as $P_{D(max)}$ at or below some *case* temperature T_{C0}, then the reciprocal of the magnitude of the slope would be θ_{JC}. Such information is often related to case temperature for diodes or transistors that are intended to be provided with heat sinks. These are metallic structures with a relatively large heat-radiating surface to which the semiconductor is attached. The derating curve of Fig. 4.6 applies to an RCA 2N2015 silicon *npn* power transistor. A correct specification such as this has led to unfortunate (but deserved) embarrassment for too many young engineers who incorrectly interpreted T_C as T_A. This curve indicates that the transistor will dissipate 150 W if $T_C \leq 25°C$, a condition that might be

Fig. 4.6 The derating curve for a 2N2015 power transistor is given in terms of the case temperature T_C. We note that $\theta_{JC} = 175/150 = 1.17°C/W$.

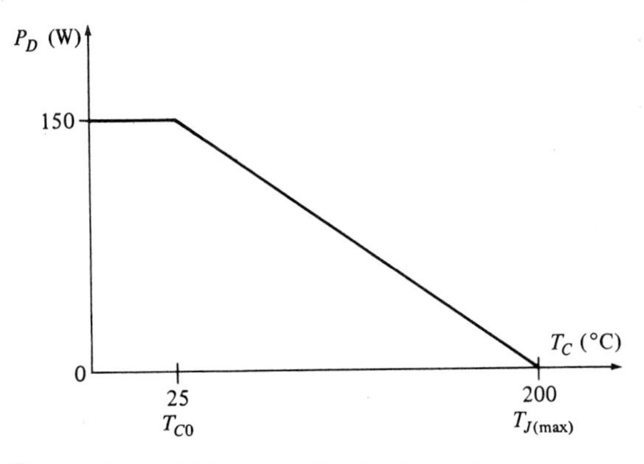

achieved if the transistor case were welded to a 500-pound anvil and placed 15 feet deep in the middle of Lake Michigan in January.

If the *ambient* temperature is 25°C and the thermal resistance between case and ambience has the relatively good value of 0.5°C/W, then for P_D = 150 W, we find that

$$T_C - T_A = 0.5(150) = 75°\text{C}$$

Thus,

$$T_C = 100°\text{C}$$

Since the derating curve shows that $\theta_{JC} = 7/6°\text{C/W}$, we have

$$T_J - T_C = (7/6)(150) = 175°\text{C}$$

It follows that $T_J = 275°\text{C}$. Since $T_{J(\text{max})} = 200°\text{C}$, ZAP!

D4.4 Data from a transistor specification sheet show that the maximum junction temperature is 175°C and the maximum allowable dissipation at any temperature is 20 W. The transistor should be derated above 25°C ambient. Assume a heat sink is used for which $\theta_{CH} = 1°\text{C/W}$ and $\theta_{HA} = 1.5°\text{C/W}$. Let P_D = 10 W in a 40°C ambient, and find (a) T_J, (b) T_C, (c) T_H.

Answers. 115°C; 65°C; 55°C

4.4 Common-collector and common-base biasing

Whatever the type of external circuit in which the bipolar junction transistor is installed, a stable operating point can be achieved by keeping the emitter current constant. The circuit we have been considering for the common-emitter arrangement does this very effectively, and the same basic circuit is employed in the less-used common-collector and common-base configurations.

For a common collector, the collector terminal is the reference for the input and output ports. It is connected to the supply voltage rather than to ground, but the ideal voltage source ensures that there can be no *signal* voltage between collector and ground. This circuit is shown in Fig. 4.7. Note that the circuit external to the transistor is *identical to that for the common-emitter* shown in Fig. 4.1*a* if $R_C = 0$. The design and analysis procedures are unchanged.

In the common-base circuit shown in Fig. 4.8*a*, two voltage sources and two resistors provide a stable operating point. Writing an equation around the left mesh, we have

$$V_{BE} + R_E(I_C + I_B) + V_{EE} = 0$$

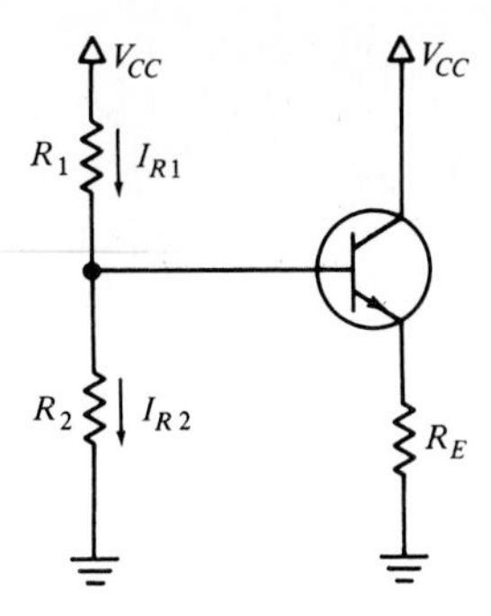

Fig. 4.7 A biasing circuit for an *npn* junction transistor in the common-collector configuration is identical to that used with the common emitter for $R_C = 0$.

Since $I_C + I_B = -I_E$,

$$I_E = \frac{V_{EE} + V_{BE}}{R_E}$$

Thus the emitter current is fixed if V_{EE} is stable and several times greater than V_{BE} (typically 0.65 V). Providing two supplies of opposite polarity with respect to ground, however, is usually uneconomical. One supply is preferable, and the circuit of Fig. 4.8*b* shows how a voltage divider may be used with the collector supply to keep the base positive with respect to the emitter. The circuit is of course identical to that used with the common-emitter earlier, and its design is accomplished by an identical procedure. The base is made common to the input and output signals by placing a large capacitor in parallel with R_2. This is often called a by-pass capacitor. For the common-emitter circuit, the by-pass capacitor appears across R_E. We shall consider the selection of these capacitors in Chapter 7.

Fig. 4.8 Two common-base biasing circuits. (*a*) A simple (but probably uneconomical) circuit using two voltage supplies. (*b*) A more practical single-supply circuit that is identical to the common-emitter circuit of Fig. 4.1*a*. For operation with a signal, a large capacitor is placed in parallel with R_2, thereby making the base a signal ground.

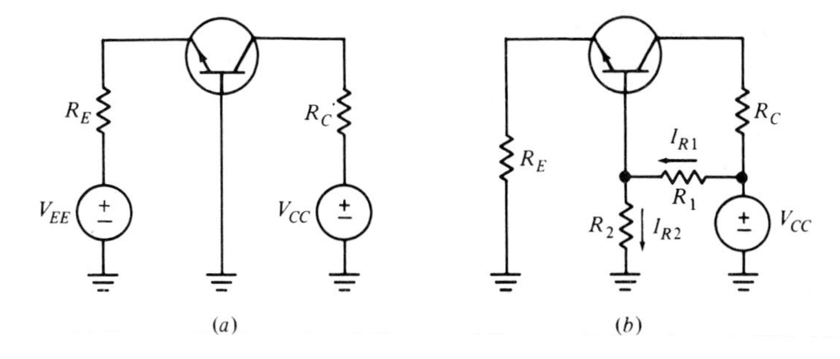

D4.5 For a silicon *npn* transistor with negligible I_{CEO}, let $V_{BE} = 0.65$ V and $\beta_{dc} = 110$. If $V_{CC} = 24$ V and $R_E = 4$ kΩ, find I_C and V_{CE} in the circuit of (a) Fig. 4.7 if $R_1 = 36$ kΩ and $R_2 = 12$ kΩ; (b) Fig. 4.8*a* if $R_C = 5$ kΩ and $V_{EE} = -6$ V; (c) Fig. 4.8*b* if $R_1 = 30$ kΩ, $R_2 = 10$ kΩ, and $R_C = 10$ kΩ.

Answers. 1.299 mA, 18.76 V; 1.325 mA, 18.02 V; 1.303 mA, 5.70 V

4.5 Integrated-circuit operating-point design

When a bipolar transistor is incorporated into an integrated circuit design, we often use techniques that are very different from those appearing in the discrete circuits of Section 4.1. The primary reason for this is that diodes and transistors require much less surface area on a chip than resistors, and large areas are very expensive to provide. This leads to our first rule of thumb for integrated circuit design: use diodes and transistors in place of resistors wherever possible, and if a resistor must be used, keep its resistance small.

One great advantage of an integrated circuit is that similar components of a circuit are made at the same time. Since they lie within several mils (1 mil = 0.001 inch) of each other, they experience the same environmental stresses. This results in highly matched parameter values, so that R_s, β_{dc}, V_{BE}, and I_{CEO} differ by only a few percent from unit to unit and behave similarly as the temperature changes. It is therefore possible to rely on matched components throughout a given circuit. Variability, however, should be expected between units made in different wafers at different times. We thus are led to a second rule of thumb: design circuits so that the performance is dependent on ratios of resistor values rather than on absolute values.

The general idea of stabilizing the emitter current and thereby stabilizing the operating point holds true for I-C designs as it does in discrete designs. However, the method by which we achieve a constant emitter current is somewhat different. Before discussing the details, let's consider an I-C diode.

An I-C diode is typically formed from a bipolar transistor by shorting the collector to the base, as illustrated in Fig. 4.9*a*. Since $V_{BC} = 0$, the transistor is operating at the boundary between the saturation and active regions, and we can use the active-region model of Fig. 4.9*b*. Note that the diode current I_D is given by

$$I_D = \beta_{dc} I_B + I_B = I_B(\beta_{dc} + 1) \qquad (16)$$

This is equal to the magnitude of the emitter current, $I_D = |I_E|$. Also,

$$V_D = I_B R_{BB} + V_0$$

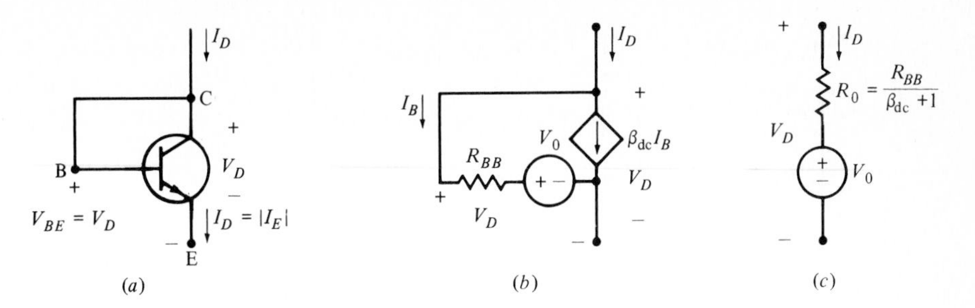

Fig. 4.9 (*a*) An integrated-circuit diode is constructed from a bipolar transistor by shorting the collector to the base on the chip. (*b*) A piecewise-linear equivalent circuit for $V_{BE} = V_D \geq 0$. (*c*) A simpler equivalent circuit for $V_{BE} = V_D \geq 0$.

and therefore

$$V_D = I_D \frac{R_{BB}}{(\beta_{dc} + 1)} + V_0$$

Thus, the equivalent circuit of Fig. 4.9*c* is applicable with

$$R_0 = \frac{R_{BB}}{(\beta_{dc} + 1)} \tag{17}$$

We conclude that shorting the base to the collector of a transistor yields a diode whose current is equal to the magnitude of the emitter current of the transistor.

To examine one of the many possible designs of an integrated circuit in which the operating point is established by using an integrated diode, let us consider the circuit shown in Fig. 4.10*a*. Two identical transistors are operating with their bases tied together and their emitters grounded; one is connected as an integrated diode, while the other is to have its operating point safely in the active region. We replace the integrated diode by $R_{BB}/(\beta_{dc} + 1)$ in series with V_0, as shown in Fig. 4.10*b*, and we also use the active-region equivalent circuit for the right-hand transistor. Equating the currents at the left node:

$$I_1 = I_D + I_{B2}$$

We may use Eq. (16) to eliminate I_D:

$$I_1 = I_{B1}(\beta_{dc} + 1) + I_{B2}$$

With the emitters and bases of the two transistors common, V_{BE} is the same, and thus the base currents are also equal; $I_{B1} = I_{B2}$. Therefore

$$I_B = \frac{I_1}{\beta_{dc} + 2}$$

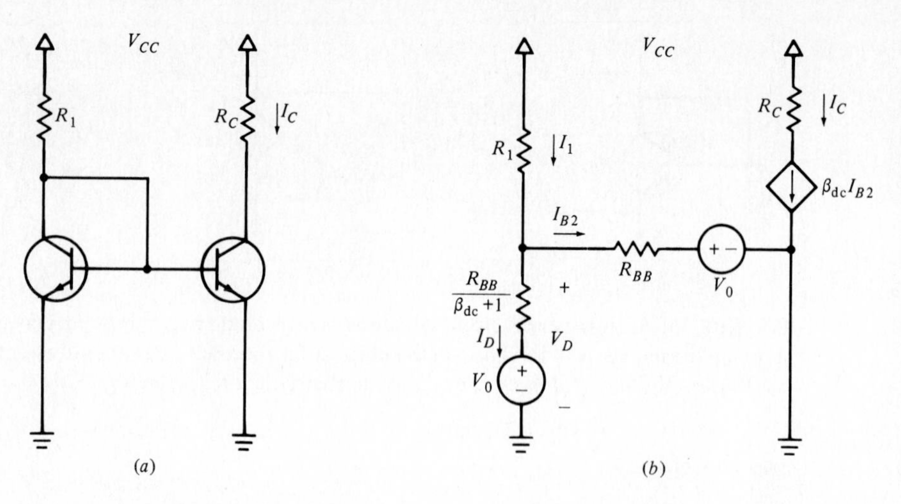

Fig. 4.10 (*a*) The left transistor is connected as an integrated diode to establish I_C for the right transistor. (*b*) The left transistor is replaced by a diode equivalent circuit, while the right unit is represented by its active-region equivalent circuit.

The collector current we wish to stabilize is $\beta_{dc}I_{B2} = \beta_{dc}I_B$. Thus,

$$I_C = \frac{\beta_{dc}}{\beta_{dc} + 2} I_1$$

Since $\beta_{dc} \gg 1$, we see that

$$I_C \doteq I_1$$

Therefore, the collector current and the operating point for the right-hand transistor will be stabilized if I_1 is held constant; moreover, the collector current will be very nearly equal to I_1.

Let us see what affects the value of I_1. From Fig. 4.10*b*, we may write

$$V_{CC} = R_1 I_1 + V_D$$

or

$$I_1 = \frac{V_{CC} - V_D}{R_1}$$

The voltage across the integrated diode V_D is approximately 0.65 V and is not apt to vary more than 0.1 V from this value. Also, V_{CC} is much larger than 0.65 V, and thus $V_{CC} - V_D$ will be constant if the supply voltage V_{CC} is constant. This is accomplished through voltage regulation in the power supply. Finally, the resistance R_1 can be held fairly constant; in any case, it is often made equal to R_C. A decrease in R_1 thus results in an increase in I_1 and I_C, but little change in $R_C I_C$ or V_{CE2}.

Thus we may hold I_C constant by keeping I_1 constant, a relatively easy task. Finally, the fact that $I_C \doteq I_1$ leads to the name of *current mirror* for the circuit of Fig. 4.10*a*; that is, the current I_C mirrors I_1.

Two additional comments may be made about the current mirror. First, the integrated diode may stabilize the current of several transistors if their bases are connected together and their emitters are grounded. All collector currents are approximately equal to I_1. Thus, several collector currents are stabilized while using only one resistor. Second, the collector supply to the right-hand transistor is essentially acting as a constant current supply with value $I_C \doteq I_1$. Thus, if we need a constant current supply for some device, it is only necessary to connect it in series with the collector of the right-hand transistor. The current mirror is a current source that is a basic building block for linear integrated circuit designs.

D4.6 Both transistors in Fig. 4.10*a* are operating with $\beta_{dc} = 100$, $V_0 = 0.65$ V, $V_{CC} = 6$ V, and $R_{BB} \doteq 0$. Let $R_1 = 2.67$ kΩ, $R_C = 2$ kΩ, and $V_{CC} = 6$ V, and find (a) I_{C2}, (b) V_{CE2}, (c) I_{C1}.

Answers. 1.964 mA; 2.07 V; 1.964 mA

4.6 JFET and depletion-mode IGFET operating-point design

This section considers the design of circuits that can provide an operating point for FETs that is fixed against variations between units and in temperature. A typical specification might be to maintain I_D within 10% of the nominal value under all possible conditions.

We have already analyzed several dc circuits containing JFETs or depletion-mode MOSFETs. For example, the *n*-channel JFET in Fig. 3.14 was furnished with both drain and gate voltage supplies, as well as gate and drain resistors, and the source was grounded. We used a nonlinear model to determine the operating point V_{GS}, I_D, and V_{DS}. The design of a similar *p*-channel circuit was considered in Fig. 3.15*a*, again using the nonlinear model. Either of these examples could apply equally well to depletion-mode MOSFETs.

In designing the circuit external to the transistor, we assumed that the parameters of the FET (I_{DSS} and V_P) were well known and stable. As we have seen with the bipolar transistor, however, variations among units of the same commercial type, temperature effects, and a combination of these two variables can cause extreme variations in the transistor parameters. With JFETs, the major variation occurs with the drain-to-source saturation current I_{DSS} and the pinch-off voltage V_P. But these are precisely the two values needed to specify the nonlinear transfer characteristic in the region beyond pinch-off.

Temperature affects V_P only slightly, while I_{DSS} decreases as temperature increases, perhaps a 25% decrease for a 100°C increase in temperature. A much greater variation in V_P and I_{DSS} occurs because of slight differences in the manufacturing process. Both these quantities are a function of the channel width; I_{DSS} is proportional to the first power of the channel width, and $|V_P|$ is proportional to its square. The first page of the data sheet for the Texas Instruments 2N3823 *n*-channel silicon JFET is shown as Fig. 4.11. From it we can see that I_{DSS} may range from 4 to 20 mA, an inconvenient five-to-one range, whereas the value of V_{GS} required to reduce I_D to 0.4 mA may be anywhere between -1 and -7.5 V. This is not quite the same as pinch-off voltage, which would be slightly more negative, but it is not difficult to calculate the pinch-off voltage from the data given. The nonlinear characteristic of a unit for which $I_{DSS} = 20$ mA and $I_D = 0.4$ mA when $V_{GS} = -7.5$ V is described by

$$I_D = \frac{I_{DSS}}{V_P^2}(V_{GS} - V_P)^2 = \frac{20}{V_P^2}(V_{GS} - V_P)^2$$

and therefore

$$0.4 = \frac{20}{V_P^2}(-7.5 - V_P)^2$$

Solving for V_P, we find that it is -8.74 V. Using data for the lower-current unit, we find its pinch-off voltage to be -1.46 V. Thus V_P may be anywhere within the six-to-one range from -1.46 to -8.74 V. Note that we have associated the more negative pinch-off voltage with the larger drain-to-source saturation current, while the less negative value of V_P goes with the smaller I_{DSS}. There is no guarantee by the manufacturer that this is the case, but it does tend to happen in practice. Thus the two limiting transfer characteristics at 25°C can be taken to be those shown in Fig. 4.12.

We may now consider the design of a biasing circuit that will provide a fixed operating point for transistors having such diverse characteristics. Suppose we begin with the fixed-bias arrangement shown in Fig. 4.13*a*. Two separate dc sources are required for this *n*-channel JFET: positive with respect to ground for the drain, and negative with respect to ground for the gate. This is probably more expensive than a circuit that has only one supply; moreover, it leads to quite different operating points for the two characteristics shown in Fig. 4.12 and repeated in Fig. 4.13*b*. Since the bias supply is negative, $I_G \doteq 0$, and therefore $V_{GS} = -2$ V. This leads to two operating points, as shown in Fig. 4.13*b*. The unit with $I_{DSS} = 20$ mA and $V_P = -8.74$ V has $I_D = 11.9$ mA. Then, when we use the nonlinear model $I_D = (I_{DSS}/V_P^2)(V_{GS} - V_P)^2$, we find that the unit with $I_{DSS} = 4$ mA and $V_P = -1.46$ V is cut off and $I_D \doteq 0$, since $V_{GS} = -2$ V. If the fixed bias were changed to -1 V, both transistors would be operating

TYPE 2N3823
N-CHANNEL EPITAXIAL PLANAR SILICON FIELD-EFFECT TRANSISTOR

SYMMETRICAL N-CHANNEL FIELD-EFFECT TRANSISTOR
FOR VHF AMPLIFIER AND MIXER APPLICATIONS

- **Low Noise Figure: $\leq$ 2.5 db at 100 Mc**
- **Low C_{rss}: $\leq$ 2 pf**
- **High y_{fs}/C_{iss} Ratio (High-Frequency Figure-of-Merit)**
- **Cross Modulation Minimized by Square-Law Transfer Characteristic**

***mechanical data**

4 LEADS $\frac{0.019}{0.016}$ DIA; 0.210/0.170; 0.030 MAX; 0.230/0.209 DIA; 0.195/0.178 DIA; 0.500 MIN; 0.100; 0.050; 0.050; 0.100; 3 – GATE; 4 – CASE; 2 – DRAIN; 1 – SOURCE; 45°; 0.048/0.028; 0.046/0.036

ALL DIMENSIONS ARE IN INCHES UNLESS OTHERWISE SPECIFIED

THE ACTIVE ELEMENTS ARE ELECTRICALLY INSULATED FROM THE CASE

ALL JEDEC TO-72† DIMENSIONS AND NOTES ARE APPLICABLE

†TO-72 outline is same as TO-18 except for addition of a fourth lead.

*** absolute maximum ratings at 25°C free-air temperature (unless otherwise noted)**

Drain-Gate Voltage	30 v
Drain-Source Voltage	30 v
Reverse Gate-Source Voltage	−30 v
Gate Current	10 ma
Continuous Device Dissipation at (or below) 25°C Free-Air Temperature (See Note 1)	300 mw
Storage Temperature Range	−65°C to + 200°C
Lead Temperature 1/16 Inch from Case for 10 Seconds	300°C

***electrical characteristics at 25°C free-air temperature (unless otherwise noted)**

	PARAMETER	TEST CONDITIONS‡	MIN	MAX	UNIT
$V_{(BR)GSS}$	Gate-Source Breakdown Voltage	$I_G = -1\ \mu a$, $V_{DS} = 0$	−30		v
I_{GSS}	Gate Cutoff Current	$V_{GS} = -20$ v, $V_{DS} = 0$		−0.5	na
		$V_{GS} = -20$ v, $V_{DS} = 0$, $T_A = 150°C$		−0.5	μa
I_{DSS}	Zero-Gate-Voltage Drain Current	$V_{DS} = 15$ v, $V_{GS} = 0$, See Note 2	4	20	ma
V_{GS}	Gate-Source Voltage	$V_{DS} = 15$ v, $I_D = 400\ \mu a$	−1	−7.5	v
$V_{GS(off)}$	Gate-Source Cutoff Voltage	$V_{DS} = 15$ v, $I_D = 0.5$ na		−8	v
$\|y_{fs}\|$	Small-Signal Common-Source Forward Transfer Admittance	$V_{DS} = 15$ v, $V_{GS} = 0$, f = 1 kc, See Note 2	3500	6500	μmho
$\|y_{os}\|$	Small-Signal Common-Source Output Admittance	$V_{DS} = 15$ v, $V_{GS} = 0$, f = 1 kc, See Note 2		35	μmho
C_{iss}	Common-Source Short-Circuit Input Capacitance	$V_{DS} = 15$ v, $V_{GS} = 0$, f = 1 Mc		6	pf
C_{rss}	Common-Source Short-Circuit Reverse Transfer Capacitance	$V_{DS} = 15$ v, $V_{GS} = 0$, f = 1 Mc		2	pf
$\|y_{fs}\|$	Small-Signal Common-Source Forward Transfer Admittance	$V_{DS} = 15$ v, $V_{GS} = 0$, f = 200 Mc	3200		μmho
$Re(y_{is})$	Small-Signal Common-Source Input Conductance	$V_{DS} = 15$ v, $V_{GS} = 0$, f = 200 Mc		800	μmho
$Re(y_{os})$	Small-Signal Common-Source Output Conductance	$V_{DS} = 15$ v, $V_{GS} = 0$, f = 200 Mc		200	μmho

NOTES: 1. Derate linearly to 175°C free-air temperature at the rate of 2 mw/C°.
2. These parameters must be measured using pulse techniques. PW = 100 msec, Duty Cycle $\leq$ 10%.

*Indicates JEDEC registered data.
‡The fourth lead (case) is connected to the source for all measurements.

Fig. 4.11 The first of four pages of data furnished by Texas Instruments for the 2N3823 transistor, reproduced with permission.

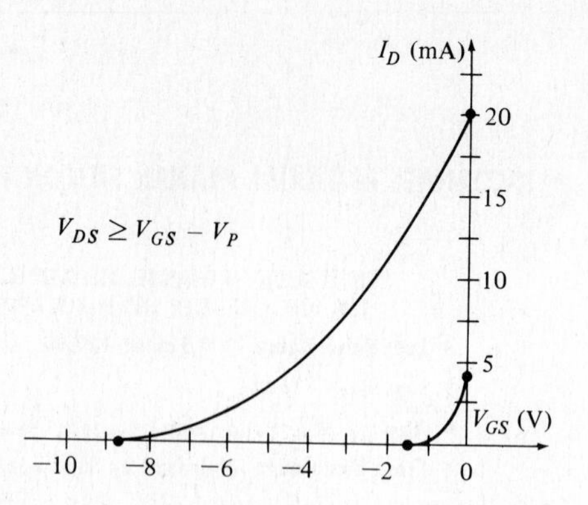

Fig. 4.12 Data given for the 2N3823 transistor in Fig. 4.11 predict these transfer characteristics for transistors having the maximum and minimum values of I_{DSS}.

beyond pinch-off; however, it is obvious that the two operating points would be completely different (I_D = 15.7 mA and 0.4 mA).

Some improvement is obtained by using self-bias, as supplied by the source resistor R_{SS} in Fig. 4.14*a*. If we again assume a negligible gate current and sum voltages around the gate mesh,

$$V_{GS} = -I_G R_G - R_{SS} I_D \doteq -R_{SS} I_D$$

This linear equation may be plotted as a straight line (Chapter 3, Eq. (16), with $V_G = 0$) on the I_D-V_{GS} axes of Fig. 4.14*b*; the load line is shown for

Fig. 4.13 (*a*) An *n*-channel JFET is operated with fixed bias, $V_{GS} = -2$ V. (*b*) The transfer characteristics for widely different 2N3823s show that one of the transistors is cut off ($I_D = 0$), while the other is at a satisfactory operating point in the saturation region.

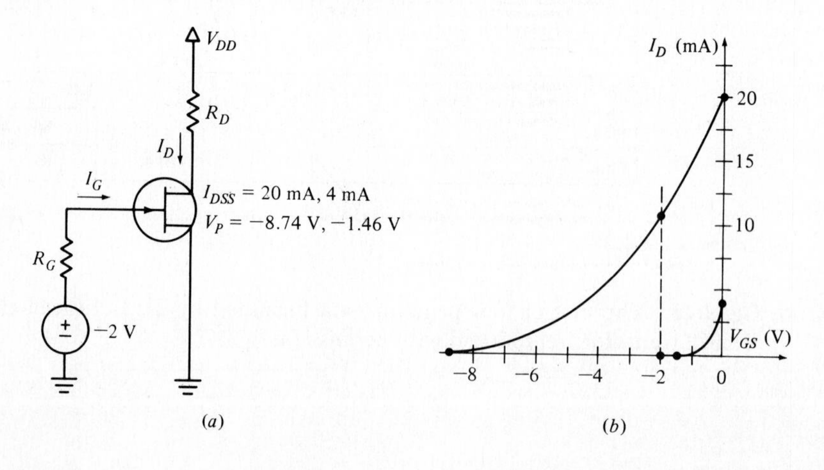

$R_{SS} = 0.5\ \text{k}\Omega$. Its equation is

$$V_{GS} = -0.5I_D$$

Since

$$I_D = \frac{20}{(-8.74)^2}[V_{GS} - (-8.74)]^2$$

we can solve the two equations simultaneously for V_{GS}, obtaining

$$V_{GS} = -3.54\ \text{V}$$

and therefore

$$I_D = 7.08\ \text{mA}$$

For a unit with $I_{DSS} = 4$ mA and $V_P = -1.46$ V, we find that $V_{GS} = -0.637$ V and $I_D = 1.27$ mA. These two values for I_D, 7.08 and 1.27 mA, are not nearly as diverse as they were with fixed bias. Smaller values of R_{SS} would produce steeper lines and even more disparate values of I_D at the operating points. Larger resistances lead to more equal currents, but they would be quite small, thus leading to reduced signal gain, as we have suggested before.

If we did not have to design for such a wide range of transistor characteristics, this self-bias arrangement could certainly be used to lead to a satisfactory range of values for I_D, and it has the virtue of simplicity. However, if no selection of R_{SS} permits a good solution to the bias problem, we have to try a different approach.

Fig. 4.14 (*a*) Self-bias is provided by a source resistor R_{SS}. (*b*) For $R_{SS} = 0.5\ \text{k}\Omega$, both operating points are beyond pinch-off, but they are quite different.

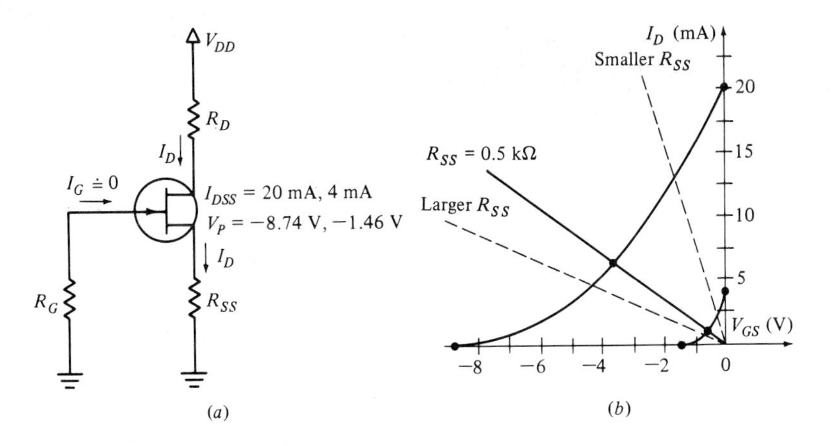

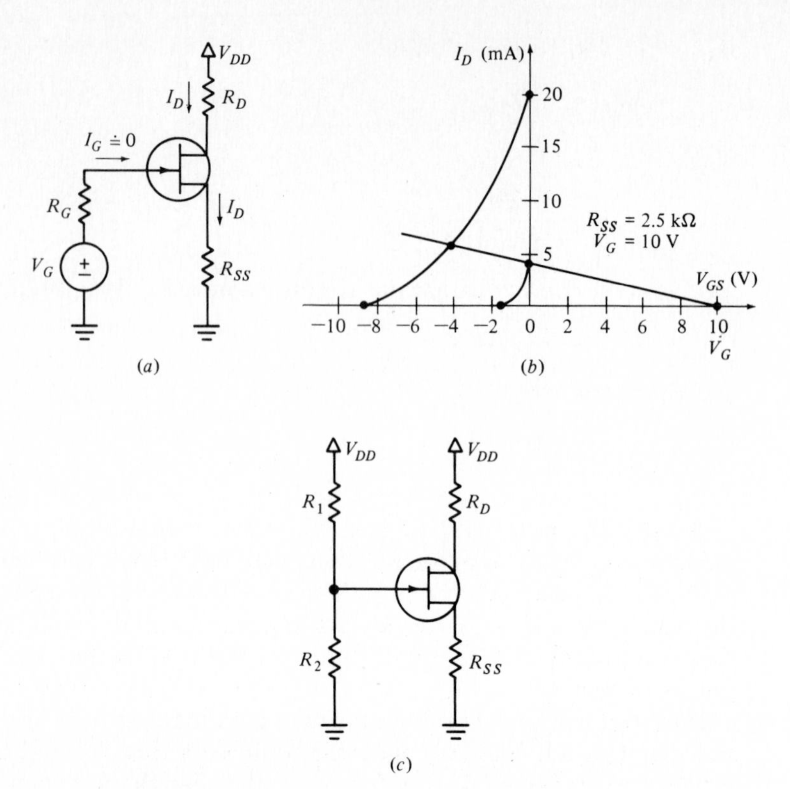

Fig. 4.15 (*a*) A biasing circuit employing a combination of fixed and self-bias. (*b*) The bias load line leads to more equal drain currents for the two transistors. (*c*) Only one dc supply is required in practice.

We would like a bias line that is horizontal and as high up the I_D axis as we can manage. Since the maximum current obtainable from the lower-current unit is 4 mA, the best we can hope for from these two very different transistors are drain currents of about 4 mA maximum. This condition is approached by using both self-bias and a positive fixed-bias supply in the gate circuit, as illustrated in Fig. 3.21*a* or 4.15*a*. Here,

$$V_{GS} = V_G - R_{SS} I_D$$

This linear equation in V_{GS} and I_D leads to a straight line on the transfer curves of Fig. 4.15*b*.

Since the operating-point value for I_D must be at the intersection of this line and the transfer characteristic, let us try for an almost horizontal line through the point $I_D = 4$ mA, $V_{GS} = 0$, the point of maximum current for the lower-current unit. The V_{GS}-axis intercept is at $V_{GS} = V_G$, so we move this point well out to the right by selecting $V_G = 10$ V. At the I_D-axis intercept,

$$0 = 10 - R_{SS}(4)$$

and $R_{SS} = 2.5$ kΩ. This bias load line is the one shown in Fig. 4.15*b*. It provides an operating point for the larger-current unit at $V_{GS} = -4.10$ V, $I_D = 5.64$ mA, as compared with the smaller unit at $V_{GS} = 0$, $I_D = 4$ mA. This is about as well as we can do with these two very different FETs, unless we make V_G even larger, as well as increasing the value of R_{SS}.

If there is no appreciable gate current, then it is easier to supply the positive voltage V_G by a voltage divider from V_{DD} than it is to provide an extra dc source. This is illustrated in Fig. 4.15*c*, and we see that

$$V_G = V_{DD}\frac{R_2}{R_1 + R_2}$$

To establish the voltage $V_G = 10$ V from a drain supply voltage $V_{DD} = 25$ V, we would set $R_1 = 1.5R_2$. Any convenient values of resistance can be used as long as they are high enough not to load the signal input to the transistor. We should also place a by-pass capacitor across R_{SS} in order to place the source at signal ground; hence the name common source.

In the depletion-mode *n*-channel IGFET, the insulated gate region prevents gate current. The transfer characteristic is parabolic for $V_{GS} < 0$, always assuming that V_{DS} is sufficiently large to provide operation in the region beyond pinch-off ($V_{DS} \geq V_{GS} - V_P$). The practical implementation of the biasing circuit is shown in Fig. 4.16*a*. Since there is no gate current,

$$V_{GS} = V_{DD}\frac{R_2}{R_1 + R_2} - R_{SS}I_D$$

The voltage divider provides the voltage V_G, the first term on the right side of the equation above, as identified on the abscissa in Fig. 4.16*b*. If operation extends into the positive V_{GS} region, as shown in Fig. 4.16*c*, the depletion-mode IGFET or MOSFET may show a smooth transition into enhancement-mode operation. The parabolic transfer characteristic continues to the right of the I_D axis.

D4.7 Let $R_1 = 700$ kΩ, $R_2 = 500$ kΩ, $V_{DD} = 24$ V, and $R_D = 2$ kΩ in the circuit of Fig. 4.15*c*. Transistor A has $V_{P(A)} = -5$ V, $I_{DSS(A)} = 12$ mA, while Transistor B has $V_{P(B)} = -1.5$ V and $I_{DSS(B)} = 4$ mA. (a) Specify the minimum value of R_{SS} so that $V_{GS} \leq 0$ for either transistor. (b) Let $R_{SS} = 3$ kΩ and determine $I_{D(A)}$, $I_{D(B)}$, $V_{DS(A)}$, and $V_{DS(B)}$.

Answers. 2.5 kΩ; 4.03 mA, 3.37 mA, 3.83 V, 7.13 V

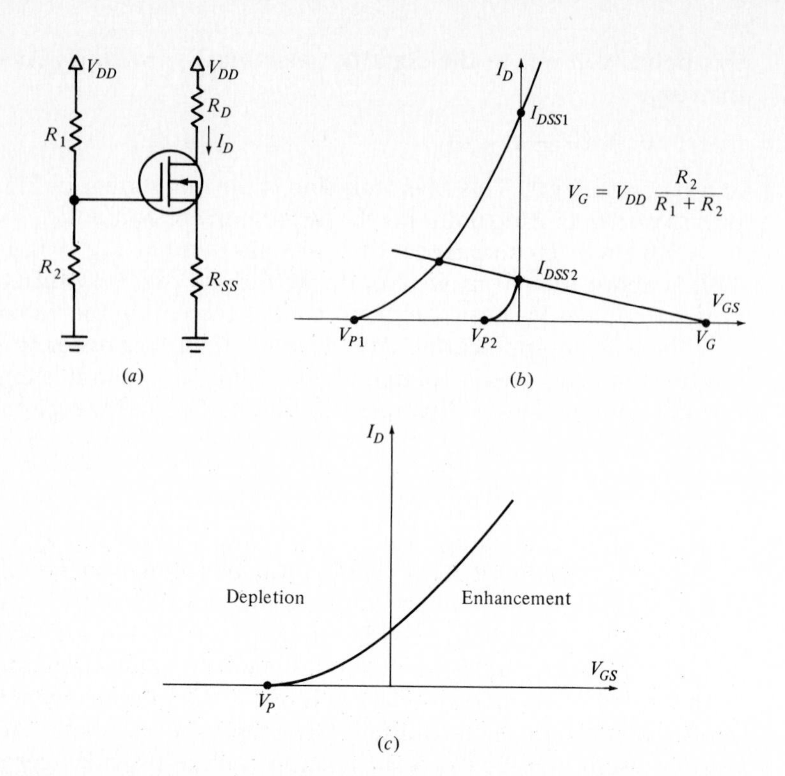

Fig. 4.16 (*a*) A practical biasing arrangement for an *n*-channel depletion-mode MOSFET. (*b*) With sufficiently large values of V_G and R_{SS}, similar values of I_D may be obtained for dissimilar units. (*c*) The transfer characteristic of a depletion-enhancement-mode MOSFET.

4.7 Enhancement-mode IGFET operating-point design

The design of biasing circuitry for an IGFET or MOSFET operating in the enhancement mode closely follows the form and principles discussed in Section 4.6. We recall that the transfer characteristic for the region beyond pinch-off is described by the constant K and the threshold voltage V_T. For the *n*-channel or *p*-channel device,

$$I_D = \frac{K}{2}(V_{GS} - V_T)^2$$

The effect of an increased operating temperature shows up as a very slight decrease in V_T and a decrease in the value of K, similar to the decrease of I_{DSS} with temperature for the JFET. Neither of these changes is apt to be as great as the unit-to-unit or wafer-to-wafer variation.

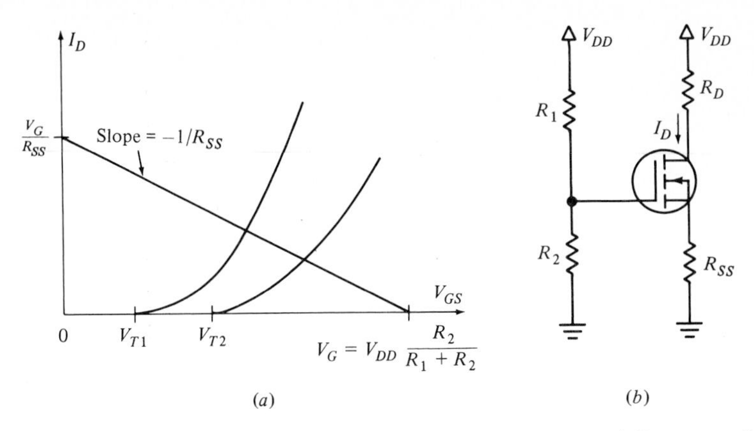

Fig. 4.17 (*a*) Similar operating points are obtained for two different *n*-channel enhancement-mode MOSFETs by using a combination of fixed and self-bias, as shown in (*b*).

Two transfer characteristics with different values of V_T and K are shown in Fig. 4.17*a*. Note that fixed bias alone would lead to a vertical biasing line and grossly unequal values of I_D. However, if we use the combination of fixed bias and self-bias shown in Fig. 4.17*b*, the load-line equation is

$$V_{GS} = V_G - R_{SS}I_D$$

where

$$V_G = V_{DD}\frac{R_2}{R_1 + R_2}$$

This is plotted on the characteristics of Fig. 4.17*a*. It is evident that the determination of suitable values for R_1, R_2, and R_{SS} follows the same procedures used for the JFETs and depletion-mode IGFETs. Using these procedures enables us to secure operating points at which the drain-current values are reasonably equal. It is also necessary that $V_{GS} \geq V_T$ for *n*-channel enhancement-mode devices.

Let's consider the design of a circuit operating with a low drain current that has the form of Fig. 4.17*b*. We want I_D to be 300 μA, plus or minus 20%. Thus, $240 \leq I_D \leq 360\ \mu$A. The drain supply voltage is specified as 10 V, and the device parameters range from $V_{T1} = 0.5$ V and $K_1 = 200$ μA/V^2 to $V_{T2} = 1$ V with $K_2 = 100\ \mu$A/V^2. Hence the limiting transfer characteristics are

$$I_{D1} = \frac{200}{2}(V_{GS1} - 0.5)^2$$

and

$$I_{D2} = \frac{100}{2}(V_{GS2} - 1)^2$$

These two transfer characteristics are shown in Fig. 4.18, with the limiting operating points defined by the maximum and minimum values of I_D indicated. By setting $I_{D1} = 360\ \mu A$, we find $V_{GS1} = 2.40$ V; when $I_{D2} = 240\ \mu A$, $V_{GS2} = 3.19$ V. These two operating points determine the load line corresponding to the smallest possible value of R_{SS}, and we find that this is

$$R_{SS} = \frac{3.19 - 2.40}{360 - 240} 10^6 = 6.61\ \text{k}\Omega$$

The intercept of the load line and the V_{GS} axis is at $V_{GS} = V_G$; we find

$$V_G = 3.19 + (240 \times 10^{-6})6.61 \times 10^3 = 4.78\ \text{V}$$

Values of R_1 and R_2 may now be selected to obtain the desired value of V_G:

$$V_G = 10 \frac{R_2}{R_1 + R_2} = 4.78$$

Arbitrarily choosing $R_2 = 100\ \text{k}\Omega$, we find that $R_1 = 109.3\ \text{k}\Omega$.

It remains to choose a value for R_D. The following chapters will show that larger resistance values lead to larger signal-gain magnitudes. However, larger values may also move the operating point into the ohmic region. Since $V_{DS} \geq V_{GS} - V_T$, we find the limiting values are $V_{DS1} \geq 2.40$

Fig. 4.18 Given the two limiting cases of the transfer characteristic for an enhancement-mode MOSFET, a drain current in the range $240 \leq I_D \leq 360\ \mu A$ is obtained for any operating point on the section of the load line extending from Q_1 to Q_2.

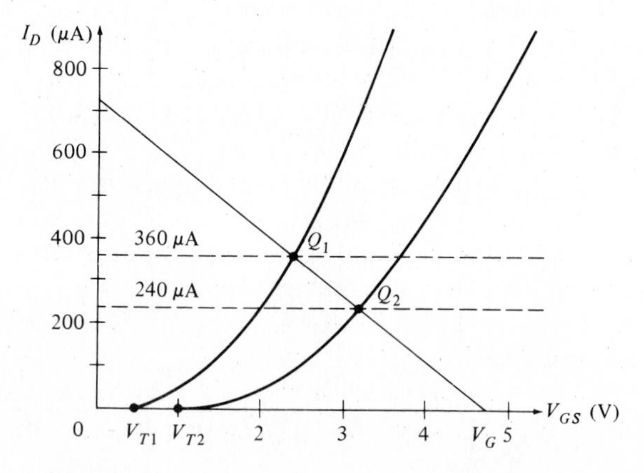

$- 0.5 = 1.90$ V, and $V_{DS2} \geq 3.19 - 1 = 2.19$ V. In turn, these show that the maximum possible voltages across the combination of R_{SS} and R_D are $V_1 = 10 - 1.90 = 8.10$ V, and $V_2 = 10 - 2.19 = 7.81$ V. Since $V_1 = 8.10$ V occurs when $I_{D1} = 360\ \mu$A, we find that the maximum permissible value of $R_{SS} + R_D$ is $8.10/0.360 = 22.5$ kΩ for this transistor, and $7.81/0.240 = 32.5$ kΩ for the other. Since either case is possible, we are restricted to $R_{SS} + R_D = 6.61 + R_D \leq 22.5$ kΩ. For safety's sake, we let $R_D = 15$ kΩ.

D4.8 Two enhancement-mode MOSFETS have $V_{T1} = 2$ V and $V_{T2} = 3$ V; when $V_{GS} = 4$ V, $I_{D1} = 20$ mA and $I_{D2} = 10$ mA. Let $R_2 = 200$ kΩ, $R_D = 5$ kΩ, and $V_{DD} = 20$ V. If $I_D = 2$ mA ± 25%, and R_{SS} is the minimum possible value, find (a) R_{SS}, (b) R_1.

Answers. 0.680; 708 kΩ

4.8 Common-drain and common-gate biasing

After discussing bias circuits for the common-emitter configuration, it was a relatively simple matter to extend the same circuit to the case of the common collector and common base. In actuality, all that had to be changed was the location of the by-pass capacitor, and we have not even shown that element in our circuit diagrams as yet.

A similar situation exists for the common-drain and common-gate FET circuits. After a quick referral to our most general biasing circuit for the common-source case, Figs. 4.15*c*, 4.16*a*, and 4.17*b*, it is a simple matter to determine that the common-drain circuit of Fig. 4.19*a* for an *n*-channel JFET differs only from the common-source circuit in having $R_D = 0$.

The common-gate JFET circuit is shown in Fig. 4.19*b*. A signal ground will be achieved at the gate by placing a large capacitor across R_2. If the temperature change is no problem and there is no great variability among the transistors to be used in the circuit, or if it is hand-tailored for a specific unit, the desired operating point may often be obtained without using R_1 and R_2 ($R_1 = \infty$, $R_2 = 0$). Doing so, however, may cause R_{SS} to be such a small value that much of the signal is shunted to ground. The use of R_1 and R_2 makes the gate positive with respect to ground and permits the use of a larger value of R_{SS}.

The circuits of Fig. 4.19 are also applicable to IGFETs and MOSFETs, either depletion- or enhancement-mode. One must of course be careful to provide the proper bias polarities and to maintain operation in the region beyond pinch-off.

D4.9 An *n*-channel enhancement-mode MOSFET is to be operated in a common-gate circuit, Fig. 4.19*b*. With $R_1 = 3R_2$, $V_{DD} = 24$ V, K

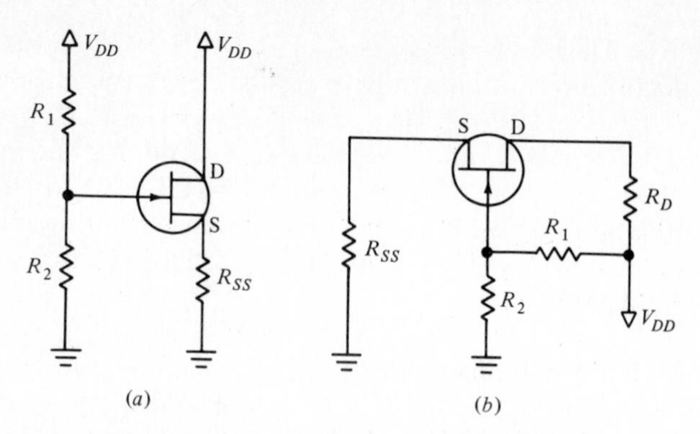

Fig. 4.19 (*a*) A general dc bias circuit for an *n*-channel JFET that is operated common drain. (*b*) The general bias circuit is shown for the common-gate arrangement. A large by-pass capacitor is used across R_2 under signal conditions.

= 8 mA/V^2, and V_T = 3 V, (a) select R_{SS} so that I_D = 5 mA. (b) Select R_D so that the voltage across it is five times V_{DS}. (c) If K doubles and R_{SS} is unchanged, by what factor does I_D increase?

Answers. 376 Ω; 3.69 kΩ; 1.145

Problems

1. Modify the design example in Section 4.1 (which uses a 2N5377 transistor) to use a smaller amount of negative feedback by selecting the voltage across R_E to be only one-quarter of V_{CE}. Maintain V_{CC} = 25 V and $I_{R1} = 50I_B$ to facilitate comparison of results. (a) Use the most accurate analysis to find I_C, I_B, and V_{CE} for the extreme cases $\beta_{dc(min)}$ = 100 and $\beta_{dc(max)}$ = 500. (b) Repeat Part (a) if $I_{R1} = 10I_B$.
2. A 2N3903 transistor (see Appendix A) is used in the circuit designed in Section 4.1 (V_{CC} = 25 V, R_C = 15 kΩ, R_E = 5 kΩ, R_2 = 11.5 kΩ, and R_1 = 38.7 kΩ). Determine I_C and V_{CE} for the minimum value of β_{dc}. Let V_{BE} = 0.65 V.
3. For the circuit shown in Fig. 4.20, let $R_C = R_2$ = 9 kΩ, and R_E = 2 kΩ. Calculate I_C and V_{CE} for (a) β_{dc} = 50, (b) β_{dc} = 500. (c) Select other values for R_C, R_E, and R_2 so that $1.9 \le I_C \le 2.1$ mA and $3.5 \le V_{CE} \le 5.5$ V for $50 \le \beta_{dc} \le 500$.
4. Let I_{CEO} = − 1 μA and V_{BE} = − 0.65 V for a *pnp* transistor operating at I_C = − 2.5 mA, V_{CE} = − 8 V. Design a circuit that will give I_C = − 2.5 mA within ± 0.1 mA for $50 \le \beta_{dc} \le 150$.
5. Let V_{BE} = 0.65 V for a 2N5376 transistor (on the 2N5377 data sheet) operating at I_C = 1 mA, V_{CE} = 5 V at 25°C. Specify values for R_1, R_2, R_C, R_E, and V_{CC} in a bias circuit if $I_{R1} = 20I_B$.

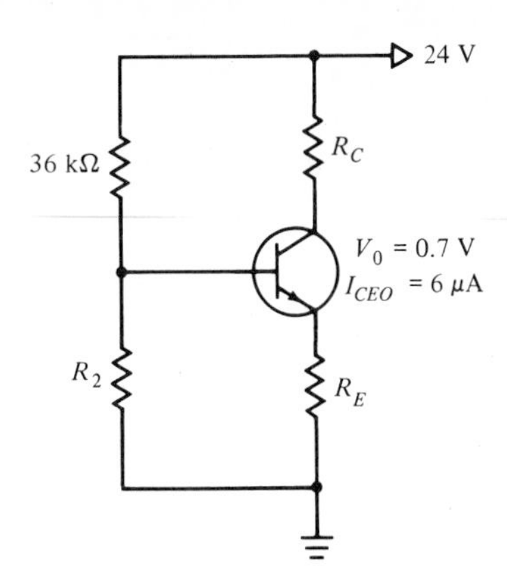

Fig. 4.20 See Problem 3.

6. Let R_1 = 50 kΩ, R_2 = 20 kΩ, R_E = 4 kΩ, R_C = 8 kΩ, and V_{CC} = 28 V in the circuit of Fig. 4.1*a*. If I_{CEO} = 8 μA, find I_C and V_{CE} if (a) β_{dc} = 30 and V_0 = 0.6 V, (b) β_{dc} = 300 and V_0 = 0.6 V, (c) β_{dc} = 30 and V_0 = 0.7 V. (d) Select other values for R_1, R_C, and R_E so that I_C = (1.5 ± 0.1) mA and V_{CE} = (9 ± 1) V if $30 \leq \beta_{dc} \leq 300$ and V_0 = 0.6 V. (e) Select other values for R_1, R_C, and R_E so that I_C = (1.5 ± 0.1) mA and V_{CE} = (9 ± 1) V if $30 \leq \beta_{dc} \leq 300$ and $0.5 \leq V_0 \leq 0.7$ V. Assume that any combinations of values for β_{dc} and V_0 may occur.
7. In the basic circuit of Fig. 4.1*a*, let R_1 = 40 kΩ, R_2 = 10 kΩ, R_E = 5 kΩ, R_C = 18 kΩ, V_{CC} = 24 V, β_{dc} = 250, V_0 = 0.65 V, and R_{BB} = 10 kΩ. Assume that I_{CEO} is negligible. (a) Find I_C and V_{CE}. (b) Increase or decrease each resistor independently by 10% to maximize I_C; calculate the percentage change in I_C for each case. (c) Increase or decrease each resistor independently by 10% to minimize V_{CE}; calculate V_{CE} for each case.
8. Let R_1 = 40 kΩ, R_2 = 10 kΩ, R_E = 5 kΩ, R_C = 15 kΩ, V_{CC} = 25 V, β_{dc} = 100, V_0 = 0.7 V, R_{BB} = 8 kΩ, and $I_{CEO} \doteq 0$ for a circuit of the form shown in Fig. 4.1*a*. (a) Find I_C. (b) Determine worst-case values for I_C if V_0, R_{BB}, and β_{dc} are each subject to a ±10% error. Assume V_{CC}, I_{CEO}, and the circuit resistance values remain constant.
9. Figure 4.21 shows a circuit similar to that found in many audio amplifiers. The signal components are not shown. Let I_C = 1 mA, V_{CE} = 5 V, β_{dc} = 50, and V_{BE} = 0.6 V for T1, while I_C = 2 mA, V_{CE} = 10 V, β_{dc} = 100, and V_{BE} = 0.65 V for T2. If V_{RE} = V_{CE} for T1, determine values for R_{E1}, R_{C1}, R_{E2}, R_{C2}, and R_F, in that sequence.
10. Let V_{CC} = 24 V, R_1 = 60 kΩ, R_2 = 20 kΩ, R_C = 5 kΩ, and R_E = 3 kΩ in Fig. 4.1*a*. At 25°C, β_{dc} ranges from 100 to 400, while V_0 lies be-

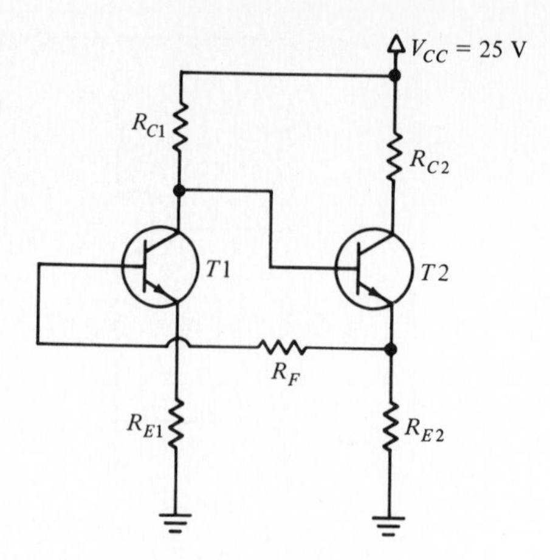

Fig. 4.21 See Problem 9.

tween 0.65 and 0.75 V. In the temperature range from -45 to 115°C, let β_{dc} equal the value at 25°C times $[1 + 0.003(T - 25)]$, where T is in degrees Celsius, and also let $V_0 = V_{0(25°C)}[1 - 0.002(T - 25)]$. Neglect I_{CEO} and find the maximum and minimum values of I_C and V_{CE} for $-45 \leq T \leq 115$°C.

11. Select new values for R_E and R_1 in the circuit of Problem 10 so that $I_{C(max)} = 1.8$ mA and $I_{C(min)} \geq 1.65$ mA. Do not make any other changes in the circuit.
12. In Fig. 4.1*a* let $R_1 = 80$ kΩ, $R_2 = 40$ kΩ, $R_E = 3$ kΩ, $R_C = 4$ kΩ, and $V_{CC} = 15$ V. The variation of β_{dc} with temperature for the transistor is given by $\beta_{dc} = \beta_{dc(25°C)}[1 + (T - 25)/450]$, where $150 \leq \beta_{dc(25°C)} \leq 350$. Also, $0.6 \leq V_{0(25°C)} \leq 0.7$, and $V_0 = V_{0(25°C)}[1 - (T - 25)/300]$, where T is in degrees Celsius. Assume I_{CEO} and R_{BB} may be neglected. Determine $I_{C(min)}$, $I_{C(max)}$, $V_{CE(min)}$, and $V_{CE(max)}$ if the unit is to operate from -50 to 150°C.
13. Specify different values for R_2 and R_E in the circuit of Problem 12 so that $1.2 \leq I_C \leq 1.4$ mA under all conditions. Other values in the circuit are unchanged.
14. A certain transistor is installed in a standard bias circuit with $R_C = R_E = 4$ kΩ, $R_1 = 50$ kΩ, $R_2 = 30$ kΩ, and $V_{CC} = 16$ V. Assume that I_{CEO} is negligible and that β_{dc} and V_0 may be represented by $\beta_{dc} = 200(T/298)^{0.7}$ and $V_0 = 0.65(T/298)^{-1.2}$ V in the temperature range of interest. Values for T are in kelvins. (a) Find I_C and V_{CE} at 25°C. (b) Find $I_{C(min)}$, $I_{C(max)}$, $V_{CE(min)}$, and $V_{CE(max)}$, and the temperature at which each occurs if $213 \leq T \leq 423$ K.
15. Using data furnished in Appendix A for the FD600 diode, determine (a) $P_{D(max)}$, T_{A0}, and $T_{J(max)}$; (b) the thermal resistance θ_{JA}. (c) What

would the junction temperature be when I_D = 200 mA for a "typical" diode operating in an ambient temperature of 50°C?

16. Refer to the data on the FD600 diode and answer the following questions: (a) How much power can the device dissipate safely in boiling water? (b) How much power can it dissipate safely at −25°C? (c) What would the junction temperature be for a "maximum" unit operating at 25°C with I_D = 200 mA?
17. A transistor data sheet states that the maximum allowable junction temperature is 160°C, the maximum allowable dissipation is 200 mW, and the transistor should be derated above T_C = 25°C. If θ_{CH} = 0.2°C/mW and θ_{HA} = 0.4°C/mW, find the ambient temperature and the temperature of the heat sink when (a) 200 mW is being dissipated and T_C = 25°C, (b) 200 mW is being dissipated and T_C = 0°C, (c) as much power as possible is being dissipated and T_C = 100°C.
18. Typical thermal data for an operational amplifier might be as follows: "Total device dissipation at T_A = 25°C is 750 mW; derate above 25°C, 5 mW/°C." What power may be dissipated safely at an ambient temperature of (a) 75°C? (b) 0°C? (c) If θ_{JC} is 0.15°C/mW, what is θ_{CA}?
19. A power-derating curve for a power transistor is shown in Fig. 4.22. If θ_{CA} = 3°C/W, find (a) θ_{JA}, (b) θ_{JC}. (c) What is the maximum safe value of P_D if T_A = 125°C? (d) How much power may be dissipated safely if T_C = 125°C?
20. A thermal equivalent circuit has θ_{HA} = 0.3°C/mW, θ_{CH} = 0.1°C/mW, and θ_{JC} = 1°C/mW for a certain transistor. (a) If the unit dissipates 0.1 W into an ambient temperature of 35°C, find T_J, T_C, and T_H. (b) If T_J = 170°C with T_A = 50°C, find T_C and P_D.
21. (a) Design a stable biasing circuit for a 2N5089 transistor if I_C = 2 ± 0.1 mA and V_{CE} = 5 ± 1 V for $-55 \le T \le 125$°C. Use V_{CC}

Fig. 4.22 See Problem 19.

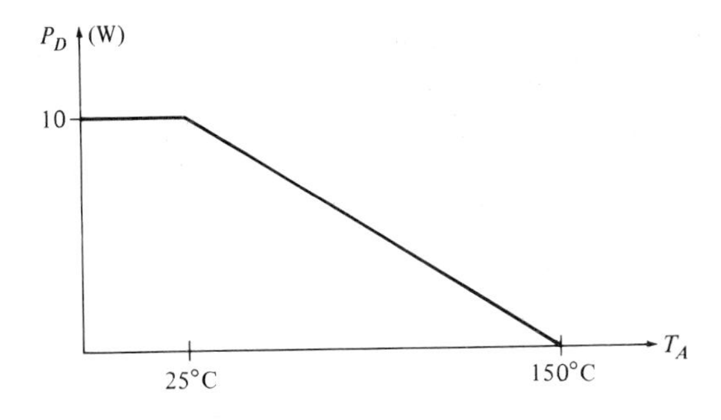

$= 18$ V and assume that $V_{BE} = 0.7$ V at 25°C and decreases linearly by 2.3 mV/°C. (b) Find the junction temperature at $T_A = -55$°C and 125°C.

22. Design a common-collector circuit that will provide a stable operating point at $I_C = 3$ mA, $V_{CE} = 4$ V, for a transistor having the parameters $80 \le \beta_{dc} \le 300$, $2 \le I_{CEO} \le 10$ μA, $0.6 \le V_0 \le 0.7$ V. Let $V_{CC} = 12$ V.
23. Select values for R_1, R_2, R_E, and V_{CC} to provide an operating point at $I_C = 0.2 \pm 0.02$ mA, $V_{CE} = 2.5 \pm 0.5$ V for a common-collector circuit using a transistor for which $100 \le \beta_{dc} \le 400$, $1 \le I_{CEO} \le 5$ μA, and $0.63 \le V_0 \le 0.73$ V. Use $V_{CC} = 6$ V.
24. Let $V_0 = -0.7$ V, $I_{CEO} = -60$ nA, and $\beta_{dc} = 60$ for a *pnp* transistor that is to be operated with $I_C = -20$ mA and $V_{CE} = -5$ V. (a) Design a common-collector bias circuit using $V_{CC} = -12$ V and $I_{R1} = 15I_B$. (b) Design a common-base circuit with $V_{CC} = -15$ V, $I_{R1} = 15I_B$, and $V_{RE} = V_{CE}$.
25. Let $V_0 = 0.7$ V, $R_{BB} = 10$ kΩ, $I_{CEO} = 0.1$ μA, and $\beta_{dc} = 50$ for an *npn* transistor that is to be operated at $I_E = -1$ mA and $V_{CB} = 4$ V in a common-base configuration. (a) Let $V_{CC} = 2.5V_{CE}$, $V_{RE} = 0.8V_{CE}$, and $R_2 = 20$ kΩ. Specify values for V_{CC}, R_1, R_E, and R_C. (b) Calculate I_E if $\beta_{dc} = 500$.
26. The two transistors used in a current mirror are characterized by $\beta_{dc} = 40$, $V_0 = 0.7$ V, and $R_{BB} = 10$ kΩ. If $R_1 = 2.5$ kΩ, $R_C = 2$ kΩ, and $V_{CC} = 7.5$ V, find V_{CE} and I_E for both transistors.
27. (a) At 25°C let both transistors in the circuit of Fig. 4.23 have $V_0 = 0.6$ V, $R_{BB} = 10$ kΩ, and $\beta_{dc} = 100$. Specify R_1 and R_C if V_{CC}

Fig. 4.23 See Problem 27.

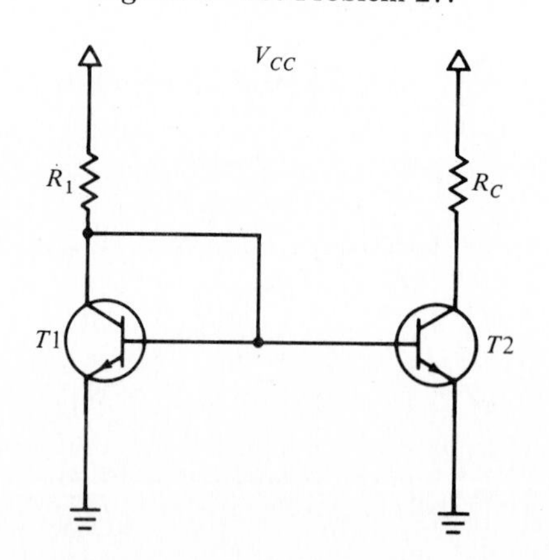

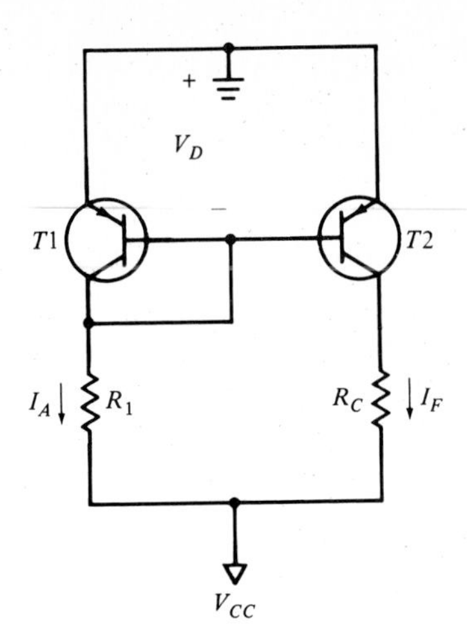

Fig. 4.24 See Problem 28.

= 9 V and the operating point for T2 is at I_{C2} = 1 mA, V_{CE2} = 5 V. (b) Assume that $dV_0/dT = -2$ mV/°C for the transistors, while β_{dc} and R_{BB} are temperature-independent. Also let both resistors have a TCR of + 1500 ppm/°C. Find I_{C2} at 75°C.

28. Two *pnp* transistors appear in the I-C shown in Fig. 4.24. (a) Assume that R_{BB} is negligibly small and derive a relationship between I_F and I_A. (b) If I_F = 1.5 mA, $V_{CC} = -6$ V, β_{dc} = 25, and $V_0 = -0.65$ V, find R_1.

29. Let V_0 = 0.65 V, R_{BB} = 10 kΩ, and β_{dc} = 100 for the *npn* transistor in Fig. 4.25, while $V_0 = -0.6$ V, R_{BB} = 8 kΩ, and β_{dc} = 25 for the

Fig. 4.25 See Problem 29.

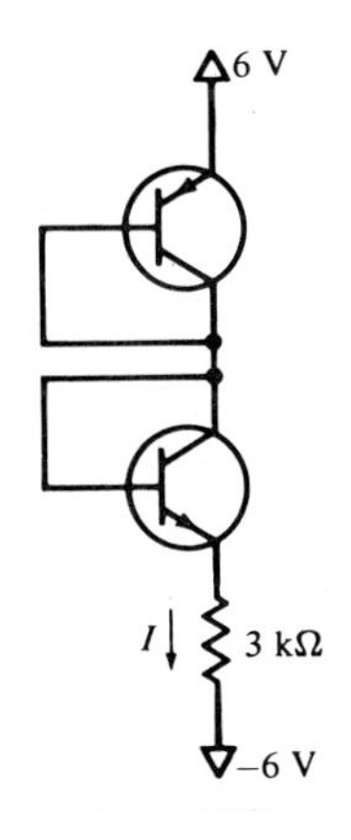

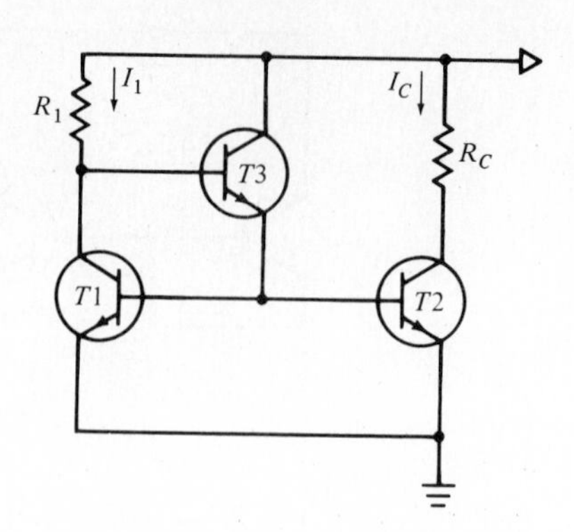

Fig. 4.26 See Problem 30.

pnp unit. (a) Determine I. (b) If $d|V_0|/dT = -2$ mV/°C for each transistor, while the TCR is 1500 ppm/°C for the resistor, calculate dI/dT in microamperes per degree Celsius.

30. (a) Neglecting the effects of R_{BB} for the three identical transistors in Fig. 4.26, show that $I_C/I_1 = \beta_{dc}/[\beta_{dc} + 2/(\beta_{dc} + 1)]$. Calculate I_C and V_{CE2} if $R_1 = 5$ kΩ, $R_C = 1$ kΩ, $V_0 = 0.65$ V, $V_{CC} = 9$ V, and $\beta_{dc} =$ (b) 50, (c) 500.
31. Element values for a JFET in a standard four-resistor bias network (Fig. 4.15*c*) are $R_1 = 150$ kΩ, $R_2 = 50$ kΩ, $R_{SS} = 3$ kΩ, and $R_D = 2.5$ kΩ. Let $V_{DD} = 16$ V. If $I_D = 5(1 + 0.25V_{GS})^2$ mA for the transistor, (a) find I_D and V_{DS}. (b) Find I_D if all four resistors decrease 20% in value.
32. The JFET in a standard bias circuit (Fig. 4.15*c*) is guaranteed to have a transfer characteristic bounded by the parabolas $I_D = 5(1 + 0.25V_{GS})^2$ and $I_D = 2(1 + 0.5V_{GS})^2$ mA. Specify values for R_1, R_2, R_{SS}, and R_D so that $2 \le I_D \le 2.5$ mA for $V_{DD} = 16$ V.
33. Let $R_2 = 40$ kΩ, $R_{SS} = 4$ kΩ, and $V_{DD} = 20$ V in the circuit shown in Fig. 4.27. (a) If $I_D = 1.28(V_{GS} + 2.5)^2$ mA, find I_D and V_{DS}. (b) By what percentage does I_D increase if V_{DD} increases 10%?
34. Let $V_{DD} = 24$ V in Fig. 4.27, and let the JFET have a transfer characteristic lying between $I_D = 1.28(V_{GS} + 2.5)^2$ and $I_D = 0.75(V_{GS} + 4)^2$ mA. Select values for R_2 and R_{SS} so that $1.8 \le I_D \le 2$ mA.
35. In Fig. 4.28, let $R_1 = \infty$ and the JFET transfer characteristic be $I_D = 2(V_{GS} - V_P)^2$ mA, where $-3 \le V_P \le -2$ V for different transistors in a given lot. (a) If $V_{DD} = 12$ V and $R_{SS} = 400$ Ω, find the maximum and minimum values of I_D and V_{DS}. (b) Select values for V_{DD} and R_{SS} so that $2.5 \le I_D \le 5$ mA and operation is beyond pinch-off.
36. Let $V_{DD} = 15$ V in Fig. 4.28 and use a JFET for which $I_D = 2(V_{GS} - V_P)^2$ mA, where $-3 \le V_P \le -2$ V for different transistors of the

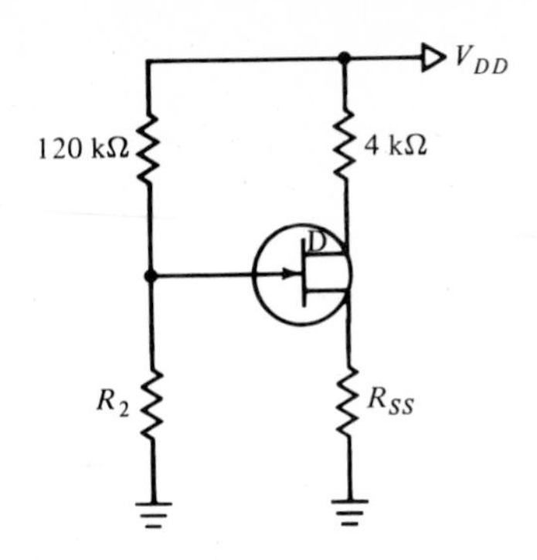

Fig. 4.27 See Problems 33 and 34.

same type. Select R_{SS} and R_1 so that $4 \le I_D \le 5$ mA and operation is beyond pinch-off.

37. For the transistor in Fig. 4.29, let $I_D = 10(0.5V_{GS} - 1)^2$ mA. The total resistance of the potentiometer is $R_1 + R_2 = 100$ kΩ, and the ratio of R_2 to R_1 may be continuously adjusted. (a) What is I_D when $R_1 = R_2$? (b) At what position of the potentiometer is the transistor at the boundary between pinch-off and the ohmic region? (c) At what position does the power dissipated by the transistor reach a maximum?
38. An n-channel enhancement-mode MOSFET has a transfer characteristic given by $I_D = 2.5b(V_{GS} - 2b)^2$ mA, where b may range from 1 to 1.5 for different units of the same type. Using $V_{DD} = 20$ V and $R_D = 1.2$ kΩ, design a bias circuit for the common-source configuration so that $9 \le I_D \le 10$ mA for any unit and operation is beyond pinch-off.
39. The transfer characteristic of an n-channel enhancement-mode MOSFET is bounded by curves for which $K_1 = 200\ \mu\text{A/V}^2$, V_{T1}

Fig. 4.28 See Problems 35 and 36.

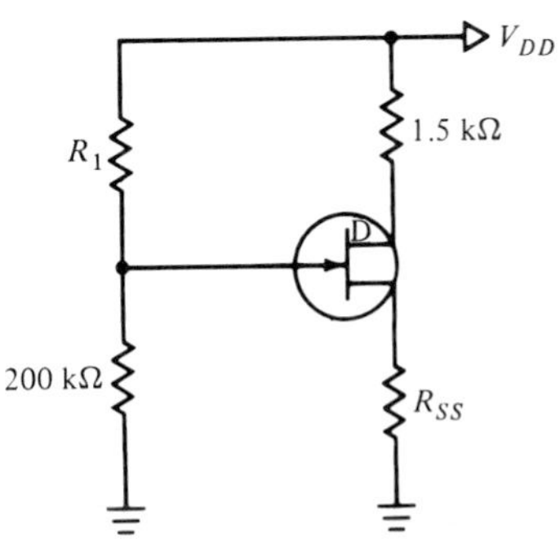

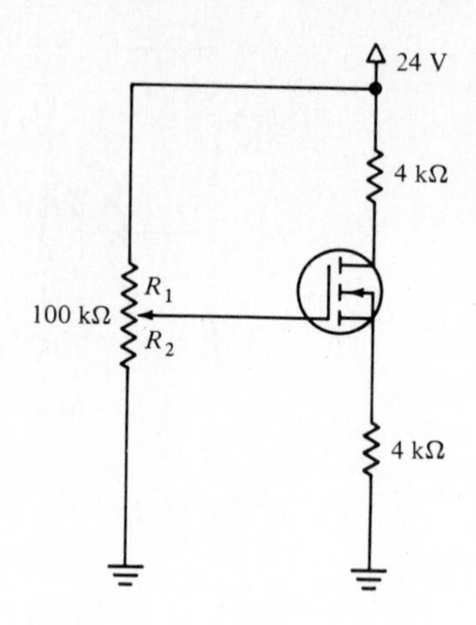

Fig. 4.29 See Problem 37.

= 0.5 V, and K_2 = 100 μA/V^2, V_{T2} = 1 V. (a) Using V_{DD} = 15 V, design a bias circuit so that I_D = 100 μA ± 20%. (b) What is the maximum possible value of R_D in your circuit that will permit saturated operation?

40. Let K = 100 μA/V^2 and V_T = 1.3 V for the transistor in Fig. 4.30. Find V_{out} if V_{in} equals (a) 5 V, (b) 0.5 V, (c) 2.5 V.
41. In Fig. 4.31, let K_1 = 200 μA/V^2 and V_{T1} = 1.5 V for T1, while K_2 = 200 μA/V^2 and V_{P2} = −4 V for T2. Calculate V_{out} for V_{in} = (a) 5 V, (b) 0.5 V.
42. In a common-drain circuit (Fig. 4.19*a*), I_D = 6 mA, $R_1 + R_2$ = 1 MΩ, V_{DD} = 15 V, and the FET dissipates a total power of 54 mW. Select values for R_1, R_2, and R_{SS} if the device is (a) an *n*-channel

Fig. 4.30 See Problem 40.

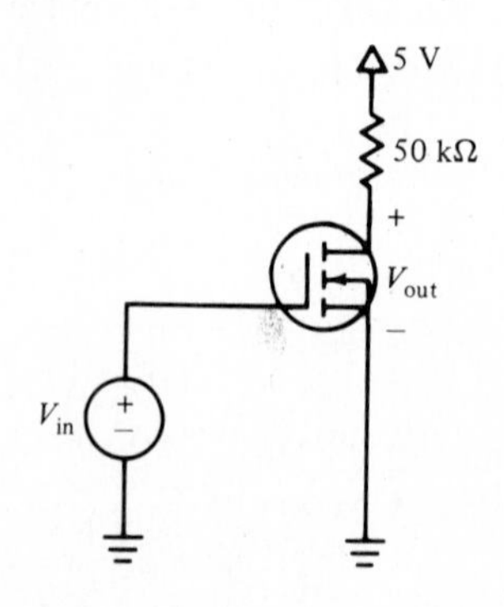

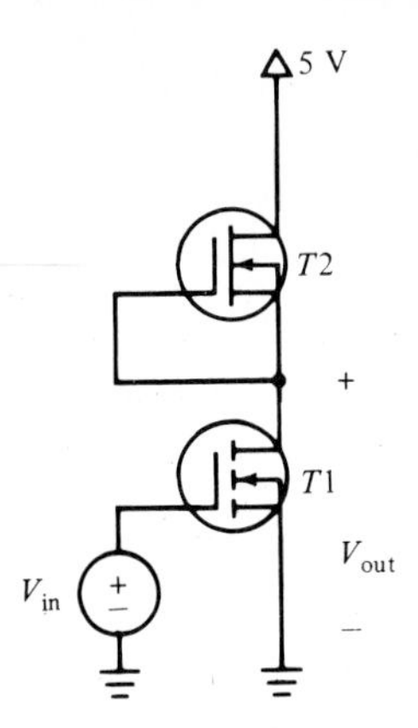

Fig. 4.31 See Problem 41.

enhancement-mode MOSFET for which $I_D = 1.28(V_{GS} - 2.5)^2$ mA; (b) an n-channel JFET having $I_D = 1.28(V_{GS} + 2.5)^2$ mA.

43. Using the common-drain circuit of Fig. 4.19a, let $V_{DD} = 20$ V and $R_1 = 1$ MΩ. The transistor is an n-channel enhancement-mode MOSFET for which $I_D = 2.5b(V_{GS} - 2b)^2$ mA in the region beyond pinch-off, where $1 \le b \le 1.5$ for different units of the same type. Select values for R_2 and R_{SS} so that the slope of the transfer characteristic is greater than 12 mA/V at the operating point for any possible transfer characteristic.

44. Let $R_1 = 150$ kΩ, $R_2 = 200$ kΩ, and $R_D = 16$ kΩ in a common-gate bias network of the form shown in Fig. 4.19b. If the transfer characteristic of the p-channel JFET is given by $I_D = -2 + 4V_{GS} - 2V_{GS}^2$ mA, specify values for V_{DD} and R_{SS} so that R_D dissipates 4 mW and the FET dissipates 2 mW.

45. The transfer characteristic of an n-channel depletion-mode MOSFET is $I_D = (V_{GS} + 2.5)^2$ mA. If the unit is used in a bias circuit similar to Fig. 4.19b, with $V_{DD} = 21$ V, $R_D = 6$ kΩ, $R_{SS} = 5$ kΩ, $R_1 = 250$ kΩ, and $R_2 = 100$ kΩ, find (a) I_D, V_{GS}, V_{DS}, and the total power dissipated in the complete circuit. To reduce I_D to exactly 1 mA, what single change of value in the given circuit should be made for the resistor (b) R_1, (c) R_2, (d) R_{SS}?

46. An n-channel JFET has $I_{DSS} = 5$ mA and $V_P = -1$ V. It is used in a circuit in which the drain is connected directly to the V_{DD} voltage supply, the gate is connected directly to the source terminal, and the source is connected to ground through $R_{SS} = 1$ kΩ. Find the voltage across R_{SS} if $V_{DD} =$ (a) 18 V, (b) 12 V, (c) 6 V. (d) Which of the following is a logical name for this circuit: source follower, gate follower, constant-current source, voltage regulator, Bruce?

5

Small-signal circuit models

We have spent enough time on dc models for the various types of transistors to enable us to find any operating point by analysis or provide any operating point by design. We must now turn our attention to the central problem of providing the desired signal response.

In this chapter we restrict the class of signals to those having such small amplitudes that the transistor characteristics may be approximated as linear in the active region. With this assumption of linearity, it is then possible to use superposition to break the complete problem into two separate parts: the already completed dc or bias problem, and the ac or small-signal problem, which we now consider. Devices operating on small signals are found in almost every electronic system, such as preamplifiers and operational amplifiers, for example, and they are therefore of major importance.

There are two quite similar models that are widely used for the bipolar transistor at low frequencies. We shall spend the major portion of our time on one known as the *hybrid-π model*, but we will also study the *h-parameter model* and investigate the relationships between the two. The h-parameter model is used primarily for bipolar transistors at low frequencies. As frequency increases beyond perhaps 10 kHz, the hybrid-π model proves to be much more convenient, both for the bipolar transistor and the FETs. The hybrid-π model is useful at both low and high frequencies. Some use is made of the *y-parameter model* for the FET, particularly in measured data provided by the manufacturers. We shall relate these data to the hybrid-π parameters.

The values of both the passive- and active-circuit elements appearing in any of the small-signal models are functions of more different variables than we would prefer. Just as the dynamic resistance of the semiconductor diode varies with the operating point, so do the dc voltages and currents have a major effect on transistor equivalent circuits.

We have already seen that temperature can effect large changes in the transistor dc characteristics; even if we are successful in maintaining a fixed operating point, the small-signal circuit parameters are still functions of temperature. The effect of frequency is also extremely important; it will force us to employ more sophisticated circuit analysis techniques than we have needed up to this time.

All these models apply only to the active region of the bipolar transistor or to the saturation region for the FET. Of course, these are also the only regions in which large signal gains can be achieved.

5.1 Bipolar transistor hybrid-π models: small-signal low-frequency

Throughout this chapter we shall limit our attention to *small-signal* models. By this we mean that the transistor never operates very far from its dc operating point. It is difficult to give a general quantitative definition of *small*, but the signal currents and voltages are often kept less than one-tenth the dc operating-point values.

In this section we consider small-signal *low-frequency* transistor models. For our purposes, a frequency is sufficiently low when all capacitive reactances associated with the transistor model are so large in comparison with other impedances that they may be considered to be open circuits. Inductive reactance is usually not a part of a transistor model. We find that capacitances in the transistor model can usually be neglected in amplifiers used in audio or instrumentation systems but must usually be included when frequency components above several hundred kilohertz are present.

A number of small-signal low-frequency bipolar transistor models have been developed and used. We shall emphasize the hybrid-π model shown in Fig. 5.1, but we shall meet the h-parameter model for the bipolar transistor in Section 5.6, and the y-parameter model will appear when we discuss high-frequency models for FETs. The model shown in Fig. 5.1 is actually only a π network (with the horizontal branch missing), and the complete hybrid-π model differs from it in having an additional series resistance in the base, as well as the missing branch from base to collector. This will become clearer when we consider the high-frequency case in the next section; all elements then appear. The model is shown in the common-emitter arrangement and *is identical for* npn *and* pnp *transistors.*

Fig. 5.1 The small-signal low-frequency hybrid-π model for both *npn* and *pnp* bipolar transistors. The resistor r_d is sometimes omitted since it is usually much larger than the external impedance levels.

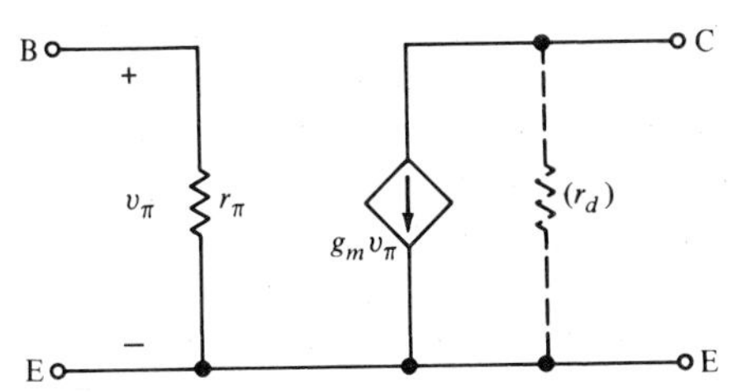

We shall investigate the range of values for these elements in the low-frequency equivalent circuit and the effect of temperature, frequency, and the choice of operating point on them later, but it might be helpful now to know that a typical set of values could be $r_\pi = 4\ \text{k}\Omega$, $g_m = 50\ \text{m}\mho$, and $r_d = 100\ \text{k}\Omega$. Note that these are signal parameters and therefore carry lower-case subscripts.

The model in Fig. 5.1 is a *linear* model; we are able to use such a model in spite of the greatly nonlinear nature of the transistor characteristics because only small-amplitude signals are present. In a sufficiently small neighborhood of the dc operating point, the characteristics are well approximated by straight lines. The dynamic resistance appearing in the ac-equivalent circuit of the diode is a similar example.

As an illustration of the use of the hybrid-π model, let us find the signal gain

$$A_V = \frac{v_o(t)}{v_s(t)}$$

for the basic common-emitter amplifier shown in Fig. 5.2*a*. Our first step is the construction of the ac-equivalent circuit. If $v_s(t) = 0.001 \cos 500t$

Fig. 5.2 (*a*) A simple common-emitter small-signal amplifier. (*b*) The low-frequency small-signal equivalent circuit.

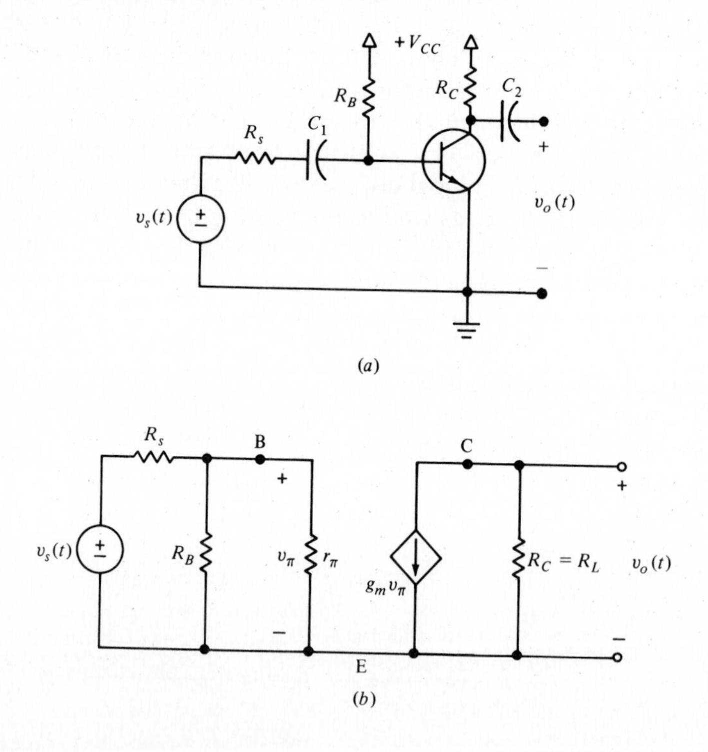

volts, we may use the small-signal low-frequency model for the transistor, because V_{BE} is typically 0.7 V and 500 rad/s is small enough to allow the transistor capacitances to be neglected. Since the total solution is arrived at by superposition, we have to replace all the dc sources by their internal impedances in order to obtain the ac equivalent of the external circuit. Thus, the transistor model is placed in an external circuit in which *all the dc voltage sources are replaced by short circuits*, as illustrated in Fig. 5.2*b*. The resistor R_B now goes from the base to signal ground (here also the emitter), while the load resistor R_C is connected between collector and signal ground. Capacitor C_1 is present to avoid having the signal source affect the dc operating point by providing a low-resistance path to ground. In a similar fashion, C_2 prevents a following amplifier stage or load from affecting V_{CE}. These capacitors are often called *dc-blocking capacitors* or *coupling capacitors*. At 500 rad/s, each should be large enough that its reactance is much less than adjacent resistance values. It may then be represented by a short circuit in the ac-equivalent circuit. We assume that r_d is large enough to be neglected.

The calculation of the voltage gain is now a straightforward circuit-analysis problem. If we indicate the parallel combination of R_B and r_π as $R_B \| r_\pi$, then voltage division gives us

$$v_\pi = v_s(t) \frac{R_B \| r_\pi}{R_B \| r_\pi + R_s}$$

The output signal voltage is

$$v_o(t) = -g_m v_\pi R_L$$

or

$$v_o(t) = -g_m R_L \frac{R_B \| r_\pi}{R_B \| r_\pi + R_s} v_s(t)$$

Thus

$$A_V = \frac{v_o(t)}{v_s(t)} = -g_m R_L \frac{R_B \| r_\pi}{R_B \| r_\pi + R_s} \tag{1}$$

Typically, we try to design the circuit so that $R_B \gg r_\pi$, thus avoiding any loss of signal in R_B. The load resistor R_C $(= R_L)$ is made as large as possible for the same reason.

D5.1 Calculate the voltage gain A_V for the amplifier of Fig. 5.2*a* if $g_m = 50$ m℧, $r_\pi = 2$ kΩ, $R_s = 100$ Ω, $R_L = 5$ kΩ, and (a) $R_B \gg r_\pi$, $r_d \gg R_L$; (b) $R_B = 10$ kΩ, $r_d \gg R_L$; (c) $R_B = 10$ kΩ, $r_d = 50$ kΩ.

Answers. −238; −236; −214

5.2 Bipolar transistor hybrid-π models: small-signal high-frequency

As the frequency increases, the voltage gain that we calculated at the lower frequencies begins to decrease as a result of the presence of internal capacitance and resistance in the bipolar transistor. A new model is needed that includes these effects; we will adopt the hybrid-π model shown in Fig. 5.3. One resistor, r_x, and two capacitors, C_μ and C_π, have been added to the low-frequency model.

The capacitor C_μ represents fairly well the collector-to-base capacitance of that reverse-biased junction; it is similar to the junction or depletion capacitance C_j of the reverse-biased diode. On the other hand, C_π is similar to the diffusion capacitance C_D present across a forward-biased diode, here represented by the emitter-base junction. We shall find that C_π is usually much larger than C_μ, a relationship that is also true for the diode capacitances. Typical values are $C_\mu = 2$ pF and $C_\pi = 200$ pF.

The resistance r_x is sometimes called the *base spreading resistance*, and it represents the resistive path between the external ohmic contact to the base (at B) and the active base region (at B'). An alternate symbol for r_x is $r_{bb'}$. A typical range of values for r_x is 20 to 200 Ω.

Other capacitors and resistors may be added to the model to provide greater accuracy, but their values are not usually given on most data sheets and they are difficult to determine experimentally.

D5.2 Let $r_x = 40\ \Omega$, $r_\pi = 2\ \text{k}\Omega$, $C_\pi = 100$ pF, $C_\mu = 2$ pF, and $g_m = 40$ m℧ for the model of Fig. 5.3. Assume that r_d is very large. Short-circuit the collector and emitter terminals and find the input admittance when ω equals (a) 0, (b) 20 Mrad/s. (c) If $Y_{in} = G_{in} + j\omega C_{in}$, what is C_{in} when $\omega =$ 20 Mrad/s?

Answers. 490 μ℧; 646 + j1948 μ℧; 97.4 pF

Fig. 5.3 The small-signal high-frequency model for the bipolar transistor contains three more elements than the low-frequency model: r_x, C_π, and C_μ.

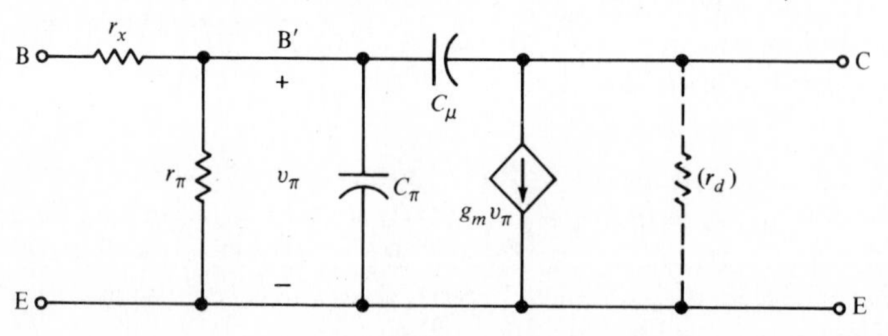

5.3 Low-frequency bipolar-model parameter values

The small-signal circuit models that were introduced in Section 5.2 contain three resistors, r_x, r_π, and r_d, two capacitors, C_π and C_μ, and one dependent-source parameter, g_m. The values of these six quantities depend on the particular operating point used for the transistor and on the temperature. In this section, we will try to discover the basic dependence of r_π and g_m on the dc operating point. We will also obtain a numerical value for r_d so that we can make an informed decision about whether or not to include it in our model. We assume room-temperature (25°C) operation at audio frequencies ($\omega < 10^4$ rad/s).

One quantity provided on transistor data sheets enters strongly into the relationships among the element values of the equivalent circuits. It is the small-signal short-circuit current gain,[1] β or h_{fe}:

$$\beta = h_{fe} = \left.\frac{i_c}{i_b}\right|_{v_{ce}=0 \text{ or } v_{CE}=\text{constant}} \tag{2}$$

where i_b and i_c are directed inward at the transistor terminals and the lower-case subscripts again identify the (small) signal components. The fact that $v_{ce} = 0$ or v_{CE} is constant simply indicates that the output terminals of the small-signal equivalent circuit are short-circuited. A large capacitor from collector to emitter in the amplifier circuit can provide this short circuit for the signal.

We shall indicate the low-frequency value of β as β_0. From the characteristics given for the 2N5088 transistor (in Appendix A), we see that h_{fe} ($= \beta_0$) may vary between 350 and 1400 at the operating point $I_C = 1$ mA, $V_{CE} = 5$ V. These values are given under "Dynamic Characteristics" and should not be confused with those for h_{FE} ($= \beta_{dc}$) given under "On Characteristics."

From the model in Fig. 5.1, we can see that

$$r_\pi = \left.\frac{v_{be}}{i_b}\right|_{v_{ce}=0} \tag{3}$$

For our model, it is not necessary to add the restriction $v_{ce} = 0$, but a more complex model might contain a dependent source in series with r_π, where the source is controlled by v_{ce}. From this same circuit of Fig. 5.1 we may also define g_m quite easily if we again short-circuit the output terminals:

$$g_m = \left.\frac{i_c}{v_{be}}\right|_{v_{ce}=0} \tag{4}$$

[1]The subscript-naming system for the h-parameters will be described in Section 5.6.

By comparing Eqs. (2), (3), and (4), we discover a simple, important relationship between r_π, g_m, and β_0:

$$g_m r_\pi = \beta_0 \tag{5}$$

We now need to find a value for either g_m or r_π, and then a value for β_0 will allow us to evaluate the remaining parameter. Consider Eq. (4) once again. This is simply the dynamic conductance (slope) of the i_C-vs.-v_{BE} characteristic for the pertinent value of v_{CE}; a similar conductance was found for the diode in Chapter 1. Up to this time we have not seen any plot of i_C vs. v_{BE}, but one can be obtained easily, since i_C is simply $\beta_{dc} i_B$ in the active region if I_{CEO} is negligible. Curves of i_C vs. v_{BE} are thus identical in form to the input characteristics i_B vs. v_{BE}, differing from them only in a change in the scale of the ordinate. Recall that these input characteristics are essentially identical once $|v_{CE}|$ is greater than about 1 V. One of these characteristics is given quite accurately by the diode equation [Eq. (1)] of Section 1.1, modified by selecting appropriate subscripts for an *npn* transistor:

$$i_C = I_{EO}(e^{qv_{BE}/nkT} - 1) \doteq I_{EO}e^{qv_{BE}/nkT}$$

The slope is

$$g_m = \left.\frac{di_C}{dv_{BE}}\right|_{\text{OP}} = \frac{i_c}{v_{be}} = \frac{q}{nkT} I_C \qquad (v_{CE} \text{ constant})$$

Since I_C is positive for an *npn* device and negative for the *pnp* unit, while g_m itself is always positive,[2] we usually write

$$g_m = \frac{q}{nkT}|I_C| = \frac{1}{nV_T}|I_C| \tag{6}$$

For $n = 1$, $T = 25°\text{C}$, we find that

$$g_m = 38.92|I_C|$$

When $|I_C|$ is 1 mA, g_m is about 40 m℧. If $\beta_0 = 200$ when $|I_C| = 1$ mA, then $r_\pi = 5\ \text{k}\Omega$.

The functional dependence of g_m on the operating point is shown clearly by Eq. (6); it is directly proportional to the magnitude of the collector current throughout the active region.

Now let us consider β_0. As a first-order result, we might say that β_0 is constant; certainly it will not increase by a factor of five as g_m would if I_C changed by that amount. However, it may change by a significant amount, as indicated by the two curves shown in Fig. 5.4. The upper graph shows a typical variation of β_0 with collector current, while the

[2] $g_m = i_c/v_{be} = \Delta i_C/\Delta v_{BE}$, and an increase in v_{BE} (either less negative for *pnp* or more positive for *npn*) always leads to an increase in i_C (less negative for *pnp* or more positive for *npn*).

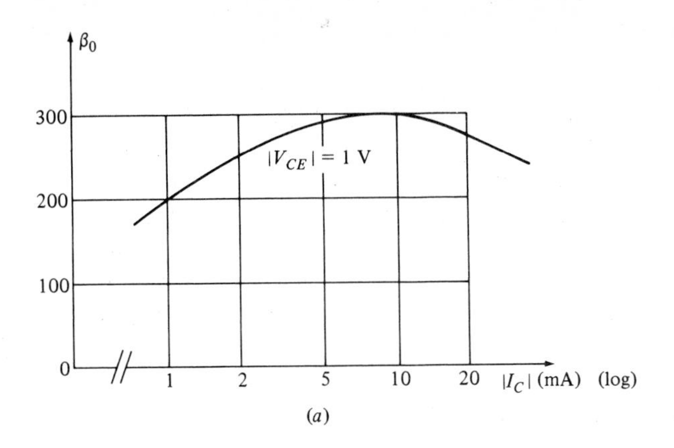

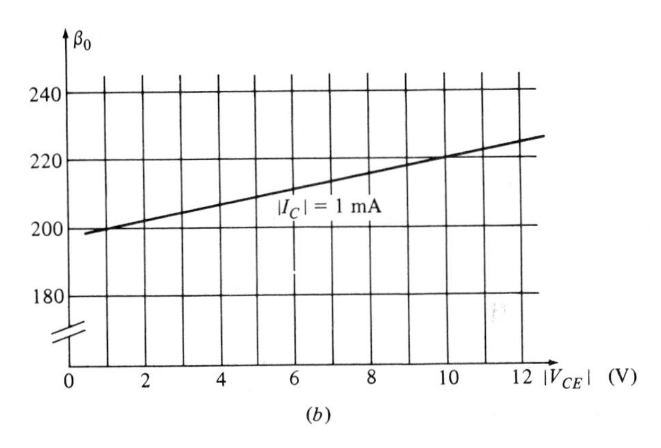

Fig. 5.4 (*a*) The variation of β_0 with $|I_C|$ is shown for $|V_{CE}| = 1$ V. (*b*) A curve of β_0 vs. $|V_{CE}|$ for $|I_C| = 1$ mA.

lower one illustrates the effect of changing V_{CE}. Both curves indicate that $\beta_0 = 200$ at $|I_C| = 1$ mA, $|V_{CE}| = 1$ V; they also show that β_0 increases from 200 to 250 as $|I_C|$ increases from 1 to 2 mA with $|V_{CE}| = 1$ V, and that β_0 increases from 200 to 208 as $|V_{CE}|$ increases from 1 to 4 V when $|I_C| = 1$ mA. Combining these results, we calculate β_0 at $|I_C| = 2$ mA, $|V_{CE}| = 4$ V as $250(208/200) = 260$.

We next consider r_d. To determine its value, we note from Fig. 5.1 that it is the resistance viewed from the output (C-E) terminals with no input signal present ($v_\pi = 0$). In our model this occurs when the input (base) is either short-circuited or open-circuited. The information commonly provided by the manufacturers is listed as the open-circuit common-emitter output admittance h_{oe}:

$$h_{oe} = \left.\frac{i_c}{v_{ce}}\right|_{i_b=0}$$

Therefore,

$$r_d = \left. \frac{v_{ce}}{i_c} \right|_{i_b=0} = \frac{1}{h_{oe}} \tag{7}$$

For a 2N5088 transistor, the range of h_{oe} is from 14 to 40 μ℧ (at $I_C = 1$ mA, $V_{CE} = 10$ V). Thus, r_d lies between 25 kΩ and 71 kΩ.

If we use graphic techniques to find a value for r_d, we see from Eq. (7) that it is the reciprocal of the slope of the output characteristic (I_C vs. V_{CE}) at the operating point. This is illustrated in Fig. 5.5. Since the tangent line is almost horizontal, it intersects the ordinate at approximately the value I_C has at the operating point. If the straight line is continued to the left, it intersects the abscissa at $V_{CE} = -V_A$, where V_A is a positive quantity known as the *Early voltage*. This intercept is approximately the same for any of the output characteristics, and therefore r_d is given by the ratio of the two intercepts:

$$r_d = \frac{V_A}{|I_C|} \tag{8}$$

Typical values of the Early voltage for integrated circuit devices and discrete transistors lie between 25 and 300 V.

The dc model developed in Chapter 3 (Fig. 3.4) includes a resistance R_0 that is also the reciprocal of the slope of an I_C vs. V_{CE} curve, and thus $r_d = R_0$ when the transistor is operating in the active region.

To illustrate the determination of the element values for the low-frequency equivalent circuit and its use in finding the gain of an amplifier, consider the circuit and data provided in Fig. 5.6. We are asked to find the amplification obtainable from this common-emitter circuit. The input signal is only 1 μV in amplitude at a frequency of 1000 rad/s, and our problem therefore involves calculations with the small-signal low-frequency equivalent circuit. *However, to find appropriate values for* r_π *and* $\mathrm{g_m}$ *we need to determine the dc operating point.* Without drawing the dc-equivalent circuit, we may follow the procedures used in Section 4.1.

Fig. 5.5 In the active region, tangents to the output characteristics all intersect the V_{CE} axis near $V_{CE} = -V_A$. Then, $r_d = V_A/|I_C|_{\text{O.P.}}$.

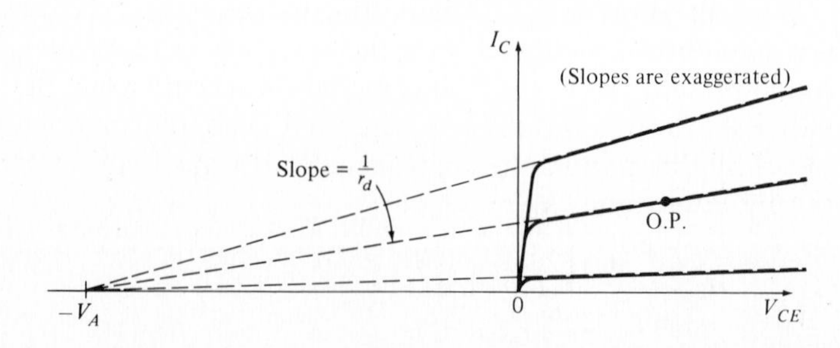

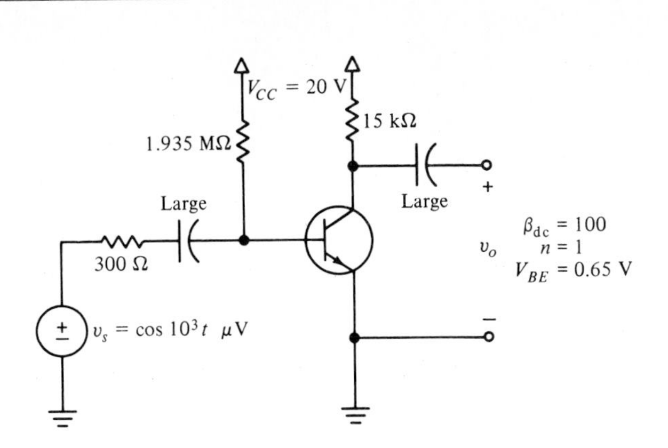

Fig. 5.6 An amplifier whose low-frequency gain is to be calculated.

The base current is

$$I_B = \frac{20 - 0.65}{1.935} = 10\ \mu\text{A}$$

This leads to

$$I_C = \beta_{dc} I_B = 100(10^{-5}) = 1\text{ mA}$$

Finally, we pin down the operating point by calculating

$$V_{CE} = 20 - 15(1) = 5\text{ V}$$

Now that we know I_C and V_{CE}, our next move is to the data sheets for the particular transistor used. Let us assume that we find $\beta_0 = 110$. Using the value of the collector current, we find g_m:

$$g_m = 38.92 \times 10^{-3} = 38.9\text{ m℧}$$

for $T = 25°\text{C}$. Then

$$r_\pi = \frac{\beta_0}{g_m} = \frac{110}{38.92} \times 10^3 = 2.83\text{ k}\Omega$$

We may now find the signal voltage gain:

$$A_V = \frac{v_o}{v_s} = -g_m R_L \frac{R_B \| r_\pi}{R_B \| r_\pi + R_s}$$

$$= -38.9 \times 10^{-3} \times 15 \times 10^3 \frac{2.83}{2.83 + 0.3} = -528$$

This short example illustrates the typical three-step procedure used in calculating the gain of an amplifier:

1. Determine the operating point so that values for the small-signal parameters may be calculated.

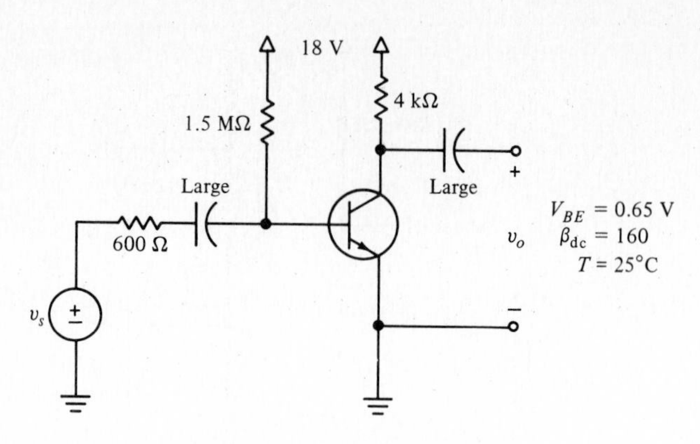

Fig. 5.7 See Problem D5.4.

2. With an appropriate model replacing the transistor, construct the ac or signal-equivalent circuit.
3. Find the gain by standard circuit analysis techniques. It can often be written by inspection from the equivalent circuit.

D5.3 If $V_A = 80$ V and the data of Fig. 5.4 apply to a certain *npn* transistor, determine values for r_π and r_d if (I_C, V_{CE}) equals (a) (5 mA, 1 V), (b) (1 mA, 5 V), (c) (2 mA, 10 V).

Answers. 1.5, 16; 5.4, 80; 3.5, 40 kΩ.

D5.4 Find the signal gain v_o/v_s for the amplifier shown in Fig. 5.7 if $\beta_0 = 180$ and (a) $n = 1$, (b) $n = 1.5$, (c) $n = 1.5$ and $V_A = 75$ V.

Answers. − 232; − 165.5; − 150.7

5.4 High-frequency bipolar-model parameter values

The high-frequency small-signal model for the bipolar transistor (Fig. 5.3) contains three resistors, two capacitors, and one dependent source. The values of these six elements must be known before any high-frequency calculations can be made from the model. The determination of suitable values for these parameters at a specified operating point is the subject of this section.

Half our problem is solved very simply, because the elements that appear in the low-frequency model maintain their same values in the high-frequency model; these are r_π, r_d, and g_m. This is one of the reasons that the hybrid-π model is so convenient.

The remaining three elements are C_μ, C_π, and the base spreading resistance r_x. We first consider the capacitance C_μ. This capacitor between the collector and the internal base region is essentially the same as the depletion capacitance of the reverse-biased collector-base junction. This depletion capacitance is customarily designated by C_{cb}, the collector-base capacitance, or C_{ob}, the common-base output capacitance. It is measured between the collector and base terminals with the emitter open-circuited. We may use results that we developed earlier for the reverse-biased diode in Chapter 1:

$$C_j = \frac{C_{j0}}{(1 - V_D/V_{bi})^N}$$

With nomenclature more suitable for the collector and base terminals of an *npn* transistor, we have

$$C_\mu = C_{ob} = C_{cb} = \frac{C_{\mu 0}}{(1 + V_{CB}/V_{bi})^N} \tag{9}$$

We recall that N ranges from ½ to ⅓, depending on the type of junction, while $|V_{bi}|$ is typically 0.8 V for silicon transistors. The value of C_μ obviously depends on V_{CB} (and V_{CE}), and $C_{\mu 0}$ is simply the limit approached by C_μ as V_{CB} approaches zero. Data may be given for C_μ, C_{ob}, or C_{cb} at one particular value of V_{CB}; we then use Eq. (9) to calculate $C_{\mu 0}$ and then the value of C_μ at any desired operating point.

For a *pnp* transistor, V_{bi} is still taken as a positive quantity, and it is therefore necessary to use $|V_{CB}|$ in place of V_{CB} in Eq. (9).

As an example of the calculation of C_μ, let us find the maximum value to be expected for C_μ with a 2N5377 transistor at $I_C = 2$ mA, $V_{CE} = 5$ V. Referring to the data sheet in Appendix A, we find that C_{cb}, the collector-base capacitance, at $V_{CB} = 10$ V, $f = 1$ MHz, and $I_E = 0$ has a maximum value of 8 pF. Using Eq. (9), we find $C_{\mu 0}$ first:

$$8 = \frac{C_{\mu 0}}{(1 + 10/V_{bi})^N}$$

We assume that V_{bi} is 0.8 V and N is ½, obtaining

$$C_{\mu 0} = 8\sqrt{1 + 12.5} = 29.4 \text{ pF}$$

At $V_{CE} = 5$ V, we have

$$V_{CB} = V_{CE} - V_{BE} = 5 - 0.65 = 4.35 \text{ V}$$

and therefore

$$C_\mu = \frac{29.4}{\sqrt{1 + 4.35/0.8}} = 11.6 \text{ pF}$$

when $V_{CE} = 5$ V. If we had assumed that $N = 1/3$, the result would have been 10.2 pF, not an important difference.

One of the most difficult parameters to determine for the high-frequency hybrid-π model is r_x or $r_{bb'}$, the base spreading resistance. It is not easy to make a direct experimental measurement of r_x; its indirect determination has been the subject of numerous investigations. The value typically ranges between 20 and 200 Ω, and it decreases somewhat as the base current I_B increases. It is less for transistors intended for high-frequency operation, as they are carefully designed to have a small value of r_x.

Occasionally, we find r_x or $r_{bb'}$ listed directly on the data sheet for a transistor designed specifically for high-frequency applications. For low-frequency devices, however, more often than not no specific data are available, and the value of r_x must be estimated.

The last element in the high-frequency hybrid-π model that we have to evaluate is C_π, the diffusion capacitance associated with the forward-biased emitter-base junction. It is similar to the diffusion capacitance of the forward-biased diode, and therefore increases as the emitter current increases. The value of C_π is found indirectly by measuring the variation of β with frequency.

In doing this, we shall begin working in the frequency domain with sinusoidal signals that are functions either of $j\omega$ or of the complex frequency s. The voltages and currents are phasors, which we represent by capital letters with lower-case subscripts. Thus $I_e = 20 + j10\ \mu$A signifies the time-domain emitter current $i_e = 22.4 \cos(\omega t + 26.6°)\ \mu$A, a signal component. If the dc component is 4 mA, then we would write

$$I_E = 4 \text{ mA} \qquad \text{(dc)}$$

$$i_E = 4 + 0.0224 \cos(\omega t + 26.6°) \text{ mA} \qquad \text{(total)}$$

$$\left.\begin{aligned} i_e &= 22.4 \cos(\omega t + 26.6°)\ \mu\text{A} \\ I_e &= 20 + j10 = 22.4\underline{/26.6°}\ \mu\text{A} \end{aligned}\right\} \qquad \text{(ac)}$$

Figure 5.8 illustrates the procedure we shall use to obtain $|\beta|$ vs. ω. The output terminals are short-circuited, since β is defined as the ratio of i_c to i_b with $v_{ce} = 0$ (or I_c to I_b with $V_{ce} = 0$); this places C_μ, C_π, and r_π in parallel and enables us to calculate V_π easily:

$$V_\pi = \frac{I_b}{1/r_\pi + j\omega(C_\pi + C_\mu)}$$

or

$$V_\pi = \frac{r_\pi I_b}{1 + j\omega(C_\pi + C_\mu)r_\pi}$$

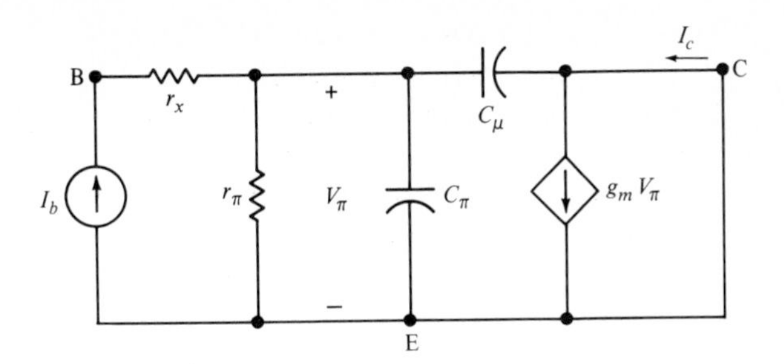

Fig. 5.8 With the output short-circuited, the frequency variation of $\beta = I_c/I_b$ may be determined.

Now, the collector current is the sum of the dependent-source current and the current through C_μ:

$$I_c = g_m V_\pi - j\omega C_\mu V_\pi = (g_m - j\omega C_\mu) V_\pi$$

The current through C_μ is so much smaller than that through the dependent source that it can be neglected at any frequency of interest. Therefore

$$I_c \doteq g_m V_\pi = \frac{g_m r_\pi I_b}{1 + j\omega(C_\pi + C_\mu) r_\pi}$$

or, since $\beta = I_c/I_b$ and $\beta_0 = g_m r_\pi$,

$$\beta = \frac{\beta_0}{1 + j\omega(C_\pi + C_\mu) r_\pi} \tag{10}$$

Thus, β is a complex quantity whose magnitude is

$$|\beta| = \frac{\beta_0}{\sqrt{1 + \omega^2 (C_\pi + C_\mu)^2 r_\pi^{\,2}}} \tag{11}$$

As $\omega \to 0$, we see that $|\beta|$ and $\beta \to \beta_0$, as they should. When ω becomes very large, however, $|\beta|$ decreases almost inversely with frequency.

There are two important frequencies that we define in terms of the performance of $|\beta|$ with frequency. The first is the frequency at which $|\beta|$ is 0.707 or $1/\sqrt{2}$ times its low-frequency value β_0. We call this frequency ω_β, and it occurs when the denominator of Eq. (10) is $1 + j1$. We have

$$\omega_\beta (C_\pi + C_\mu) r_\pi = 1$$

so that

$$\omega_\beta = \frac{1}{(C_\pi + C_\mu) r_\pi} = 2\pi f_\beta \tag{12}$$

This frequency, expressed in either radians per second or hertz, is called the *beta cutoff frequency*. A typical value for f_β is 1 MHz.

The value of C_π may be given in terms of ω_β:

$$C_\pi = \frac{1}{\omega_\beta r_\pi} - C_\mu \tag{13}$$

Thus if we know the beta cutoff frequency, or can determine it from the data given, we can find a value for C_π.

Another special frequency is more apt to be available, however—that at which the value of $|\beta|$ drops to unity. It is designated ω_T and is called the *gain-bandwidth product*, for a reason that we shall discover in succeeding chapters. A bipolar transistor designed for use at high frequencies probably has a value of 500 MHz or greater for f_T. From Eq. (11), we see that as ω increases beyond ω_β, $|\beta|$ will be equal to unity when

$$\omega_T(C_\pi + C_\mu)r_\pi \doteq \beta_0 = g_m r_\pi$$

or

$$\omega_T = \frac{\beta_0}{(C_\pi + C_\mu)r_\pi} = \frac{g_m}{C_\pi + C_\mu} \qquad (\omega_T \gg \omega_\beta) \tag{14}$$

Solving Eq. (14) for C_π, we have a relationship that we shall often use to obtain a value for C_π:

$$C_\pi = \frac{g_m}{\omega_T} - C_\mu \tag{15}$$

We also note from a comparison of Eq. (14) with Eq. (12),

$$\omega_T = \beta_0 \omega_\beta \tag{16}$$

Figure 5.9 illustrates one form in which manufacturers provide data on f_T at various operating points. At constant I_C, note that an increase in V_{CE} yields a larger value for f_T and hence a smaller value of C_π. At a constant value of V_{CE}, say 2 V, f_T is largest at about 2 mA, decreasing for both larger and smaller values of I_C.

Some data sheets do not give the value of either ω_β or ω_T. Instead, $|\beta|$ is given at some frequency sufficiently greater than ω_β that we may assume that $|\beta|$ is inversely proportional to frequency. On logarithmic scales, this inverse relationship shows up as a straight line, as is illustrated in Fig. 5.10. One of the problems at the end of the chapter suggests that Eq. (11) may be used to show that this proportionality between $|\beta|$ and $1/\omega$ is in error by less than 5.5% when $\omega > 3\omega_\beta$. At such frequencies, Eq. (11) then leads to

$$|\beta| = \frac{\beta_0}{\omega(C_\pi + C_\mu)r_\pi} \qquad (\omega > 3\omega_\beta)$$

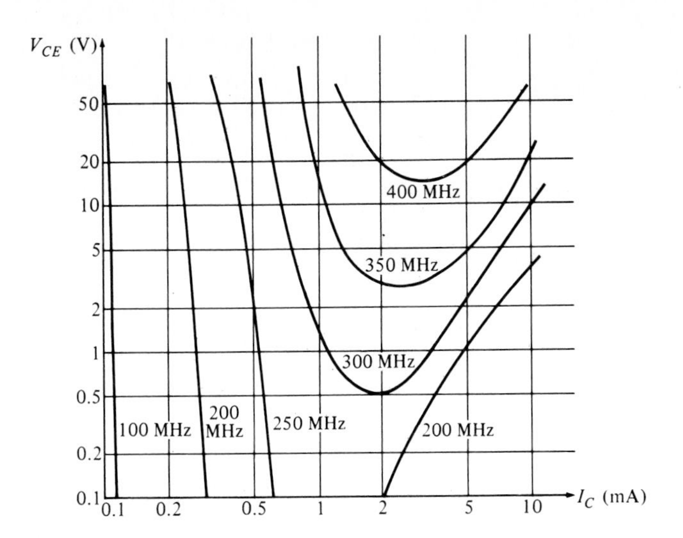

Fig. 5.9 Loci of constant values for f_T are shown as functions of I_C and V_{CE} for a typical *npn* transistor.

Comparing this with Eq. (13), we have

$$\omega|\beta| = \omega_T \qquad (\omega > 3\omega_\beta) \tag{17}$$

Thus, a knowledge of the value of $|\beta|$ at some frequency ω that is at least three times greater than ω_β automatically provides us with a value for ω_T as their product.

As an example illustrating the use of the gain-bandwidth product to find C_π, let us estimate the maximum value of C_π that might be expected for a 2N5377 transistor operating at $I_C = 0.5$ mA, $V_{CE} = 5$ V. From the data sheet, we note that $|\beta|$ $(= |h_{fe}|)$ lies between 3 and 15 at $I_C = 0.5$

Fig. 5.10 A plot of log $|\beta|$ vs. log ω shows that $|\beta| \doteq \beta_0$ for $\omega \ll \omega_\beta$, and $|\beta| = 1$ at $\omega = \omega_T$.

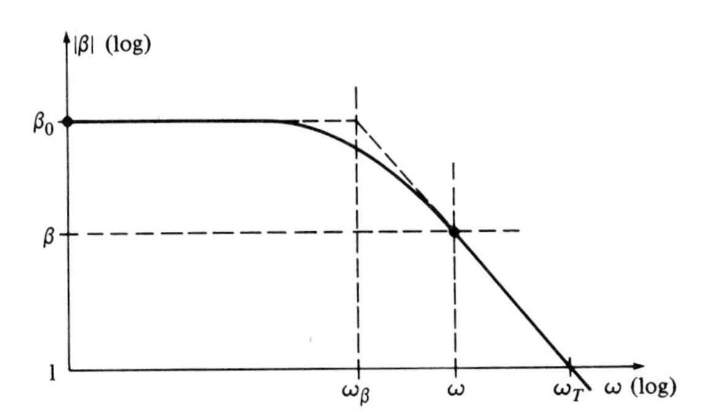

mA, $V_{CE} = 5$ V, and $f = 10$ MHz. Thus, from Eq. (17), f_T will lie between 30 and 150 MHz at this operating point. Since we are asked for the maximum value of C_π and Eq. (15) shows that C_π increases as ω_T decreases, we use the minimum gain-bandwidth product,

$$\omega_T = 2\pi f_T = 2\pi(30 \times 10^6) = 6\pi 10^7 \text{ rad/s}$$

We also need a value for g_m. Using Eq. (6) and assuming that $n = 1$ and $T = 25°$C, we have

$$g_m = 38.9 I_C = 19.45 \text{ m℧}$$

The sum of C_π and C_μ is therefore

$$C_\pi + C_\mu = \frac{g_m}{\omega_T} = \frac{19.45 \times 10^{-3}}{6\pi 10^7} = 103 \text{ pF}$$

The value of C_μ given on the data sheet is 8 pF at $V_{CB} = 10$ V; an earlier example in this section showed that a reasonable value at $V_{CE} = 5$ V is about 11 pF. Hence

$$C_\pi = 103 - 11 = 92 \text{ pF}$$

It is also possible to calculate the β-cutoff frequency from Eq. (16) if we select the minimum value of β_0, which is consistent with our earlier assumptions. From the specifications, we have $h_{fe(\min)} = 100$, and

$$\omega_\beta = \frac{\omega_T}{\beta_0} = \frac{6\pi 10^7}{100} = 6\pi 10^5 \text{ rad/s}$$

or $f_\beta = 0.3$ MHz. This frequency is useful in two ways. First, when $f > 3f_\beta$ or $f > 0.9$ MHz, $\omega|\beta| = \omega_T$ with good accuracy. Calculating ω_T from the data provided at 10 MHz is thus valid. Second, high-frequency effects are quite noticeable at ω_β, and therefore the high-frequency equivalent circuit should be used whenever $\omega > \omega_\beta/3$, or for frequencies greater than about 100 kHz for the 2N5377.

Before concluding this section, let us continue with this example by completing the high-frequency hybrid-π equivalent circuit. We have g_m, C_π, and C_μ, and we still need values for the three resistances, r_π, r_d, and r_x. Knowing β_0 and g_m, we find r_π:

$$r_\pi = \frac{\beta_0}{g_m} = \frac{100}{0.01945} = 5.14 \text{ k}\Omega$$

The output resistance r_d is probably large enough to neglect, particularly in high-frequency circuits where the impedance level of the external elements is apt to be low. However, let us check this assumption. We have seen that r_d is the reciprocal of h_{oe}, a commonly given bit of data, so we inspect the 2N5377 data sheet, finding that $h_{ob(\max)} = 0.2\ \mu\mho$ at $I_C = 1$ mA but no value is given for h_{oe}. These h-parameters are related by

$$h_{oe} = (h_{fe} + 1)h_{ob}$$

Therefore

$$h_{oe} = (100 + 1)0.2 \doteq 20\ \mu\mho$$

and

$$r_d = \frac{1}{h_{oe}} = 50\ \text{k}\Omega \qquad (I_C = 1\ \text{mA})$$

Using Eq. (8), we see that $V_A = 50$ V, and therefore $r_d = 100$ kΩ at $I_C = 0.5$ mA.

Finally, we find no data for r_x. Since this is not a high-frequency transistor, we should not select too small a value for r_x; we arbitrarily pick $r_x = 50\ \Omega$. The resultant small-signal high-frequency hybrid-π equivalent circuit for a 2N5377 transistor operating with $I_C = 0.5$ mA, $V_{CE} = 5$ V is shown in Fig. 5.11. The element values have been rounded off, an indication that we have used an approximation here and there, as well as an occasional guess.

The operating point for which this model applies happens to be the only point at which the data sheet provides a high-frequency value for β. We might well ask, how does the equivalent circuit differ for another operating point, say $I_C = 2$ mA, $V_{CE} = 8$ V?

The transconductance g_m is proportional to $|I_C|$, so a new value is easily calculated, $g_m = 38.9 \times 2 \times 10^{-3} = 77.8$ m℧.

The resistance r_π is found from g_m and β_0, and β_0 is obtained at the new operating point from curves provided on the data sheet, if any, or by assuming that β_0 is constant in lieu of other information. No curves are given for the 2N5377, so we maintain $\beta_0 = 100$, and therefore $r_\pi = 100/0.0778 = 1.285$ kΩ.

Fig. 5.11 A suitable high-frequency hybrid-π model for a 2N5377 transistor operating at $I_C = 0.5$ mA, $V_{CE} = 5$ V.

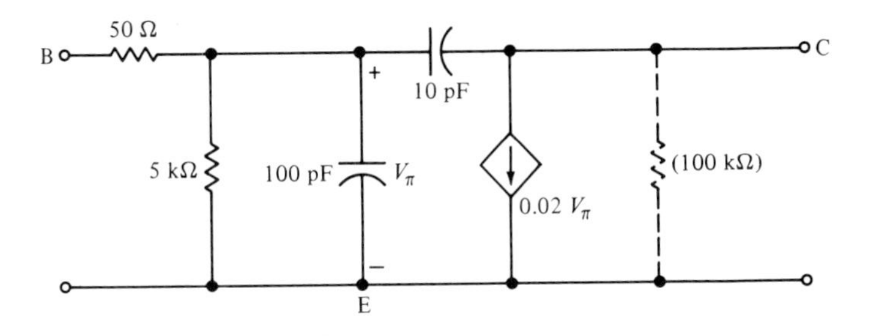

At $I_C = 2$ mA, the resistance r_d decreases to 25 kΩ, since it is inversely proportional to the collector current, and the current has doubled.

No data were given for r_x, and the only noticeable effect of a change in operating point is a slight decrease for large $|I_C|$. Since 2 mA is not very large, we keep $r_x = 50$ Ω.

The two capacitances remain. The collector-to-base capacitance C_μ varies with V_{CE}, and we have already determined $C_{\mu 0}$. Therefore

$$C_\mu = \frac{29.4}{\sqrt{1 + (8 - 0.65)/0.8}} = 8.53 \text{ pF}$$

This represents a slight decrease, in this particular case hardly worth changing in our rounded-off equivalent circuit.

Finally, we turn to C_π. Using Eq. (15), we see that it depends on g_m, ω_T, and C_μ, so we need to find ω_T at the new operating point. Curves are provided on some data sheets, showing f_T vs. $|I_C|$ for several values of V_{CE}, such as those given in Fig. 5.9. For small currents, f_T is proportional to $|I_C|$; for larger currents, the rate of increase is less than the first power, and eventually decreases. The gain-bandwidth product also increases slightly with V_{CE}. Since I_C is increasing from 0.5 to 2 mA while V_{CE} goes from 5 to 8 V, we might well let ω_T increase from $6\pi 10^7$ to $18\pi 10^7$ rad/s, a factor of three. Then

$$C_\pi = \frac{g_m}{\omega_T} - C_\mu = \frac{77.8 \times 10^{-3}}{18\pi 10^7} - 8.5 \times 10^{-12}$$

$$= 137.3 - 8.5 = 128.8 \text{ pF}$$

Thus we might wish to make several changes in the high-frequency equivalent circuit of Fig. 5.11 as the operating point moves from 0.5 mA, 5 V, to 2 mA, 8 V: decreasing r_π from 5 to 1.25 kΩ, increasing C_π from 100 to 125 pF, and increasing g_m from 0.02 to 0.08 ℧.

D5.5 Calculate values of C_π for a typical 2N5088 with V_{CE} equal to (a) 8 V; (b) 5 V; (c) 2.5 V. To check answers given below, let $N = 0.5$, $V_{bi} = 0.8$ V, and $V_{BE} = 0.65$ V. (d) Find the three corresponding values from the curves given on the data sheet in Appendix A and compare them with the values calculated above.

Answers. 1.52; 1.91; 2.66; 1.7, 2.0, 2.65 pF

D5.6 If $\beta_0 = 100$ and $f_\beta = 0.1$ MHz, calculate $|\beta|$ at (a) 10 MHz, (b) 1 MHz, (c) 0.2 MHz, (d) 0.01 MHz.

Answers. 1.000; 10 (9.95); 44.7; 100 (99.5)

D5.7 Use Figs. 11 and 12 of the data sheets in Appendix A to find a typical value for C_π at 25°C for a 2N5088 transistor at the operating point (a) $I_C = 0.4$ mA, $V_{CE} = 5$ V, (b) 5 mA, 5 V, (c) 0.4 mA, 15 V, (d) 5 mA, 15 V.

Answers. 15.1; 52.3; 14.3; 45.7 pF

5.5 The effect of temperature on bipolar-model parameters

In designing bias circuits to provide a specified operating point, the two problems that caused us the most trouble were the variation present among transistors of the same type and the effect of changing temperature. These same two problems exist for the signal models and may be treated in a similar manner by proper modeling and careful circuit design. In this section we shall discuss the variation with temperature of the six parameters appearing in the high-frequency hybrid-π model. The effect of unit-to-unit variability is considered as a worst-case design problem in the next chapter.

We can use Eq. (6) to calculate the transconductance, g_m, of a bipolar transistor with good accuracy:

$$g_m = \frac{q}{nkT}|I_C|$$

It thus varies inversely with temperature for a fixed collector current, and we may write

$$g_m \propto T^{-1}$$

For bipolar transistors, β_0 increases with temperature. The form of the variation is approximately as T^a, where T is the absolute temperature and a may be as large as 2.5 for a silicon *npn* transistor. A value for β_0 is most often specified at 25°C, and additional data at other temperatures cannot always be found. In lieu of other information, a square-law relationship between β_0 and T is a reasonable assumption. Thus

$$\beta_0 \propto T^2 \qquad \text{(for Si)}$$

Some data sheets list β_0 at 25°C and at 125°C. If it is assumed that $\beta_0 \propto T^a$, then this provides enough information to calculate β_0 at any temperature. For example, if $\beta_0 = 200$ at 25°C and 340 at 125°C, then we can obtain an equality by taking the ratio of two proportionalities:

$$\frac{\beta_0(T_1)}{\beta_0(T_2)} = \frac{K(T_1)^a}{K(T_2)^a} = \left(\frac{T_1}{T_2}\right)^a \tag{18}$$

For our example,

$$\frac{200}{340} = \left(\frac{273.16 + 25}{273.16 + 125}\right)^a$$

Thus, $0.588 = 0.749^a$, and we may take natural logarithms to find

$$\ln 0.588 = a(\ln 0.749) \qquad \text{or} \qquad a = 1.835$$

Thus we have determined the proportionality for this particular transistor to be $\beta_0 \propto T^{1.835}$; since $\beta_0 = 200$ at 25°C, $\beta_0(T) = 200(T/298.16)^{1.835}$. Thus at 80°C, this transistor has a predicted β_0 of $200(353.16/298.16)^{1.835} = 273$.

The effect of temperature on r_π is discovered by expressing it in terms of g_m and β_0:

$$r_\pi = \frac{\beta_0}{g_m} = \frac{nkT\beta_0}{q|I_C|}$$

Assuming that I_C is held constant, we therefore see the direct effect of T as a linear factor in the numerator and an indirect relationship through β_0. If we make the simplest assumption that $\beta_0 \propto T^2$ for silicon transistors, then

$$r_\pi \propto T^3 \qquad (\text{for } \beta_0 \propto T^2 \text{ for Si})$$

Under the more general assumption that $\beta_0 \propto T^a$, we would obtain other powers of T. For the particular example used above in which $\beta_0 \propto T^{1.835}$, we see that $r_\pi \propto T^{1.835+1}$, or $r_\pi \propto T^{2.835}$.

The collector-to-base output capacitance C_μ is almost independent of temperature. Any change that arises comes from a decrease in V_{bi} as T increases; this effect is usually negligible.

The diffusion-related capacitance C_π is given approximately by the ratio of g_m to ω_T. The transconductance varies inversely with temperature, and the gain-bandwidth product decreases with temperature, although

Fig. 5.12 See Problem D5.8.

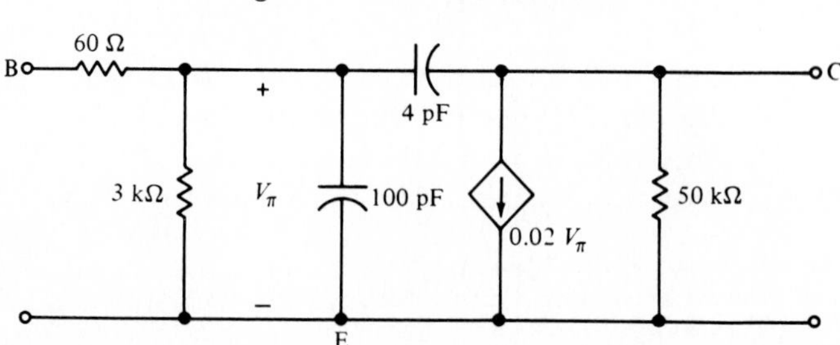

not in the same way for silicon and germanium units. The net result is that C_π increases as about the square root of temperature for silicon transistors:

$$C_\pi \propto T^{1/2} \qquad \text{(for Si)}$$

We find that the base spreading resistance increases as temperature increases and the output resistance r_d decreases as the temperature rises.

The information that we have accumulated on the effect of the operating point and temperature on the hybrid-π element values is collected in Table 5.1.

D5.8 The hybrid-π model of an *npn* silicon transistor is shown in Fig. 5.12 for a temperature of 25°C. If it is known that $\beta_0 = 120$ at 125°C, determine suitable element values at a temperature of −35°C.

Answers. 50 Ω; 1.4 kΩ; 90 pF; 4 pF; 25 m℧; 60 kΩ

5.6 Bipolar transistor *h*-parameter model

The hybrid-π model is applicable to both bipolar transistors and FETs at any frequency from near dc to the UHF region. Some data sheets, however, list values of the *h*-parameters for bipolar transistors at low frequencies. The *h* stands for "hybrid," but because of possible confusion between hybrid parameters and hybrid-π parameters, we will use only the single letter, saying "aitch" quietly to ourselves.[3]

The reason for using *h*-parameters is the convenience with which they may be measured at audio frequencies, typically 1 kHz. They provide an accurate model if the frequency is low enough that the interelectrode capacitances may be neglected.

Figure 5.13 shows a general linear two-port network with the signal currents I_1 and I_2 entering the input and output ports respectively, and the signal voltages V_1 and V_2 with positive references at the terminals where the currents enter. Two equations may be written expressing any two of these variables in terms of the remaining two, so there are thus six different ways of naming dependent and independent variables. One method leads to the *y*-parameters, which we shall study in Section 5.8, and another to the *h*-parameters, defined by these equations:

$$V_1 = h_{11}I_1 + h_{12}V_2 \qquad (19a)$$

$$I_2 = h_{21}I_1 + h_{22}V_2 \qquad (19b)$$

[3] It looks better if the lips do not move.

Table 5.1 Hybrid-π Element Values for Bipolar Transistors

Parameter	Basic formula	Typical variation with T (K), $\lvert I_C \rvert$, and $\lvert V_{CE} \rvert$			Typical range
		T^a, a is	$\lvert I_C \rvert^a$, a is	$\lvert V_{CE} \rvert^a$, a is	
g_m	$\dfrac{q \lvert I_C \rvert}{nkT}$	$= -1$	$= +1$	0	1–500 m℧
β_0	$\left.\dfrac{I_c}{I_b}\right\rvert_{V_{ce}=0}$	$\doteq +2$ (Si)	>0, small $\lvert I_C \rvert$ <0, large $\lvert I_C \rvert$ or assume $= 0$	>0	50–1500
r_π	$\dfrac{\beta_0}{g_m}$	$\doteq +3$ (Si)	$= -1$	>0	0.1–75 kΩ
C_μ	$\dfrac{C_{\mu 0}}{(1 + \lvert V_{CB}/V_{bi} \rvert)^N}$	0	0	<0	0.1–20 pF
ω_T	$\beta_0 \omega_\beta = \lvert \beta \rvert \omega$ for $\omega > 3\omega_\beta$	$= -3/2$ (Si)	$\doteq 1$, small $\lvert I_C \rvert$ <1, large $\lvert I_C \rvert$	>0	10–5000 Mrad/s
C_π	$\dfrac{g_m}{\omega_T} - C_\mu$	$= 1/2$ (Si)	<1 and >0	<0	15–500 pF
r_x ($r_{bb'}$)	$\doteq 50\ \Omega$	>0	0 <0, large $\lvert I_C \rvert$	0	10–500 Ω
r_d	$\dfrac{1}{h_{oe}} = \dfrac{V_A}{\lvert I_C \rvert}$	<0	-1	>0, small $\lvert I_C \rvert$ <0, large $\lvert I_C \rvert$	20–100 kΩ

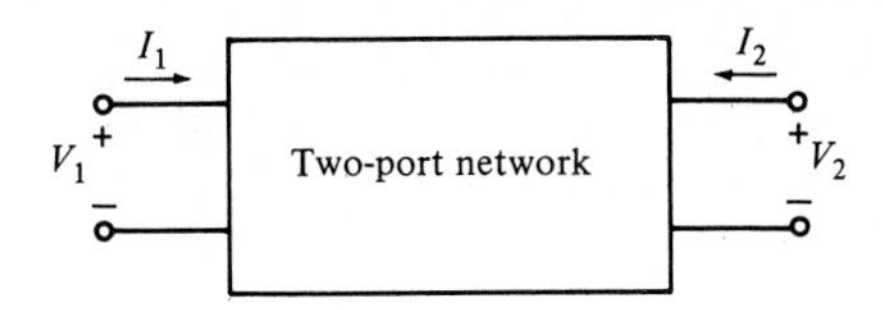

Fig. 5.13 A general linear two-port network is shown with the traditional reference polarities for the input and output currents and voltages.

From an inspection of these equations, each h-parameter may be expressed as a ratio of two quantities with either an open circuit at the input ($I_1 = 0$) or a short circuit at the output ($V_2 = 0$).

Equations (19*a*) and (19*b*) may be used term by term to form an equivalent circuit, as shown in Fig. 5.14*a*. When it is applied to a bipolar transistor, the terminology is modified to indicate which of the three terminals is common to both the input and output. For the common-emitter configuration, Eqs. (19*a*) and (19*b*) become

$$V_{be} = h_{ie}I_b + h_{re}V_{ce} \tag{20a}$$

$$I_c = h_{fe}I_b + h_{oe}V_{ce} \tag{20b}$$

Fig. 5.14 (*a*) An equivalent circuit for the h-parameters as defined for a general two-port network. (*b*) The low-frequency h-parameter model for a bipolar transistor connected in the common-emitter configuration.

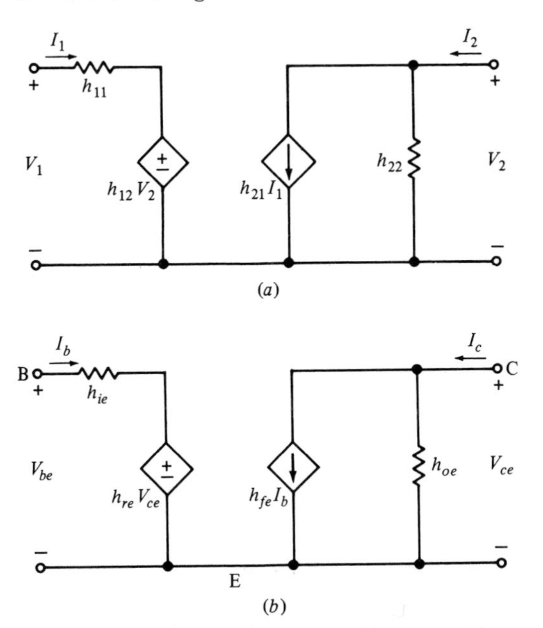

The subscripts are descriptive in that they refer to input, reverse, forward, and output, as well as identifying which terminal is common.

These equations lead to the h-parameter model for the bipolar transistor, Fig. 5.14b. Compare Eqs. (20a) and (20b) to a corresponding set written for the low-frequency hybrid-π model:

$$V_{be} = r_\pi I_b = h_{ie} I_b + h_{fe} V_{ce} \tag{21a}$$

Fig. 5.15 The data sheets for the 2N5088 Motorola transistor include curves for h_{ie}, h_{re}, h_{fe}, and h_{oe} vs. I_C (Figs. 3, 4, 5, and 6 respectively) and one curve showing the variation with V_{CE}. Note in Fig. 5 that h_{fe} for Unit 4 is less than the specified minimum (350 at 1 mA, 5 V), and, we trust, would never leave the factory.

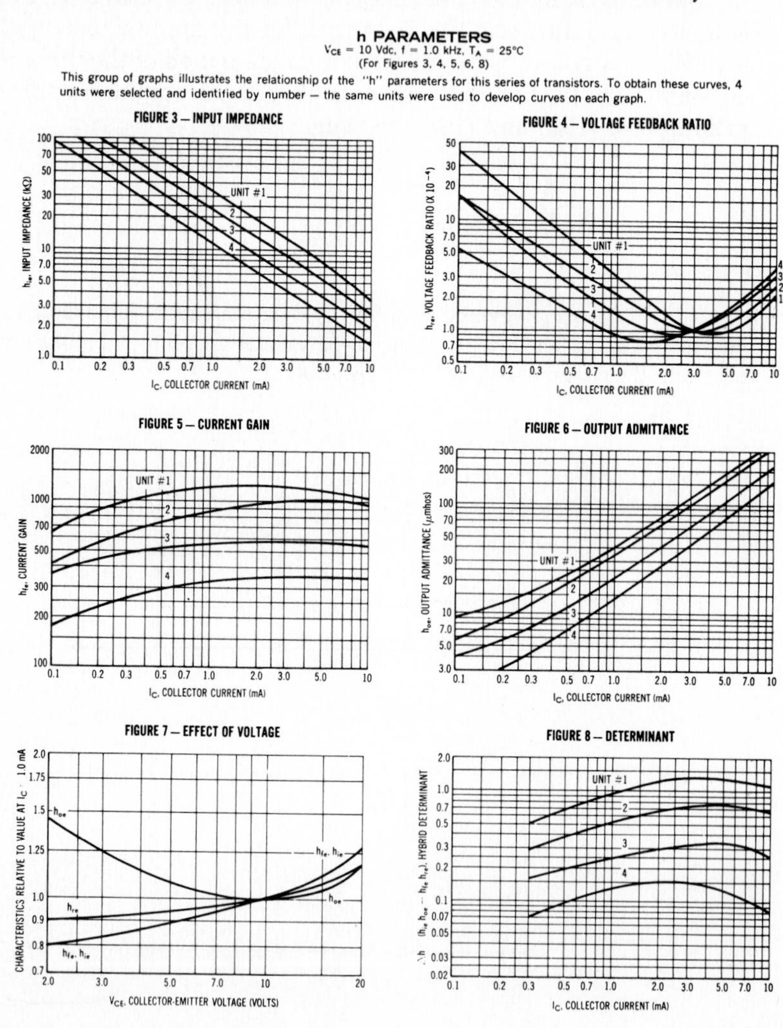

$$I_c = g_m V_{be} + \left(\frac{1}{r_d}\right) V_{ce} = g_m r_\pi I_b + \left(\frac{1}{r_d}\right) V_{ce} = h_{fe} I_b + h_{oe} V_{ce} \tag{21b}$$

It is easy to see that $h_{fe} = g_m r_\pi = \beta_0$, $h_{ie} = r_\pi$, $h_{oe} = 1/r_d$, and $h_{re} = 0$. Typical values of the h-parameters for a silicon transistor operating at $I_C = 1$ mA, $V_{CE} = 5$ V are $h_{ie} = 2.5$ kΩ, $h_{re} = 4 \times 10^{-5}$, $h_{fe} = 100$, and $h_{oe} = 20$ μ℧. These values change with operating point and temperature, just as the values do in the hybrid-π model. One page of the data sheets for the 2N5088 transistor is reproduced as Fig. 5.15; note that it gives curves (their Figs. 3, 4, 5, and 6) showing the variation of all four parameters with dc collector current, and one figure (their Fig. 7) illustrating the changes in the four parameters with V_{CE}. The data apply to an audio frequency of 1 kHz. The effect of temperature can be determined from the relationships between h- and hybrid-π parameters.

D5.9 (a) Let $h_{ie} = 2$ kΩ, $h_{re} = 0$, $h_{fe} = 80$, and $h_{oe} = 50$ μ℧ in Fig. 5.14*b*. Find the ratio V_{ce}/V_s if a source for which $V_s = 1$ mV and $R_s = 600$ Ω is connected to the input and a 4-kΩ load is connected to the output. (b) Find the ratio if $h_{re} = 5 \times 10^{-4}$, all other values remaining the same.

Answers. −102.6; −97.6

5.7 FET models: small-signal low-frequency

The field-effect transistor is also modeled well by a hybrid-π equivalent circuit. When interelectrode capacitances can be neglected, we may use the *small-signal low-frequency model*, which is particularly simple, as shown in Fig. 5.16. Only a single value, g_m, need be specified.

We may define g_m by inspecting Fig. 5.16. It is obvious that $i_d = g_m v_\pi$, but we usually specify that the output (drain-source) terminals are short-circuited since the high-frequency model may contain other elements in parallel with the dependent source. Thus

$$g_m = \left.\frac{i_d}{v_\pi}\right|_{v_{ds}=0 \text{ or } v_{DS}=\text{constant}}$$

The drain current i_d and gate-to-source voltage v_{gs} or v_π are signal quantities, and their ratio is once more given by the slope of the dc transfer characteristic at the specified operating point in the region beyond pinch-off. This type of analysis began with the ac-equivalent circuit of the diode in Chapter 1. Thus

$$g_m = \left.\frac{di_D}{dv_{GS}}\right|_{OP} \tag{22}$$

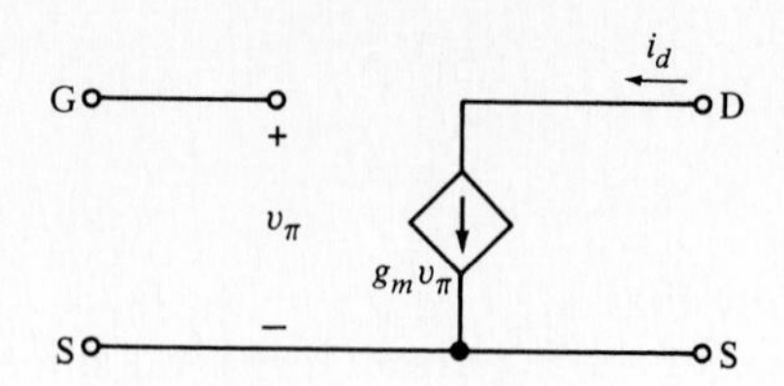

Fig. 5.16 The small-signal low-frequency hybrid-π model for an n-channel or p-channel JFET, IGFET, or MOSFET has a very simple form.

For the n-channel JFET, we use Eq. (10) from Chapter 3, with I_D and V_{GS} replaced by i_D and v_{GS}:

$$i_D = \frac{I_{DSS}}{V_P^2}(v_{GS} - V_P)^2$$

This equation is independent of V_{DS}, provided that the transistor is operating in the region beyond pinch-off. Then

$$g_m = \left.\frac{di_D}{dv_{GS}}\right|_{OP} = \left.\frac{2I_{DSS}}{V_P^2}(v_{GS} - V_P)\right|_{V_{GS}}$$

so that

$$g_m = \frac{2I_{DSS}}{V_P^2}(V_{GS} - V_P) \tag{23}$$

The value of g_m will vary with the operating point, since Eq. (23) shows that it is a linear function of V_{GS}. The maximum value in the range $V_P \le V_{GS} \le 0$ is g_{m0}, which occurs when $V_{GS} = 0$. The minimum value is zero, occurring when $V_{GS} = V_P$. In terms of signal performance, we can see that any JFET will produce the greatest gain when V_{GS} is near zero and the drain current is large; the gain approaches zero as $V_{GS} \to V_P$ and $I_D \to 0$.

The values of g_m and g_{m0} for a p-channel JFET are also positive quantities. For example, I_{DSS} in Eq. (23) is negative, while V_P is positive with $V_{GS} \le V_P$. Depletion-mode IGFETs, either n-channel or p-channel, also have positive values of g_m and g_{m0} that we determine by identical formulas.

Enhancement-mode devices have slightly different forms of the nonlinear transfer characteristic. For an n-channel enhancement-mode IGFET,

$$i_D = \frac{K}{2}(v_{GS} - V_T)^2 \qquad \text{for } V_{DS} \ge V_{GS} - V_T \text{ and } V_{GS} \ge V_T$$

$$g_m = \left.\frac{di_D}{dv_{GS}}\right|_{OP} = \left.K(v_{GS} - V_T)\right|_{OP}$$

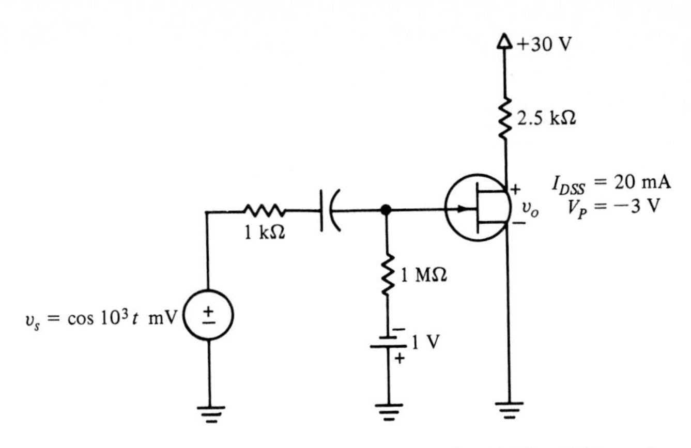

Fig. 5.17 An amplifier for which the voltage gain $[v_o(t)]/[v_s(t)]$ is to be calculated.

Thus,

$$g_m = K(V_{GS} - V_T) \tag{24}$$

We note that $g_m = 0$ when $V_{GS} = V_T$ and that g_m increases linearly as V_{GS} increases. Larger signal gains therefore occur as V_{GS} and I_D increase.

A *p*-channel enhancement-mode IGFET also has a positive g_m because K and V_T are negative and $V_{GS} \leq V_T$.

As an illustration of the use of the low-frequency model, suppose we calculate the voltage gain v_o/v_s for the amplifier shown in Fig. 5.17. We note that 1 mV is small-signal and 10^3 rad/s is low-frequency, so we shall use the model of Fig. 5.16. To find g_m, we need to determine the dc operating point. This is a simple single-stage amplifier and the nonlinear model is easily applicable. Since $V_{GS} = -1$ V,

$$I_D = \frac{I_{DSS}}{V_P^2}(V_{GS} - V_P)^2 = \frac{20}{9}(-1 + 3)^2 = 8.89 \text{ mA}$$

As a precaution, we calculate V_{DS}:

$$V_{DS} = 30 - 8.89(2.5) = 7.78 \text{ V}$$

Since V_{DS} is greater than $V_{GS} - V_P = 2$ V, we are operating safely beyond pinch-off. Thus

$$g_m = \frac{2I_{DSS}}{V_P^2}(V_{GS} - V_P) = \frac{40}{9}(-1 + 3) = 8.89 \text{ m℧}$$

We therefore have a signal model for the JFET with $g_m = 8.89$ m℧. The complete signal circuit is shown in Fig. 5.18. It is obtained by replac-

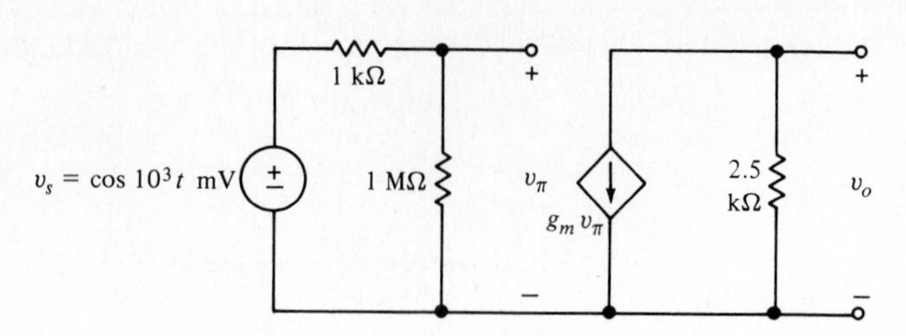

Fig. 5.18 The ac-equivalent circuit of the amplifier in Fig. 5.17; $g_m = 8.89$ m℧.

ing the dc voltage sources and large capacitors with short circuits. We then use voltage division to obtain

$$v_\pi = \frac{1000}{1001} v_s$$

while

$$v_o = -g_m v_\pi 2500 = -8.89 \frac{1000}{1001} (2.5 v_s)$$

and

$$A_V = \frac{v_o}{v_s} = -22.2$$

Note that an increase in g_m or R_D causes $|A_V|$ to increase.

D5.10 Determine g_m for (a) a *p*-channel JFET with $I_{DSS} = -8$ mA, $V_P = 1.6$ V, and $V_{GS} = 0.2$ V; (b) an *n*-channel depletion-mode MOSFET with $I_{DSS} = 5$ mA, $V_P = -2$ V, and $I_D = 4$ mA; (c) an *n*-channel enhancement-mode MOSFET having $V_T = 1.5$ V, $K = 4$ mA/V^2, and $V_{GS} = 4$ V.

Answers. 8.75; 4.47; 10 m℧

D5.11 In Fig. 5.17, replace the 1-V battery with a short circuit and the 2.5-kΩ resistor with a 1.2-kΩ resistor. Find the signal power dissipated in (a) $R_s = 1$ kΩ, (b) $R_2 = 1$ MΩ, (c) $R_D = 1.2$ kΩ, (d) the transistor, (e) $V_{DD} = 30$ V.

Answers. 0.499 fW; 0.499 pW; 106 nW; −106 nW; 0

5.8 FET models: small-signal high-frequency

In the frequency range where the reactances of the interelectrode capacitances are comparable to the resistance levels in the external circuit, the small-signal FET model must be changed to the *high-frequency hybrid-π model* shown in Fig. 5.19. Note that there is no series resistance at the input, so the model is actually only a π network. The resistor r_d and capacitor C_{ds} are sometimes omitted, since their impedances should be considerably larger than that of the external load resistor. The major high-frequency effects are caused by the two capacitances C_{gs} and C_{gd}. This model is simple and accurate up to about 500 MHz. At higher frequencies, the element values in Fig. 5.19 become frequency-dependent. This high-frequency hybrid-π model is a particularly important model because the FET is widely used at UHF and VHF.

The high-frequency parameters for FETs are often measured and listed in data sheets as short-circuit y-parameters for a two-port network. These admittance parameters are defined for the general two-port of Fig. 5.13 as follows:

$$I_1 = y_{11}V_1 + y_{12}V_2$$

$$I_2 = y_{21}V_1 + y_{22}V_2$$

Note that the parameters may be found for a general two-port by applying a 1-V signal to one port, short-circuiting the other port, and measuring a phasor current.

The notation used for the short-circuit admittance parameters when they are applied to an FET is shown in Fig. 5.20 for the *common-source connection*. The subscript system is similar to that used with the h-parameters; thus y_{fs} is the forward short-circuit transfer admittance for an FET having the source (signal) terminal common to both input and output. It would be measured with the drain short-circuited to the source (at signal frequencies by a large capacitor).

For the model of Fig. 5.19, including r_d and C_{ds}, we may relate the element values and the y-parameters by first short-circuiting the right port:

$$y_{is} = j\omega C_{is} = j\omega(C_{gs} + C_{gd}) \tag{25a}$$

Fig. 5.19 The small-signal high-frequency model for an FET typically includes C_{gs} and C_{gd}, while C_{ds} and r_d are sometimes omitted.

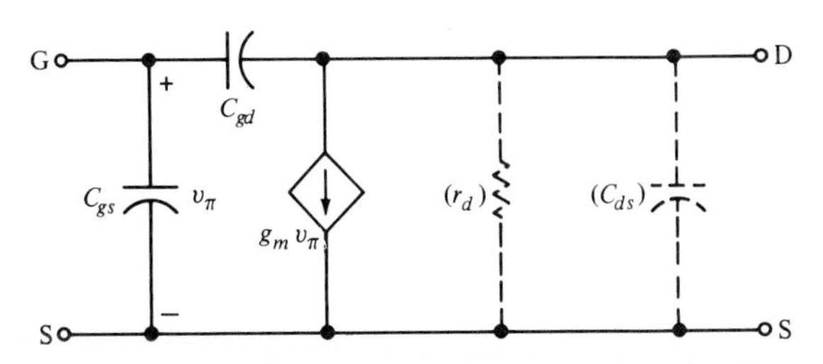

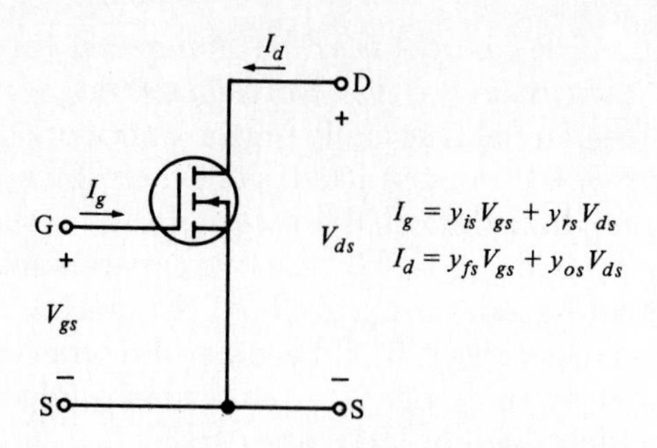

Fig. 5.20 The short-circuit admittance parameters for an FET in common-*source* configuration carry a second subscript *s* and a first subscript signifying input, reverse, forward, or output.

where we have defined an equivalent capacitance,

$$C_{is} = C_{gs} + C_{gd} \tag{25b}$$

Also,

$$y_{fs} = g_{fs} + jb_{fs} = g_m - j\omega C_{gd} = g_{fs} - j\omega C_{fs} \tag{26a}$$

$$C_{fs} = C_{gd} \tag{26b}$$

$$g_{fs} = \text{Re}(y_{fs}) = g_m \tag{26c}$$

With the gate and source short-circuited,

$$y_{rs} = -j\omega C_{rs} = -j\omega C_{gd} \tag{27a}$$

$$C_{rs} = C_{gd} \tag{27b}$$

and

$$y_{os} = g_{os} + j\omega C_{os} = \frac{1}{r_d} + j\omega(C_{gd} + C_{ds}) \tag{28a}$$

$$C_{os} = C_{gd} + C_{ds} \tag{28b}$$

$$g_{os} = \text{Re}(y_{os}) = \frac{1}{r_d} \tag{28c}$$

The four capacitances, C_{is}, C_{fs}, C_{rs}, and C_{os}, are sometimes designated C_{iss}, C_{fss}, C_{rss}, and C_{oss} on data sheets. The second *s* is a reminder that there is a (signal) short circuit at one of the ports.

Several facts are evident from the relationships above. For example, at low frequencies, y_{fs} approaches g_m, the low-frequency transconductance. Also, C_{gd} is found from the imaginary part of either y_{fs} or y_{rs}; knowing its value, we find that Eqs. (25*b*) and (28*b*) give C_{gs} and C_{ds}.

To illustrate the use of this collection of formulas in finding the hybrid-π element values from y-parameters given on specification sheets, let us determine the high-frequency model for a typical 2N3823 transistor at $V_{GS} = 0$ and $V_{DS} = 15$ V for a frequency of 10 MHz. A portion of one page of the data sheets for this transistor is shown in Fig. 5.21, wherein we see that curves are given for Re(y), Im(y), and C for most of the parameters as functions of frequency and V_{GS}. Their Figs. 6 through 9 all apply to $V_{GS} = 0$, $V_{DS} = 15$ V, and $T = 25°$C.

Beginning with their Fig. 6, we see that the real part of y_{is} is negligible, while $C_{is} = 4.8$ pF. The remaining curve, Im(y_{is}), is simply ωC_{is} plotted for a logarithmic frequency scale.

In their Fig. 7, we note that Re(y_{fs}) is independent of frequency, so we set

$$g_m = \text{Re}(y_{fs}) = 4.7 \text{ m℧}$$

At a frequency of 10 MHz, the imaginary part has a magnitude of about 0.1 m℧, which is negligible.

From their Fig. 8, it is evident that Re(y_{rs}) is negligible, while the imaginary part provides a capacitance $C_{rs} = 1.4$ pF.

Finally, y_{os} from their Fig. 9 is mostly imaginary, so we use $C_{os} = 1.8$ pF over the frequency range up to 300 MHz. The small value of Re(y_{os}) looks as if it were less than 0.1 m℧, and thus $r_d > 10$ kΩ.

The elements in the model are therefore

$$C_{gs} = C_{is} - C_{rs} = 4.8 - 1.4 = 3.4 \text{ pF}$$

$$C_{gd} = C_{rs} = 1.4 \text{ pF}$$

$$g_m = 4.7 \text{ m℧}$$

$$C_{ds} = C_{os} - C_{rs} = 1.8 - 1.4 = 0.4 \text{ pF}$$

$$r_d > 10 \text{ k}\Omega$$

How constant may we expect these values to remain as the operating point is changed? Let us agree first that operation will always be beyond pinch-off. We already know that g_m is a linear function of V_{GS}, and we may neglect any change in r_d. The three capacitances vary in a different manner for JFETs and MOSFETs. For a MOSFET, the values of C_{gs} and C_{gd} are relatively constant with changing V_{GS} or V_{DS} The capacitances are mainly determined by the thickness and type of the insulating oxide layer, as well as the area and placement of the metallic gate electrode. Changes in the depletion or enhancement of the channel therefore have only a minor effect.

The JFET, however, operates by virtue of the creation of a depletion region and a reverse-biased junction between gate and channel. Thus any increase in the reverse bias voltage between gate and channel should cause

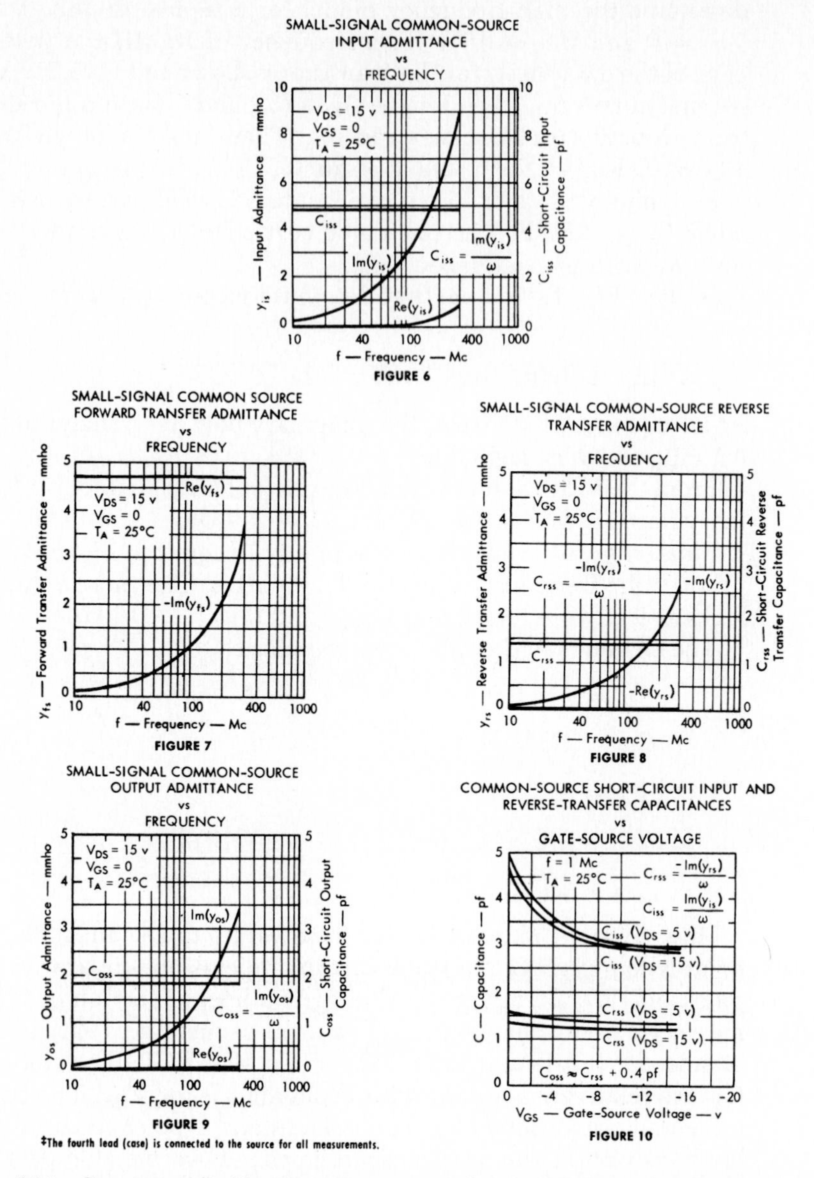

Fig. 5.21 Curves supplied by Texas Instruments for its 2N3823 n-channel JFET. The values apply to the high-frequency small-signal y-parameter model.

a decrease in capacitance, as it would in the reverse-biased diode. Since the depletion region is larger between gate and drain than it is between gate and source, C_{gd} is less than C_{gs}. Moreover, any reduction in the dimensions of the channel (by increasing $|V_{GS}|$ or V_{DS} for an n-channel JFET) has less effect at the drain end than at the source end. Thus, C_{gd} is relatively insensitive to V_{GS} and V_{DS}, while C_{gs} exhibits a variation with V_{GS} that is very similar to the junction capacitance of the reverse-biased semiconductor diode. Referring again to the 2N3823 data in Fig. 5.21, their Fig. 10 shows that C_{rss} ($= C_{gd}$) is small and fairly constant; it does decrease slightly as V_{GS} or V_{DS} increases. On the other hand, C_{iss} ($= C_{gs} + C_{gd}$) decreases much more markedly with V_{GS}, but only a small amount with V_{DS}.

D5.12 Given a Type 2N3823 transistor operating at 1 MHz, determine typical values for (a) C_{gs} at $V_{GS} = 0$, $V_{DS} = 10$ V, (b) C_{gs} at $V_{GS} = -4$ V, $V_{DS} = 10$ V, (c) C_{ds} at $V_{GS} = 0$, $V_{DS} = 10$ V, (d) C_{gd} at $V_{GS} = 0$, $V_{DS} = 10$ V.

Answers. 3.4; 2.2; 0.4; 1.4 pF

D5.13 According to curves published by Motorola for its 2N4223 n-channel silicon JFET, a possible set of y-parameter values at 100 MHz is (in m℧) $y_{is} = 0.2 + j2.5$, $y_{rs} = -0.01 - j0.65$, $y_{fs} = 3.1 - j0.65$, $y_{os} = 0.05 + j0.8$. Determine suitable values for (a) C_{gd}, (b) g_m, (c) C_{gs}, (d) r_d, (e) C_{ds}.

Answers. 1.0 pF; 3.1 m℧; 3.0 pF; 20 kΩ; 0.3 pF

5.9 The effect of temperature on FET-model parameters

The only important effect that is caused by a change in operating temperature on either of the small-signal FET models is in the value of g_m. We may recall from Section 4.6 that an increase in temperature for a JFET or depletion-mode MOSFET results in a very slight decrease in the pinch-off voltage $|V_P|$ and a more significant decrease in the drain saturation current $|I_{DSS}|$, perhaps a 25% decrease for a 100°C increase in temperature. The effect of temperature on an enhancement-mode device was mentioned in Section 4.7. It is similar to the behavior of the JFET, for V_T decreases slightly and K decreases significantly as T increases.

The net result in either case is that g_m, which is proportional to the quotient of $|I_{DSS}|$ and $|V_P|$, varies as T^a, where a is roughly -1.5 and T is the absolute temperature. An example of this variation is shown in Fig. 5.22, which shows curves supplied for the 2N3823 JFET. For a typical unit with

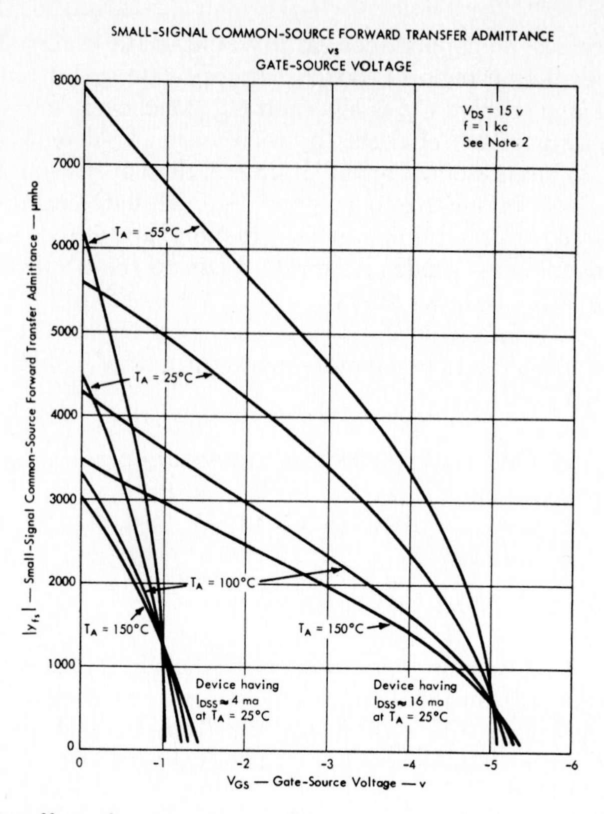

Fig. 5.22 The effect of temperature and V_{GS} on g_m or $|y_{fs}|$ is shown for a Texas Instruments 2N3823 FET.

$I_{DSS} \doteq 16$ mA at 25°C, we can see that g_m or $|y_{fs}|$ at $V_{GS} = 0$, $V_{DS} = 15$ V, decreases from 7.9 to 3.5 m℧ as the ambient temperature increases from -55 to 150°C. This corresponds to the value $a = -1.23$. For the lower-current devices, the magnitude of a is found to be a little less. We can also see that g_m is almost a linear function of V_{GS}.

Table 5.2 summarizes the information we have on the effect of operating point and temperature on the parameter values in the low- or high-frequency small-signal models for the several varieties of FET.

D5.14 Use Fig. 5.22 to specify g_m for a 2N3823 transistor having $I_{DSS} = 4$ mA when $T = 25$°C if it is operating with $V_{DS} = 15$ V and (a) $V_{GS} = 0$, $T = 25$°C, (b) $V_{GS} = -0.5$ V, $T = -55$°C, (c) $V_{GS} = -0.5$ V, $T = 150$°C.

Answers. 4.5; 4.3; 2.3 m℧

Table 5.2 Hybrid-π Element Values for FETs

Parameter	Basic formula	Typical variation with T(K), $\lvert V_{GS}\rvert$, and $\lvert V_{DS}\rvert$			Typical range
		T^a, a is	$\lvert V_{GS}\rvert$	$\lvert V_{DS}\rvert^a$, a is	
JFET					
g_m	$\frac{2I_{DSS}}{V_P^2}(V_{GS} - V_P)$	$\doteq -1.5$	$\propto (V_{GS} - V_P)$	0	0.5–20 m℧
C_{gs}	$\frac{C_{gs0}}{(1 + \lvert V_{GS}/V_{bi}\rvert)^N}$	$\doteq 0$	$\propto \left(1 + \left\lvert \frac{V_{GS}}{V_{bi}} \right\rvert\right)^{-N}$	< 0	2–20 pF
C_{gd}	. . .	$\doteq 0$	$\propto (1 + \lvert V_{GS}\rvert)^a$, $a < 0$	< 0	0.3–5 pF
IGFET, Depletion-mode					
g_m	$D(V_{GS} - V_P)$	$\doteq -1.5$	$\propto (V_{GS} - V_P)$	0	0.5–20 m℧
C_{gs}	. . .	0	0	$\doteq 0$	2–20 pF
C_{gd}	. . .	0	0	$\doteq 0$	0.1–5 pF
IGFET, Enhancement-mode					
g_m	$K(V_{GS} - V_T)$	$\doteq -1.5$	$\propto (V_{GS} - V_T)$	0	0.5–20 m℧
C_{gs}	. . .	0	0	$\doteq 0$	2–20 pF
C_{gd}	. . .	0	0	$\doteq 0$	0.1–5 pF

Problems

1. Include r_d in the model of Fig. 5.2*b* and then derive a new version of Eq. (1).
2. Values applying to the small-signal equivalent circuit of Fig. 5.2*b* are $R_B = 40$ kΩ, $r_\pi = 3$ kΩ, $g_m = 50$ m℧, $r_d = 80$ kΩ, and $R_C = 5$ kΩ. A source $0.1u(t)$ mV in series with 500 Ω is connected to the input, and a resistor R_L is placed across the output. The values of C_1 and C_2 are large. Find $v_o(t)$ if R_L equals (a) 2 kΩ, (b) 20 kΩ.
3. In Fig. 5.2*a*, let $R_s = 100$ Ω, $R_B = 400$ kΩ, $R_C = 5$ kΩ, and $V_{CC} = 15$ V. Both capacitors are large. The *npn* transistor has a negligible I_{CEO}, $\beta_{dc} = 50$, and $V_0 = 0.6$ V. (a) Find I_C. If $g_m = 36I_C$, $r_\pi = 1.4/I_C$ (kΩ, I_C in mA), $r_d = 40$ kΩ, and $v_s(t) = 2 \cos 500t$ mV, find (b) $v_o(t)$, (c) $v_{CE}(t)$.
4. In Fig. 5.23, let $v_s(t) = 2 \cos 10^4 t$ mV, $\beta_{dc} = 200$, $V_0 = -0.7$ V, and $I_{CEO} \doteq 0$. (a) Find I_C. (b) Find V_{CE}. (c) If $g_m = 32|I_C|$ and $r_\pi = 2.5$ kΩ, find $v_{CE}(t)$.
5. The small-signal hybrid-π equivalent circuit for the transistor in Fig. 5.24 contains $r_\pi = 5$ kΩ and $g_m = 32$ m℧. Some of the dc biasing circuitry is unimportant at the operating frequency and is not shown. Find $v_o(t)$ if $v_s(t) = 10 \cos 10^4 t$ mV.
6. A bipolar transistor is modeled at high frequencies by $r_\pi = 4$ kΩ, $C_\pi = 50$ pF, $r_x = 100$ Ω, $g_m = 25$ m℧, $C_\mu = 5$ pF, and $r_d = \infty$. (a) If the model is short-circuited at the C-E terminals, calculate $|I_c/I_b|$ as a function of ω and sketch the results on log-log paper for $10^5 \le \omega \le 10^9$ rad/s. (b) At what value of ω is this ratio $1/\sqrt{2}$ times its low-frequency value?
7. The high-frequency equivalent circuit of a *pnp* transistor amplifier is shown in Fig. 5.25. Find the input impedance seen at B′-E if the output is left open-circuited.
8. Elements appearing in the high-frequency equivalent circuit of a junction transistor are $r_x = 100$ Ω, $r_\pi = 2$ kΩ, $C_\pi = 100$ pF, $C_\mu = 5$

Fig. 5.23 See Problem 4.

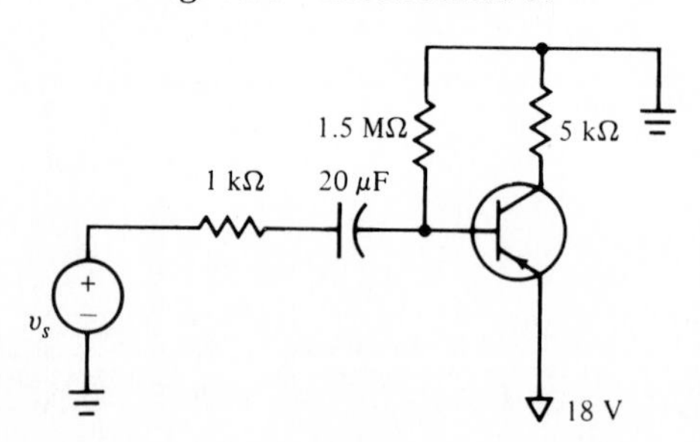

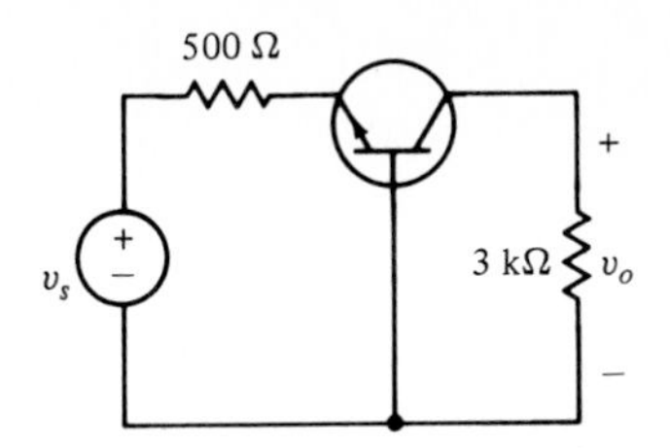

Fig. 5.24 See Problem 5.

pF, and $g_m = 100$ m℧. At 10^7 rad/s, determine the impedance viewed at the C-E terminals if a 1.9-kΩ resistor is connected between the B-E terminals.

9. The transistor of Fig. 5.2*a* and *b* has $n = 1$ with $\beta_0 = 100$ at $I_C = 0.1$ mA, and $\beta_0 = 200$ at $I_C = 1$ mA. Calculate the voltage gain if $R_B = 40$ kΩ, $R_s = 1$ kΩ, $R_L = 8$ kΩ, and I_C equals (a) 0.1 mA, (b) 1 mA.
10. The transistor of Fig. 5.2 has $n = 1$ and a β_0 given by Fig. 5.4. If $R_s = 1$ kΩ, $R_B = 50$ kΩ, and $R_L = 10$ kΩ, calculate A_V at the operating point (a) $I_C = 1$ mA, $V_{CE} = 1$ V, (b) $I_C = 10$ mA, $V_{CE} = 10$ V.
11. Let the transistor used in the amplifier of Fig. 5.2 have $\beta_0 = 100$ at $I_C = 1$ mA and $\beta_0 = 150$ at 5 mA. Given $n = 1$, an Early voltage $V_A = 50$ V, and $R_s = 2$ kΩ, $R_B = 25$ kΩ, and $R_L = 4$ kΩ, calculate the magnitude of the voltage gain at I_C equals (a) 1 mA, (b) 5 mA.
12. Calculate g_m, r_π, and A_V for the amplifier shown in Fig. 5.6 if (a) V_{CC} is reduced to 15 V and $\beta_0 = 105$; (b) V_{CC} is increased to 30 V and $\beta_0 = 110$.
13. For the small-signal low-frequency amplifier shown in Fig. 5.6, change R_B to 1.8 MΩ and n to 1.3, and let $\beta_0 = 90$. (a) Find I_C, g_m, r_π, and A_V. (b) A voltage gain of exactly −400 is now desired, and this can be achieved by changing the value of only one external circuit element. What is this new value for R_L? (c) for R_B?

Fig. 5.25 See Problem 7.

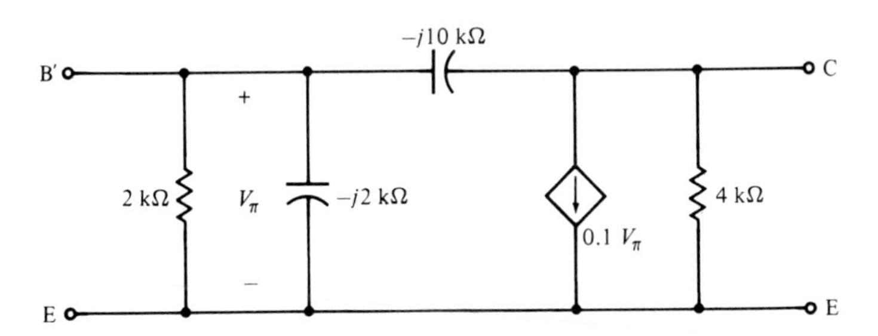

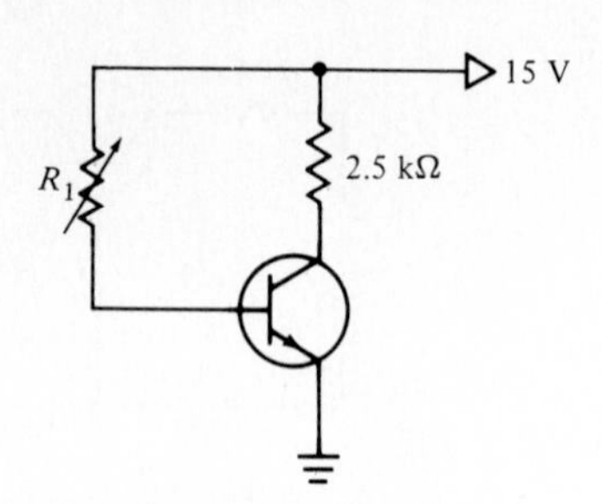

Fig. 5.26 See Problem 15.

14. A 2N5088 transistor, similar to Unit 2 on the data sheets in Appendix A, is used in a stabilized bias circuit with $R_1 = 60$ kΩ, $R_2 = 30$ kΩ, $R_E = 2$ kΩ, $R_C = 3$ kΩ, and $V_{CC} = 12$ V. Assuming operation at 25°C, let $V_{BE} = 0.75$ V and determine values for β_{dc}, V_0, I_C, V_{CE}, β_0, r_π, g_m, n, and r_d.
15. The transistor in Fig. 5.26 has $\beta_{dc} = 200$, $V_0 = 0.75$ V, $\beta_0 = 225$, and $n = 1.25$. What value of R_1 will cause (a) $g_m = 100$ m℧? (b) $r_\pi = 3$ kΩ? (c) $V_{CE} = 0.4$ V?
16. (a) If $V_{bi} = 0.8$ V and $C_{\mu 0} = 12$ pF for a certain *pnp* transistor and $N = 0.42$, determine C_μ when $V_{CB} = -6$ V. (b) What is V_{CB} when $C_\mu = 10$ pF?
17. Assume that the gain-bandwidth curves shown in Fig. 5.9 apply to an *npn* transistor having $n = 1.1$, $N = 0.5$, $V_{BE} = 0.65$ V, $V_{bi} = 0.8$ V, and $C_\mu = 2$ pF when $V_{CE} = 1$ V. Calculate C_π at the following values for I_C and V_{CE}: (a) 0.5 mA, 2 V; (b) 1 mA, 20 V; (c) 2 mA, 20 V.
18. Knowing that $|\beta| = 3$ at 100 MHz and $\beta_0 = 100$, (a) calculate ω_β and ω_T; (b) find $|\beta|$ at 10 MHz.
19. Consider the 2N5088 listed in the "Dynamic Characteristics" on the data sheets in Appendix A as having $\beta_0 = 350$ at $I_C = 1$ mA, $V_{CE} = 5$ V. Assuming that $C_{cb} = 1$ pF at $V_{CE} = 15$ V, use Figs. 11 and 12 of the data sheets to calculate C_π at the following values for I_C and V_{CE}: (a) 1 mA, 15 V; (b) 7 mA, 5 V; (c) 7 mA, 15 V.
20. Let $N = 1/3$, $V_{BE} = 0.65$ V, and $V_{bi} = 0.75$ V for a silicon transistor. If $C_{ob} = 6$ pF at $V_{CE} = 5$ V, and $\beta_0 = 200$, (a) find C_μ, g_m, and r_π when $I_C = 1.5$ mA, $V_{CE} = 10$ V. (b) If the gain-bandwidth product for this transistor is 6×10^8 rad/s at $I_C = 1.5$ mA, $V_{CE} = 10$ V, determine C_π and $|\beta|$ at 20 MHz.
21. Find the output impedance of the equivalent circuit shown in Fig. 5.27. Let $r_x = 100$ Ω and place a short circuit across the input terminals.

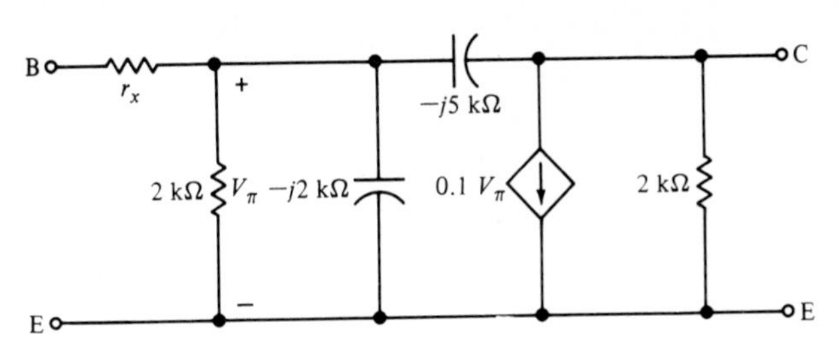

Fig. 5.27 See Problem 21.

22. The high-frequency equivalent circuit for a certain transistor at a specified operating point is shown in Fig. 5.28. Find (a) β_0, (b) f_β, (c) f_T, (d) $|\beta|$ at 20 MHz, (e) $|\beta|$ at 2 kHz, (f) β at 2 MHz.
23. At $I_C = 1$ mA, $V_{CE} = 10$ V, a certain transistor data sheet shows that $C_\mu = 3$ pF, $\beta_0 = 200$, and $\omega_T = 500$ Mrad/s. (a) Assume $n = 1$, $V_{bi} = 0.8$ V, and $V_{BE} = 0.65$ V, and find g_m, r_π, C_π, and ω_β. (b) Estimate values for C_μ, β_0, ω_T, C_π, and ω_β if the operating point is changed to 2 mA, 5 V.
24. The resistors in the amplifier of Fig. 5.6 may be assumed to have temperature coefficients of 1200 ppm/°C at 25°C, while the transistor has a β_{dc} that is proportional to $T^{1.5}$. (a) Estimate V_{BE} and β_{dc} at 50°C. (b) Find the operating point at 50°C. (c) Determine a new low-frequency small-signal equivalent circuit and predict the voltage gain at 50°C. Assume $\beta_0 = 110$ at 25°C.
25. If it is assumed that $\beta_0 = \beta_{dc}$ for the *pnp* transistor of Fig. 5.29, determine values for the low-frequency hybrid-π equivalent circuit of a typical transistor for which $V_A = 25$ V, $n = 1.2$, and I_C equals (a) -1 μA, (b) -100 μA.
26. For the transistor of Fig. 5.29, make the assumption that $\beta_{dc} \propto T^{1.8}$ and determine values of $\beta_{dc(max)}$ and $\beta_{dc(min)}$ at $I_C = -100$ μA for T equals (a) -50°C, (b) 125°C.

Fig. 5.28 See Problems 22 and 28.

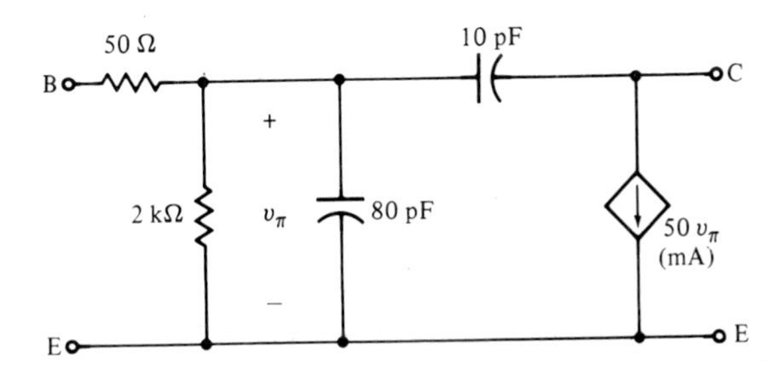

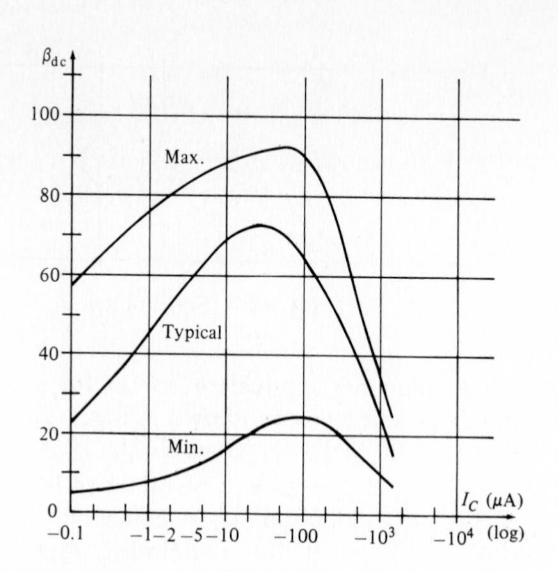

Fig. 5.29 The variation of β_{dc} with I_C is shown for a *pnp* transistor used in an integrated circuit; $V_{CE} = -10$ V and $T = 25°$C. See Problems 25 and 26.

27. An *npn* silicon bipolar transistor has a β_0 of 200 at $I_C = 1.8$ mA, $V_{CE} = 6$ V, and $T = 25°$C. If $\beta_0 \propto T^{1.7} I_C^{0.2} V_{CE}^{0.4}$, plot a locus of those operating points on I_C-V_{CE} axes for which (a) $\beta_0 = 200$ at 25°C, (b) $\beta_0 = 200$ at 100°C.

28. The high-frequency parameter values for the transistor of Fig. 5.28 vary as T^a, where the values for a are given below. Assuming that the given circuit is valid at 25°C, (a) construct an equivalent circuit useful at 125°C; (b) calculate new values for β_0, f_β, and f_T, and compare them with the values for 25°C.

Parameter:	g_m	r_π	C_μ	C_π	r_x
a:	-1	3	0	$-1/3$	1.4

29. The small-signal high-frequency equivalent circuit of a silicon *npn* transistor contains $r_x = 50\ \Omega$, $r_\pi = 2$ kΩ, $C_\pi = 50$ pF, $C_\mu = 5$ pF, $g_m = 30$ m℧, and $r_d = 20$ kΩ when $I_C = 1$ mA, $V_{CE} = 4$ V, and $T = 25°$C. For the sake of this problem, modify Table 5.1 by changing every $\doteq$ to $=$, > 0 to 0.25, < 0 to -0.25, < 1 to 0.75, and large $|I_C|$ to $|I_C| > 1$ mA. (a) Find accurate values for the six parameters above at $I_C = 2$ mA, $V_{CE} = 10$ V, $T = 125°$C. (b) The high-frequency performance of a bipolar transistor amplifier begins to deteriorate at a frequency given very roughly by $(R_s C_\pi + g_m R_L R_s C_\mu)^{-1}$ rad/s, where R_s and R_L are the source and load resistance respectively. Assume that $R_s = R_L = 1$ kΩ, and calculate this frequency at 25°C. (c)

In what direction should T, I_C, and V_{CE} be changed to increase this frequency?

30. If $h_{ie} = 2\ \text{k}\Omega$, $h_{re} = 10^{-4}$, $h_{fe} = 120$, and $h_{oe} = 40\ \mu\mho$, draw an equivalent circuit similar to the hybrid-π, but include an additional element to account for h_{re}. Give all element values.
31. The transistor in Fig. 5.2*a* has h parameters $h_{ie} = 2.5\ \text{k}\Omega$, $h_{re} = 0$, $h_{fe} = 60$, and $h_{oe} = 2 \times 10^{-5}\ \mho$. If $R_s = 1\ \text{k}\Omega$, $R_B = 16\ \text{k}\Omega$, and $R_C = 4\ \text{k}\Omega$, find (a) $v_o(t)/v_s(t)$, (b) $v_o(t)/v_{be}(t)$.
32. Let $h_{ie} = 2.5\ \text{k}\Omega$, $h_{re} = 4 \times 10^{-5}$, $h_{fe} = 100$, and $h_{oe} = 20\ \mu\mho$ for the transistor of Fig. 5.30 at its established operating point. Find (a) V_o, (b) I_{in}.
33. (a) Determine the hybrid-π parameters and the value of n for a 2N5088 transistor similar to Unit 3 on the data sheets in Appendix A at $I_C = 2$ mA, $V_{CE} = 10$ V. (b) Find h_{fe} at $I_C = 0.2$ mA, $V_{CE} = 4$ V.
34. Calculate h_{ie}, h_{re}, h_{fe}, and h_{oe} for a common-emitter amplifier having (a) $r_\pi = 2\ \text{k}\Omega$, $g_m = 40\ \text{m}\mho$, and $r_d = 100\ \text{k}\Omega$; (b) $I_C = 0.8$ mA, $n = 1.25$, $\beta_0 = 70$, and $V_A = 60$ V.
35. Determine values for the four common-base h parameters h_{ib}, h_{rb}, h_{fb}, and h_{ob} for a transistor having $r_\pi = 1\ \text{k}\Omega$, $r_d = 10\ \text{k}\Omega$, and $g_m = 10\ \text{m}\mho$.
36. Using data from Appendix A, specify a typical value of g_m for the 2N3823 transistor if $I_{DSS} = 16$ mA, $V_{GS} = -2$ V, and T equals (*a*) $-55°$C, (*b*) $25°$C, (*c*) $100°$C. (*d*) Repeat parts (*a*), (*b*), and (*c*) if $V_{GS} = -1$ V.
37. An n-channel enhancement-mode MOSFET with $V_T = 2$ V and $K = 5\ \text{mA/V}^2$ is operating in saturation with a gate-source voltage $v_{GS} = 3.6 + 0.02 \cos 1200t$ V. Find i_D.

Fig. 5.30 See Problem 32.

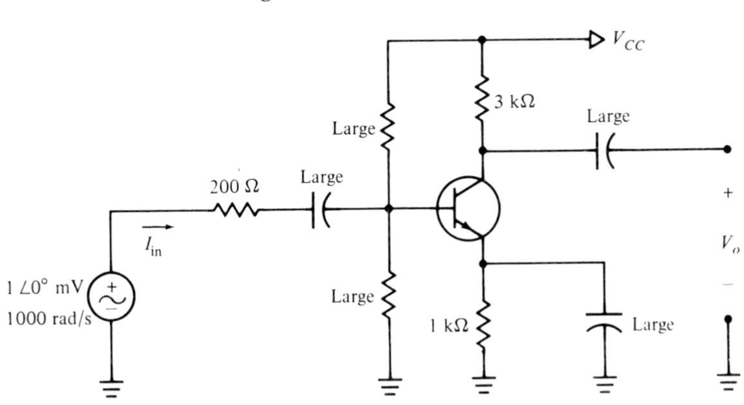

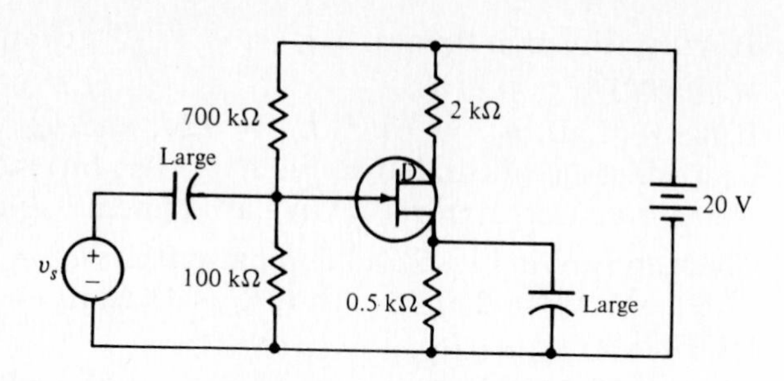

Fig. 5.31 See Problem 38.

38. Let $V_P = -2.5$ V and $I_{DSS} = 10$ mA for the JFET in Fig. 5.31. If $v_s = 0.01 \sin 2000t$ V, assume that the capacitors are open circuits to dc and short circuits to ac, and calculate v_{DS}.
39. In Fig. 5.32, assume that all three capacitors are very large and that the operating frequency is low. If $I_D = 10(0.4V_{GS} + 1)^2$ mA in saturation, determine R_{SS} so that $V_o/V_s = -8$.
40. The transistor in the amplifier circuit of Fig. 5.33 has $I_{DSS} = -8$ mA and $V_P = 3$ V. (a) Find I_D, V_{DS}, and g_m. (b) Assume V_s represents a small low-frequency signal with an amplitude of 1 mV, and find the amplitude of the signal voltage across R_D.
41. Let V_s in Fig. 5.34 represent a small-amplitude low-frequency signal such that V_o is 20 times V_s in amplitude. Find the necessary value for R_{SS} if $I_D = 15$ mA when $V_{GS} = 4$ V and $V_T = 2$ V.
42. Calculate the four y-parameters at 250 MHz if $C_{gs} = 3.6$ pF, $C_{gd} = 0.9$ pF, $C_{ds} = 0.2$ pF, $r_d = 12$ kΩ, and $g_m = 8$ m℧.
43. Use the information in Fig. 5.21 to develop a high-frequency equivalent circuit for a 2N3823 transistor at 50 MHz if $V_{DS} = 15$ V and V_{GS} equals (a) 0, (b) -2 V with $V_P = -4$ V.

Fig. 5.32 See Problem 39.

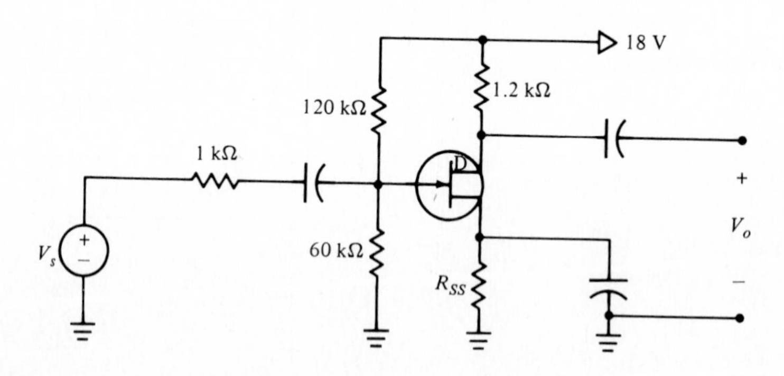

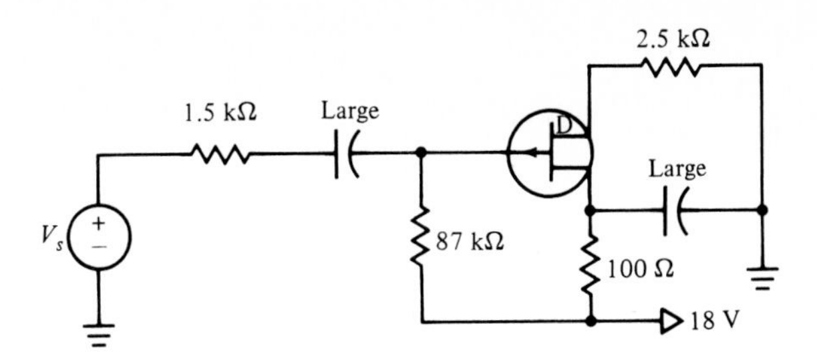

Fig. 5.33 See Problem 40.

44. Data for the Motorola 2N4223 n-channel silicon JFET show that typical values for the y-parameters at 300 MHz are $y_{is} = 1.6 + j8.3$ m℧, $y_{rs} = -j1.6$ m℧, $y_{fs} = 3.1 - j1.6$ m℧, and $y_{os} = 0.1 + j3$ m℧. Determine suitable values for (a) C_{gd}, (b) g_m, (c) C_{gs}, (d) r_d, (e) C_{ds}. (f) What element do their data suggest should also be included in the equivalent circuit?
45. Figure 5.35 shows curves of y-parameter values vs. f for the RCA 40673 n-channel depletion-mode IGFET. Let $f = 200$ MHz and determine values for C_{gs}, C_{gd}, C_{ds}, r_d, g_m, and r_g (gate to source). Note that g_m is provided with a phase angle for increased accuracy in the high-frequency model.
46. Test data for an RCA 3N152 n-channel depletion-mode MOSFET at 200 MHz show that $y_{is} = 0.44 + j7$, $y_{rs} = -j0.19$, $y_{fs} = 6.8 - j2$, and $y_{os} = 0.32 + j1.9$, all in m℧. Calculate C_{gs}, C_{gd}, C_{ds}, g_m, and r_g (gate to source). Use y_{rs} to determine C_{gd}, and use this value and y_{fs} to calculate a magnitude and angle for g_m.

Fig. 5.34 See Problem 41.

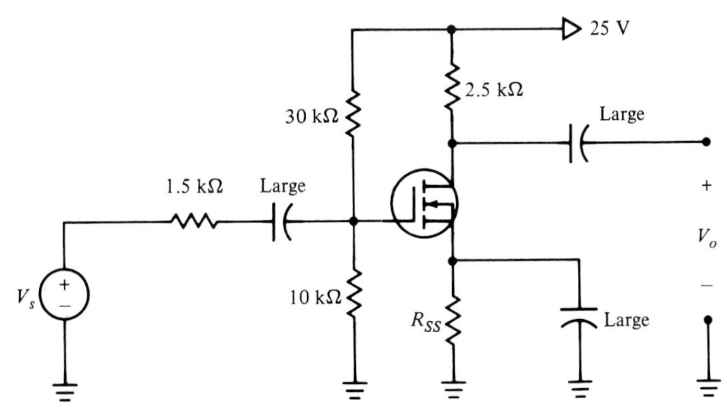

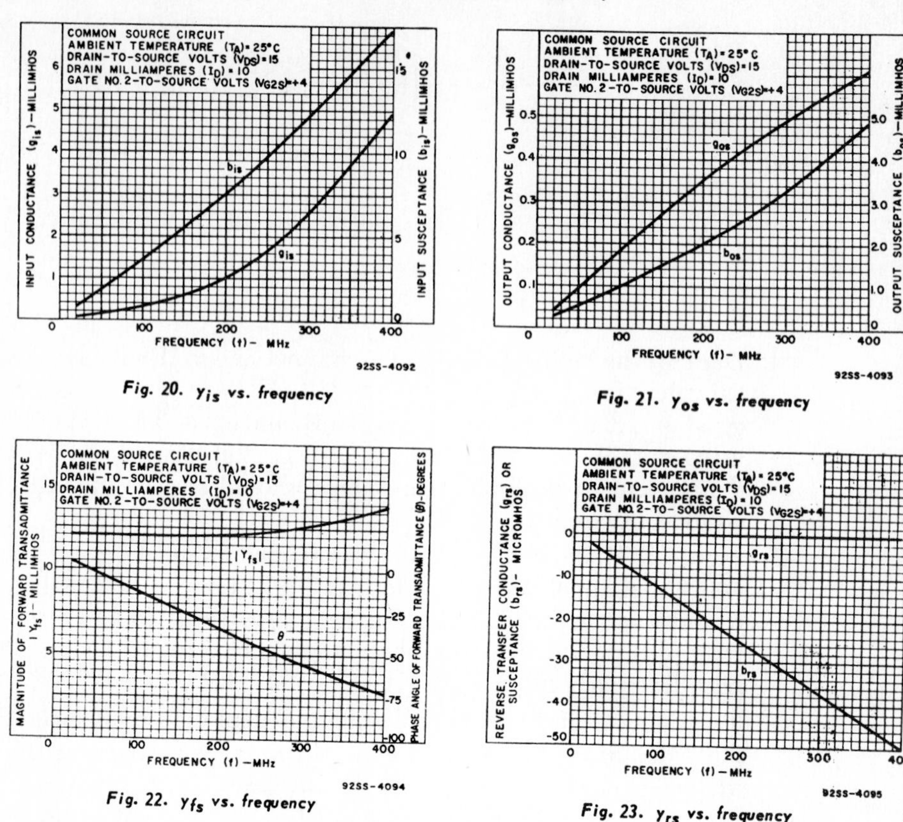

Fig. 5.35 A portion of the data provided by RCA for the 40673 n-channel depletion-mode MOSFET. See Problem 45.

47. (a) Use data from the $-55°C$ and $150°C$ curves of the lower-current transistor in Fig. 5.22 to establish values for a and K in the equation $g_m = KT^a$, with $V_{GS} = 0$. (b) Use your results to calculate values for 25°C and 100°C, and compare them with data from the curves.
48. For a certain JFET operating at $T = 25°C$, $V_{GS} = -0.5$ V, $V_{DS} = 10$ V, and $V_P = -2.5$ V, the high-frequency equivalent circuit contains $C_{gs} = 7$ pF, $C_{gd} = 1$ pF, and $g_m = 8$ m℧. Estimate new values for these three quantities at (a) $T = 125°C$, $V_{GS} = -0.5$ V, $V_{DS} = 10$ V, (b) $T = 125°C$, $V_{GS} = -0.5$ V, $V_{DS} = 5$ V.
49. An n-channel JFET ($V_P = -2$ V, $I_{DSS} = 15$ mA, $V_{bi} = 0.8$ V, $N = 0.5$, and $V_{GS} = -1$ V), an n-channel depletion-mode MOSFET ($V_P = -2$ V, $I_{DSS} = 15$ mA, $V_{GS} = -1$ V), and an n-channel enhancement-mode MOSFET ($V_T = 2$ V, $K = 7.5$ mA/V^2, $V_{GS} = 3$ V)

are all operating at 25°C, $V_{DS} = 10$ V with $C_{gs} = 10$ pF and $C_{gd} = 2.5$ pF. (a) Find I_D and g_m for each device. (b) In Table 5.2, replace $\doteq$ by $=$ and <0 by -0.25, and find g_m, C_{gs}, and C_{gd} for each device if $T = 125$°C, $V_{DS} = 5$ V, and V_{GS} becomes 1 V more positive for each transistor.

50. A p-channel enhancement-mode MOSFET is operating at 25°C with $g_m = 3.5$ m℧. Assuming that g_m decreases to 2.4 m℧ when the temperature increases to 100°C and the operating point remains constant, estimate g_m at -40°C.

6

Single-stage amplifiers at mid-frequencies

Let us pause for a moment to reflect on our progress. After studying dc models in Chapter 3, we used them in Chapter 4 to establish operating points that were safely located in the active region for bipolar transistors and in the region beyond pinch-off for the FETs. With this accomplished, we were able to consider small-signal models in Chapter 5 and discuss the relationship of the parameter values to the dc operating point and to temperature. We even calculated the voltage gain for the simple common-emitter and common-source amplifiers.

In this chapter we will consider the mid-frequency analysis and design of single-stage amplifiers, those containing only one transistor. The extension of this work to higher and lower frequencies is discussed in Chapter 7, and the problems associated with multistage amplifiers are addressed in Chapter 8.

Amplifier analysis is considered first, after which we will illustrate several designs. In analyzing an amplifier, we seek values for the voltage, current, and power gain, and for the input and output impedance. It turns out that once we determine the voltage gain, it is easy to find values of the current and power gain.

Amplifiers in the common-emitter and common-source configuration yield the largest power gain and are therefore widely used. The common-collector amplifier is often employed when its large input impedance or small output impedance is advantageous. The common-drain FET amplifiers have similar characteristics. Common-base and common-gate amplifiers have a low value of input impedance and a large output impedance. All of these amplifiers are considered at mid-frequencies in this chapter.

6.1 The common-emitter amplifier: analysis

The general single-stage common-emitter amplifier is shown in Fig. 6.1*a*. Its dc bias circuit is obtained in Fig. 6.1*b* by replacing all capacitors with open circuits. A suitable dc-equivalent circuit from Section 3.1 could be

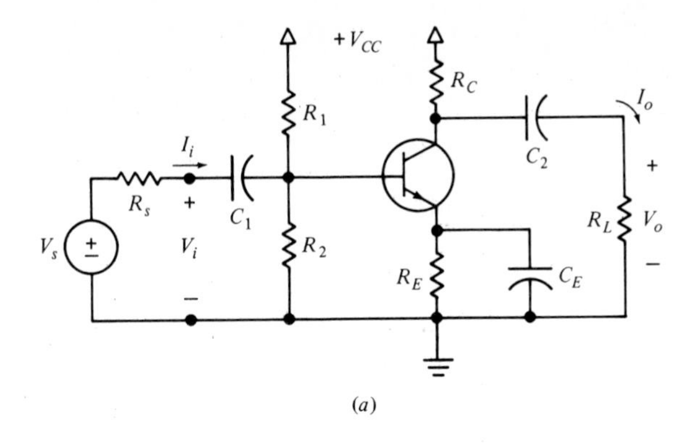

(a)

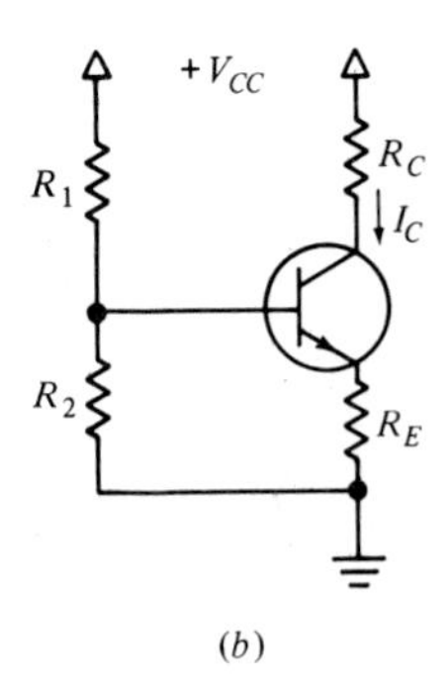

(b)

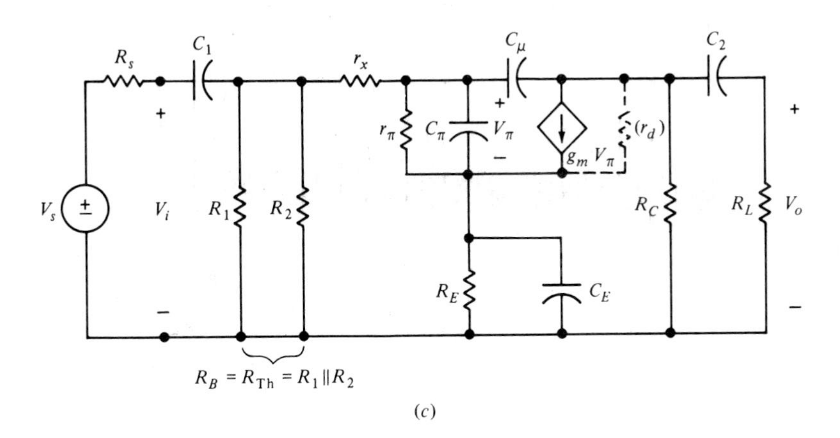

(c)

Fig. 6.1 (*a*) The general single-stage common-emitter amplifier circuit. (*b*) The dc biasing circuit. (*c*) The complete small-signal equivalent circuit.

used in place of the transistor, if a dc analysis is desired. The complete small-signal equivalent circuit appears in Fig. 6.1*c*, and it is obviously not simple. Its analysis is accomplished by simplifying the circuit in a manner appropriate to a specified range of operating frequencies. We define the *mid-frequencies* to be those frequencies for which the *large capacitors* (C_1, C_2, and C_E in this circuit) have capacitive reactances that are much smaller than the associated resistances, while the *small capacitors* (here C_π and C_μ) have reactances much larger than the resistances with which they are associated. Thus we replace C_1, C_2, and C_E by short circuits and C_π and C_μ by open circuits. The resultant mid-frequency equivalent circuit is shown in Fig. 6.2*a*. Note that all analytical results obtained from it will be frequency-independent, since it contains no capacitors. This procedure is appropriate for almost all broadband amplifiers. In the following chapter we will discover how to determine what range of frequencies the mid-frequencies extend over.

As the frequency of the applied signal decreases well below the mid-frequency range, it is necessary to add C_1, C_2, and C_E to the low-frequency equivalent circuit, as shown in Fig. 6.2*b*. The high-frequency equivalent includes C_π and C_μ and appears in Fig. 6.2*c*. These latter two circuits will be considered in Chapter 7. Note that the base spreading resistance r_x would appear in series with r_π in the mid- and low-frequency equivalents, but since it is usually much less than r_π, it is often neglected in those frequency ranges.

We now analyze the mid-frequency equivalent of Fig. 6.2*a*. The voltage gain that we can measure most easily is the ratio of the amplifier signal output voltage V_o to the signal voltage V_i that is present at the output terminals of the signal generator (including its source resistance). We call it A_{Vi}, the voltage gain with respect to the amplifier input:

$$A_{Vi} = \frac{V_o}{V_i} \tag{1}$$

By inspection of the mid-frequency equivalent, we see that $V_\pi = V_i$, and thus

$$V_o = -g_m V_\pi (R_C \| R_L) = -g_m V_i (R_C \| R_L)$$

so that

$$A_{Vi} = -g_m (R_C \| R_L) \tag{2}$$

If r_d is not very much larger than R_C and R_L, then the factor in parentheses becomes $r_d \| R_C \| R_L$. Another important gain is V_o / V_s, easily obtained from Eq. (2) and the equivalent circuit. We let

$$R_B = R_{\text{Th}} = R_1 \| R_2$$

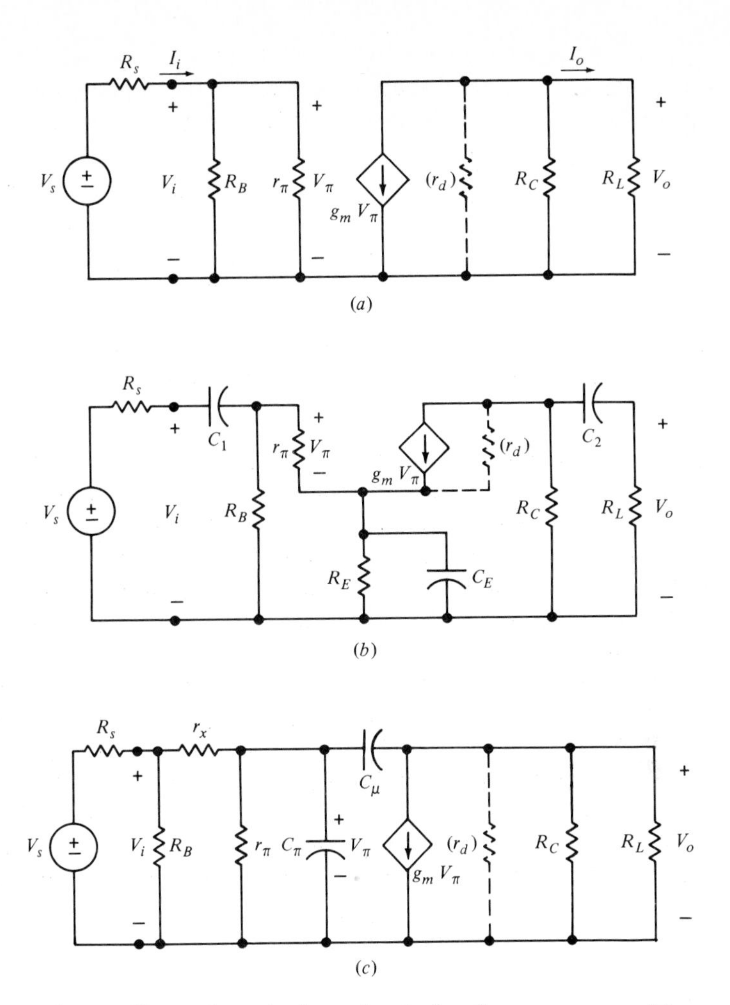

Fig. 6.2 The small-signal equivalent circuit for the common-emitter amplifier shown in Fig. 6.1*c* is simplified for use at (*a*) mid-frequencies, (*b*) low frequencies, and (*c*) high frequencies.

and then

$$V_i = V_s \frac{(r_\pi \| R_B)}{R_s + (r_\pi \| R_B)}$$

Therefore

$$A_{Vs} = \frac{V_o}{V_s} = \frac{V_o}{V_i} \times \frac{V_i}{V_s} = A_{Vi} \frac{V_i}{V_s} = A_{Vi} \frac{(r_\pi \| R_B)}{R_s + (r_\pi \| R_B)}$$

$$= -g_m (R_C \| R_L) \frac{(r_\pi \| R_B)}{R_s + (r_\pi \| R_B)} \tag{3}$$

This gain is a measure of the amplification of the open-circuit source voltage (the Thévenin-equivalent voltage of the signal source), for example the voltage appearing at the unloaded terminals of the pickup cartridge in a hi-fi system. The two voltage gains tend to be equal as $R_B \| r_\pi$ becomes much greater than R_s.

Current gain, power gain, input impedance, and output impedance are also important parameters in describing amplifier performance. Any or all of them may be specified as criteria for a design. We define the input impedance as the impedance offered to the signal source:

$$Z_i = \frac{V_i}{I_i} \tag{4}$$

By direct inspection of Fig. 6.2*a*, we see that it is given at mid-frequencies by

$$Z_i = r_\pi \| R_B \tag{5}$$

Having the input impedance, it is now quite easy to find the current gain in terms of the voltage gain:

$$A_I = \frac{I_o}{I_i} = \frac{V_o/R_L}{V_i/Z_i} = A_{Vi}\frac{Z_i}{R_L} \tag{6}$$

The power gain is commonly defined as the ratio of the power delivered to the load R_L to the power supplied to the input terminals of the amplifier. Assuming our signals are expressed as rms quantities, we have

$$A_P = \frac{V_o I_o}{V_i I_i} = A_{Vi}A_I = (A_{Vi})^2\frac{Z_i}{R_L} \tag{7}$$

Therefore, the voltage gain A_{Vi} is seen to be of central importance.

Just as Z_i may be considered to be the Thévenin impedance viewed at the input terminals, the output impedance Z_o is the Thévenin impedance at the output of the amplifier with all the *independent signal* sources set to zero and R_L removed. In Fig. 6.2*a* we set $V_s = 0$; thus V_π must be zero, and the dependent current generator $g_m V_\pi$ is also zero. Therefore

$$Z_o = R_C \qquad \text{(or } R_C \| r_d\text{, if necessary)} \tag{8}$$

Let us solidify our gains by obtaining values for all these quantities in the case of the amplifier shown in Fig. 6.3. We require values for the voltage, current, and power gain, the input and output impedance, and the actual rms output voltage. Since 0.1 mV is small and 2 kHz is neither too high nor too low,[1] we assume that the mid-frequency small-signal equivalent circuit of Fig. 6.2*a* is applicable. To calculate the required quantities, we need to determine values of g_m, r_d, and r_π; this in turn requires that

[1] The definition of "too" appears in Chapter 7.

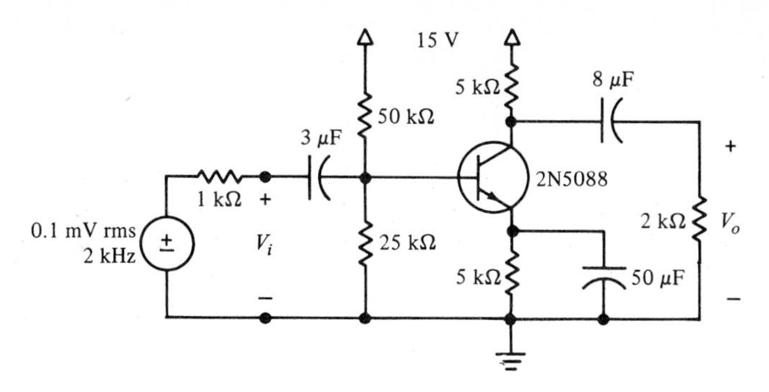

Fig. 6.3 A common-emitter amplifier for which the voltage gain, current gain, power gain, input impedance, output impedance, and output voltage are to be determined.

the dc operating point be identified. The problem thus divides itself into three parts: (1) finding the dc operating point; (2) determining hybrid-π parameter values for the small-signal equivalent circuit; and (3) calculating the required gains and impedances directly from the equivalent circuit. We shall go through the analysis using the extreme values of β_{dc} and β_0 for the 2N5088, obtained from the data sheets in Appendix A.

Let us work with the lower-gain unit first. To find β_{dc} from the data sheets, we need to know I_C. However, to calculate I_C, we need to know β_{dc}—a vicious circle indeed. But the presence of R_E suggests that the collector current should be rather insensitive to β_{dc}, so we probably do not need to know β_{dc} with great accuracy. We begin by noting that 15 V distributed across R_E, R_C, and the transistor leads to a collector current of $(15 - V_{CE})/(5 + 5)$, which is certainly less than 1.5 mA. If $I_C = 1$ mA and $V_{CE} = 5$ V, the data sheets give $\beta_{dc(min)} = 350$. Following the procedure of Section 4.1, we calculate $R_{Th} = R_B = 16.67$ kΩ, $V_{Th} = 5$ V. If we let $V_{BE} = 0.65$ V,

$$I_C = \frac{350(5 - 0.65)}{16.67 + 351(5)} = 0.859 \text{ mA}$$

$$V_{CE} = 15 - 5(0.859) - 5(0.862) = 6.40 \text{ V}$$

These values may be used in conjunction with the transistor specifications to give an improved value of $\beta_{dc(min)}$. From Fig. 9 on the data sheet we estimate a multiplying factor of about 1.2 for this operating point, and we apply it to the reference value of 300 at $I_C = 0.1$ mA, $V_{CE} = 5$ V. This leads to $\beta_{dc(min)} = 360$, just about the same as the value used earlier. Thus we conclude that the operating point is at $I_C = 0.86$ mA, $V_{CE} = 6.4$ V.

Moving into the second part of the analysis, we assume $n = 1$ and have

$$g_m = 38.92|I_C| = 38.92(0.86) = 33.5 \text{ m℧}$$

To find r_π, we need a value for β_0, which usually cannot be found precisely. From the data sheet, we find that $h_{fe} = \beta_0 = 350$ at $I_C = 1$ mA, $V_{CE} = 5$ V. Since our operating point is in that vicinity, let's assume $\beta_0 = 350$, and therefore

$$r_\pi = \frac{\beta_0}{g_m} = \frac{350}{33.5} = 10.45 \text{ k}\Omega$$

Note also that r_d is about 100 kΩ, and quite negligible when in parallel with 5 kΩ and 2 kΩ.

The last step involves the gain and impedance calculations. We have, in order,

$$A_{Vi} = -g_m(R_C \| R_L) = -33.5(5\|2) = -47.9$$

$$Z_i = r_\pi \| R_B = 10.45\|16.67 = 6.42 \text{ k}\Omega$$

$$A_{Vs} = A_{Vi}\frac{(r_\pi \| R_B)}{R_s + (r_\pi \| R_B)} = -47.9\left(\frac{6.42}{7.42}\right) = -41.4$$

$$A_I = A_{Vi}\frac{Z_i}{R_L} = -47.9\left(\frac{6.42}{2}\right) = -154$$

$$Z_o = 5 \text{ k}\Omega$$

$$A_P = A_{Vi}A_I = (-47.9)(-154) = 7380$$

$$V_o = A_{Vs}V_s = -41.4(0.1) = -4.14 = 4.14\underline{/180^\circ} \text{ mV rms}$$

In order to see the effect of a higher-gain unit on these performance parameters, we now use $\beta_{dc(max)} = 1050$, and find

$$I_C = \frac{1050(5 - 0.65)}{16.67 + 1051(5)} = 0.866 \text{ mA}$$

$$V_{CE} = 15 - 5(0.866) - 5(0.867) = 6.34 \text{ V}$$

$$g_m = 38.92(0.866) = 33.7 \text{ m}\mho$$

From the data sheet, $\beta_0 \doteq 1400$ at 1 mA and 5 V, and we let

$$\beta_0 = 1400 \qquad (I_C = 0.866 \text{ mA}, V_{CE} = 6.34 \text{ V})$$

$$r_\pi = \frac{1400}{33.7} = 41.5 \text{ k}\Omega$$

$$A_{Vi} = -33.7(5\|2) = -48.1$$

$$Z_i = 41.5\|16.67 = 11.9 \text{ k}\Omega$$

$$A_{Vs} = -48.1\left(\frac{11.9}{12.9}\right) = -44.4$$

$$A_I = -48.1\left(\frac{11.9}{2}\right) = -286$$

$$A_P = (-48.1)(-286) = 13{,}800$$

$$Z_o = 5 \text{ k}\Omega$$

$$V_o = -44.4(0.1) = -4.44 \text{ mV}$$

We note that both voltage gains, the voltage output, and the output impedance have stayed quite constant, particularly for this large change in β_{dc} and β_0. The output current V_o/R_L is also relatively unchanged. However, β is a measure of current gain, and it is therefore not surprising to find that the current gain and power gain have increased. The increase is by virtue of a decrease in I_i (Z_i increases), since I_o is almost unchanged.

D6.1 In Fig. 6.2*a*, let $V_s = 2\angle 0°$ mV rms, $R_s = 300\ \Omega$, $R_B = 25$ kΩ, $\beta_0 = 400$, $g_m = 80$ m℧, $R_C = 4$ kΩ, and $R_L = 6$ kΩ. Find (a) r_π, (b) A_{Vs}, (c) A_I, (d) the average signal power delivered to the load.

Answers. 5 kΩ; -179.1; -133.3; 21.4 μW

6.2 The common-emitter amplifier: design

Let us now consider the problem of designing a small-signal common-emitter amplifier to meet a set of mid-frequency performance specifications, typically input impedance, output impedance, and voltage gain, although current, power gain, or dc-supply voltage might be given as alternate requirements to be met. Suppose the problem is stated as follows:

> Design a common-emitter amplifier using a 2N5089 transistor to provide a voltage gain $|V_o/V_s| \geq 200$, between a small-signal voltage source having a resistance of 500 Ω and a load $R_L = 5$ kΩ. It is also specified that $Z_i \geq 5$ kΩ.

The first step in the design process is the selection of a suitable circuit topology. Figure 6.4*a* is a stable-operating point version of the common-emitter circuit, and we show $R_s = 500\ \Omega$ and $R_L = 5$ kΩ, as specified. In thinking through the design procedure, we should see that the performance criteria must be satisfied by an appropriate small-signal equivalent circuit but that the parameter values in that circuit are either determined by the dc-equivalent circuit (Fig. 6.4*b*) or actually appear as part of it. Only R_E and the three capacitors will not appear in the small-signal circuit. Let us think first in terms of some reasonable small-signal parameter values and then see whether or not a dc-equivalent circuit can be designed to provide them. If not, we must modify our original small-signal values and try again.

The gain requirement is

$$|A_{Vs}| = \left|\frac{V_o}{V_s}\right| = g_m(R_C \| R_L)\frac{Z_i}{R_s + Z_i} \geq 200$$

If $Z_i \geq 5\ \text{k}\Omega$, and $R_s = 0.5\ \text{k}\Omega$,

$$0.9 < \frac{Z_i}{R_s + Z_i} < 1$$

We try 0.9, which will require the largest value for g_m (or R_C). Next we select a moderate value for R_C so that V_{CC} will not be too large, say $R_C = R_L = 5\ \text{k}\Omega$. Thus

$$g_m > \frac{200}{2500(0.9)} = 88.9\ \text{m}℧$$

and therefore

$$I_C > \frac{88.9}{38.9} = 2.285\ \text{mA}$$

The voltage drop across R_C is seen to be 11.42 V; if we use a 27-V supply, then there are about 15.6 V to divide between V_{RE} and V_{CE}. So let's round these numbers off a little on the safe side and try to set the operating point at $I_C = 2.4$ mA, $V_{CE} = 7.5$ V, and therefore $V_{RE} = 7.5$ V.

We complete the design of the bias circuit by the procedure given in Section 4.1, finding

$$R_E = \frac{7.5}{2.4} = 3.125\ \text{k}\Omega$$

$$I_{B(\text{max})} = \frac{2.4}{480} = 5\ \mu\text{A}$$

since the data sheets indicate that $\beta_{\text{dc(min)}} \doteq 480$ at $I_C = 2.4$ mA. We next select a current through R_1:

$$I_{R1} = 50 I_B = 250\ \mu\text{A}$$

Then

$$R_2 = \frac{V_{RE} + V_0}{I_{R1} - I_B} = \frac{8.15}{0.245} = 33.3\ \text{k}\Omega$$

$$R_1 = \frac{27 - 8.15}{0.250} = 75.4\ \text{k}\Omega$$

$$R_{\text{Th}} = \frac{75.4(33.3)}{108.7} = 23.1\ \text{k}\Omega$$

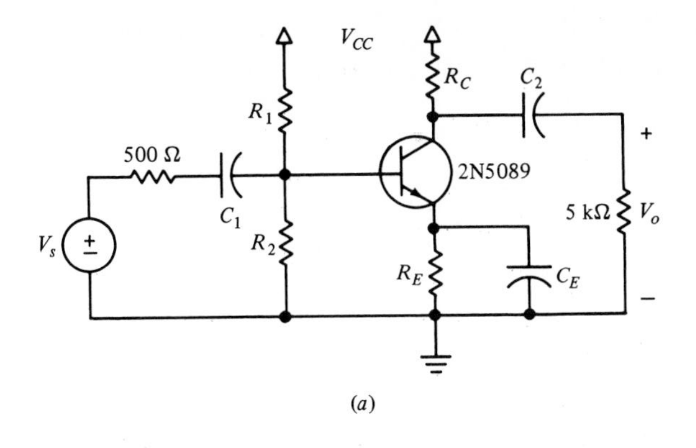

(a)

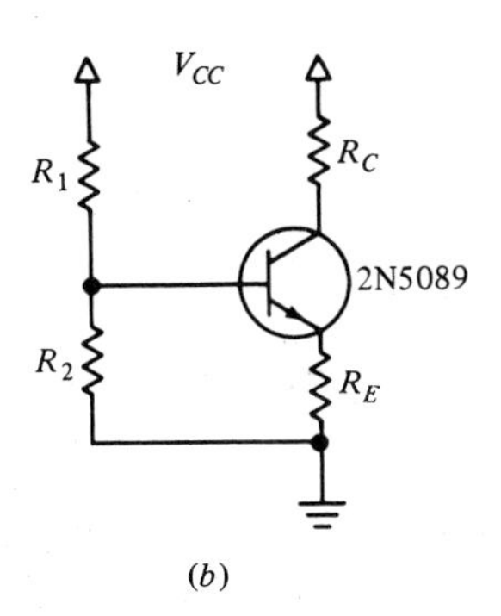

(b)

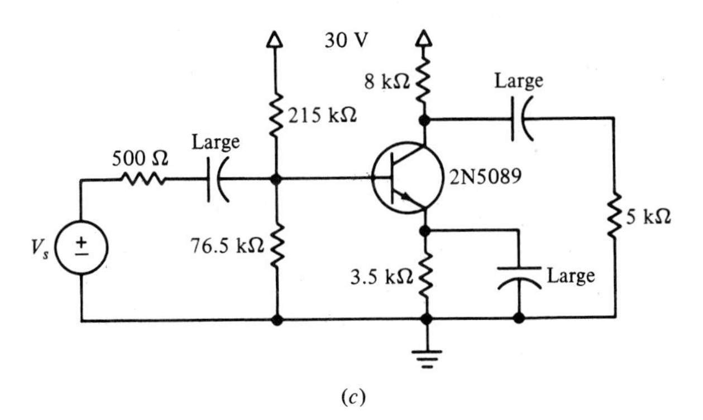

(c)

Fig. 6.4 (*a*) The circuit selected to provide $|V_o/V_s| \geq 200$, $Z_i \geq 5$ kΩ, at mid-frequencies between the given source and load. (*b*) The dc-equivalent circuit. (*c*) The final design gives $V_o/V_s = -219$, $Z_i = 5.24$ kΩ.

Finally, we assume that C_1, C_2, and C_E are very large and produce small reactances in the frequency range of interest. We will assign specific values to these capacitors in Chapter 7.

This completes the preliminary design. We must now carry out an analysis for the lowest-gain unit to see whether or not we have met all the specifications. We check to see that no errors have been made in the dc circuit; of course they haven't, and $I_C = 2.4$ mA, $V_{CE} = 7.5$ V. It follows that $g_m = 38.92(2.4) = 93.4$ m℧. The tabulated value of $\beta_{0(min)}$ is 450 for $I_C = 1$ mA, and the data sheet indicates that β_0 changes only slightly with I_C in this range; 450 is therefore a good choice. Thus, $r_\pi = 450/93.4 = 4.82$ kΩ. Then

$$Z_i = r_\pi \| R_B = 4.82 \| 23.1 = 3.99 \text{ k}\Omega$$

which is less than 5 kΩ, and we see that we have already missed the specifications; Z_i is too low. We need to increase r_π and perhaps R_B as well.

To increase r_π, we must decrease g_m, which will drop the gain below permissible levels unless $(R_C \| R_L)$ can be increased a little. Let us try $I_C = 2$ mA, $g_m = 77.8$ m℧, and $R_C = 8$ kΩ. We increase V_{CC} to 30 V because of the greater voltage drop across R_C, and then let $V_{CE} = 7$ V, $V_{RE} = 7$ V. The new bias design leads to $R_E = 3.5$ kΩ, $I_{B(max)} = 2/480 = 4.17$ μA. With $I_{R1} = 250$ μA, we find that $R_2 = 7.65/0.2458 = 31.1$ kΩ, $R_1 = 22.35/0.250 = 89.4$ kΩ, and $R_{Th} = 23.1$ kΩ. Therefore

$$r_\pi = \frac{450}{77.8} = 5.78 \text{ k}\Omega$$

and

$$Z_i = 4.63 \text{ k}\Omega$$

This is still too low, but at least we have raised r_π above 5 kΩ. It now seems likely that slightly higher values for R_1 and R_2 will lead to victory.

Let us try $I_{R1} = 25 I_{B(max)} = 104$ μA, and then

$$R_2 = \frac{7.65}{0.100} = 76.5 \text{ k}\Omega$$

$$R_1 = \frac{22.35}{0.104} = 215 \text{ k}\Omega$$

$$R_{Th} = R_B = 56.4 \text{ k}\Omega$$

$$Z_i = 5.24 \text{ k}\Omega$$

$$A_{Vs} = -77.8(8\|5)\frac{5.24}{5.74} = -219$$

These results are satisfactory, and the final values are indicated on Fig. 6.4*c*; we could even raise I_{R1} slightly if we wished and achieve a little better sensitivity against a change in β_{dc} or in temperature. The design should be checked for the higher-β transistors, but this is left as a problem for the eager learner. We expect a larger voltage gain and input impedance.

It is hoped that at least one important point has been made in the course of this example: there are few, if any, design problems that have a unique answer, and it is often easier to design and redesign than it is to solve the problem all at once. The use of a sequence of simple formulas, plus a little experience, enables us to gauge the effect of each variable on the amplifier performance and to change parameter values intelligently.

D6.2 Determine the percentage increase in $|V_o/V_s|$ for the amplifier of Fig. 6.4*c* if there is a 10% increase in the value of (a) R_L, (b) R_2, (c) β_0.

Answers. 5.9%; 7.0%; 0.7%

6.3 Common-base and common-collector amplifiers

The bipolar transistor is used most often as an amplifier when it is in the common-emitter configuration. It turns out that this form produces the largest power and current gain, good voltage gain, and intermediate values for the input and output impedance. Certain applications, however, require a very high or very low input or output impedance. Such specifications often lead to the use of the common-collector or common-base amplifier.

We first investigate the common-base amplifier at mid-frequencies and show that it has low input impedance with good voltage gain.

A typical circuit for the common-base amplifier is shown in Fig. 6.5*a*. It should be noted that the dc-equivalent circuit has the identical form of that for the common-emitter amplifier; this dc circuit was considered in Chapter 4. The mid-frequency ac-equivalent circuit appears in Fig. 6.5*b*; we shall study the various performance parameters under the assumption that r_d is very large and may be considered infinite.

We first see that $V_\pi = -V_i$, and we therefore have the voltage gain directly from the equivalent circuit. Since

$$V_o = -g_m V_\pi (R_C \| R_L)$$

we have

$$A_{Vi} = \frac{V_o}{V_i} = \frac{V_o}{-V_\pi} = \frac{-g_m V_\pi (R_C \| R_L)}{-V_\pi} = g_m (R_C \| R_L) \qquad (9)$$

Note the lack of a 180° phase reversal in A_{Vi} as compared to the common-emitter circuit.

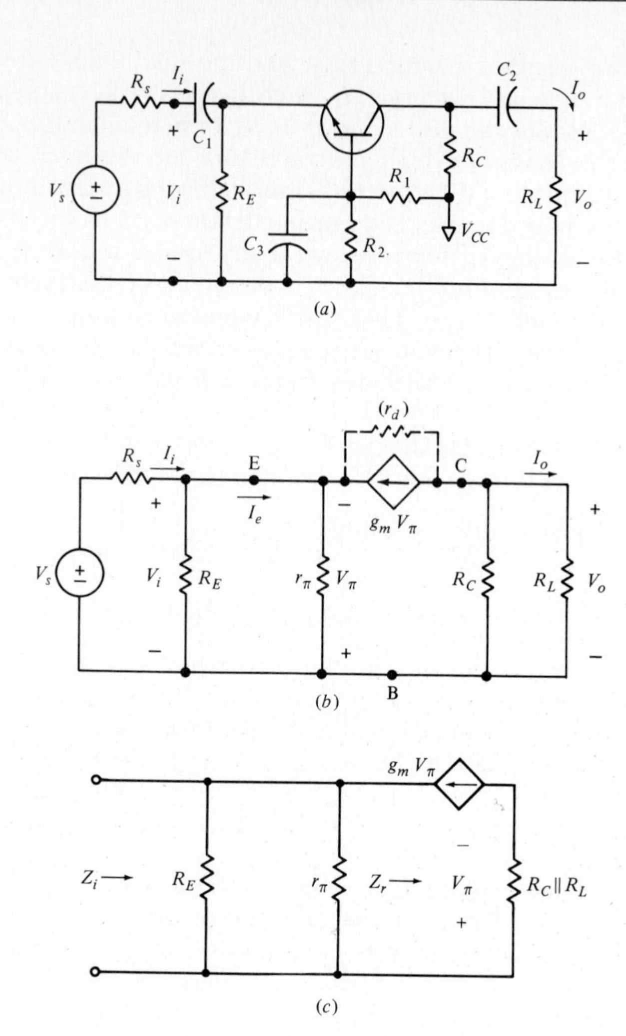

Fig. 6.5 (*a*) A single-stage common-base amplifier circuit. (*b*) The mid-frequency equivalent circuit. (*c*) Z_i is the parallel combination of R_E, r_π, and Z_r.

The input impedance is $Z_i = V_i/I_i$. This is obtained most easily by finding the parallel combination of R_E, r_π, and the impedance seen to the right of r_π, indicated as Z_r in Fig. 6.5*c*. The voltage at these terminals is $-V_\pi$, and the current flowing to the right in the upper conductor is the dependent-source current $-g_m V_\pi$. From Ohm's law,

$$Z_r = \frac{-V_\pi}{-g_m V_\pi} = \frac{1}{g_m}$$

Making the parallel combination, we have

$$Z_i = R_E \| r_\pi \| \frac{1}{g_m} \tag{10}$$

Since g_m usually lies between 10 and 100 m℧, Z_i is certainly quite small for the common-base amplifier, usually less than 100 Ω. It is interesting to note the result of combining r_π and $1/g_m$ in parallel,

$$r_\pi \| \frac{1}{g_m} = \frac{r_\pi/g_m}{r_\pi + 1/g_m} = \frac{r_\pi}{\beta_0 + 1}$$

This is the input impedance seen at the emitter-base terminals of Fig. 6.5*b*. Its value is the parallel resistance r_π reduced by the factor $\beta_0 + 1$. This factor will appear in a number of the results for the common-base and common-collector circuits. The reason it appears is that the dependent-source current, $g_m V_\pi$ or $(\beta_o/r_\pi)V_\pi$, is exactly β_0 times as large as the current in r_π, V_π/r_π. Their sum is the current marked I_e in Fig. 6.5*b*, and

$$I_e = -\frac{V_\pi}{r_\pi} - \beta_0 \frac{V_\pi}{r_\pi} = -\frac{V_\pi}{r_\pi}(\beta_0 + 1)$$

Knowing the input impedance, we may now find the voltage gain with respect to the signal source:

$$A_{Vs} = \frac{V_o}{V_s} = A_{Vi} \frac{Z_i}{R_s + Z_i} \tag{11}$$

This gain can be much less than A_{Vs} whenever $Z_i \ll R_s$.

The current gain is also found easily:

$$A_I = \frac{I_o}{I_i} = \frac{V_o/R_L}{V_i/Z_i} = A_{Vi} \frac{Z_i}{R_L} \tag{12}$$

Let us substitute the pertinent expressions from Eqs. (9) and (10) into Eq. (12):

$$A_I = g_m(R_C \| R_L) \frac{R_E \| r_\pi \| \dfrac{1}{g_m}}{R_L} = \frac{g_m R_C}{R_C + R_L}\left(R_E \| r_\pi \| \frac{1}{g_m}\right)$$

Since the term in parentheses must be less than $1/g_m$, A_I must be less than $R_C/(R_C + R_L)$, which is less than unity. The current gain for the common-base amplifier is therefore actually equivalent to a slight attenuation. We note that the output and input currents are in phase, as is the voltage gain.

The power gain from the amplifier input to the load is

$$A_P = \frac{V_o I_o}{V_i I_i} = A_{Vi} A_I = A_{Vi}^2 \frac{Z_i}{R_L} \tag{13}$$

This result has the same form as that for the common-emitter, but the numerical value is likely to be much lower because the input impedance is usually much smaller than the load resistance.

The only remaining parameter to be found is the Thévenin-equivalent output impedance, which we find by returning to Fig. 6.5*b*, removing R_L, and looking into R_C across the collector-base terminals with V_s set to zero, a short circuit. The open-circuit voltage there is $-g_m V_\pi R_C$, while the short-circuit current is $-g_m V_\pi$. The control voltage V_π is unchanged for an open-circuit or short-circuit termination. The ratio gives the output impedance

$$Z_o = \frac{V_{oc}}{I_{sc}} = \frac{-g_m V_\pi R_C}{-g_m V_\pi} = R_C \tag{14}$$

This is unchanged from the common-emitter result. The value is as large as we wish, and it is independent of the transistor parameters.

Comparing these results with those for the common-emitter, we find that the common-base amplifier has comparable voltage gain and output impedance, much lower input impedance, current gain less than unity, and modest power gain. Most of its applications take advantage of the lower value of Z_i and the lack of 180° phase shift in the voltage and current gain.

The common-base amplifier circuits shown in Fig. 6.6 leads to an ac-equivalent that has almost the same element values as the common-emitter circuit of Fig. 6.3 analyzed in Section 6.1. Only the source resistance has been decreased to a more appropriate low value. The dc operating point is the same as it was, and g_m and r_π are also the same, 33.5 m℧ and 10.45 kΩ respectively. Using these results and the other element values, we

Fig. 6.6 A common-base amplifier containing element values similar to those appearing in the common-emitter circuit of Fig. 6.3.

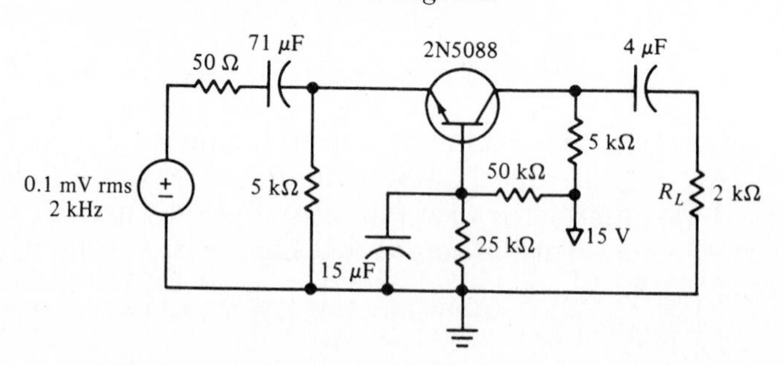

may calculate and compare the performance parameters of this common-base unit with the common-emitter results obtained earlier and given in parentheses:

$$A_{Vi} = 47.9 \qquad (-47.9)$$

$$Z_i = 29.6\ \Omega \qquad (6.42\ \mathrm{k}\Omega)$$

$$A_{Vs} = 17.8 \qquad (-41.4)$$

$$A_I = 0.709 \qquad (-154)$$

$$A_P = 33.9 \qquad (7380)$$

$$Z_o = 5\ \mathrm{k}\Omega \qquad (5\ \mathrm{k}\Omega)$$

The relative magnitudes bear out the general comparison made above.

The common-collector amplifier, often called an *emitter-follower*, is illustrated in Fig. 6.7*a*; its mid-frequency equivalent circuit is shown in Fig. 6.7*b*. If r_d must be included in the analysis, it appears in parallel with R_E in the equivalent circuit. Note also that there is no by-pass capacitor in parallel with R_E, because the signal voltage is present at that point.

We again require expressions for the six performance criteria. The results are simplified somewhat if we agree to let R_L' represent R_E and R_L in parallel as an effective load resistance:

$$R_L' = R_E \| R_L$$

Let us begin by determining the input impedance. Looking at Fig. 6.7*b*, we can see that it is R_B in parallel with the series combination of r_π and the impedance Z_r looking to the right at the emitter-collector terminals, shown in Fig. 6.7*c*. We first find Z_r by squirting a current I_b into the base terminals. Thus $V_\pi = r_\pi I_b$, while $g_m V_\pi = g_m r_\pi I_b = \beta_0 I_b$. Therefore the currents I_b and $\beta_0 I_b$ combine in the effective load resistance R_L':

$$V_o = (I_b + \beta_0 I_b) R_L'$$

and

$$Z_r = \frac{V_o}{I_b} = (\beta_0 + 1) R_L'$$

We see that the resistance connected to the emitter appears in the base circuit multiplied by a factor of $\beta_0 + 1$. This is a general result worth remembering. Note that the higher resistance appears on the base side where the current is smaller.

We now have

$$Z_i = \frac{V_i}{I_i} = R_B \| [r_\pi + (\beta_0 + 1) R_L'] \tag{15}$$

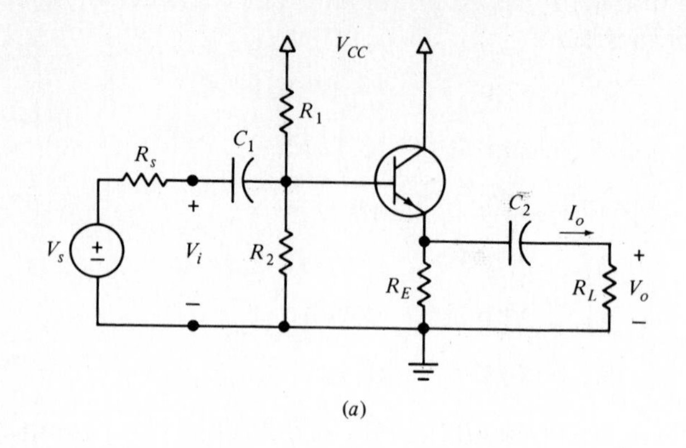

(a)

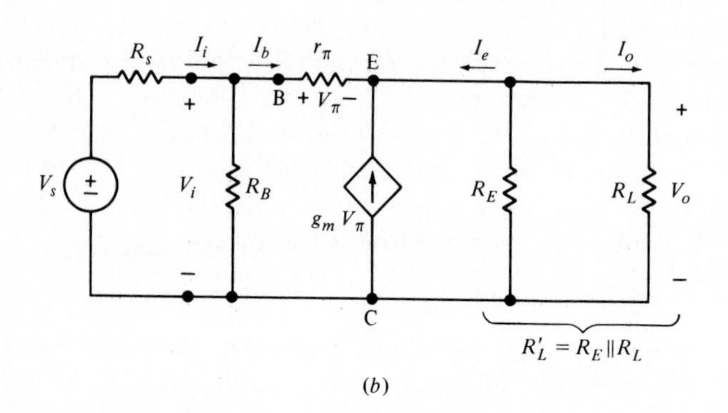

(b)

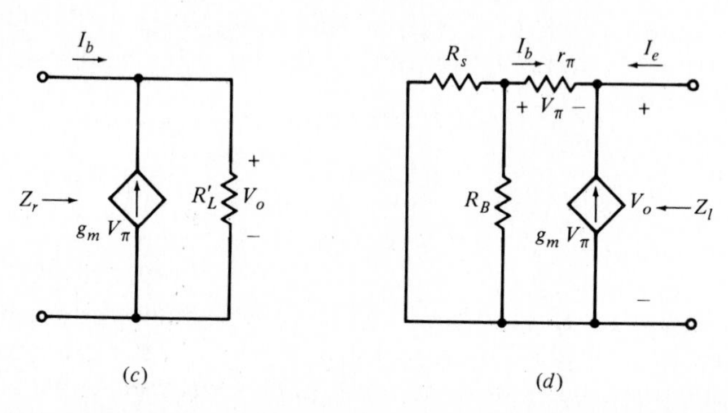

(c) (d)

Fig. 6.7 (*a*) A common-collector amplifier circuit, often called an emitter-follower. (*b*) The mid-frequency equivalent circuit. (*c*) The definitions of Z_r and (*d*) of Z_l.

Both R_B and $(\beta_0 + 1)R_L'$ can be quite large; our first discovery is that the common-collector configuration can have a large input impedance and is often used for that reason. In practice, the magnitude of Z_i is often limited by R_B.

Knowing Z_r, we can use voltage division to help us obtain the voltage gain A_{Vi}:

$$V_o = V_i \frac{Z_r}{r_\pi + Z_r}$$

or

$$A_{Vi} = \frac{V_o}{V_i} = \frac{Z_r}{r_\pi + Z_r} = \frac{(1 + \beta_0)R_L'}{r_\pi + (1 + \beta_0)R_L'} \tag{16}$$

This is certainly less than unity, although it is often as large as 0.9. Note, however, that there is no phase reversal for the voltage gain.

More voltage-divider manipulations give us A_{Vs}:

$$A_{Vs} = \frac{V_o}{V_s} = \frac{V_o}{V_i}\frac{V_i}{V_s} = A_{Vi}\frac{V_i}{V_s} = A_{Vi}\frac{Z_i}{R_s + Z_i} \tag{17}$$

If the right-hand expression is written out in terms of the element values, it is so long that it is frightening. One does better to calculate A_{Vi} and Z_i first. The two voltage gains turn out to be almost equal in magnitude, since Z_i is usually much larger than R_s. In general, we have $A_{Vs} < A_{Vi} < 1$.

As for the current gain, it is once again easy to find in terms of the voltage gain:

$$A_I = \frac{I_o}{I_i} = \frac{V_o/R_L}{V_i/Z_i} = A_{Vi}\frac{Z_i}{R_L} \tag{18}$$

Since the input impedance is apt to be much larger than the load impedance, the current gain has a satisfactorily large value.

We again take the power gain as

$$A_P = \frac{V_o I_o}{V_i I_i} = A_{Vi}A_I = A_{Vi}^2\frac{Z_i}{R_L} \tag{19}$$

The last parameter to be determined is the output impedance, which is the impedance seen looking to the left from R_L with R_L removed and $V_s = 0$. This in turn is R_E in parallel with the impedance seen looking to the left of R_E. This latter impedance is given by the ratio of V_o to I_e, and we call it Z_l, Fig. 6.7*d*. As we have already seen, the dependent-source current is $\beta_0 I_b$; therefore

$$I_e = -(\beta_0 + 1)I_b$$

But I_b flows through the series combination of r_π and $(R_B \| R_s)$, and we have

$$V_o = -I_b[r_\pi + (R_B \| R_s)]$$

Therefore

$$Z_l = \frac{V_o}{I_e} = \frac{r_\pi + (R_B \| R_s)}{\beta_0 + 1}$$

Note that the impedance seen at the emitter is the impedance connected to the base, including r_π, *divided* by $(\beta_0 + 1)$. Impedance levels increase by $(\beta_0 + 1)$ times going from emitter to base, and they decrease by a factor of $(\beta_0 + 1)$ from base to emitter.

Combining Z_l and R_E in parallel, we have

$$Z_o = R_E \left\| \left[\frac{r_\pi + (R_B \| R_s)}{\beta_0 + 1} \right] \right. \tag{20}$$

The factor of $\beta_0 + 1$ causes the bracketed resistance to be of the order of $1/g_m$; combining this in parallel with R_E does not lower the value appreciably. Thus we find that the output impedance of the common-collector is similar in value to the input impedance of the common-base; each is apt to lie in the range from 10 to 500 Ω.

Figure 6.8 shows an emitter-follower that may be compared with the common-emitter and common-base amplifiers of Figs. 6.3 and 6.6 respectively. The dc-equivalent circuit differs in lacking any collector resistance, but this does not affect the value of I_C, and it merely increases V_{CE} from 6.4 to 10.7 V. This has no appreciable effect on g_m, β_0, or r_π, so we can use the values we found earlier: $g_m = 33.5$ m℧ and $r_\pi = 10.45$ kΩ. We obtain the gains and impedance values given below; the corresponding results are given for the other two connections to provide easy comparison.

	C-E	C-B	C-C
$Z_i =$	6.42 kΩ	29.6 Ω	16.4 kΩ
$A_{Vi} =$	−47.9	47.9	0.980
$A_{Vs} =$	−41.4	17.8	0.922
$A_I =$	−154	0.709	7.91
$A_P =$	7380	33.9	7.74
$Z_o =$	5 kΩ	5 kΩ	32.3 Ω

The common-collector or emitter-follower amplifier may be characterized as having high input impedance, low output impedance, and moderate values of current and power gain.

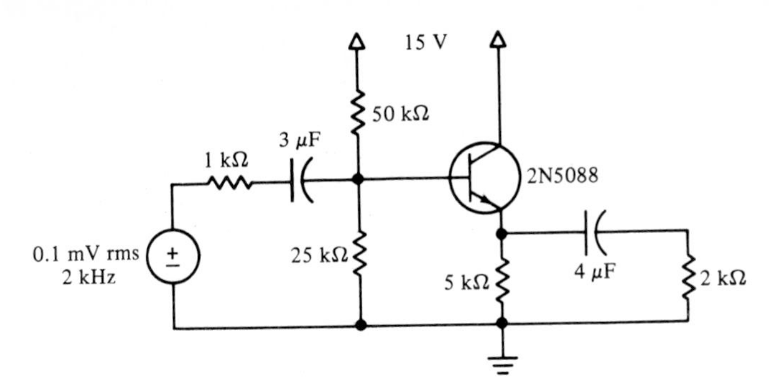

Fig. 6.8 A common-collector amplifier circuit having element values similar to those of the common-emitter of Fig. 6.3 and the common-base of Fig. 6.6.

D6.3 The common-base amplifier shown in Fig. 6.5*a* contains these elements: $R_s = 40\ \Omega$, $R_E = 4\ \mathrm{k}\Omega$, $R_C = R_L = 6\ \mathrm{k}\Omega$, $R_1 = 50\ \mathrm{k}\Omega$, $R_2 = 30\ \mathrm{k}\Omega$, and three large capacitors. If $\beta_0 = 100$ and $g_m = 50$ m℧, find (a) Z_i, (b) Z_o, and (c) A_{Vs}.

Answers. 19.7 Ω; 6 kΩ; 49.5

D6.4 Element values for the common-collector circuit of Fig. 6.7*a* are $R_s = 3\ \mathrm{k}\Omega$, $R_1 = 50\ \mathrm{k}\Omega$, $R_2 = 30\ \mathrm{k}\Omega$, $R_E = 4\ \mathrm{k}\Omega$, $R_L = 6\ \mathrm{k}\Omega$, and two large capacitors. If $\beta_0 = 100$ and $g_m = 50$ m℧, find (a) Z_i, (b) Z_o, (c) A_{Vs}.

Answers. 17.4 kΩ; 44.9 Ω; 0.846

6.4 The common-source amplifier: mid-frequency analysis

We now leave the mid-frequency version of the bipolar-transistor single-stage amplifier and begin considering the performance of the field-effect transistors under similar conditions. The most widely used form is the common-source configuration. Figure 6.9*a* illustrates a typical arrangement for a JFET. The resistor R_{SS} at the source terminal (the JFET source, not the voltage source) and the divider composed of R_1 and R_2 provide operating-point stability over a wide range of temperature and transfer characteristics. The three large capacitors, C_1, C_2, and C_{SS}, are assumed to have reactances at the operating frequencies that are small compared with R_B, R_L, and R_{SS} respectively, where R_B is the parallel combination of R_1 and R_2, as usual. The mid-frequency equivalent circuit then becomes that shown in Fig. 6.9*b*. Note that the interelectrode capacitances C_{gs}, C_{gd}, and C_{ds} may be replaced by open circuits in this frequency range, since their reactances are much larger than all the resistance values.

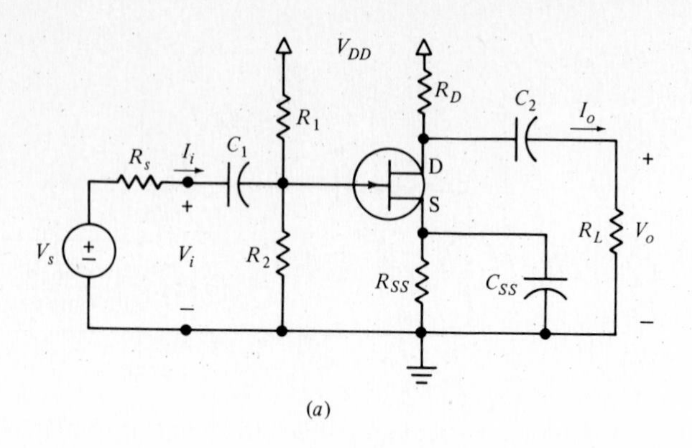

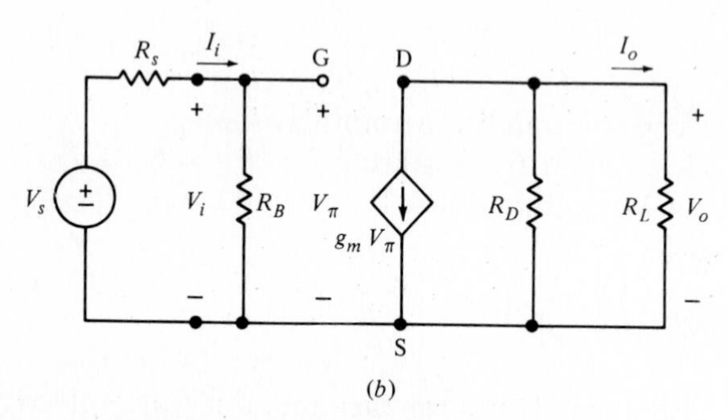

Fig. 6.9 (*a*) A typical common-source single-stage amplifier. (*b*) The equivalent circuit in the mid-frequency range; $R_B = R_1 \| R_2$.

The two voltage gains are obtained by inspection of Fig. 6.9*b*:

$$V_o = -g_m V_\pi (R_D \| R_L) = -g_m V_i (R_D \| R_L)$$

so that

$$A_{Vi} = \frac{V_o}{V_i} = -g_m (R_D \| R_L) \tag{21}$$

and

$$A_{Vs} = \frac{V_o}{V_s} = \frac{R_B}{R_s + R_B} A_{Vi} = -\frac{R_B}{R_s + R_B}(g_m)(R_D \| R_L) \tag{22}$$

Since R_B can be made arbitrarily large, $A_{Vi} \doteq A_{Vs}$ for FET amplifiers.

The input impedance is simply R_B:

$$Z_i = \frac{V_i}{I_i} = R_B \tag{23}$$

and is not a function of any transistor parameters. The current gain depends on Z_i; hence

$$A_I = \frac{I_o}{I_i} = \frac{V_o/R_L}{V_i/Z_i} = A_{Vi}\frac{Z_i}{R_L} = A_{Vi}\frac{R_B}{R_L} \tag{24}$$

This is not a particularly important performance parameter for FET amplifiers. We can make the current gain almost as large as we choose by increasing the value of R_B, since R_1 and R_2 serve only as a potential divider for the gate ($I_G = 0$). The power gain is also wholly under our control by varying the size of R_B:

$$A_P = \frac{V_o I_o}{V_i I_i} = A_{Vi}A_I = A_{Vi}^2\frac{R_B}{R_L} \tag{25}$$

Finally, the output impedance presented to the load R_L is simply

$$Z_o = R_D \tag{26}$$

As an example of a typical set of values for a common-source FET amplifier, let us analyze the circuit shown in Fig. 6.10. For the sake of simplicity, let us assume that the transistor is a unit with $V_P = -3$ V and $I_{DSS} = 9$ mA. From the transfer characteristic,

$$I_D = \frac{9}{(-3)^2}[V_{GS} - (-3)]^2 \tag{27}$$

and the gate-to-source-loop equation,

$$V_{GS} = 20\frac{0.3}{1.7 + 0.3} - (1)I_D \tag{28}$$

we can solve for the operating point by substituting Eq. (27) into Eq. (28), or vice versa, finding that $I_D = 4$ mA and $V_{GS} = -1$ V. Therefore $V_{DS} = 20 - 4(3 + 1) = 4$ V, and the device is operating in the region beyond

Fig. 6.10 A typical common-source amplifier circuit for which the mid-frequency performance parameters are to be determined.

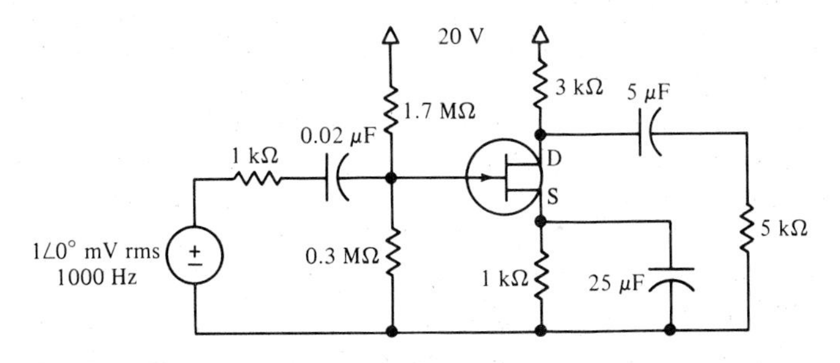

pinch-off, $V_{DS} \geq -1 - (-3)$. The transconductance at this operating point is

$$g_m = \frac{2(9)}{(-3)^2}(-1 + 3) = 4 \text{ m℧}$$

The several gains and impedances are

$$A_{Vi} = -4(3 \| 5) = -7.5$$

$$Z_i = R_B = 1.7 \| 3 = 255 \text{ k}\Omega$$

$$A_{Vs} = -7.5 \frac{255}{256} = -7.47$$

$$A_I = -7.5 \frac{255}{5} = -382$$

$$A_P = (-7.5)(-382) = 2870$$

$$Z_o = 3 \text{ k}\Omega$$

The source is supplying only 3.9 nA rms, or an average power of 3.9 pW.

As a different type of example, consider the circuit shown in Fig. 6.11. This network is known as a *phase splitter*, for it develops equal but opposite output voltages, $V_{o1} = -V_{o2}$, from a single input V_s. It may be used in place of a transformer having a single primary winding and two identical secondary windings (or a single center-tapped secondary). This particular circuit uses an n-channel enhancement-mode MOSFET, but any type of transistor can be used with the appropriate bias circuitry.

From Fig. 6.11b, we see that $V_{o1} = -g_m V_\pi R_D$ and $V_{o2} = g_m V_\pi R_{SS}$. It is apparent that $V_{o1} = -V_{o2}$ whenever $R_{SS} = R_D$. To find the voltage gains, we note that the voltage between gate and ground is

$$V_s \frac{R_B}{R_B + R_s} = V_\pi + g_m V_\pi R_{SS}$$

Solving for V_π yields

$$V_\pi = \frac{V_s}{1 + g_m R_{SS}} \frac{R_B}{R_B + R_s}$$

Since $V_{o1} = -g_m V_\pi R_D$ and $V_{o2} = g_m V_\pi R_{SS}$, the two gains are

$$\frac{V_{o1}}{V_s} = -\frac{g_m R_D}{1 + g_m R_{SS}} \frac{R_B}{R_B + R_s}$$

and

$$\frac{V_{o2}}{V_s} = +\frac{g_m R_{SS}}{1 + g_m R_{SS}} \frac{R_B}{R_B + R_s}$$

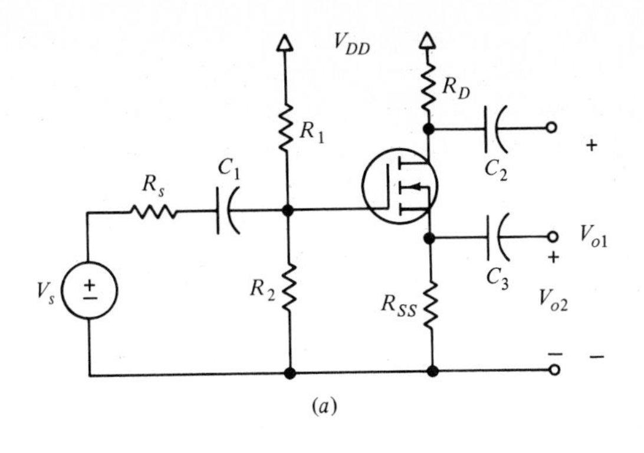

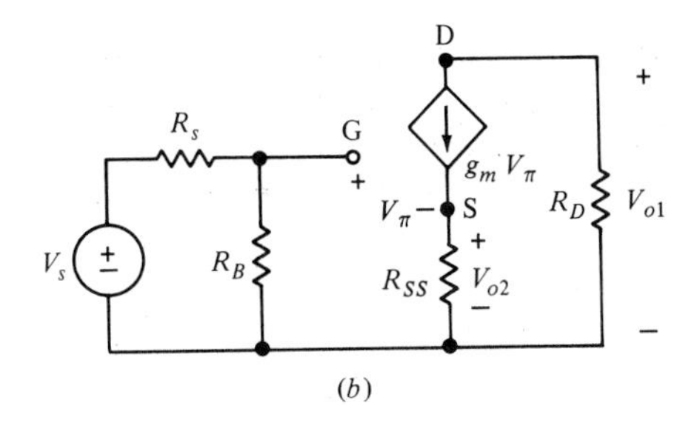

Fig. 6.11 (*a*) A phase-splitter circuit using an *n*-channel enhancement-mode MOSFET. (*b*) The mid-frequency equivalent circuit; $V_{o1} = -V_{o2}$.

Usually $R_B \gg R_s$, while $g_m R_D$ and $g_m R_{SS}$ range from 1 to 100. Thus each of the two output voltages is slightly less than the open-circuit source voltage. The phase-splitting is therefore accomplished without any voltage amplification, but current and power gains greater than unity are possible.

D6.5 Element values for a C-S amplifier are $R_s = 800\ \Omega$, $R_1 = 750\ \text{k}\Omega$, $R_2 = 150\ \text{k}\Omega$, $R_D = 4\ \text{k}\Omega$, $R_L = 6\ \text{k}\Omega$, $R_{SS} = 2.5\ \text{k}\Omega$, $V_{DD} = 24$ V, $I_{DSS} = 8$ mA, and $g_{m0} = 8$ m℧. Assuming operation at mid-frequencies where C_1, C_2, and C_{SS} can be neglected, find (a) A_{Vi}, (b) A_{Vs}, (c) A_I, (d) A_P, (e) Z_i, (f) Z_o.

Answers. -9.60; -9.54; -200; 1920; 125 kΩ; 4 kΩ

6.5 The common-source amplifier: design

The design problem we shall use to illustrate a suitable procedure may be stated as on the following page.

> Design a common-source amplifier to provide an input impedance greater than 1 MΩ and a voltage gain $|A_{Vi}| \geq 20$ betweeen a 10-kΩ load and a source developing 1 mV at 2 kHz with $R_s = 1$ kΩ.

To these conditions let us also add the constraint that we must utilize an available *n*-channel depletion-mode MOSFET having a range of pinch-off voltages $-2 \leq V_P \leq -0.5$ V, a corresponding range of drain-to-source saturation currents $10 \geq I_{DSS} \geq 4$ mA, and a range of g_{m0} from 10 to 16 m℧. A circuit of the form shown in Fig. 6.12*a* is selected, and the two limiting dc transfer characteristics are sketched as solid curves in Fig. 6.12*b*. We see that we can achieve a maximum value for g_m, $g_{m0} = 2(10)/2$, or 10 m℧, for the unit with the higher I_{DSS}, while $g_{m0} = 2(4)/0.5 = 16$ m℧ for the smaller I_{DSS} variety. Note that since there is a range of values for V_P, other values of g_{m0} may be found associated with a specific value of I_{DSS}. Two such curves that will be investigated later as possible worst cases are shown as broken lines in Fig. 6.12*b*.

As we did with the common-emitter design, it is helpful to obtain some insight into the required value of g_m by making a rough estimate from the gain equation:

$$|A_{Vi}| = g_m(R_D \| 10) \geq 20$$

If we let $R_D = 10$ kΩ, then $g_m \geq 4$ m℧, a value that should be easy to obtain. Suppose that we identify the two operating points at which $g_m = 4$ m℧ on the limiting transfer characteristics. For the upper curve,

$$4 = 5(V_{GS} + 2)$$

$$V_{GS} = -1.2 \text{ V}$$

and

$$I_D = 2.5(-1.2 + 2)^2 = 1.6 \text{ mA}$$

The lower unit requires

$$4 = 32(V_{GS} + 0.5)$$

$$V_{GS} = -0.375 \text{ V}$$

and

$$I_D = 16(-0.375 + 0.5)^2 = 0.25 \text{ mA}$$

These two operating points are identified in Fig. 6.12*b*. Any operating point at a higher value of drain current is satisfactory, since g_m would be larger, and therefore any bias line passing through or above both points will lead to a sufficiently high value of g_m. To simplify the circuit, we let R_1 be infinite, so that the bias load-line equation is

$$V_{GS} = -R_{SS} I_D$$

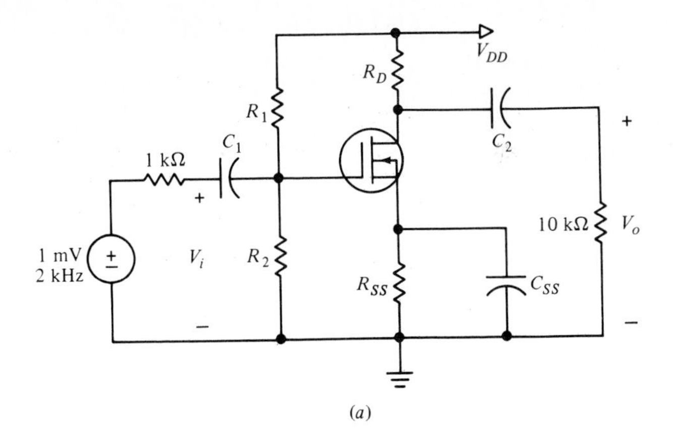

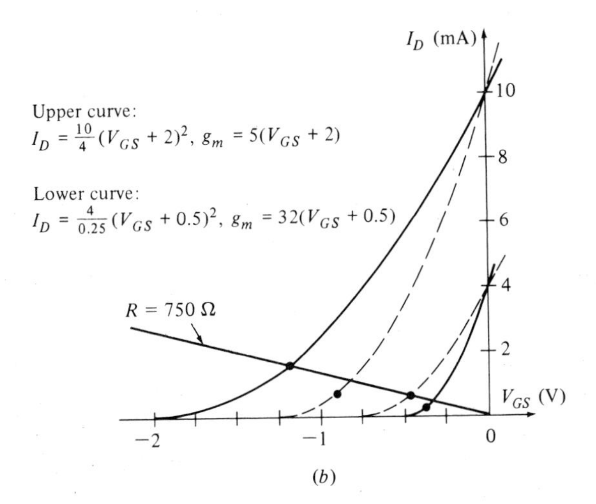

Fig. 6.12 (*a*) The topology selected for a common-source amplifier design that must give $|A_{Vi}| \geq 20$ for a range of MOSFETs. (*b*) The two extremes of the transfer characteristics are shown, along with the final choice of load line. The broken lines are transfer characteristics that are investigated later.

To pass through the point -1.2 V, 1.6 mA, we must have $R_{SS} = 1.2/1.6$ or 750 Ω. A few more calculations show that the intersection with the lower curve is at -0.333 V, 0.444 mA. This is above the lower limit of the possible operating points on that curve and is therefore quite satisfactory.

Two other possible worst cases should also be investigated. One is the transfer characteristic that is obtained for $I_{DSS} = 10$ mA and $g_{m0} = 16$ m℧; therefore $V_P = -2I_{DSS}/g_{m0} = -1.25$ V. This is shown as a broken line in Fig. 6.12*b*. The other occurs when $I_{DSS} = 4$ mA and $g_{m0} = 10$ m℧, so that $V_P = -0.8$ V; this is also indicated in Fig. 6.12*b*. The points at which $g_m = 4$ m℧ are identified and are seen to lie safely on or below the 750-Ω bias line.

The maximum possible value of I_D is 1.6 mA, so our selection of R_D as 10 kΩ leads to $V_{RD} = 16$ V. Since $V_{RSS} = 0.75(1.6) = 1.2$ V, a supply voltage $V_{DD} = 20$ V will leave $V_{DS} = 2.8$ V and ensure operation beyond pinch-off.

We complete the design by selecting $R_2 = R_B = 1$ MΩ, thus having the values $R_D = 10$ kΩ, $R_{SS} = 750$ Ω, $R_1 = \infty$, $R_2 = 1$ MΩ, and $V_{DD} = 20$ V. Checking our design through analysis, we obtain the following values for the two different units:

I_{DSS} =	10 mA	4 mA
V_P =	− 2 V	− 0.5 V
I_D =	1.6 mA	4/9 mA
V_{GS} =	− 1.2 V	− 1/3 V
V_{DS} =	2.8 V	15.22 V
A_{Vi} =	− 20	− 26.7
Z_i =	1 MΩ	1 MΩ

Both devices meet specifications, and we assume that transistors with intermediate characteristics will provide values of gain whose magnitudes are greater than 20.

D6.6 A common-source amplifier is operating with $R_s = 1$ kΩ, $R_1 = 600$ kΩ, $R_2 = 150$ kΩ, $R_D = 4.5$ kΩ, $R_L = 9$ kΩ, $R_{SS} = 2$ kΩ, $V_{DD} = 20$ V, and $I_D = 2.5(V_{GS} + 2)^2$ mA. If it is desired to reduce V_{DS} to 3 V, what new value should be selected for (a) R_D, (b) R_2, (c) R_1?

Answers. 4.80; 162.1; 555 kΩ

6.6 Common-gate and common-drain amplifiers

The common-gate amplifier is shown in Fig. 6.13*a*. As we found in Section 4.7, the analysis or design problems associated with the dc operating point are identical to those of the common-source circuit. Capacitors C_1, C_2, and C are large, and they approximate short circuits for the signal at the mid-frequencies. Again remembering the discussion in Chapter 4, we may be able to achieve a suitable design without using R_1 and R_2 ($R_1 = \infty$, $R_2 = 0$), depending on the range of temperature expected, the variability of the FET parameters, and the desired input impedance.

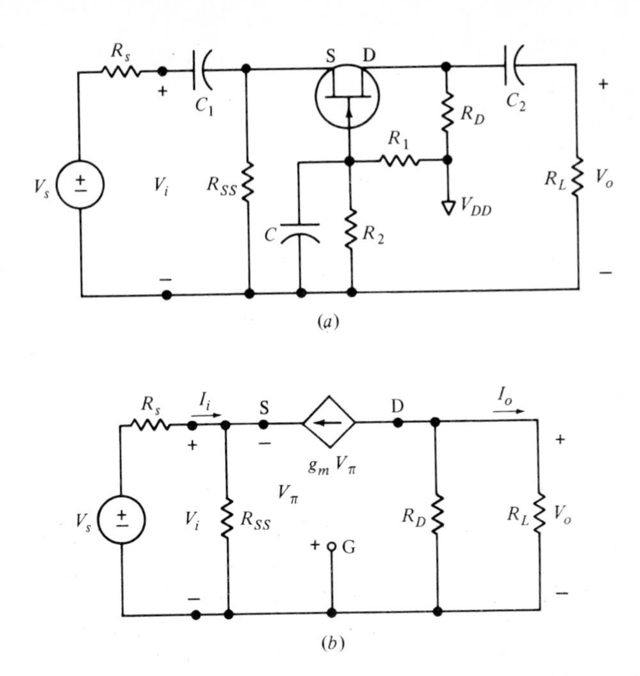

Fig. 6.13 The general circuit diagram of a common-gate amplifier is shown in (*a*) and its mid-frequency equivalent in (*b*).

The performance characteristics of the common-gate amplifier are similar to those of the common-base circuit: low input impedance, high output impedance, and good voltage gain. As a matter of fact, the similarity is so great, as shown by a comparison of the mid-frequency equivalents in Fig. 6.13*b* and 6.5*b*, that we may use all the common-base results by letting $R_E \to R_{SS}$, $R_C \to R_D$, and $r_\pi \to \infty$. We have

$$A_{Vi} = g_m(R_D \| R_L) \tag{29}$$

$$Z_i = R_{SS} \left\| \frac{1}{g_m} \right. \tag{30}$$

$$A_{Vs} = A_{Vi} \frac{Z_i}{R_s + Z_i} \tag{31}$$

$$A_I = A_{Vi} \frac{Z_i}{R_L} \tag{32}$$

$$A_P = A_{Vi}^2 \frac{Z_i}{R_L} \tag{33}$$

$$Z_o = R_D \tag{34}$$

(29)–(34): common-gate

Before considering a numerical example, let us complete the FET-amplifier picture by presenting the common-drain circuit of Fig. 6.14*a* and its mid-frequency equivalent in Fig. 6.14*b*. By analogy with the common-collector (emitter-follower) circuit, we may conclude that the common-drain amplifier or *source-follower* will be blessed with a low output impedance, the price being a voltage gain slightly less than unity. The voltage and current gains will not show a phase reversal.

The gains and impedances for the source-follower could be found from those pertaining to the emitter-follower by comparing the equivalent circuits of Figs. 6.14*b* and 6.7*b*, replacing R_E by R_{SS} and β_0 by r_π/g_m and then letting r_π become infinite. However, the educational benefit would be minimal, and it is easy to derive the formulas anew from Fig. 6.14*b*. We let

$$R_L' = R_{SS} \| R_L$$

and then

$$V_o = g_m V_\pi R_L' = g_m (V_i - V_o) R_L'$$

since $V_\pi = V_i - V_o$. Therefore

$$A_{Vi} = \frac{V_o}{V_i} = \frac{g_m R_L'}{1 + g_m R_L'} \tag{35}$$

and

$$Z_i = \frac{V_i}{I_i} = R_B \tag{36}$$

so that

$$A_{Vs} = \frac{V_o}{V_s} = A_{Vi} \frac{R_B}{R_B + R_s} = \frac{g_m R_L'}{1 + g_m R_L'} \frac{R_B}{R_B + R_s} \tag{37}$$

Moreover

$$A_I = \frac{V_o/R_L}{V_i/Z_i} = A_{Vi} \frac{R_B}{R_L} \tag{38}$$

and

$$A_P = A_{Vi}^2 \frac{R_B}{R_L} \tag{39}$$

Both the current and power gains can be made as large as we wish by increasing the magnitude of R_B. Only the output impedance remains to be found, and Fig. 6.14*b* shows this to be R_{SS} in parallel with the impedance seen looking to the left into the source-drain terminals. The current into the source is $-g_m V_\pi$ and the source-to-drain voltage is V_o. But V_π

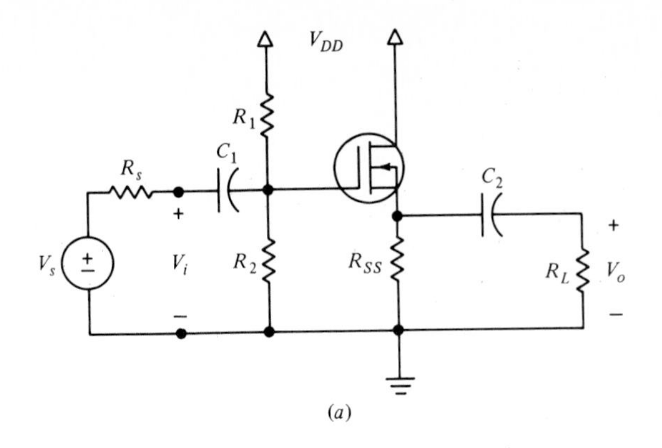

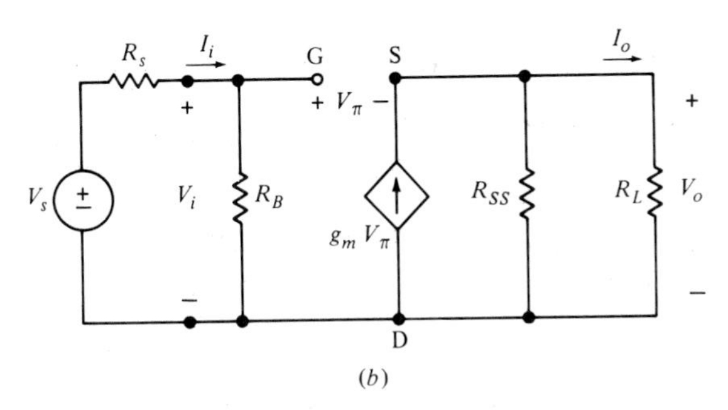

Fig. 6.14 (a) A representative common-drain amplifier, or source-follower. (b) The mid-frequency equivalent circuit; $R_B = R_1 \| R_2$.

$= -V_o$, since V_s and V_i are both zero. Therefore R_{SS} is in parallel with a resistance of $V_o/[-g_m(-V_o)] = 1/g_m$, and

$$Z_o = R_{SS} \left\| \frac{1}{g_m} \right. \tag{40}$$

Since g_m is of the order of 10 m℧, it is apparent that the output impedance is of the order of 100 Ω or less.

To gain some quantitative feeling for the relative values of these parameters, we can analyze C-G and C-D amplifiers that are similar to the C-S circuit of Fig. 6.10 discussed earlier. The two comparable circuits are shown in Fig. 6.15. The only change in the dc operating point is an increased value of V_{DS} for the common-drain circuit, which does not affect the earlier 4-m℧ value of g_m. The analytical results for the three amplifiers are tabulated on the following page. They may also be compared with the performance of typical bipolar amplifiers, as tabulated at the end of Section

	C-S	C-G	C-D
Z_i =	255 kΩ	200 Ω	255 kΩ
A_{Vi} =	−7.5	7.5	0.769
A_{Vs} =	−7.47	6	0.766
A_I =	−382	0.3	39.2
A_P =	2870	2.25	30.2
Z_o =	3 kΩ	3 kΩ	200 Ω

6.3. Comparing these three types of FET amplifiers, we note that the C-S circuit is characterized by moderate voltage gain with phase reversal and a high input impedance, the C-G configuration by moderate voltage gain without phase reversal and a low input impedance, and the C-D arrangement by a high input impedance, a low output impedance, and a voltage gain less than unity without phase reversal.

Fig. 6.15 (*a*) A common-gate amplifier and (*b*) a common-drain amplifier that have element values comparable to the common-source amplifier of Fig. 6.10.

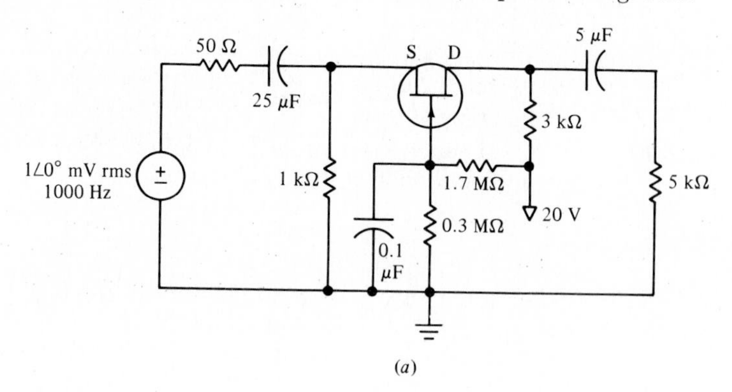

(*a*)

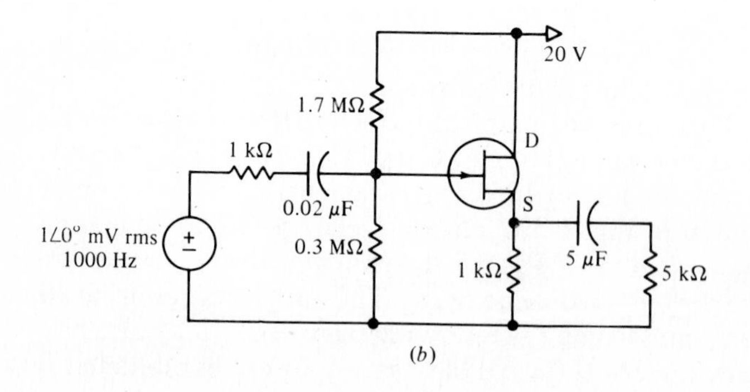

(*b*)

Compared with bipolar-transistor amplifiers, the FET circuits generally provide less gain but much greater input impedance. As we shall see in the next chapter, FETs maintain their gain and high input impedance at high frequencies, a fact which accounts for their widespread use in that frequency range.

D6.7 Element values for the common-gate amplifier of Fig. 6.13*a* are $R_s = 40\ \Omega$, $R_{SS} = 2.5\ \text{k}\Omega$, $R_D = 4\ \text{k}\Omega$, $R_L = 6\ \text{k}\Omega$, $R_1 = 750\ \text{k}\Omega$, $R_2 = 150\ \text{k}\Omega$, $V_{DD} = 24$ V, and three large capacitors. Let $I_{DSS} = 8$ mA and $g_{m0} = 8\ \text{m℧}$. Find (a) A_{Vi}, (b) A_{Vs}, (c) A_I, (d) A_P, (e) Z_i, (f) Z_o.

Answers. 9.6; 8.16; 0.364; 3.49; 227 Ω; 4 kΩ

D6.8 Element values for the common-drain amplifier of Fig. 6.14*a* are $R_s = 800\ \Omega$, $R_{SS} = 2.5\ \text{k}\Omega$, $R_L = 1\ \text{k}\Omega$, $R_1 = 750\ \text{k}\Omega$, $R_2 = 150\ \text{k}\Omega$, $V_{DD} = 24$ V, and two large capacitors. Let $I_{DSS} = 8$ mA and $g_{m0} = 8\ \text{m℧}$. Find (a) A_{Vi}, (b) A_{Vs}, (c) A_I, (d) A_P, (e) Z_i, (f) Z_o.

Answers. 0.741; 0.736; 92.6; 68.6; 125 kΩ; 227 Ω

6.7 Amplifier analysis in terms of *h*-parameters

The *h*-parameters were introduced in Section 5.6. They are only used at low- or mid-frequencies for bipolar transistors. Figure 6.16*a* shows the signal variables we used previously for a common-emitter amplifier. We have again

$$V_{be} = h_{ie} I_b + h_{re} V_{ce}$$

$$I_c = h_{fe} I_b + h_{oe} V_{ce}$$

Figure 6.16*b* shows the source and load connected to the transistor and uses the input and output variables we have applied to amplifiers. Let us replace V_{be} by V_i and V_{ce} by V_o, and write

$$V_i = h_{ie} I_b + h_{re} V_o \tag{41}$$

$$I_c = h_{fe} I_b + h_{oe} V_o \tag{42}$$

At the output,

$$V_o = -I_c (R_C \| R_L) = -I_c R_L' = -\frac{I_c}{Y_L'} \tag{43}$$

where

$$Y_L' = \frac{1}{R_C} + \frac{1}{R_L}$$

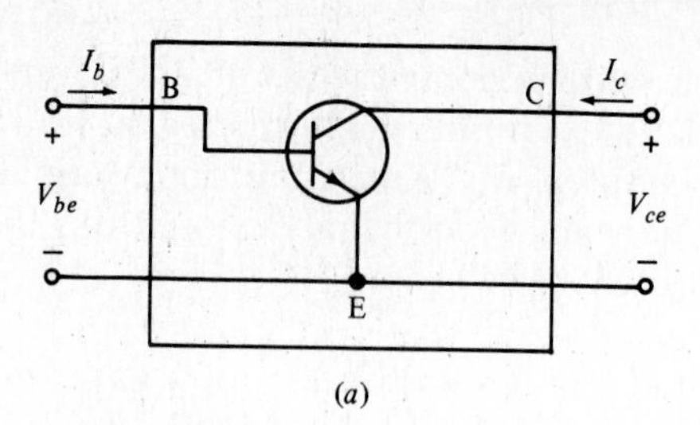

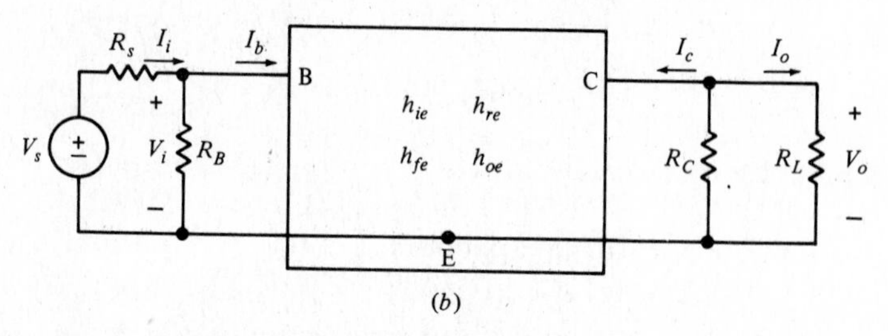

Fig. 6.16 (*a*) The signal variables used with a bipolar transistor in common-emitter configuration are shown. (*b*) Source and load are added to the basic transistor circuit.

We would like to determine again the input and output impedances and the various gains for the C-E amplifier. We begin with the input impedance. By eliminating V_o and I_c from these equations, it is not too difficult to find the input impedance at the base-emitter terminals:

$$Z_{be} = \frac{V_i}{I_b} = h_{ie} - \frac{h_{re}h_{fe}}{h_{oe} + Y_L'} \tag{44}$$

Having this input impedance, we may now move closer to the source and give the input impedance presented to it:

$$Z_i = \frac{V_i}{I_i} = R_B \| Z_{be} \tag{45}$$

Next we tackle the current gain, I_o/I_i. The voltage V_o is eliminated from Eqs. (42) and (43), yielding

$$\frac{I_c}{I_b} = \frac{h_{fe}Y_L'}{h_{oe} + Y_L'}$$

Current division is then used at the input,

$$\frac{I_b}{I_i} = \frac{R_B}{R_B + Z_{be}}$$

and at the output,

$$\frac{I_o}{I_c} = -\frac{R_C}{R_C + R_L}$$

Multiplying these last three expressions together, we have the current gain

$$A_I = \frac{I_o}{I_i} = \frac{I_b}{I_i}\frac{I_c}{I_b}\frac{I_o}{I_c} = \left(\frac{R_B}{R_B + Z_{be}}\right)\left(\frac{h_{fe}Y_L'}{h_{oe} + Y_L'}\right)\left(-\frac{R_C}{R_C + R_L}\right) \tag{46}$$

With these results, it is now easy to write an expression for the voltage gains:

$$A_{Vi} = \frac{V_o}{V_i} = \frac{I_o R_L}{I_i Z_i} = A_I \frac{R_L}{Z_i} \tag{47}$$

and

$$A_{Vs} = \frac{V_o}{V_s} = A_{Vi}\frac{Z_i}{R_s + Z_i} \tag{48}$$

For the power gain,

$$A_P = \frac{V_o I_o}{V_i I_i} = A_{Vi}A_I \tag{49}$$

For completeness, we may also obtain the output impedance. We set $V_s = 0$ and relate the two input variables,

$$V_i = -I_b(R_s \| R_B)$$

We then use this result with Eqs. (41) and (42) to eliminate both of these input variables. The net result is

$$I_c = \left[h_{oe} - \frac{h_{re}h_{fe}}{h_{ie} + (R_s \| R_B)}\right]V_o$$

Looking into the collector-emitter terminals, we therefore see the admittance,

$$Y_{ce} = \frac{I_c}{V_o} = h_{oe} - \frac{h_{re}h_{fe}}{h_{ie} + (R_s \| R_B)}$$

The output admittance or output impedance is obtained by including R_C in parallel:

$$Y_o = \frac{1}{Z_o} = \frac{1}{R_C} + h_{oe} - \frac{h_{re}h_{fe}}{h_{ie} + (R_s \| R_B)} \tag{50}$$

These formulas are more complicated and their derivation is more difficult than those we have been using for the hybrid-π parameters, but the present treatment is also more general. The transistor open-circuit output admittance, $h_{oe} = 1/r_d$, is included, as is the reverse voltage feedback ratio h_{re}. In the hybrid-π model, r_d was usually neglected ($h_{oe} = 0$) and h_{re} was assumed to be zero throughout our entire discussion. This last assumption is usually quite appropriate.

D6.9 A common-emitter amplifier using a 2N5088 transistor at low current levels might have these h parameters: $h_{ie} = 50$ kΩ, $h_{re} = 5 \times 10^{-4}$, $h_{fe} = 400$, $h_{oe} = 5$ μ℧. If $R_s = 10$ kΩ, $R_B = 500$ kΩ, and $R_C = R_L = 20$ kΩ, find (a) Z_i, (b) A_I, (c) A_{Vi}, (d) A_{Vs}, (e) A_P, (f) Z_o.

Answers. 43.9 kΩ; −173.8; −79.2; −64.5; 13,760; 19.36 kΩ

D6.10 Repeat Problem D6.9, except let $h_{re} = 0$.

Answers. 45.5 kΩ; −173.2; −76.2; −62.5; 13,190; 18.18 kΩ

Problems

1. For the amplifier of Fig. 6.1a, let $R_s = 1$ kΩ, $R_1 = 40$ kΩ, $R_2 = 30$ kΩ, $R_C = R_L = 6$ kΩ, $R_E = 4$ kΩ, and $V_{CC} = 12$ V. Assume the transistor is a 2N5377 operating at mid-frequencies with $V_{BE} = 0.65$ V and $n = 1$. (a) For the minimum-gain unit, find A_{Vi}, A_{Vs}, A_I, A_P, Z_i, and Z_o. (b) Repeat for the maximum-gain unit.
2. If it is assumed that the transistor in the amplifier of Fig. 6.3 may be represented by an ac-equivalent circuit comprised of $r_\pi = 10$ kΩ and $g_m = 40$ m℧, find the signal power dissipated in each of the six external resistors.
3. Change R_2 from 25 to 30 kΩ in Fig. 6.3 and find A_{Vi}, A_{Vs}, A_I, A_P, Z_i, and Z_o for the lower-gain unit. Use $V_{BE} = 0.65$ V.
4. Parameter values for the mid-frequency equivalent circuit of a C-E amplifier are $V_s = 5\angle 0°$ mV, $R_s = 0.5$ kΩ, $R_B = 10$ kΩ, $r_\pi = 4$ kΩ, $g_m = 50$ m℧, $R_C = 4$ kΩ, and $R_L = 6$ kΩ. Find the signal power supplied by each of the two sources and the signal power dissipated in each of the five resistors.
5. A C-E amplifier has the following parameter values: $R_s = 600$ Ω, $R_B = 24$ kΩ, $R_C = 10$ kΩ, $g_m = 52$ m℧, and $\beta_0 = 130$. The value of R_L is selected so that it will receive a maximum signal power. Calculate A_{Vs} and A_P.
6. In Fig. 6.1a, let $R_s = 500$ Ω, $R_1 = 50$ kΩ, $R_2 = 30$ kΩ, $R_C = 4$ kΩ, $R_E = 3$ kΩ, $R_L = 6$ kΩ, and $V_{CC} = 25$ V. The transistor is an npn unit with $V_0 = 5/8$ V, $R_{BB} = 6.25$ kΩ, $I_{CEO} \doteq 0$, $\beta_{dc} = 200$, $n = 1.25$, and

$\beta_0 = 215$. If $v_s(t) = 2 \cos \omega t$ mV and the capacitors are very large, find $v_o(t)$.

7. Design a common-emitter amplifier, using 2N5088 Unit 3, to provide a mid-frequency voltage gain $|V_o/V_s| \geq 100$, between a small-signal voltage source having a resistance of 500 Ω and a load $R_L = 5$ kΩ. It is also required that $|Z_i| \geq 10$ kΩ and $V_{CC} \leq 18$ V.
8. Design a common-emitter amplifier to provide a gain $|A_{Vs}| \geq 200$ between a source having $R_s = 0.3$ kΩ and a load $R_L = 6$ kΩ. The input impedance is to be greater than 1 kΩ and $V_{CC} \leq 20$ V. Facts about the transistor are: $150 \leq \beta_{dc} \leq 400$, $\beta_0 = \beta_{dc}$, $n = 1$, $0.6 \leq V_{BE} \leq 0.7$ V, $I_{CEO} \doteq 0$, $R_{BB} = 5$ kΩ, $r_d \doteq \infty$, and $V_A = 150$ V.
9. In the small-signal equivalent circuit of Fig. 6.2*a*, let $R_s = 1$ kΩ and $R_L = 5$ kΩ. Now select values for R_B, R_C, and g_m so that $|A_{Vs}| \geq 50$, $|Z_i| \geq 2.5$ kΩ, $Z_o = 4$ kΩ, and $R_B \leq 25$ kΩ. Assume that $90 \leq \beta_0 \leq 300$.
10. (a) Design a common-emitter amplifier using a 2N5376 transistor that will provide a mid-frequency voltage gain $|A_{Vs}|$ of at least 100 between a source having an internal resistance of 150 Ω and a capacitively coupled load resistance of 3 kΩ. Use $V_{CC} = 24$ V. (b) What is the maximum voltage gain that may be expected from this amplifier?
11. A common-emitter amplifier has a voltage source with $R_s = 100$ Ω, a load $R_L = 8$ kΩ, and a 12-V collector supply voltage. Select values for R_1, R_2, R_C, and R_E so that $g_m \geq 50$ m℧ and $|A_{Vs}| \geq 100$ if the transistor parameters are $120 \leq \beta_{dc} \leq 360$, $\beta_0 = 0.9\beta_{dc}$, $0.6 \leq V_0 \leq 0.7$ V, and $n = 1$, and I_{CEO} is negligible.
12. Figure 6.17 illustrates an integrated-circuit form of a common-emitter amplifier using two identical transistors. (a) Show that the voltage gain $V_o/V_i = -g_m R_L$. (b) Show that the input impedance at the V_i terminals is $(r_\pi/2) \| R_1 \| (1/g_m)$.
13. In the circuit of Fig. 6.17, let $V_{CC} = 15$ V, $R_1 = 12$ kΩ, and $R_L = 8$ kΩ, while $\beta_{dc} = 100$, $\beta_0 = 110$, $V_{BE} = 0.65$ V, and $I_{CEO} \doteq 0$ for

Fig. 6.17 See Problems 12 and 13.

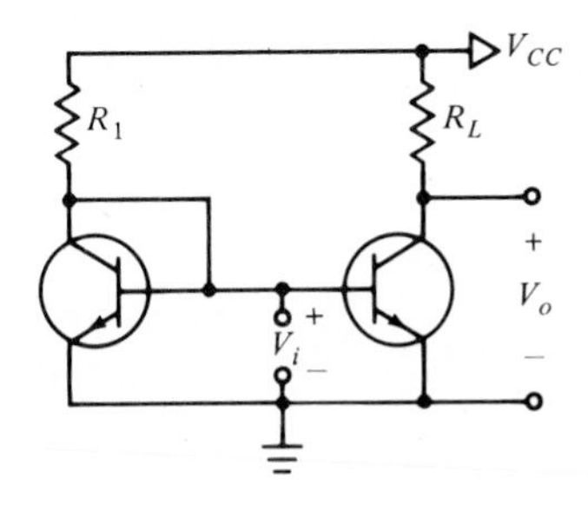

both transistors. (a) Calculate V_o/V_i, using information from Problem 12 if needed. (b) If a source V_s in series with $R_s = 1$ kΩ and a large capacitor is connected to the V_i terminals, find the voltage gain V_o/V_s.

14. If r_d ($= 1/g_d$) is included in the ac-equivalent of the common-base amplifier, (a) show that the appropriate expression for A_{Vi} is $[(g_d + g_m)(R_C \| R_L)]/[1 + g_d(R_C \| R_L)]$. (b) Show that Z_o is $R_C \| [r_d + (1 + r_d g_m)(R_E \| R_s \| r_\pi)]$.
15. Element values in the common-base amplifier circuit of Fig. 6.5*a* are $R_s = 50$ Ω, $R_E = 4$ kΩ, $R_2 = 23.1$ kΩ, $R_1 = 66.7$ kΩ, and $R_C = R_L = 6$ kΩ. The capacitors are large and $V_{CC} = 20$ V. Let $\beta_{dc} = \beta_0 = 100$, $V_0 = 0.65$ V, and $n = 1$, and find A_{Vi}, A_{Vs}, A_I, A_P, Z_i, and Z_o.
16. A C-B amplifier (Fig. 6.5*a*) contains $R_s = 80$ Ω, $R_E = 3$ kΩ, $R_2 = 30$ kΩ, $R_1 = 70$ kΩ, $R_C = 5$ kΩ, $R_L = 7$ kΩ, $V_{CC} = 18$ V, and three large capacitors. Transistor parameters are $V_0 = 0.7$ V, $n = 1.2$, $\beta_{dc} = 120$, and $\beta_0 = 125$. Calculate A_{Vs} and Z_i.
17. Design a common-base amplifier circuit to work between a source having $R_s = 80$ Ω and a load $R_L = 10$ kΩ that will provide $Z_i \leq 25$ Ω, $Z_o \geq 5$ kΩ, and $A_{Vs} \geq 30$. Use $V_{CC} = 18$ V and an *npn* transistor for which $\beta_{dc} = 100$, $V_{BE} = 0.65$ V, and $\beta_0 = 90$.
18. Design a common-base amplifier that will produce $A_{Vs} \geq 100$ with $R_s = 25$ Ω and $R_L = 10$ kΩ. Use $V_{CC} = 24$ V and a 2N5088 transistor.
19. The transistor in the common-collector circuit of Fig. 6.7*a* has $\beta_{dc} = 100$, $V_0 = 0.65$ V, $R_{BB} = 5$ kΩ, $n = 1$, and $\beta_0 = 120$. Circuit values include $R_s = 1$ kΩ, $R_1 = 90$ kΩ, $R_2 = 30$ kΩ, $R_E = 4$ kΩ, $R_L = 6$ kΩ, and $V_{CC} = 20$ V. Find A_{Vi}, A_{Vs}, A_I, A_P, Z_i, and Z_o.
20. In Fig. 6.7*a*, let the transistor be a 2N5377 having the "minimum" characteristics, and let $R_s = 800$ Ω, $R_1 = 70$ kΩ, $R_2 = 30$ kΩ, $R_E = 3$ kΩ, $R_L = 7$ kΩ, and $V_{CC} = 18$ V. Calculate expected values of A_{Vi}, A_I, Z_i, and Z_o.
21. Design an emitter follower using a 2N5376 transistor and a 15-V power supply so that $Z_i \geq 50$ kΩ for $R_s = 4.7$ kΩ and $R_L = 1.2$ kΩ. Assume $V_{BE} = 0.65$ V, $n = 1$, and $R_{BB} = 0$.
22. Design a common-collector amplifier using a 2N5377 transistor and a 12-V power supply so that $Z_o \leq 10$ Ω for $R_s = 750$ Ω and $R_L = 1.5$ kΩ. Assume $V_{BE} = 0.65$ V, $n = 1$, $R_{BB} = 0$, and $\beta_0 \propto (I_C)^{-0.1}$.
23. The mid-frequency equivalent circuit of a certain common-source amplifier contains the elements $R_s = 2.5$ kΩ, $R_1 = 750$ kΩ, $R_2 = 250$ kΩ, $g_m = 8$ m℧, $R_D = 4$ kΩ, and $R_L = 8$ kΩ. (a) Calculate A_{Vs}. (b) If $v_s(t) = 50 \cos \omega t$ mV, find the signal current amplitude in each of the five resistors listed above.

24. The transfer characteristic of the transistor in Fig. 6.18 is $I_D = -0.5(V_{GS} + 2)^2$ mA. What amplitude is required for I_s in order to produce a signal voltage amplitude of 1 V at the output?
25. Circuit element values in the common-source amplifier of Fig. 6.9*a* are $R_s = 0.5$ kΩ, $R_1 = 200$ kΩ, $R_2 = 50$ kΩ, $R_D = 6$ kΩ, $R_L = 12$ kΩ, and $V_{DD} = 30$ V. The capacitors are large. The transistor is an *n*-channel JFET with $I_{DSS} = 4.5$ mA and $V_P = -1.5$ V. (a) Let $R_{SS} = 2$ kΩ and find A_{Vs}. (b) For what value of R_{SS} is the transistor operating at the border between the ohmic and saturation regions? (c) Find A_{Vs} for this condition.
26. Exchange the transistor in Problem 25 for an *n*-channel enhancement-mode MOSFET for which $K = 4.5$ mA/V^2 and $V_T = 2$ V. Rework Parts a, b, and c.
27. A common-source amplifier utilizes a *p*-channel depletion-mode MOSFET having $I_{DSS} = -10$ mA and $V_P = 2.5$ V. Circuit values include $R_s = 1$ kΩ, $R_1 = 500$ kΩ, $R_D = 5$ kΩ, $R_{SS} = 2$ kΩ, $R_L = 15$ kΩ, $V_{DD} = -25$ V, and three large capacitors. (a) Find A_{Vs} if $R_2 = 125$ kΩ. (b) Find the maximum value of R_2 that allows operation in the region beyond pinch-off. (c) Find A_{Vs} for this value of R_2.
28. Modify the design of the C-S amplifier of Section 6.5 so that $|A_{Vi}| \geq 25$. Maintain $R_D = 10$ kΩ, $R_1 = \infty$, $R_2 = 1$ MΩ, $R_s = 1$ kΩ, and $R_L = 10$ kΩ. The same transistor with its range of possible values for I_{DSS}, V_P, and g_m should be used.
29. Let $V_P = -2.5$ V and $I_{DSS} = 8$ mA for the JFET of Fig. 6.9*a*. Also, $R_s = 1$ kΩ, $R_L = 4$ kΩ, $R_1 = \infty$, and $V_{DD} = 20$ V. Select values for R_2, R_D, and R_{SS} so that $|A_{Vi}| \geq 7.5$, $Z_i \geq 1$ MΩ, $V_{DS} \geq 5$ V, and $I_D \leq 6$ mA.
30. Let the transfer characteristic of the *n*-channel JFET in Fig. 6.9*a* be represented by $I_D = 12.5(1 + 0.3V_{GS})^2$ mA. Element values are R_s

Fig. 6.18 See Problem 24.

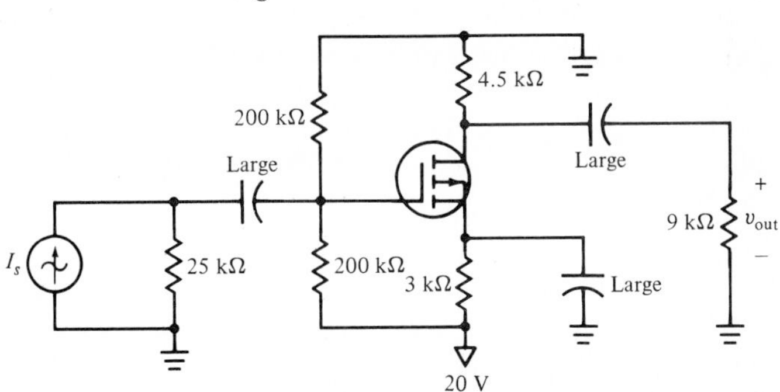

$= 2\ \text{k}\Omega$, $R_1 = \infty$, $R_2 = 98\ \text{k}\Omega$, and $R_L = 6\ \text{k}\Omega$. (a) Choose values for V_{DD}, R_D, and R_{SS} so that $V_{DD} \leq 21.5$ V, $|A_{Vs}| \geq 9.8$, and $V_{DS} = 3(V_{GS} - V_P)$, a constraint that maintains operation safely in the region beyond pinch-off. (b) Let $V_{DD} = 19$ V and select R_{SS} and R_D so that $|A_{Vs}| \geq 10$. (The restriction that $V_{DS} = 3(V_{GS} - V_P)$ no longer applies.)

31. Element values in Fig. 6.9*a* are $R_s = 1\ \text{k}\Omega$, $R_1 = \infty$, $R_L = 4\ \text{k}\Omega$, and $V_{DD} = 20$ V. The transistors have $I_{DSS} = 10$ mA and a pinch-off voltage that may range from -3 to -2 V. Select values for R_2, R_D, and R_{SS} so that $|A_{Vi}| \geq 8$ and $7 \leq I_D \leq 8$ mA.
32. A certain type of *n*-channel JFET has a transfer characteristic defined by $I_D = (8/b)(V_{GS} + b)^2$ mA, where $1 \leq b \leq 2$. Design a C-S amplifier using transistors of this type that will provide a gain $|A_{Vs}| \geq 10$ if $R_s = 500\ \Omega$ and $R_L = 5\ \text{k}\Omega$. Keep $V_{DD} \leq 10$ V.
33. The transistor in Fig. 6.15*a* is replaced by a unit for which $V_P = -2.5$ V, $I_{DSS} = 8$ mA. Determine Z_i, A_{Vi}, A_{Vs}, A_I, A_P, and Z_o.
34. In Fig. 6.13*a*, let $R_s = 100\ \Omega$, $R_{SS} = 1.5\ \text{k}\Omega$, $R_2 = 500\ \text{k}\Omega$, $R_1 = 1\ \text{M}\Omega$, $R_D = 3.5\ \text{k}\Omega$, $R_L = 7\ \text{k}\Omega$, and $V_{DD} = 40$ V. The *n*-channel JFET is replaced by an *n*-channel enhancement-mode MOSFET having the transfer characteristic $I_D = 2.5(V_{GS} - 2)^2$ mA. Find Z_i and A_{Vs} at mid-frequencies.
35. Design a common-gate amplifier to provide a mid-frequency voltage gain $|V_o/V_s| \geq 10$. Let $R_s = 50\ \Omega$, $R_L = 5\ \text{k}\Omega$, $I_{DSS} = 6$ mA, and $V_P = -0.8$ V.
36. Specify values for R_{SS} and g_m in a C-G amplifier having $R_s = 30\ \Omega$, $R_D = 4\ \text{k}\Omega$, and $R_L = 6\ \text{k}\Omega$ if $V_o/V_s = 20$ and $Z_i = 20\ \Omega$ at mid-frequencies.
37. The transistor of Fig. 6.15*b* is replaced by a unit for which $V_P = -2.5$ V and $I_{DSS} = 8$ mA. Determine Z_i, A_{Vi}, A_{Vs}, A_I, A_P, and Z_o.
38. The common-drain amplifier of Fig. 6.15*b* uses a 2N3823 transistor for which $I_{DSS} = 9$ mA and $V_P = -3$ V. The performance of the circuit is tabulated near the end of Section 6.6. Repeat the analysis, except let r_d be 50 kΩ instead of infinite.
39. Design a common-drain amplifier circuit using an *n*-channel enhancement-mode transistor having $I_D = 8(V_{GS} - 2.5)^2$ mA, $V_{DD} = 18$ V, and $R_2 = \infty$ so that $Z_i = 100\ \text{k}\Omega$ and $Z_o = 100\ \Omega$. The value of R_L is not specified.
40. In Fig. 6.14*a*, let $R_s = 1\ \text{k}\Omega$, $R_1 = 100\ \text{k}\Omega$, $R_{SS} = R_L = 2\ \text{k}\Omega$, and $V_{DD} = 20$ V. (a) Select R_2 so that $I_D = 8$ mA if $I_D = 2.5(V_{GS} + 2)^2$ mA. (b) Calculate Z_i, Z_o, and A_{Vs} for this value of R_2.
41. A common-emitter amplifier has the parameter values $h_{ie} = 2\ \text{k}\Omega$, $h_{fe} = 80$, $h_{oe} = 50\ \mu\mho$, $R_B = 100\ \text{k}\Omega$, $R_s = 1\ \text{k}\Omega$, $R_C = 3\ \text{k}\Omega$, and R_L

$= 6\ \text{k}\Omega$. (a) Find Z_i, Z_o, and A_{Vs} if $h_{re} = 0$. (b) Repeat for $h_{re} = 5 \times 10^{-4}$.

42. The transistor in a C-E amplifier is a 2N5088 with the parameters of Unit 3 on the data sheets in Appendix A. Let the operating point be at $I_C = 1.5$ mA, $V_{CE} = 4$ V, with $R_s = 100\ \Omega$, $R_B = 20\ \text{k}\Omega$, $R_C = 6\ \text{k}\Omega$, and $R_L = 10\ \text{k}\Omega$. Calculate Z_i, Z_o, A_{Vi}, and A_{Vs}.
43. Parameters of a common-emitter amplifier are $R_s = 200\ \Omega$, $R_B = 16\ \text{k}\Omega$, $R_C = 5\ \text{k}\Omega$, $R_L = 10\ \text{k}\Omega$, $h_{ie} = 3\ \text{k}\Omega$, $h_{fe} = 150$, and $h_{oe} = 80\ \mu\mho$. Plot Z_i vs. h_{re} (log scale), $10^{-5} \leq h_{re} \leq 10^{-2}$.
44. The common-emitter h parameters of the transistor used in the common-base circuit of Fig. 6.5a are $h_{ie} = 2.5\ \text{k}\Omega$, $h_{re} = 10^{-3}$, $h_{fe} = 60$, and $h_{oe} = 0$. If $R_s = 1\ \text{k}\Omega$, $R_E = 2\ \text{k}\Omega$, and $R_C = R_L = 5\ \text{k}\Omega$, find A_{Vi} and Z_i.

7

Single-stage amplifiers at low and high frequencies

Now that we have become adept at analyzing and designing single-stage transistor amplifiers in the mid-frequency range where the coupling and by-pass capacitors are effectively short circuits, while the interelectrode capacitances behave as open circuits, we need to consider similar analysis and design problems at lower and higher frequencies, where such simplifying assumptions are inappropriate. Suitable equivalent circuits for the transistors themselves were studied in Chapter 5, and these were installed in typical external circuitry at the beginning of Chapter 6.

In this chapter, we shall install an appropriate hybrid-π equivalent circuit for the transistor into an external circuit containing a source, a load, and a stable bias network. The high-frequency case is considered first, since it is the more important problem in practice. We will look at the FET in its three configurations, following this with the bipolar transistor in its three configurations, again at high frequencies.

Amplification at the lower frequencies for the FET and bipolar transistors is the next exciting topic, and we will find that both the analysis and design problems are easier. The chapter concludes with the analysis and design of a common-emitter amplifier at low, mid-, and high frequencies.

7.1 Frequency response

The circuits that we shall be analyzing and designing in this chapter all contain one or more reactive elements. Inductors appear infrequently in most circuits because of their size and weight and the difficulty of building them in integrated-circuit form. Capacitive impedances, $1/j\omega C$, thus appear regularly in our circuit equations, leading to expressions for voltage and current gain that are complex-valued; that is, they have a magnitude and a phase angle. A curve showing the magnitude of the voltage gain of an amplifier as a function of frequency enables us to judge how well the various frequency components in the signal are amplified. For a good stereo system, we may want uniform amplification over a frequency range from 20 to 20,000 Hz, whereas the telephone company finds that 300 to 3000 Hz is adequate. However, the video amplifier in a television receiver needs a bandwidth from 30 to 4,000,000 Hz to be able to reproduce a good-quality picture. Phase information is less important for audio applications, since the ear is not sensitive to phase difference, at least until the difference is large enough to be sensed as a time delay, say a few tenths of a second. Phase distortion is important in video applications such as amplifiers in cathode-ray oscilloscopes and TV receivers. We shall devote the major portion of our efforts to magnitude response.

Information about the frequency performance of an amplifier can be obtained in many forms. One method is experimental, but we shall assume that we still have a paper amplifier that must be analyzed before it is built. An accurate plot of gain magnitude vs. frequency is probably the most useful form for presenting frequency response. The frequency response of a typical amplifier designed to work well over a broad frequency range is illustrated by Fig. 7.1. The magnitude of the gain $|A|$ is plotted to a logarithmic scale as a function of frequency in hertz or in radians per second, also on a logarithmic scale. The use of logarithmic scales enables us to show a range of several decades for both variables.

We see from the curve that the gain magnitude is essentially constant over a *mid-frequency range*, a term which we shall define shortly. The gain falls off at lower frequencies where the reactances of the large coupling and bypass capacitors are too great, and at higher frequencies where the reactances of the interelectrode capacitances of the transistor are too small.

The high-frequency case will be considered first. Here, the fundamental analysis problem is the determination of the *upper half-power frequency* f_H (or ω_H), defined as the higher of the two frequencies in Fig. 7.1, at which a specified voltage or current gain magnitude has decreased to $1/\sqrt{2}$ times its mid-frequency value. The basic design problem involves selecting an appropriate value for this frequency and then specifying a suitable external circuit and transistor to meet the design requirements. Both bipolar and field-effect transistors are considered.

At low frequencies, we define f_L (or ω_L), the *lower half-power frequency*, as the lower frequency at which the gain magnitude is $1/\sqrt{2}$ times the mid-frequency value. The design problem is concerned with the specification of suitable values for the capacitors so that the desired low-frequency performance is obtained.

We may now define the *mid-frequency range* as all those frequencies that are greater than ten times ω_L and less than $0.1\omega_H$, or $10\omega_L < \omega < 0.1\omega_H$. If $\omega_H < 100\omega_L$, there is no mid-frequency range.

Finally, we define the *bandwidth* of an amplifier as the frequency difference between the upper and lower half-power frequencies, $\omega_H - \omega_L$ or $f_H - f_L$.

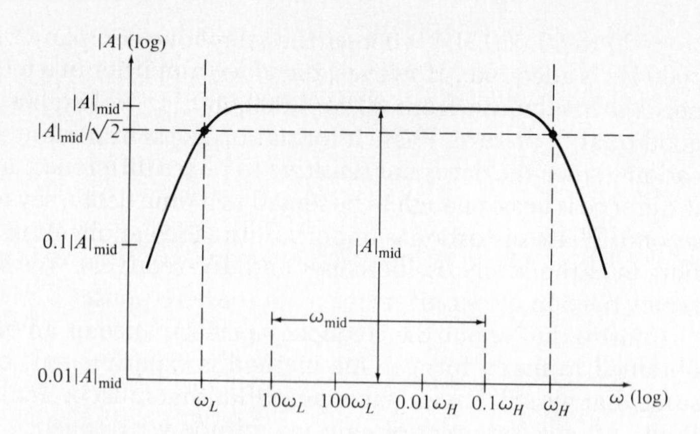

Fig. 7.1 The frequency response of a typical broadband amplifier is shown with logarithmic frequency and amplitude scales. At the lower and upper half-power frequencies ω_L and ω_H, $|A| = |A|_{mid}/\sqrt{2}$.

The frequency response problem can be simplified by using two analysis techniques that we have not yet emphasized. The first technique involves looking carefully at the behavior of two simple RC circuits as frequency varies. For each case we shall identify a frequency we call the half-power, corner, or break frequency, and we shall also find out how to express the gain magnitude in decibels. The second technique invokes the use of the digital computer and special circuit analysis programs that are widely available. With these we can obtain the amplifier performance at a number of frequencies. The programs are usually referred to as computer-aided-design (CAD) programs, although they perform an analysis, not a design. The engineer must accomplish the design and merely use CAD to check his work; if changes are needed, the designer, not the computer, has to decide what to change and how much.

Several CAD circuit simulator programs are available, such as ECAP (Electronic Circuit Analysis Program), ECAP II, CORNAP (Cornell Network Analysis Program), SNAP (Symbolic Network Analysis Program), SPICE (Simulation Program with Integrated Circuit Emphasis), SPICE2, and others too numerous to mention.

7.2 Decibels and break frequencies

In plotting response curves for a broadband amplifier, we often find that we wish to show a thousand-to-one (three-decade) range of some gain $|A|$, and a range of six or seven decades in frequency. Thus, both magnitude and frequency should be plotted on logarithmic scales. Since the phase angle changes over a smaller range, it is plotted with a linear angle scale and a logarithmic frequency scale.

The magnitude scale is commonly given in decibels. We define the *decibel* (dB) as a dimensionless unit equal to 20 times the common logarithm (base ten) of the magnitude of the gain,

$$|A|_{\text{dB}} = 20 \log |A| \qquad \text{(decibels)} \tag{1}$$

Since the gain is a function of $j\omega$, this may be written as

$$|A|_{\text{dB}} = 20 \log |A(j\omega)| \tag{2}$$

The definition of gain in dB, as given by Eqs. (1) or (2), is only applicable to a voltage gain or to a current gain, and not to power gain.

Several consequences of the use of logarithms to the base ten are worth recalling:

1. Unity gain corresponds to zero dB.
2. Negative gain in dB indicates attenuation.
3. Doubling the gain results in an increase of 6 dB (6.0206 dB).
4. Ten times the gain corresponds to an increase of 20 dB.

We first consider a circuit that illustrates the loss of gain that occurs at high frequencies. Figure 7.2*a* contains a series resistor R_S, a parallel resistor R_P, and a parallel capacitor C_P. We let $A_V = V_o/V_s$ and obtain it by first replacing everything to the left of C_P with its Thévenin equivalent, Fig. 7.2*b*. By voltage division, we have

$$\frac{V_o}{V_s\left(\dfrac{R_P}{R_P + R_S}\right)} = \frac{\dfrac{1}{j\omega C_P}}{(R_S \| R_P) + \dfrac{1}{j\omega C_P}} = \frac{1}{1 + j\omega C_P(R_S \| R_P)}$$

Thus

$$A_V = \frac{V_o}{V_s} = \frac{R_P}{R_P + R_S} \times \frac{1}{1 + j\omega C_P(R_S \| R_P)} \tag{3}$$

Fig. 7.2 (*a*) A circuit used to illustrate high-frequency performance when one break frequency is present. (*b*) The Thévenin equivalent appears with C_P.

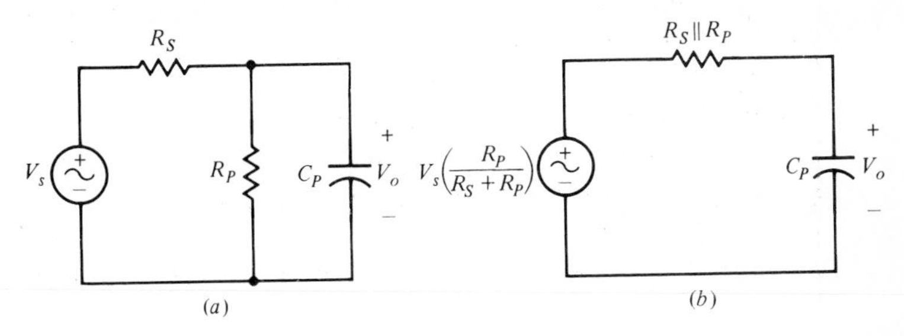

Before we investigate a plot of $|A_V|_{\text{dB}}$ vs. ω for this expression, let us rewrite it as

$$A_V = \frac{R_P}{R_P + R_S} \times \frac{1}{1 + j\dfrac{\omega}{\omega_H}} \tag{4}$$

where

$$\omega_H = \frac{1}{(R_S \| R_P) C_P} \tag{5}$$

We call ω_H the *break frequency* or *corner frequency*. Equation (4) shows that when $\omega = \omega_H$, $|1 + j(\omega/\omega_H)| = \sqrt{2}$, and the magnitude of A_V is reduced to $1/\sqrt{2}$ times the value it has when ω is very small.

Now let us try a plot of $|A_V|_{\text{dB}}$ vs. ω for Eq. (4) and the network of Fig. 7.2*a* or *b*. When ω is much less than ω_H, $|A_V|$ approaches the asymptotic value $R_P/(R_P + R_S)$. Thus

$$|A_V|_{\text{dB}} \doteq 20 \log \frac{R_P}{R_P + R_S} \qquad (\omega \ll \omega_H) \tag{6}$$

This constant value is indicated on Fig. 7.3. Note that it is negative.

When ω is much greater than ω_H, $|1 + j(\omega/\omega_H)|$ approaches $|j(\omega/\omega_H)|$ or ω/ω_H, and $|A_V|$ approaches

$$\frac{R_P}{R_P + R_S} \times \frac{1}{\omega/\omega_H}$$

Thus

$$|A_V|_{\text{dB}} \doteq 20 \log \left[\frac{R_P}{R_P + R_S} \times \frac{1}{\omega/\omega_H} \right] \qquad (\omega \gg \omega_H)$$

Fig. 7.3 The voltage gain $|A_V|_{\text{dB}}$ for the network of Fig. 7.2*a* is plotted vs. ω. At low frequencies, the gain approaches $20 \log[R_P/(R_P + R_S)]$; at high frequencies, it decreases with a slope of -20 dB/decade.

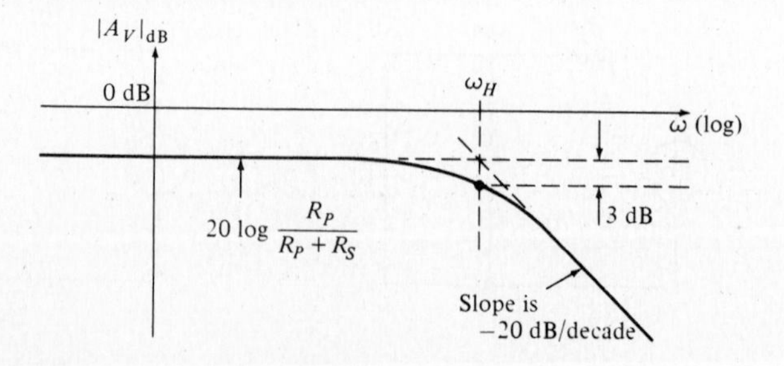

or

$$|A_V|_{\text{dB}} \doteq 20 \log \frac{R_P}{R_P + R_S} - 20 \log \frac{\omega}{\omega_H} \qquad (\omega \gg \omega_H) \tag{7}$$

The first term of Eq. (7) is the constant asymptotic value that $|A_V|_{\text{dB}}$ approaches at low frequencies. The second term is zero when $\omega = \omega_H$, -20 dB for $\omega = 10\omega_H$, -40 dB for $\omega = 100\omega_H$, and so forth. That is, it has a slope of -20 dB/decade. This straight line is also an asymptote that the true curve approaches more closely as ω becomes much greater than ω_H. At $\omega = \omega_H$, Eq. (7) shows that the asymptote has a value of $20 \log [R_P/(R_P + R_S)]$, the same as the constant value of the low-frequency asymptote. However, we used Eq. (4) earlier to show that the true value is $20 \log [R_P/(R_S + R_P)] - 20 \log \sqrt{2} = 20 \log [R_P/(R_P + R_S)] - 3$ dB. The exact value, therefore, lies 3 dB *below* the intersection of the low- and high-frequency asymptotes. This is also apparent from Fig. 7.3.

In summary, we see that the low- and high-frequency asymptotes intersect at $\omega = \omega_H$, the corner or break frequency, indicating the basis for these names. The true curve lies 3 dB below the intersection. Also, the corner frequency is given by

$$\omega_H = \frac{1}{(R_S \| R_P)C_P} = \frac{1}{R_{\text{Th}}C_P} \tag{8}$$

where R_{Th} is the Thévenin-equivalent resistance of the two-terminal network to which C_P is connected. We will use this technique extensively in the analysis of amplifiers at high frequencies.

We next consider the low-frequency performance of an amplifier by investigating the simple circuit of Fig. 7.4. It is not necessary to derive the Thévenin equivalent seen by C_S, since the circuit is already in that form. The clockwise current is $V_s/(R_P + R_S + 1/j\omega C_S)$. Multiplying this by R_P gives the output voltage:

$$V_o = V_s \frac{R_P}{R_P + R_S + \dfrac{1}{j\omega C_S}}$$

Thus

$$A_V = \frac{V_o}{V_s} = \frac{R_P}{R_P + R_S + \dfrac{1}{j\omega C_S}}$$

or

$$A_V = \frac{R_P}{R_P + R_S} \frac{1}{1 + \dfrac{1}{j\omega C_S(R_P + R_S)}} \tag{9}$$

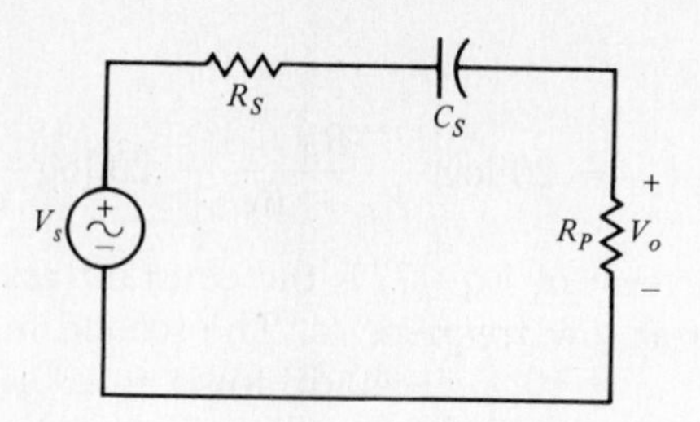

Fig. 7.4 This circuit has one break frequency, easily located at $\omega_L = 1/(R_P + R_S)C_S$ by using the Thévenin equivalent resistance seen by C_S.

We simplify this expression by identifying the corner frequency:

$$\omega_L = \frac{1}{(R_P + R_S)C_S} \tag{10}$$

so that we may write

$$A_V = \frac{R_P}{R_P + R_S} \times \frac{1}{1 - j\dfrac{\omega_L}{\omega}} \tag{11}$$

This is quite similar to Eq. (4) for the high-frequency response, but the frequency ratio containing the break frequency is inverted. Here we have ω_L/ω; in the high-frequency case we had ω/ω_H.

It follows that $|A_V|_{dB}$ approaches the constant asymptotic value of 20 log $|R_P/(R_P + R_S)|$ at *high* frequencies where $\omega \gg \omega_L$. This creates the right-hand portion of the curve shown in Fig. 7.5.

When $\omega \ll \omega_L$, $|1 - j(\omega_L/\omega)|$ is approximated by ω_L/ω, and thus when ω decreases by a factor of ten, $|A_V|_{dB}$ decreases by 20 dB. We thus estab-

Fig. 7.5 The voltage gain $|A_V|_{dB}$ for the network of Fig. 7.4 is plotted against ω. At high frequencies, the gain approaches 20 log$[R_P/(R_P + R_S)]$; at low frequencies, it has a slope of +20 dB/dec.

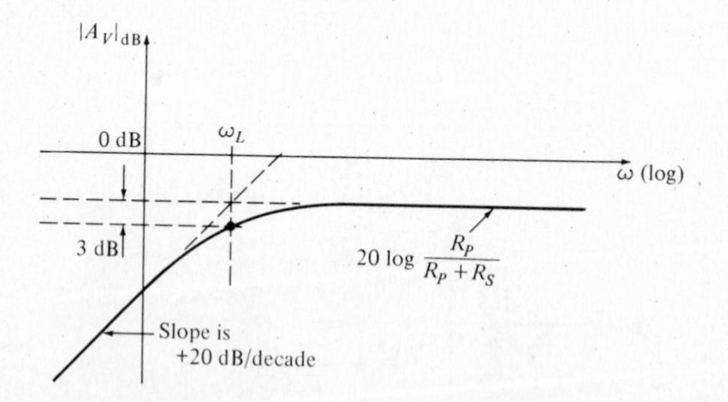

lish a low-frequency asymptote with a slope of 20 dB/dec. This asymptote also appears in Fig. 7.5.

The two straight lines intersect at the corner frequency, and the true curve once again lies 3 dB directly below the intersection. Also, the corner frequency is found by $1/(R_{\mathrm{Th}}C)$, where C is the appropriate capacitance and R_{Th} is the Thévenin-equivalent resistance it sees:

$$\omega_L = \frac{1}{R_{\mathrm{Th}}C_S} = \frac{1}{(R_P + R_S)C_S}$$

Later in this chapter, we shall begin to apply these concepts to the common-emitter amplifier. The signal-equivalent circuit for this amplifier, given in Fig. 6.1*c*, is repeated as Fig. 7.6. We note that it contains five capacitors and therefore suspect that any gain expression might have five factors in its denominator, each of the form $[1 - j(\omega_L/\omega)]$ or $[1 + j(\omega/\omega_H)]$. Our suspicions are all too true, and the exact determination of the five corner frequencies requires finding the roots of a fifth-degree polynomial, not a pleasant task. This will not be the method we follow.

Single factors of this form are also found in the numerator for some types of amplifier circuits,but the corner frequencies are usually so far below ω_L or so much above ω_H that they have little effect on the analysis.

To ease our plight, we assume that the three low-frequency corners are caused by C_1, C_2, and C_E, and that C_π and C_μ may be considered open circuits at these frequencies. Similarly, the two high-frequency corners are produced by C_π and C_μ, while C_1, C_2, and C_E are effectively short circuits at these high frequencies. For broadband amplifiers, these are excellent assumptions. We are thus led to the low- and high-frequency equivalent circuits of Figs. 6.2*b* and *c*. One has three capacitors and provides a cubic equation to tax our patience, while the other gives a simple quadratic.

Fig. 7.6 The complete small-signal equivalent circuit of a C-E amplifier, a repeat of Fig. 6.1*c*.

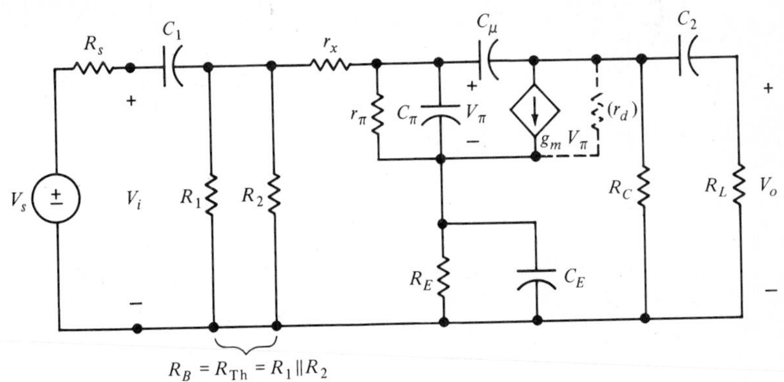

Even these procedures represent too much work, however, especially when multistage amplifiers are considered. We shall analyze the common-source amplifier by introducing an approximate method in the following section in which each capacitor is treated by itself. This technique will be extended to multistage amplifiers in Chapter 8.

D7.1 Find $|A|_{dB}$ if A equals (a) 26, (b) -260, (c) $-20 + j8$. Find $|A|$ if $|A|_{dB}$ equals (d) 26 dB, (e) 36.3 dB, (f) -8 dB.

Answers. 28.3; 48.3; 26.7; 19.95; 65.3; 0.398

D7.2 Element values in Fig. 7.2*a* are $R_S = 2$ kΩ, $R_P = 6$ kΩ, and $C_P = 80$ pF. (a) Find ω_H. Find $|A_V|_{dB}$ at ω equals (b) 10^5 rad/s, (c) 5×10^6 rad/s, (d) 10^7 rad/s, (e) 10^8 rad/s, (f) 10^9 rad/s.

Answers. 8.33 Mrad/s; -2.50 dB; -3.83 dB; -6.37 dB; -24.1 dB; -44.1 dB

D7.3 The circuit of Fig. 7.4 contains $R_S = 300$ Ω, has a corner at $\omega_L = 100$ rad/s, and shows a constant value of $|A_V|_{dB} = -2$ dB at high frequencies. Find (a) R_P, (b) C_S, (c) $|A_V|_{dB}$ at 40 rad/s.

Answers. 1159 Ω; 6.86 μF; -10.60 dB

7.3 High-frequency FET response: common-source

The field-effect transistor is often used instead of a bipolar transistor in the range from 100 MHz to 10 GHz because it introduces less noise into the signal and gives excellent high-frequency performance at small signal levels. However, it does not provide as large a mid-frequency voltage gain as the bipolar transistor. Also, FET amplifiers are often used when large input impedances are required.

The high-frequency model for the common-source amplifier of Fig. 7.7*a* is shown in Fig. 7.7*b*. Again we let R_B represent $R_1 \| R_2$, while $R_L' = R_D \| R_L$.

In analyzing amplifier circuits in the mid-frequency range, we found that the large coupling and by-pass capacitors had relatively small reactances and could be replaced by short circuits, while the small interelectrode capacitances had extremely large reactances and were treated as open circuits. However, when higher-frequency signals are applied, this latter assumption becomes less valid, because the reactance values produced by the small capacitors are of the same order of magnitude as the impedances of the other circuit components. Once again, we have a gain function that is frequency-dependent, and we are interested in calculating ω_H, the upper 3-dB frequency.

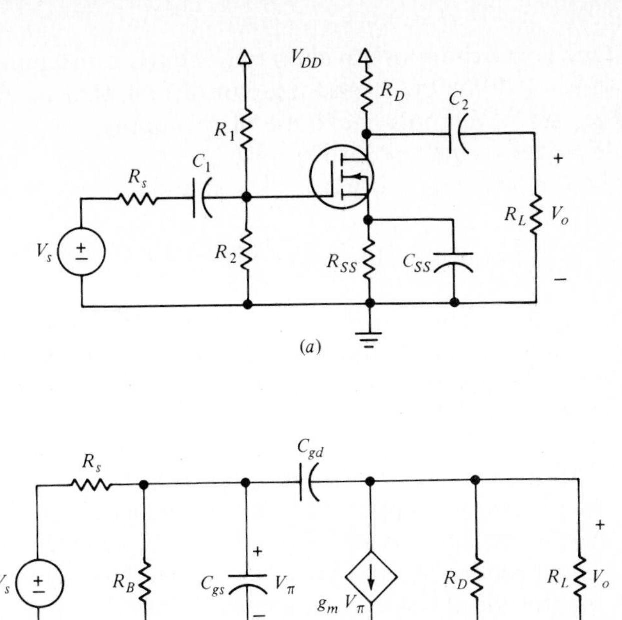

Fig. 7.7 (*a*) A typical FET common-source amplifier. (*b*) The high-frequency equivalent circuit.

Before we establish a value for ω_H, it will be helpful to consider the general approach for any high-frequency equivalent circuit. Each one will contain two small capacitances, C_{gs} and C_{gd}, or C_π and C_μ, and each gain function will contain a quadratic in its denominator that may be factored to display two corner frequencies. Thus, at any frequency $\omega \gg \omega_L$, a range that includes the mid- and high-frequencies, we will obtain a gain expression of the form

$$A_V = A_{V(\text{mid})} \frac{1}{(1 + j\omega/\omega_A)(1 + j\omega/\omega_B)} \tag{12}$$

Since ω_A and ω_B are high-frequency corners, they both are above the mid-frequency region, and very much greater than any low-frequency corner. Note that the gain approaches $A_{V(\text{mid})}$ when $\omega \ll \omega_A$ and $\omega \ll \omega_B$.

Although both the FET and bipolar high-frequency equivalent circuits can be analyzed exactly to determine expressions for ω_A and ω_B, the results seem to be functions of every element in the circuit and they are too complicated to be very useful. Instead, we settle for less accuracy but much more information concerning the factors that limit high-frequency performance.

Our approximation involves two related assumptions. The first is that ω_A and ω_B differ by at least a factor of five; that is, either $\omega_A/\omega_B \geq 5$, or $\omega_B/\omega_A \geq 5$. We now make use of this approximation to find the upper half-power frequency for Eq. (12).

At $\omega = \omega_H$, $|A_V| = |A_{V(\text{mid})}|/\sqrt{2}$, and

$$\left|\left(1 + j\frac{\omega_H}{\omega_A}\right)\left(1 + j\frac{\omega_H}{\omega_B}\right)\right| = \sqrt{2} \tag{13}$$

Let us assume that $\omega_B \gg \omega_A$, so that the second factor in the expression above may be approximated by unity, while the first factor must be $1 + j1$. Therefore, $\omega_H \doteq \omega_A$. Similarly, if ω_A is larger, then $\omega_H \doteq \omega_B$. Problem 7 at the end of the chapter shows that if $\omega_A = 1$ and $\omega_B = 5$ Mrad/s, then an accurate value for ω_H is 0.964 Mrad/s.

We thus conclude that if the two corner frequencies differ by at least a factor of five, the upper half-power frequency may be set equal to the smaller of the two corner frequencies with good accuracy.

The true value of ω_H is always less than the approximate value because the magnitude of the factor we neglected in Eq. (13) must be greater than unity. A lower bound for ω_H may be determined by expanding the left side of Eq. (13):

$$\left|1 - \left(\frac{\omega_H}{\omega_A}\right)\left(\frac{\omega_H}{\omega_B}\right) + j\omega_H\left(\frac{1}{\omega_A} + \frac{1}{\omega_B}\right)\right| = \sqrt{2}$$

and neglecting the product term. It follows that the imaginary term is unity and

$$\omega_H \doteq \frac{1}{1/\omega_A + 1/\omega_B} \tag{14}$$

Since this is a lower bound for the upper half-power frequency ω_H, it is a conservative estimate of that value. That is, if we calculate such a lower bound as our estimate for ω_H, we can be assured that the exact value will be larger.

The second assumption we make enables us to express ω_A and ω_B in terms of the two capacitances in the high-frequency equivalent circuit and the Thévenin-equivalent resistances the capacitances face. We are particularly interested in the smaller of the two frequencies ω_A and ω_B. At this lower frequency, we assume one of the two capacitors controls the response, while the other may be approximated by an open circuit. We thus let $\omega_A = 1/R_{A(\text{Th})}C_A$ and $\omega_B = 1/R_{B(\text{Th})}C_B$.

For an FET, we have

$$\omega_A = \frac{1}{R_{gs}C_{gs}}$$

and

$$\omega_B = \frac{1}{R_{gd}C_{gd}}$$

where R_{gs} and R_{gd} are Thévenin-resistance values that still must be determined from the high-frequency equivalent circuit.

Using these values for ω_A and ω_B in Eq. (14), we have

$$\omega_H = \frac{1}{R_{gs}C_{gs} + R_{gd}C_{gd}} \tag{15}$$

In practice, the RC products are called *open-circuit time constants*, since one of the capacitors is replaced by an open circuit. Thus, we let $\tau_{gs} = R_{gs}C_{gs}$ and $\tau_{gd} = R_{gd}C_{gd}$, so that

$$\omega_H = \frac{1}{\tau_{gs} + \tau_{gd}} = \frac{1}{\tau}$$

or

$$\tau = \tau_{gs} + \tau_{gd} = \frac{1}{\omega_H} \tag{16}$$

This completes our procedural discussion, and we may summarize it briefly before applying it to the common-source circuit. The steps to be taken to find ω_H for an FET are:

1. Determine the Thévenin resistance faced by each capacitor alone, with the other replaced by an open circuit.
2. Calculate the two time constants $\tau_{gs} = R_{gs}C_{gs}$ and $\tau_{gd} = R_{gd}C_{gd}$.
3. Add the two time constants to obtain $\tau = \tau_{gs} + \tau_{gd}$.
4. Let $\omega_H = 1/\tau$.

Although this procedure results in an approximate value for ω_H, we can guarantee that the true value is greater than $1/(\tau_{gs} + \tau_{gd})$.

One final important point should be made about this approach of determining τ_{gs} and τ_{gd} independently. The larger time constant is the most important in determining ω_H, and thus we can see where a design might be changed to affect ω_H the most.

We now try out the open-circuit time-constant method on the high-frequency small-signal equivalent circuit of the common-source amplifier, Fig. 7.7*b*. Letting C_{gd} be an open circuit, we see that $R_{gs} = R_s \| R_B = R_s'$. Therefore

$$\tau_{gs} = R_s'C_{gs} = (R_s \| R_B)C_{gs} \tag{17}$$

Next, C_{gs} is replaced by an open circuit and we determine R_{gd} as the Thévenin resistance viewed from the terminals of C_{gd} (gate and drain). To find R_{gd}, let us squirt 1 A into the gate terminal, as shown in Fig. 7.8. Therefore $V_\pi = R_s'$, $g_m V_\pi = g_m R_s'$, and Kirchhoff's current law shows that the upward current in R_L' must be $1 + g_m V_\pi = 1 + g_m R_s'$. The gate-to-drain voltage V_{gd} is therefore $V_{gd} = R_s' + (1 + g_m R_s')R_L' = R_L' + (1 + g_m R_L')R_s'$. Dividing this by the 1-A input current gives $V_{gd}/1 = R_{gd} = R_L' + (1 + g_m R_L')R_s'$, and thus

$$\tau_{gd} = R_{gd}C_{gd} = [R_L' + (1 + g_m R_L')R_s']C_{gd} \tag{18}$$

Adding Eqs. (17) and (18), we have

$$\tau = \tau_{gs} + \tau_{gd} = 1/\omega_H$$

In Eq. (18), note that the term $g_m R_L'$ is exactly equal to $|A_{Vi(\text{mid})}|$ for a common-source amplifier, as we found in Chapter 6. Thus, $1 + g_m R_L' = 1 + |A_{Vi(\text{mid})}|$, a value that often is as great as 20. Hence τ_{gd} is usually greater than τ_{gs} for a common-source amplifier, even though C_{gs} may be several times larger than C_{gd}. The high-frequency response of the C-S amplifier is therefore usually determined by the $g_m R_L' R_s' C_{gd}$ product, and it is necessary to decrease one or more of these factors to achieve a significant decrease in τ and a consequent increase in ω_H.

As a numerical example, let us calculate some of these values for a typical common-source amplifier having $R_s = 300\ \Omega$ (the impedance of most FM and TV antennas), $R_{SS} = 1\ \text{k}\Omega$, $R_B = 100\ \text{k}\Omega$, $R_D = R_L = 5\ \text{k}\Omega$, $g_m = 10\ \text{m℧}$, $C_{iss} = 7$ pF, and $C_{rss} = 2$ pF. We have $C_{gd} = C_{rss} = 2$ pF, $C_{gs} = C_{iss} - C_{rss} = 5$ pF, and therefore

$$\tau_{gs} = (R_s \| R_B)C_{gs} = (300 \| 10^5)5 \times 10^{-12} = 1.496 \text{ ns}$$

Also,

$$\begin{aligned}\tau_{gd} &= [R_L' + (1 + g_m R_L')R_s']C_{gd} \\ &= [2500 + (1 + 0.01 \times 2500)(300 \| 10^5)]2 \times 10^{-12} \\ &= 20.6 \text{ ns}\end{aligned}$$

Therefore

$$\tau = \tau_{gs} + \tau_{gd} = 22.1 \text{ ns}$$

and

$$\omega_H = 1/\tau = 45.3 \text{ Mrad/s or } 7.22 \text{ MHz}$$

This, of course, is only an approximate result, but we can state definitely that $\omega_H > 45.3$ Mrad/s.[1]

Higher values of ω_H and greater amplifier bandwidths are achieved by decreasing the overall time constant τ. For our example, $\tau = 22.1$ ns, of

[1] An exact solution gives $\omega_H = 48.40$ Mrad/s.

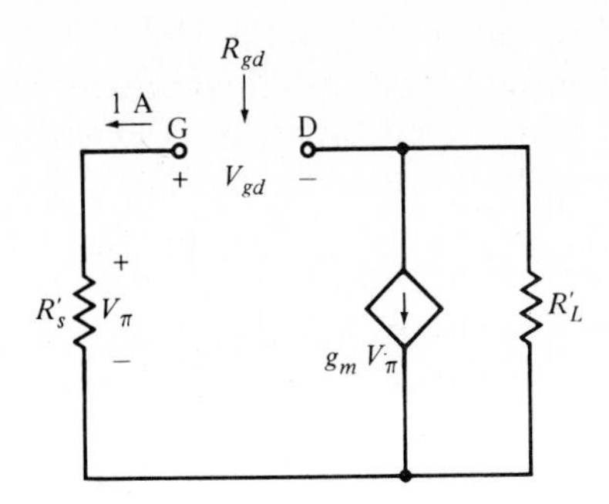

Fig. 7.8 A simplified circuit used to demonstrate that $R_{gd} = V_{gd}/1 = R_s' + (1 + g_m R_s') R_L'$.

which the larger component is $\tau_{gd} = 20.6$ ns. In turn, Eq. (18) shows that τ_{gd} is composed of three terms: $R_L' C_{gd} = 5$ ns, $R_s' C_{gd} = 0.6$ ns, and $g_m R_L' R_s' C_{gd} = |A_{Vi(\text{mid})}| R_s' C_{gd} = 15.0$ ns. Thus, the only effective way to increase ω_H is to reduce the product $g_m R_L' R_s' C_{gd}$. Since $R_s' = R_s \| R_B$ and R_B is much larger than R_s, a reduction of R_s' must be effected by a reduction in R_s. This is usually not possible, since R_s is often a specified value. The capacitance C_{gd} can only be changed appreciably by changing transistor types. Thus, increasing bandwidth is usually achieved by reducing $g_m R_L'$, the mid-frequency gain. The product of gain and bandwidth tends to be fairly constant, and we may therefore exchange gain for bandwidth. This is a technique commonly used in very wideband amplifiers, such as operational amplifiers.

One last comment about the open-circuit time-constant method of determining ω_H deserves to be made. It is readily extended to multistage amplifiers where more than two capacitances appear in the high-frequency equivalent circuit. We shall do so in the following chapter.

D7.4 The high-frequency equivalent circuit of a C-S amplifier includes $R_s = 500\ \Omega$, $R_1 = 600$ kΩ, $R_2 = 200$ kΩ, $R_D = 4$ kΩ, $R_L = 6$ kΩ, $C_{gs} = 5.2$ pF, $C_{gd} = 1.2$ pF, and $g_m = 12$ m℧. Calculate (a) τ_{gs}, (b) τ_{gd}, (c) τ, (d) an approximate value for ω_H, (e) lower and upper bounds for ω_H.

Answers. 2.59 ns; 20.7 ns; 23.3 ns; 42.9 Mrad/s; 42.9 and 48.3 Mrad/s

7.4 Miller-effect capacitance

In the previous section we saw that C_{gd} had a much greater effect in establishing a value for ω_H than did C_{gs}, in spite of the fact that C_{gs} was larger. The mathematical reason for this is evident in Eqs. (17) and (18), which show that C_{gs} is associated with a relatively small resistance, $R_s' \doteq R_s$, while C_{gd} is joined with a Thévenin resistance that is approximately $|A_{Vi(\text{mid})}|R_s'$. Thus, the ratio of the Thévenin resistances is of the order of $|A_{Vi(\text{mid})}|$, the magnitude of the voltage gain. Hence τ_{gd} is several times as large as τ_{gs}.

Now let us try to achieve a better physical understanding of the reason that C_{gd} has such a controlling importance in any high-gain amplifier having a 180° phase reversal, such as the common-source or common-emitter amplifier.

Figure 7.7*b* shows the small-signal high-frequency equivalent circuit of a C-S amplifier, and Fig. 7.9 repeats that part of the circuit to the right of C_{gs}. The voltage across C_{gd} is $V_\pi - V_o$, and therefore

$$I_g = j\omega C_{gd}(V_\pi - V_o)$$

We next let $V_o = A_{Vi}V_\pi$ and approximate A_{Vi} by its mid-frequency value, $-|A_{Vi(\text{mid})}|$,

$$V_o \doteq -|A_{Vi(\text{mid})}|V_\pi$$

Therefore

$$I_g = j\omega C_{gd}(V_\pi + |A_{Vi(\text{mid})}|V_\pi)$$

The input admittance at the left-hand terminals in Fig. 7.9 is

$$Y_{\text{in}} = \frac{I_g}{V_\pi} = j\omega C_{gd}(1 + |A_{Vi(\text{mid})}|) = j\omega C_{\text{Miller}}$$

where C_{Miller} is known as the *Miller-effect capacitance* and is defined by:

$$C_{\text{Miller}} = C_{gd}(1 + |A_{Vi(\text{mid})}|) = C_{gd}(1 + g_m R_L') \tag{19}$$

The Miller capacitance is C_{gd} multiplied by a factor slightly larger than the magnitude of the voltage gain. Thus, if $C_{gd} = 2$ pF and $A_{Vi(\text{mid})} = -20$, $C_{\text{Miller}} = 42$ pF. The effect of C_{gd} is thus greatly magnified.

Fig. 7.9 A portion of the high-frequency small-signal equivalent circuit of a common-source amplifier is used to show that $C_{\text{Miller}} = (1 + g_m R_L')C_{gd}$.

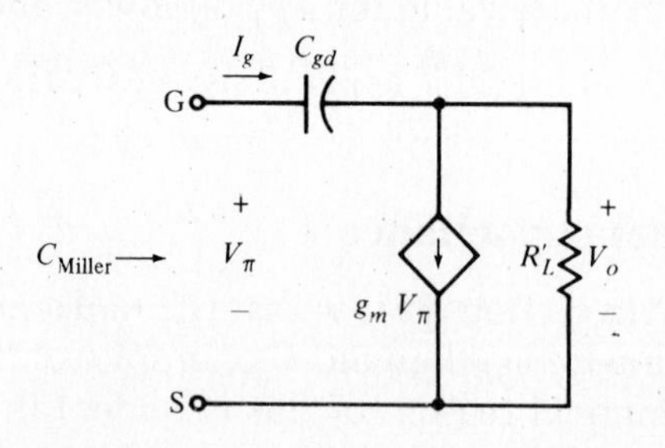

The total input capacitance seen at the gate-source terminals must also include C_{gs}:

$$C_{\text{in}} = C_{gs} + C_{\text{Miller}}$$

A physical explanation for the effective increase in the size of C_{gd} is possible with the aid of Fig. 7.9. As the voltage V_π increases, I_g increases as it begins to charge C_{gd} to the higher potential. However, the gain of the amplifier is causing V_o to *decrease*, and to a much greater extent than V_π is increasing. Hence the potential difference across C_{gd} is much greater than V_π, and a much greater charging current I_g must flow. That is, C_{gd} requires a much larger charging current than its size suggests. In some ways, trying to charge C_{gd} is like trying to fill a barrel when there's a large hole in the bottom. You have to pour a lot faster than you think you should.

We may safely draw a general conclusion from this special result for a C-S amplifier. If a capacitor is connected between the input and output of a high-gain amplifier with 180° phase reversal, then the capacitance is effectively augmented by a factor equal to the magnitude of the voltage gain. Thus, the C-E amplifier will also show a large Miller effect, but the C-G and C-D amplifiers will not, since they have no phase reversal.

D7.5 A common-source amplifier has $R_s = 250\ \Omega$, $R_B = 120\ \text{k}\Omega$, $R_D = 3\ \text{k}\Omega$, $R_L = 6\ \text{k}\Omega$, $g_m = 11\ \text{m℧}$, $C_{gs} = 4\ \text{pF}$, and $C_{gd} = 1\ \text{pF}$. Find (a) C_{Miller}, (b) C_{in}, (c) ω_H.

Answers. 23 pF; 27 pF; $114.5 < \omega_H < 129.2$ Mrad/s

7.5 High-frequency FET response: common-gate and common-drain

We now consider the FET used in the common-gate configuration. The common-gate amplifier is used extensively in high-frequency applications because of its low input capacitance and high upper half-power frequency. It is used in the radio-frequency amplifiers of most good-quality television, FM, and communication receivers, since its low input impedance can be made to match the antenna or lead-in cable.

The high-frequency equivalent circuit for the common-gate amplifier is shown in Fig. 7.10. There are two capacitors, C_{gs} appearing across the input and C_{gd} across the output. We shall find the high-frequency response by calculating the open-circuit time constants.

The resistance faced by C_{gs} consists of R_s in parallel with the input resistance of the C-G amplifier, $Z_i = R_{SS} \| 1/g_m$, Eq. (30) of Chapter 6. Therefore

$$R_{gs} = R_s \| R_{SS} \| 1/g_m$$

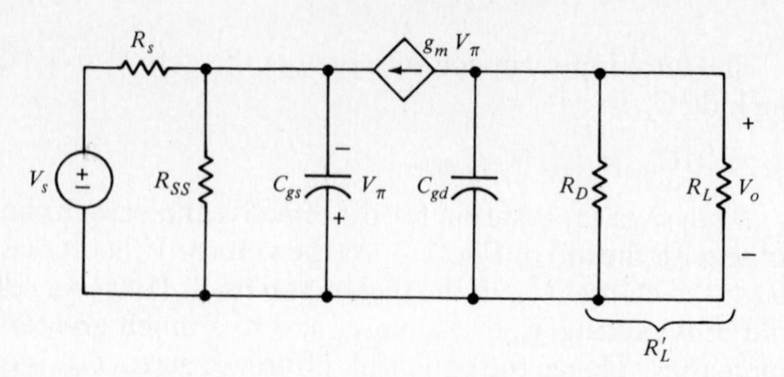

Fig. 7.10 The high-frequency model for the common-gate amplifier circuit.

and

$$\tau_{gs} = R_{gs}C_{gs} = (R_s \| R_{SS} \| 1/g_m)\, C_{gs} \tag{20}$$

At the output, C_{gd} faces R_L in parallel with R_D:

$$R_{gd} = R_L \| R_D = R_L'$$

$$\tau_{gd} = R_L' C_{gd} \tag{21}$$

Then, we again let

$$\tau = \tau_{gs} + \tau_{gd}$$

and

$$\omega_H \doteq 1/\tau$$

Once again, this estimate gives us a lower bound on ω_H. The upper bound is either $1/\tau_{gs}$ or $1/\tau_{gd}$, whichever is smaller. Here, we see that R_{gs} is of the order of $1/g_m$, a small resistance. Thus τ_{gs} is normally much smaller than τ_{gd}, and it follows that $1/\tau < \omega_H < 1/\tau_{gd}$.

As a numerical example, let us find ω_H for a typical common-gate amplifier having $R_s = 300\ \Omega$, $R_{SS} = 1\ \text{k}\Omega$, $R_D = R_L = 5\ \text{k}\Omega$, $g_m = 10\ \text{m}\mho$, $C_{iss} = 7$ pF, and $C_{rss} = 2$ pF. We have $C_{gd} = C_{rss} = 2$ pF, $C_{gs} = C_{iss} - C_{rss} = 5$ pF, and

$$\tau_{gs} = (0.3 \| 1 \| 0.1)5 = 0.349 \text{ ns}$$

$$\tau_{gd} = (5 \| 5)2 = 5 \text{ ns}$$

$$\tau = 5.35 \text{ ns}$$

$$\omega_H \doteq 187 \text{ Mrad/s}$$

There is no Miller-effect capacitance present in the common-gate circuit, since the voltage gain shows no phase reversal. Thus, ω_H is larger than it would be for the same device operating common source. This accounts for the use of the common-gate amplifier at high frequencies where a larger input capacitance is most troublesome.

The final FET circuit we need to inspect at high frequencies is the common-drain amplifier or source-follower, whose high-frequency model appears in Fig. 7.11*a*. This amplifier has a very high input impedance, since the input capacitance is quite small; a very low output impedance, a characteristic that enables the source-follower to drive a low-impedance transmission line efficiently at high frequencies; and a voltage gain less than unity with no phase reversal.

Now let us estimate ω_H by calculating the open-circuit time constants. From Fig. 7.11*a*, we have

$$R_{gd} = R_s' = R_s \| R_B$$

and

$$\tau_{gd} = R_{gd} C_{gd} = (R_s \| R_B) C_{gd} \tag{22}$$

To determine R_{gs}, we remove C_{gd} and find the Thévenin resistance faced by C_{gs}, as indicated in Fig. 7.11*b*. Let us squirt 1 A into the gate terminal so that $R_{gs} = V_\pi/1$. Then,

$$V_\pi = R_s' + R_L'(1 - g_m V_\pi)$$

Fig. 7.11 (*a*) The high-frequency model of the source-follower or common-drain amplifier. (*b*) With 1 A injected into the gate terminal and C_{gd} open-circuited, $R_{gs} = V_\pi$.

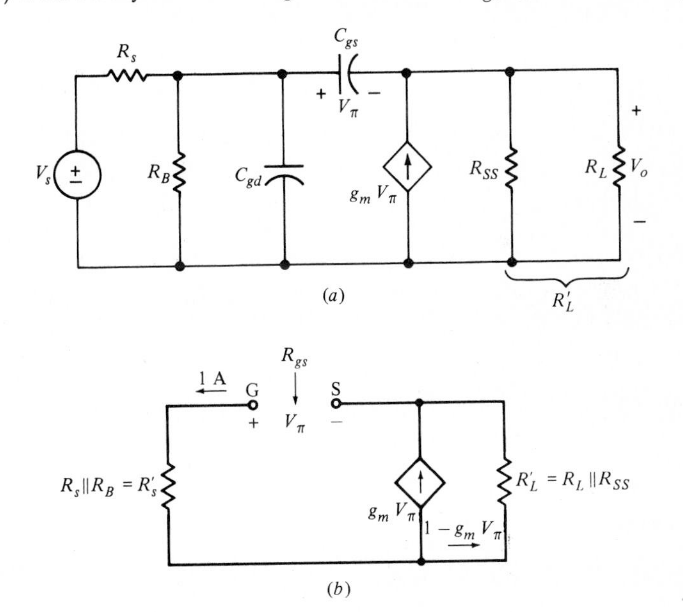

and

$$R_{gs} = \frac{V_\pi}{1} = \frac{R_s' + R_L'}{1 + g_m R_L'}$$

Thus

$$\tau_{gs} = R_{gs} C_{gs} = \frac{(R_s' + R_L') C_{gs}}{1 + g_m R_L'} \tag{23}$$

There is also a zero of the response at $\omega_z = g_m / C_{gs}$. This frequency is usually very much greater than ω_H and has little effect on our approximations.

If we let $R_s = 2\ \text{k}\Omega$, $R_B = 100\ \text{k}\Omega$, $R_{SS} = 1\ \text{k}\Omega$, $R_L = 250\ \Omega$, $g_m = 10\ \text{m℧}$, $C_{iss} = 7$ pF, and $C_{rss} = 2$ pF, then $C_{gs} = 5$ pF, $C_{gd} = 2$ pF, and

$$R_{gd} = R_s' = R_s \| R_B = 2 \| 100 = 1.961\ \text{k}\Omega$$

$$\tau_{gd} = R_{gd} C_{gd} = 1.961(2) = 3.92\ \text{ns}$$

$$R_{gs} = \frac{R_s' + R_L'}{1 + g_m R_L'} = \frac{(2 \| 100) + (1 \| 0.250)}{1 + 10(1 \| 0.250)} = 0.720\ \text{k}\Omega$$

$$\tau_{gs} = R_{gs} C_{gs} = 3.60\ \text{ns}$$

$$\tau = \tau_{gs} + \tau_{gd} = 7.52\ \text{ns}$$

$$\omega_H \doteq 1/\tau = 132.9\ \text{Mrad/s}$$

$$\omega_z = g_m / C_{gs} = 10^{-2}/5 \times 10^{-12} = 2000\ \text{Mrad/s}$$

and $132.9 < \omega_H < 255$ Mrad/s. An exact analysis of this circuit shows that $\omega_H = 136.70$ Mrad/s; the approximation $\omega_H \doteq 1/(\tau_{gs} + \tau_{gd})$ is again excellent.

D7.6 An n-channel depletion-mode MOSFET with $I_D = 4(V_{GS} + 2)^2$ mA is operated as a C-G amplifier with $V_{DD} = 35$ V and $V_{DS} = 7$ V. If $R_1 = 200\ \text{k}\Omega$, $R_2 = 50\ \text{k}\Omega$, $R_{SS} = 2\ \text{k}\Omega$, $R_s = 0.5\ \text{k}\Omega$, $R_L = 20\ \text{k}\Omega$, $C_{gs} = 4$ pF, and $C_{gd} = 1.6$ pF, find (a) τ_{gs}, (b) τ_{gd}. (c) Estimate ω_H.

Answers. 0.381 ns; 6.40 ns; 147.5 Mrad/s

D7.7 Element values for the model of Fig. 7.11*a* are $R_s = 2\ \text{k}\Omega$, $R_B = 98\ \text{k}\Omega$, $R_{SS} = 800\ \Omega$, $R_L = 200\ \Omega$, $g_m = 8\ \text{m℧}$, $C_{gs} = 2$ pF, and $C_{gd} = 0.8$ pF. Find (a) τ_{gs}, (b) τ_{gd}. (c) Estimate ω_H.

Answers. 1.860 ns; 1.568 ns; 292 Mrad/s

7.6 High-frequency bipolar response: common-emitter

We now leave the field-effect transistor temporarily while we consider the high-frequency performance of the common-emitter amplifier. At these higher frequencies, capacitors C_1, C_2, and C_E in Fig. 7.6 have extremely small reactances, and they therefore appear as short circuits in the high-frequency equivalent circuit of Fig. 7.12*a*. Note that there are two resistors present, r_x and r_π, that have no counterparts in the common-source high-frequency equivalent.

We shall again use the open-circuit time-constant method to determine a value for ω_H. We begin by finding $\tau_\pi = R_s'C_\pi$, where R_s' is the Thévenin-equivalent resistance faced by C_π with C_μ replaced by an open circuit, as shown in Fig. 7.12*b*. We combine resistances in series and parallel and obtain

$$R_s' = r_\pi \| [r_x + (R_B \| R_s)] \tag{24}$$

Therefore

$$\tau_\pi = \{r_\pi \| [r_x + (R_B \| R_s)]\} C_\pi \tag{25}$$

Fig. 7.12 (*a*) The high-frequency equivalent circuit of the common-emitter amplifier. (*b*) By removing C_μ and letting $V_s = 0$, the Thévenin resistance seen by C_π is found to be $R_s' = r_\pi \| [r_x + (R_B \| R_s)]$.

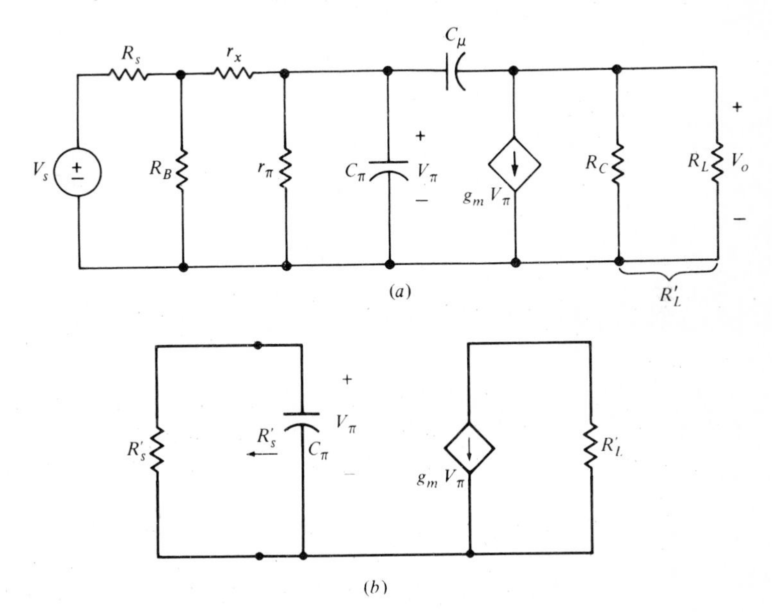

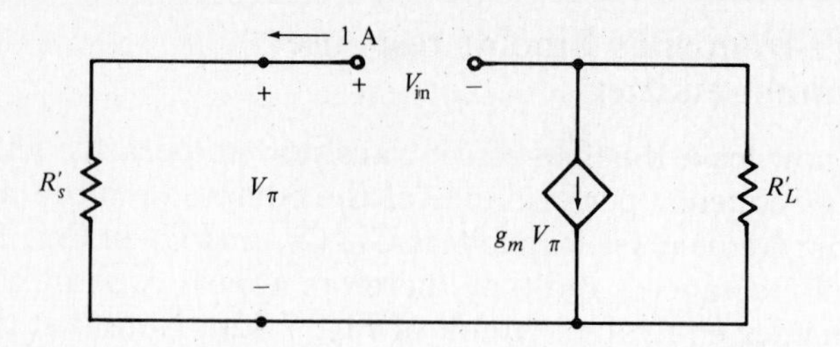

Fig. 7.13 Letting $C_\pi = 0$ and $V_s = 0$, the value of R_μ is obtained by squirting in 1 A and calculating $R_\mu = V_{in}/1$.

To calculate the time constant τ_μ, we remove (open-circuit) C_π, set $V_s = 0$, and seek the Thévenin resistance facing C_μ, as indicated in Fig. 7.13. Injecting 1 A into the equivalent circuit, we see that $V_\pi = R_s' \times 1$. Therefore, the upward current in R_L' is $1 + g_m V_\pi = 1 + g_m R_s'$, and the input voltage is

$$V_{in} = V_\pi + R_L'(1 + g_m R_s') = R_s' + R_L'(1 + g_m R_s')$$
$$= R_L' + R_s'(1 + g_m R_L')$$

It follows that the Thévenin resistance is $R_\mu = V_{in}/1$, or

$$R_\mu = R_L' + R_s'(1 + g_m R_L') \tag{26}$$

and

$$\tau_\mu = R_\mu C_\mu = [R_L' + R_s'(1 + g_m R_L')]C_\mu \tag{27}$$

Then, $\tau = \tau_\pi + \tau_\mu$, and

$$\omega_H \doteq \frac{1}{\tau} = \frac{1}{\tau_\pi + \tau_\mu} \tag{28}$$

Since $g_m R_L' \doteq |A_{Vi(mid)}|$, generally quite a large value, we see that the largest contribution to τ_μ is $g_m R_L' R_s' C_\mu$, and that this term will control ω_H. As an example, let us calculate ω_H for a C-E amplifier having

$$g_m = 33.5\ \text{m℧} \qquad r_x = 20\ \Omega$$
$$r_\pi = 8.78\ \text{k}\Omega \qquad C_\pi = 17.25\ \text{pF}$$
$$\beta_0 = 294 \qquad C_\mu = 1.8\ \text{pF}$$
$$R_s = 1\ \text{k}\Omega \qquad R_C = 5\ \text{k}\Omega$$
$$R_B = 16.67\ \text{k}\Omega \qquad R_L = 2\ \text{k}\Omega$$

We find

$$R_s' = 8.78\|[0.02 + (1\|16.67)] = 0.868 \text{ k}\Omega$$

$$R_L' = 5\|2 = 1.429 \text{ k}\Omega$$

$$R_\mu = 1.429 + 0.868(1 + 33.5 \times 1.429) = 43.8 \text{ k}\Omega$$

$$\tau_\pi = R_s'C_\pi = 0.868(17.25) = 14.98 \text{ ns}$$

$$\tau_\mu = R_\mu C_\mu = 43.8(1.8) = 78.9 \text{ ns}$$

$$\tau = 14.98 + 78.9 = 93.9 \text{ ns}$$

$$\omega_H \doteq 1/93.9 \times 10^{-9} = 10.65 \text{ Mrad/s}$$

$$f_H \doteq 1.695 \text{ MHz}$$

$$|A_{Vs(\text{mid})}| = 40.8, \text{ or } 32.2 \text{ dB } (40.7, \text{ or } 32.2 \text{ dB, including } r_x)$$

Since $\tau_\mu > \tau_\pi$, and $1/\tau_\mu = 10^9/78.9 = 12.67$ Mrad/s, we may state definitely that $10.65 < \omega_H < 12.67$ Mrad/s. The use of a CAD program locates ω_H at 10.7 Mrad/s and provides data for the solid curve of Fig. 7.14. The broken-line response curve is an approximation based on a single

Fig. 7.14 The results of the CAD analysis of the amplifier described in Section 7.6 are shown in the solid line for $C_\pi = 17.25$ pF, $C_\mu = 1.8$ pF, and $r_x = 20\ \Omega$. The single-break-frequency asymptotic diagram is shown as a broken line.

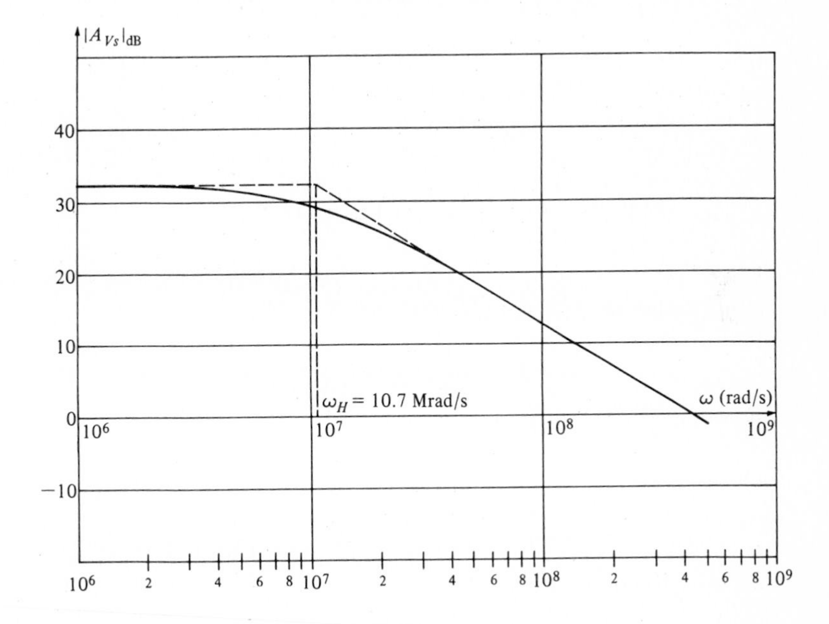

break frequency located at 10.65 Mrad/s, using the open-circuit time-constant method. The agreement is excellent.

The commanding effect that C_μ has on ω_H may also be attributed to the Miller effect. By direct analogy to the C-S amplifier and Eq. (19), we have

$$C_{\text{Miller}} = C_\mu(1 + g_m R'_L) \tag{29}$$

and

$$C_{\text{in}} = C_\pi + C_\mu(1 + g_m R'_L) \tag{30}$$

For this amplifier,

$$C_{\text{Miller}} = 1.8(1 + 33.5 \times 1.429) = 87.9 \text{ pF}$$

$$C_{\text{in}} = 17.25 + 87.9 = 105.2 \text{ pF}$$

Using this capacitance in association with the equivalent input resistance R'_s, we obtain a time constant

$$\tau_{\text{in}} = R'_s C_{\text{in}} = 0.868(105.2) = 91.3 \text{ ns}$$

We also obtain an approximate upper half-power frequency of $10^9/91.3 = 10.95$ Mrad/s. This also shows excellent agreement with the exact value of 10.7 Mrad/s.

An increase in ω_H for an amplifier such as this is usually obtained by reducing $g_m R'_L$ and the mid-frequency gain. As we found for the FET common-source circuit, the product of gain and bandwidth tends to be fairly constant, and thus a decrease in gain can lead to an increase in bandwidth.

In order to increase the bandwidth without an accompanying decrease in gain, we must increase the gain-bandwidth product. As a very rough approximation, $|A_{V(\text{mid})}| \doteq g_m R'_L$ and $\omega_H \doteq 1/g_m R'_L R'_s C_\mu$. Thus, the product $1/R'_s C_\mu$ may be increased by decreasing R'_s, largely governed by R_s, which is often not under our control; or decreasing C_μ by increasing $|V_{CB}|$; or selecting a different transistor with a smaller C_μ and a larger ω_T.

D7.8 In an effort to increase the bandwidth of the amplifier used as an example above, R_C is changed from 5 to 2 kΩ. As a result, V_{CE} changes and leads to the new values, $C_\pi = 16$ pF, $C_\mu = 1.5$ pF. Neglecting r_x at mid-frequencies, calculate (a) $|A_{Vs(\text{mid})}|$, (b) τ_π, (c) τ_μ, (d) ω_H.

Answers. 28.5; 13.9 ns; 46.4 ns; 16.6 Mrad/s

D7.9 A C-E amplifier has $R_s = 1$ kΩ, $R_C = 3$ kΩ, $g_m = 40$ m℧, $r_\pi = 5$ kΩ, $R_B = 15$ kΩ, $r_x = 50$ Ω, $C_\pi = 25$ pF, and $C_\mu = 2.5$ pF. Let $\omega_H = 1/(\tau_\pi + \tau_\mu)$ and calculate the gain-bandwidth product if R_L equals (a) 10 kΩ, (b) 3 kΩ, (c) 1 kΩ.

Answers. 330; 313; 272 Mrad/s

7.7 High-frequency bipolar response: common-base and common-collector

Common-base

In some applications where the large input capacitance caused by the Miller effect must be avoided, and a low value of input resistance is desired, a common-base configuration is desirable. Examples may be found in the two-stage cascode amplifier described in the next chapter, or an input amplifier driven by a 50- or 75-Ω transmission line.

The high-frequency equivalent circuit of the common-base amplifier is shown in Fig. 7.15. Let us first assume that r_x is so small that we may set it equal to zero. With this assumption, it can be seen that the dependent current source isolates the network in two parts, thus avoiding any interaction between the two capacitors C_π and C_μ. Then, the equivalent resistance seen by C_μ with C_π open-circuited is $R_L' = R_C \| R_L$, and

$$\tau_\mu = R_L' C_\mu \qquad (r_x = 0) \tag{31}$$

The open-circuit time constant τ_π depends on the product of C_π and the resistance given by the parallel combination of R_s and the input resistance of the common-base amplifier, Eq. (10) in Section 6.3, repeated here as

$$Z_i = R_E \| r_\pi \| \frac{1}{g_m} \tag{32}$$

Thus

$$\tau_\pi = \left(R_s \| R_E \| r_\pi \| \frac{1}{g_m}\right) C_\pi \qquad (r_x = 0) \tag{33}$$

Having τ_π and τ_μ, we may easily obtain

$$\tau = \tau_\pi + \tau_\mu$$

Fig. 7.15 The high-frequency equivalent of the common-base amplifier circuit.

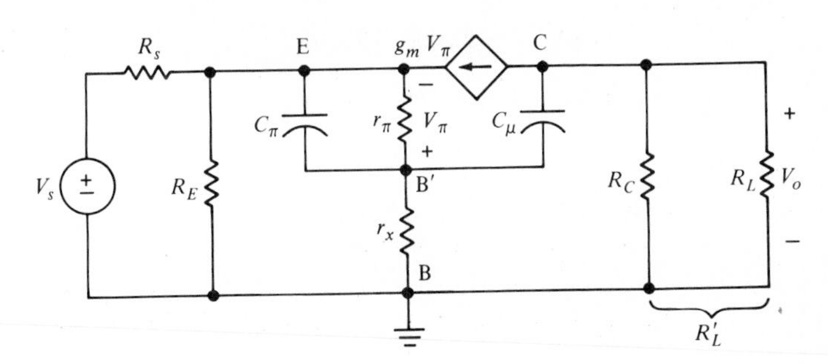

and

$$\omega_H \doteq \frac{1}{\tau} = \frac{1}{R_L' C_\mu + \left(R_s \| R_E \| r_\pi \middle\| \frac{1}{g_m}\right) C_\pi} \qquad (r_x = 0) \tag{34}$$

Since $1/g_m$ is usually quite small, τ_π tends to be much less than τ_μ, and ω_H is approximately equal to $1/\tau_\mu$.

In Eq. (31), note that C_μ is *not* multiplied by the voltage gain. The absence of any Miller effect capacitance is due to two reasons. First, there is no phase reversal in the common-base amplifier, and second, no capacitance is present between input and output in the model used.

We therefore conclude that C-B amplifiers tend to have larger bandwidths than C-E amplifiers do.

If r_x is not negligibly small, then it is necessary to include it in the equivalent circuit when the Thévenin resistances faced by C_π and C_μ are determined. These two derivations are not impossibly difficult, but an overwhelming student demand for the opportunity to do them as homework problems[2] forces us to include only the results:

$$R_\pi = r_\pi \left\| \frac{R_s' + r_x}{1 + g_m R_s'} \right. \tag{35}$$

where

$$R_s' = R_s \| R_E$$

and

$$R_\mu = r_x + R_L' + \frac{r_x(\beta_0 R_L' - r_x)}{r_x + r_\pi + (\beta_0 + 1)R_s'} \tag{36}$$

Both resistances are larger than they would be if r_x were zero; it follows that τ_π, τ_μ, and τ will also be larger. Hence ω_H and the bandwidth decrease as r_x increases.

As a numerical example, let a C-B amplifier have the following parameter values:

$R_s = 0.1\ \text{k}\Omega \qquad g_m = 38\ \text{m}\mho$

$R_E = 1\ \text{k}\Omega \qquad r_\pi = 6\ \text{k}\Omega$

$R_C = 10\ \text{k}\Omega \qquad C_\pi = 20\ \text{pF}$

$R_L = 10\ \text{k}\Omega \qquad C_\mu = 2\ \text{pF}$

[2] See Problems 24 and 25.

Table 7.1 The Effect of r_x on High-Frequency Common-Base Performance

	Calculated				CAD:
r_x, Ω	τ_π, ns	τ_μ, ns	τ, ns	ω_H, Mrad/s	ω_H, Mrad/s
0	0.407	10.00	10.41	96.8	99.8
50	0.629	14.34	14.97	66.8	68.5
100	0.851	18.67	19.52	51.2	52.2
600	3.02	61.1	64.1	15.60	15.68

We first let $r_x = 0$, obtaining

$$\tau_\mu = (10 \| 10)2 = 10 \text{ ns}$$

$$\tau_\pi = (0.1 \| 1 \| 6 \| 1/38)20 = 0.407 \text{ ns}$$

$$\tau = 10.41 \text{ ns}$$

$$\omega_H \doteq 10^9/10.41 = 96.1 \text{ Mrad/s}$$

$$f_H \doteq 15.29 \text{ MHz}$$

Note that the value obtained for ω_H depends almost entirely on τ_μ.

To see the effect that r_x has on ω_H, we use Eqs. (35) and (36) to calculate R_π and R_μ for a few values of r_x. Having these, we determine the open-circuit time constants and the upper half-power frequencies. The results are shown in Table 7.1, along with "exact" data obtained with a CAD program. Note the serious reduction in ω_H that occurs when r_x assumes its maximum value of 600 Ω.

In general, the computer agrees quite closely with our calculations, which speaks well for the computer, although it insists that a more exact value for ω_H when $r_x = 0$ is 99.8 Mrad/s.

Common-collector

Before leaving high-frequency bipolar transistor performance, we should look briefly at the emitter-follower or common-collector circuit. The high-frequency equivalent appears in Fig. 7.16. The feedback-capacitance role is now filled by C_π, while C_μ appears across the input. The voltage gain is less than unity, and there is no 180° phase shift. Thus, the Miller effect is unimportant. An analysis of the equivalent circuit, a task that can easily be accomplished in three or four hours, shows that C_{in} is only slightly larger than C_μ, the smaller of the two transistor capacitances. The input capacitance is thus quite small. Also, the input resistance is very large, as we discovered in Section 6.3. When the impedance

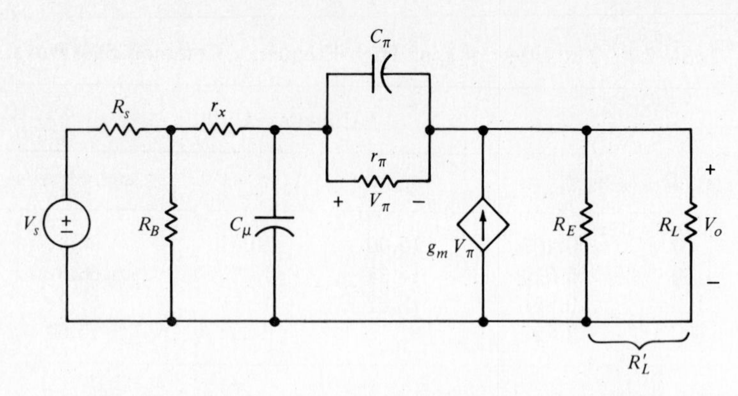

Fig. 7.16 The high-frequency equivalent of the common-collector amplifier.

of the source is high, and that of the load is low, the emitter-follower often provides high-frequency performance superior to that obtainable from a comparable common-emitter circuit.

Let us estimate ω_H by making use of the open-circuit time-constant method and Fig. 7.16. To obtain R_μ, we open-circuit C_π, set $V_s = 0$, and view the network from the terminals of C_μ. Looking to the left, we see $R_s' = r_x + (R_B \| R_s)$. The equivalent resistance looking to the right is $r_\pi + (\beta_0 + 1)R_L'$, where we have applied results obtained for the common-collector amplifier in Section 6.3. Thus,

$$R_\mu = R_s' \| [r_\pi + (\beta_0 + 1)R_L'] \tag{37}$$

and

$$\tau_\mu = R_\mu C_\mu \tag{38}$$

where

$$R_s' = r_x + (R_B \| R_s) \tag{39}$$

and

$$R_L' = R_L \| R_E \tag{40}$$

To find R_π, we remove C_μ, set V_s equal to zero once again, and look at the network from the terminals of C_π, as shown in Fig. 7.17. For simplicity, we again make use of R_s' and R_L', Eqs. (39) and (40) above. The current in r_π is V_π/r_π, that in R_s' is $I_{\text{in}} - (V_\pi/r_\pi)$, and $I_{\text{in}} - (V_\pi/r_\pi) - g_m V_\pi$ flows upward in R_L'. Equating voltages, we have

$$V_\pi = R_s'\left(I_{\text{in}} - \frac{V_\pi}{r_\pi}\right) + R_L'\left(I_{\text{in}} - \frac{V_\pi}{r_\pi} - g_m V_\pi\right)$$

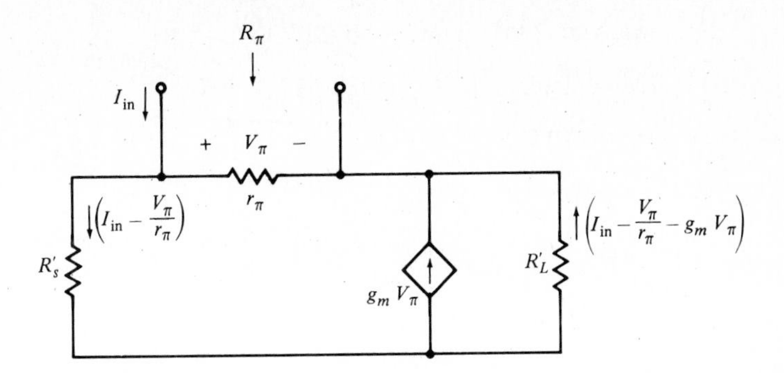

Fig. 7.17 To calculate R_π in Fig. 7.16, we open-circuit C_μ and set $V_s = 0$; then $R_\pi = V_\pi / I_{in}$.

Therefore

$$I_{in}(R'_s + R'_L) = V_\pi\left(1 + \frac{R'_s}{r_\pi} + \frac{R'_L}{r_\pi} + g_m R'_L\right)$$

so that

$$\frac{I_{in}}{V_\pi} = \frac{1}{R_\pi} = \frac{1}{r_\pi} + \frac{1 + g_m R'_L}{R'_s + R'_L}$$

or

$$R_\pi = r_\pi \left\| \frac{R'_s + R'_L}{1 + g_m R'_L} \right. \tag{41}$$

We now have

$$\tau_\pi = R_\pi C_\pi \tag{42}$$

and

$$\omega_H \doteq \frac{1}{\tau}$$

where

$$\tau = \tau_\mu + \tau_\pi$$

Note that r_x enters only into R'_s; as r_x increases, R'_s will increase, thus causing both τ_μ and τ_π to increase and ω_H to decrease.

As an example, let us select $R_B = 40\ \text{k}\Omega$, $r_x = 100\ \Omega$, and $R_C = 0$, of course, and all other values identical with those of the C-B amplifier used earlier in this section:

$$R_s = 0.1\ \text{k}\Omega \qquad r_\pi = 6\ \text{k}\Omega$$

$$R_E = 1\ \text{k}\Omega \qquad C_\pi = 20\ \text{pF}$$

$$R_L = 10\ \text{k}\Omega \qquad C_\mu = 2\ \text{pF}$$

$$g_m = 38\ \text{m℧}$$

We find that

$$R_s' = 0.1 + (0.1 \| 40) = 0.1998\ \text{k}\Omega$$

$$R_L' = 1 \| 10 = 0.909\ \text{k}\Omega$$

$$\beta_0 = 228$$

$$R_\mu = 0.1998 \| (6 + 229 \times 0.909) = 199.6\ \Omega$$

$$\tau_\mu = 0.399\ \text{ns}$$

$$R_\pi = 6 \left\| \frac{0.1998 + 0.909}{1 + 38(0.909)} \right. = 6 \| 0.0312 = 31.0\ \Omega$$

$$\tau_\pi = 0.621\ \text{ns}$$

$$\tau = 1.020\ \text{ns}$$

$$\omega_H \doteq 981\ \text{Mrad/s}$$

A computer simulation of this C-C amplifier leads to $\omega_H = 2001$ Mrad/s, and we are therefore off by approximately a factor of two. Since ω_H is really much higher than we estimate, we err on the conservative side. Although a better estimation would be welcome, we would probably be wasting our time making additional calculations. Let us calculate the gain-bandwidth product ω_T for this transistor at this operating point:

$$\omega_T = \frac{g_m}{C_\pi + C_\mu} = \frac{38 \times 10^{-3}}{(20 + 2)10^{-12}} = 1727\ \text{Mrad/s}$$

Thus, we will have a bandwidth that is comparable to the gain-bandwidth product; the transistor is operating close to its limiting frequency. We might also note that the gain-bandwidth product and the bandwidth are comparable because the gain of the C-C amplifier is approximately unity.

A more serious problem with high-frequency analysis in many cases is that the hybrid-π model does not represent the transistor accurately above a few hundred megahertz. In such cases a much more sophisticated model has to be considered, and this is a topic we leave for more advanced treatments.

D7.10 A common-base amplifier has $R_s = 50\ \Omega$, $R_E = R_L = R_C = 2\ \text{k}\Omega$, $g_m = 20\ \text{m℧}$, $\beta_0 = 40$, $C_\mu = 2\ \text{pF}$, and $C_\pi = 30\ \text{pF}$. Find ω_H if r_x equals (a) 0, (b) 50 Ω.

Answers. 366; 220 Mrad/s

D7.11 A common-collector amplifier has $R_s = R_E = R_L = 2\ \text{k}\Omega$, $R_B = 20\ \text{k}\Omega$, $g_m = 20\ \text{m℧}$, $\beta_0 = 40$, $C_\mu = 2\ \text{pF}$, and $C_\pi = 30\ \text{pF}$. (a) Find ω_T. Find $1/(\tau_\mu + \tau_\pi)$ if r_x equals (b) 0, (c) 50 Ω.

Answers. 625; 137.7; 134.8 Mrad/s

7.8 Low-frequency FET response

We now turn our attention to the low-frequency end of the amplifier response curve. A typical *n*-channel enhancement-mode MOSFET amplifier connected in the common-source configuration is shown in Fig. 7.18*a*. We recall that the term *common-source* is used because at mid-frequencies, the source capacitor C_{SS} acts as a short circuit to the signal and the source is common to one terminal of both the output and the input. The low-frequency equivalent circuit shown in Fig. 7.18*b* indicates that we have three capacitors, the two coupling capacitors C_1 and C_2 and the source by-pass capacitor C_{SS}. The presence of these capacitors causes a reduction in current and voltage gain with frequency in the low-frequency range. Our goal is to determine the variation of the magnitude

Fig. 7.18 (*a*) A typical common-source FET amplifier. (*b*) The low-frequency equivalent circuit.

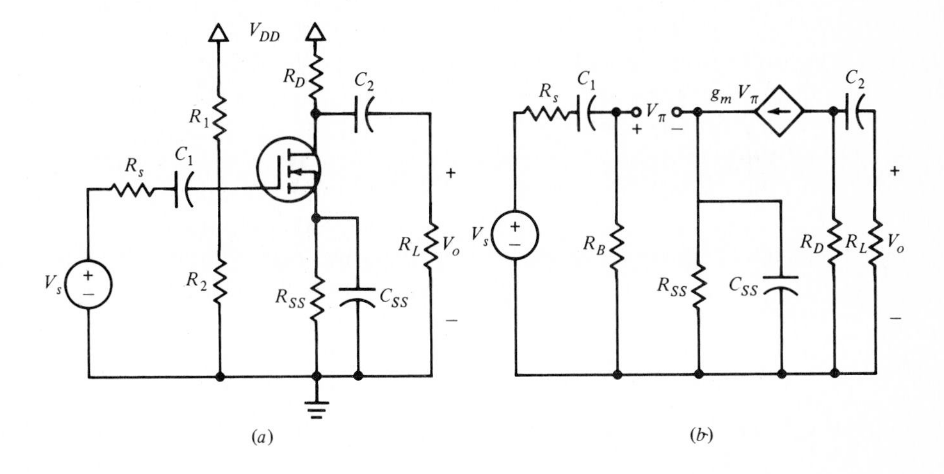

and phase of $A_{Vs}(j\omega)$ with frequency below the mid-frequency range. To a large extent, this is accomplished by finding the lower half-power frequency ω_L.

There are three methods of determining ω_L that we might use. The first is a brute-force analysis of the circuit to yield an expression for $A_{Vs}(j\omega)$. With three capacitors in the circuit, the result is a fraction in which the denominator is a cubic polynomial in $j\omega$. When ω is much greater than ω_L, $A_{Vs}(j\omega) = A_{Vs(\text{mid})}$. We are seeking a value of ω_L such that $|A_{Vs}(j\omega_L)| = (1/\sqrt{2})|A_{Vs(\text{mid})}|$. This may be found by calculation and careful plotting, by the use of a "SOLVE" routine on a good hand-held calculator, or by trial-and-error methods. The complete circuit analysis is usually not a simple process, however, and we shall try to avoid it.

A second method involves the use of a digital computer to analyze the specific circuit at a sequence of frequency values. Either plotting the results or inspecting the tabulated data will reveal the details of the response curve or the value of ω_L. Any CAD program may be used for this purpose. However, this method is better suited to analysis than it is to design, for the only means of improving the results is by an educated guess, or simple trial and error to modify the circuit, followed by a repetition of the computer analysis.

The third method is approximate, but it is often very effective, particularly in design situations. It is based on the assumption that ω_L is determined primarily by only one of the three capacitors, and that the remaining two may be treated as effective *short circuits* at ω_L. Thus three simple analyses must be made, each leading to a response curve characterized by a single corner frequency. If we find that one of these three break frequencies is clearly much larger than each of the other two, say by a factor of five to ten, then the two lower break frequencies have a relatively small effect on the response curve in the neighborhood of ω_L, and ω_L is only slightly greater than the frequency of the highest corner frequency. The error is usually less than 10%. If one of the break frequencies is not considerably larger than both of the other two, then we either must fall back on one of the two methods described earlier or accept a greater error in our estimate of ω_L.

Let us apply this third method to the C-S amplifier of Fig. 7.18*b* and determine expressions for the break frequencies that result when each capacitor acts alone, assuming the other two are both short-circuited. As we found in Section 7.2, the break frequency is given by $1/(R_{\text{Th}}C)$, where R_{Th} is the Thévenin resistance the capacitor faces.

We first consider C_1 with C_{SS} and C_2 replaced by short circuits. With $V_s = 0$, the resistance faced by C_1 is $R_s + R_B$, and we have

$$\omega_{C1} = \frac{1}{(R_s + R_B)C_1} \tag{43}$$

Next, C_1 and C_{SS} become short circuits, and we have C_2 acting with the series combination $R_D + R_L$, since the current source goes to zero when V_s is set equal to zero. Thus,

$$\omega_{C2} = \frac{1}{(R_D + R_L)C_2} \tag{44}$$

Finally, C_1 and C_2 are short-circuited with $V_s = 0$, and we look at the network from the terminals of C_{SS}, as indicated in Fig. 7.19. Since there is no voltage across R_s or R_B, then $V_{in} = -V_\pi$. Therefore, $I_{in} = (V_{in}/R_{SS}) - g_m V_\pi = V_{in}(1/R_{SS} + g_m)$, and it follows that

$$\frac{V_{in}}{I_{in}} = \frac{1}{1/R_{SS} + g_m} = R_{SS} \left\| \frac{1}{g_m} \right.$$

Thus

$$\omega_{CSS} = \frac{1}{[R_{SS} \| (1/g_m)]C_{SS}} \tag{45}$$

Let us obtain some numerical values for these three corner frequencies in the case of a common-source amplifier having $R_s = 300\ \Omega$, $R_B = 100\ \text{k}\Omega$, $R_D = R_L = 5\ \text{k}\Omega$, $R_{SS} = 2\ \text{k}\Omega$, $g_m = 10\ \text{m℧}$, $C_1 = 0.2\ \mu\text{F}$, $C_2 = 1\ \mu\text{F}$, and $C_{SS} = 10\ \mu\text{F}$. We find

$$\omega_{C1} = \frac{1}{(R_s + R_B)C_1} = \frac{1000}{100.3(0.2)} = 49.9\ \text{rad/s}$$

$$\omega_{C2} = \frac{1}{(R_D + R_L)C_2} = \frac{1000}{10(1)} = 100\ \text{rad/s}$$

$$\omega_{CSS} = \frac{1}{[R_{SS} \| (1/g_m)]C_{SS}} = \frac{1}{(2 \| 0.1)10} = 1050\ \text{rad/s}$$

Fig. 7.19 The resistance faced by C_{SS} with $V_s = 0$ and C_1 and C_2 short-circuited is $V_{in}/I_{in} = R_{SS} \| (1/g_m)$.

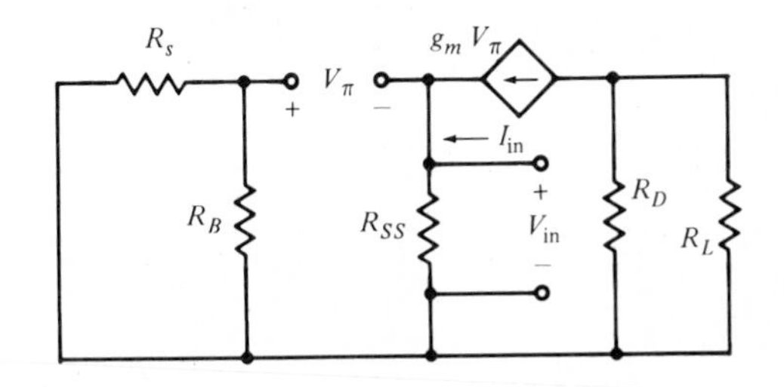

Since ω_{CSS} is much greater than both ω_{C1} and ω_{C2}, we estimate that $\omega_L \doteq \omega_{CSS} = 1050$ rad/s, or 167 Hz. The true value, obtained by an exact analysis of the circuit, is 1059.4 rad/s.

The design of a common-source circuit for a desired low-frequency performance is easily accomplished by selecting either ω_{C1}, ω_{C2}, or ω_{CSS} as ω_L, or slightly less than ω_L to be safe, and then placing the remaining two corners at frequencies less than one-tenth of that value. The controlling frequency is usually ω_{CSS} because the resistance $1/g_m$ is such a small value.

The common-gate amplifier, and its low-frequency model, are shown in Fig. 7.20*a*, *b*. Considering C_1 by itself, and remembering that the input resistance of the C-G amplifier was found to be $R_{SS} \| 1/g_m$ in Section 6.6, we find

$$\omega_{C1} = \frac{1}{[R_s + (R_{SS} \| 1/g_m)]C_1} \tag{46}$$

The resistance seen by C_2 is $R_D + R_L$, and therefore

$$\omega_{C2} = \frac{1}{(R_D + R_L)C_2} \tag{47}$$

Fig. 7.20 (*a*) A typical common-gate amplifier circuit for an *n*-channel enhancement-mode MOSFET. (*b*) The low-frequency equivalent circuit.

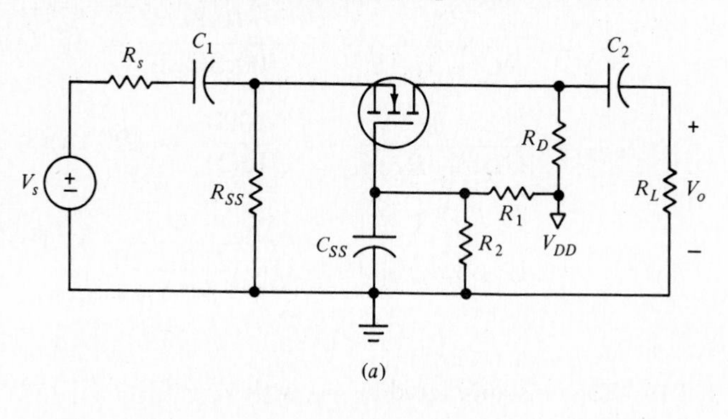

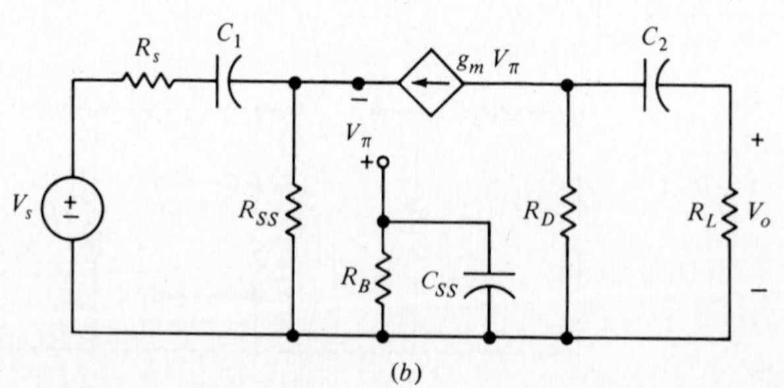

Finally, the resistance faced by C_{SS} is $R_B = R_1 \| R_2$, and

$$\omega_{CSS} = \frac{1}{R_B C_{SS}} \tag{48}$$

Since R_B is usually very large, either C_1 or C_2 will control ω_L, with the other corner frequency less than one-tenth that value. Since $1/g_m$ is small, C_1 is a good choice for the controlling capacitor in the common-gate circuit.

The remaining common-drain configuration is shown in Fig. 7.21*a*, with its low-frequency model in Fig. 7.21*b*. The two capacitors provide the break frequencies,

$$\omega_{C1} = \frac{1}{(R_s + R_B)C_1} \tag{49}$$

and, since the output impedance of the C-D amplifier was found to be $R_{SS} \| 1/g_m$ in Section 6.6,

$$\omega_{C2} = \frac{1}{[R_L + (R_{SS} \| 1/g_m)]C_2} \tag{50}$$

It usually happens that ω_L is determined by C_2 because the output impedance of the common-drain, $R_{SS} \| 1/g_m$, is so small.

Fig. 7.21 (*a*) A MOSFET common-drain amplifier, and (*b*) its low-frequency equivalent circuit.

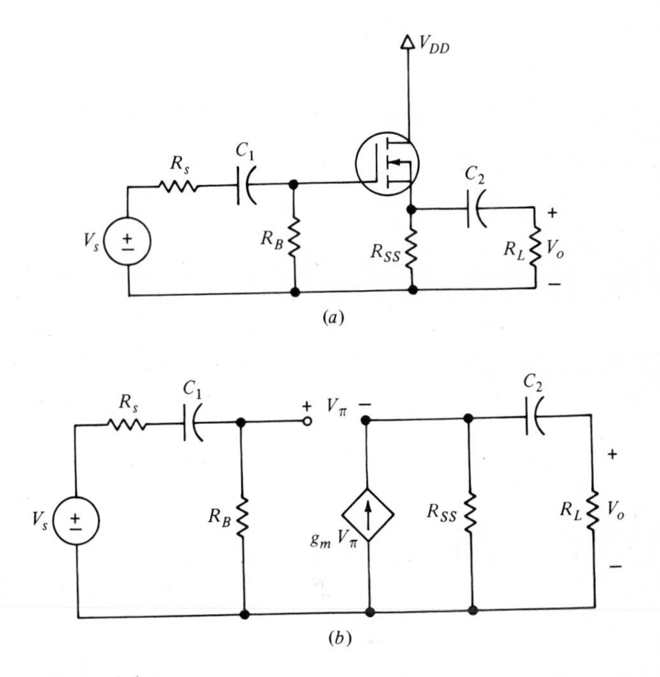

D7.12 The common-source amplifier used as a numerical example in this section ($R_s = 300\ \Omega$, $R_B = 100\ \text{k}\Omega$, $R_D = R_L = 5\ \text{k}\Omega$, $R_{SS} = 2\ \text{k}\Omega$, $g_m = 10\ \text{m℧}$) is now operated as a C-G amplifier with $C_1 = 5\ \mu\text{F}$, $C_2 = 2\ \mu\text{F}$, and $C_{SS} = 0.2\ \mu\text{F}$. Find (a) ω_{C1}, (b) ω_{C2}, (c) ω_{CSS}. (d) Estimate ω_L.

Answers. 506; 50; 50; 506

D7.13 A common-drain amplifier has the circuit parameters $R_s = R_L = 300\ \Omega$, $R_B = 100\ \text{k}\Omega$, $R_{SS} = 2\ \text{k}\Omega$, $g_m = 10\ \text{m℧}$, $C_1 = 0.5\ \mu\text{F}$, and $C_2 = 10\ \mu\text{F}$. Find (a) ω_{C1}, (b) ω_{C2}. (c) Estimate ω_L.

Answers. 19.9; 253; 253 rad/s

7.9 Low-frequency bipolar response

Common emitter

The analysis and design of bipolar circuits at low frequencies is quite similar to the procedures used for FET amplifiers. There will be one additional element in every equivalent circuit, however—the resistance r_π—and this may require a few more equations each time a Thévenin-equivalent resistance is found.

A standard common-emitter amplifier is shown in Fig. 7.22*a*, and its low-frequency equivalent circuit appears as Fig. 7.22*b*. Note that C_π and C_μ have been replaced by open circuits, and r_x is neglected since it is now in series with r_π and may be considered to be a part of r_π, if desired.

We shall again find three corner frequencies by considering each capacitor as acting alone with the other two replaced by short circuits.

In Fig. 7.23*a*, C_2 and C_E are considered to be short circuits, and we seek the Thévenin resistance presented to C_1. It is obviously $R_s + (R_B \| r_\pi)$, where $R_B = R_1 \| R_2$, as usual. Thus

$$\omega_{C1} = \frac{1}{[R_s + (R_B \| r_\pi)]C_1} \tag{51}$$

We next consider only C_2, with C_1 and C_E short-circuited, as represented by the equivalent circuit of Fig. 7.23*b*. The Thévenin resistance is $R_C + R_L$, so that

$$\omega_{C2} = \frac{1}{(R_C + R_L)C_2} \tag{52}$$

Finally, with both C_1 and C_2 replaced by short circuits, and V_s set equal to zero, the equivalent circuit of Fig. 7.24 results. The resistance offered to C_E will be obtained as the ratio of V_{in} to I_{in}.

The downward current in R_E is V_{in}/R_E, the current to the left in r_π is $-V_\pi/r_\pi$, and the current to the right in the dependent current source is

$-g_m V_\pi$. Hence

$$I_{\text{in}} = \frac{V_{\text{in}}}{R_E} - \frac{V_\pi}{r_\pi} - g_m V_\pi = \frac{V_{\text{in}}}{R_E} - \frac{V_\pi}{r_\pi}(\beta_0 + 1)$$

We also see that

$$V_{\text{in}} = -\frac{V_\pi}{r_\pi}[r_\pi + (R_s \| R_B)]$$

so that

$$-\frac{V_\pi}{r_\pi} = \frac{V_{\text{in}}}{r_\pi + (R_s \| R_B)}$$

Eliminating V_π in the first equation, we have

$$I_{\text{in}} = V_{\text{in}}\left[\frac{1}{R_E} + \frac{\beta_0 + 1}{r_\pi + (R_s \| R_B)}\right]$$

Fig. 7.22 (*a*) A typical common-emitter amplifier circuit. (*b*) The low-frequency equivalent circuit.

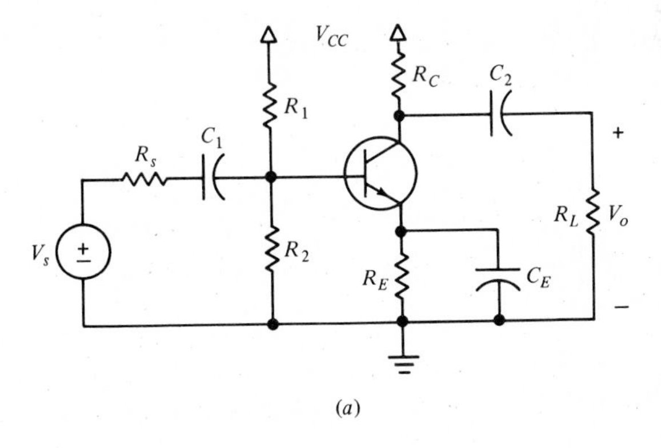

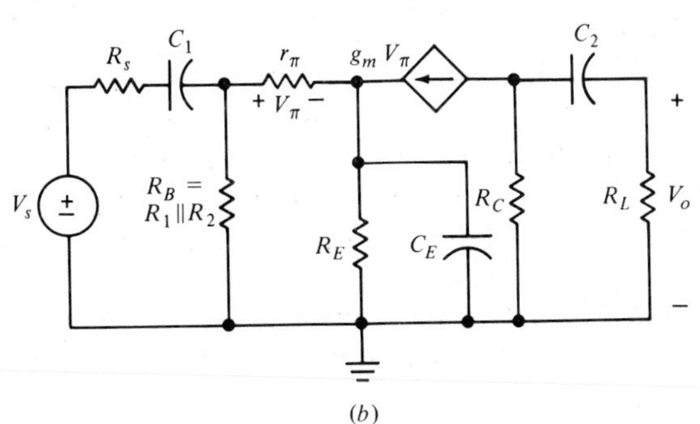

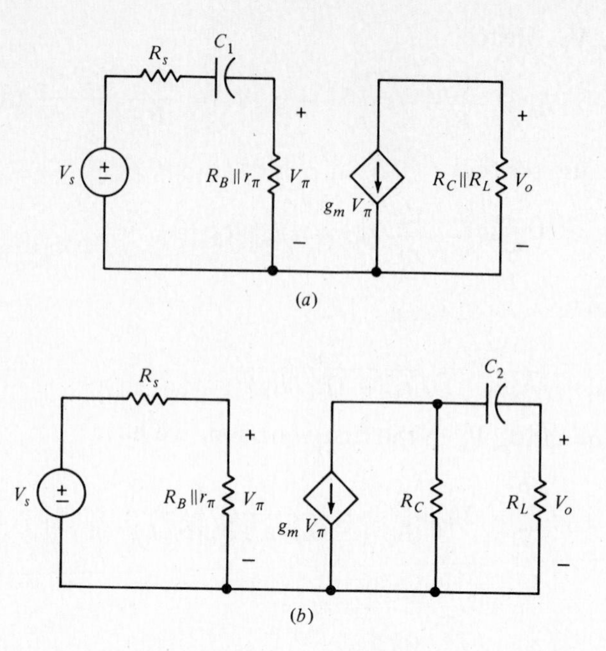

Fig. 7.23 The common-emitter low-frequency equivalent circuit of Fig. 7.22 is simplified by assuming that the controlling capacitance is (a) C_1 and (b) C_2.

If we designate this Thévenin-equivalent resistance as R_{eq}, then

$$\frac{1}{R_{eq}} = \frac{I_{in}}{V_{in}} = \frac{1}{R_E} + \frac{\beta_0 + 1}{r_\pi + (R_s \| R_B)}$$

or

$$R_{eq} = R_E \left\| \frac{r_\pi + (R_s \| R_B)}{\beta_0 + 1} \right. \tag{53}$$

We therefore obtain the third break frequency as

$$\omega_{CE} = \frac{1}{R_{eq} C_E} \tag{54}$$

An alternate approach to finding R_{eq} uses the results obtained for the common-base and common-collector amplifiers in Section 6.3. We found then that resistance values increased by a factor of $\beta_0 + 1$ when we viewed emitter loads from the base, whereas they decreased by the same factor as our viewpoint moved from the base to the emitter. In Fig. 7.24, our viewpoint is the emitter, and we see R_E in parallel with

$$\frac{r_\pi + (R_s \| R_B)}{\beta_0 + 1}$$

Thus Eq. (53) follows.

A detailed analysis of the low-frequency C-E equivalent circuit also shows that the gain expressions have a factor $1 - j(\omega_z/\omega)$ in the numerator, where

$$\omega_z = \frac{1}{R_E C_E} \tag{55}$$

When $\omega \gg \omega_z$, this corner has little effect, since $1 - j(\omega_z/\omega) \doteq 1$. This is usually the case.

If we examine Eq. (53) more closely, we may recall that R_{eq} is the output impedance of the common-collector amplifier, Eq. (20) in Chapter 6. Note that its value does not depend on the value of R_C, since that resistance is in series with a current source and is therefore a superfluous element with respect to the emitter circuit. Equation (53) also indicates that if R_s is small compared with r_π and if R_E is no smaller than about 1 kΩ, then $R_{\text{eq}} \doteq 1/g_m$. This is often a useful rule of thumb in making a quick estimate of R_{eq}.

Collecting these results, we are now prepared to calculate values for ω_{C1}, ω_{C2}, ω_{CE}, and ω_z and then make an estimate of the lower half-power frequency ω_L. If one of the three corners ω_{C1}, ω_{C2}, or ω_{CE} is at least ten times greater than both of the other two (and also ω_z), then we may select ω_L as being equal to that greatest corner frequency.

As an example of the determination of the low-frequency performance of a common-emitter amplifier, let us analyze the circuit whose mid-fre-

Fig. 7.24 The equivalent circuit of Fig. 7.22*b* with C_1, C_2, and V_s replaced by short circuits. The Thévenin resistance faced by C_E is $R_{\text{eq}} = V_{\text{in}}/I_{\text{in}}$.

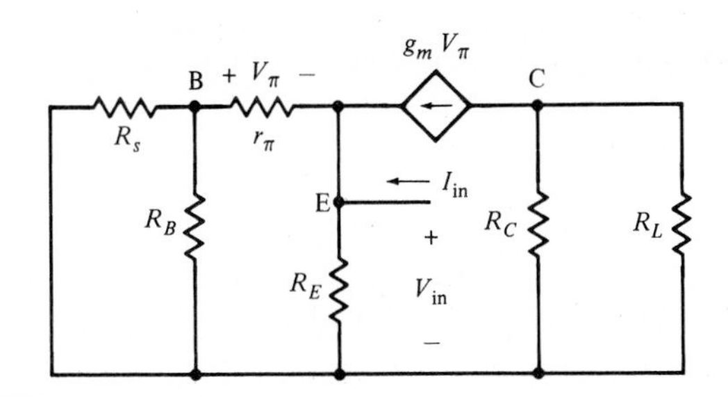

quency performance we determined in Section 6.1. The parameters are:

$$\begin{aligned} g_m &= 33.5 \text{ m℧} & R_B &= 16.67 \text{ k}\Omega \\ r_\pi &= 10.45 \text{ k}\Omega & R_C &= 5 \text{ k}\Omega \\ \beta_0 &= 350 & R_L &= 2 \text{ k}\Omega \\ C_1 &= 3\ \mu\text{F} & R_E &= 5 \text{ k}\Omega \\ C_2 &= 8\ \mu\text{F} & R_s &= 1 \text{ k}\Omega \\ C_E &= 50\ \mu\text{F} \end{aligned}$$

We first assume that C_2 and C_E are short circuits and use Eq. (51) to determine ω_{C1}:

$$\omega_{C1} = \frac{10^3}{[1 + (16.67 \| 10.45)]3} = 44.9 \text{ rad/s}$$

Short-circuiting C_1 and C_E, Eq. (52) leads to

$$\omega_{C2} = \frac{10^3}{(5 + 2)8} = 17.86 \text{ rad/s}$$

The third situation arises with C_1 and C_2 replaced by short circuits. From Eq. (53),

$$R_{eq} = 5 \left\| \frac{10.45 + (1 \| 16.67)}{1 + 350} \right. = 32.5\ \Omega$$

while the approximation $R_{eq} \doteq 1/g_m$ gives 29.9 Ω. We now employ Eq. (54) to obtain

$$\omega_{CE} = \frac{10^6}{32.5 \times 50} = 616 \text{ rad/s}$$

The break frequency of the zero is located from Eq. (55):

$$\omega_z = \frac{10^3}{5 \times 50} = 4 \text{ rad/s}$$

We note that ω_{CE} is the break frequency that controls the low-frequency response since it is more than ten times greater than any of the other corner frequencies. Thus, $\omega_L \doteq \omega_{CE} = 616$ rad/s.

As a check on these conclusions, the computer may be used to calculate the low-frequency response data. The results are shown as the solid curve in Fig. 7.25. The mid-frequency gain was found previously, with the value $A_{Vs(\text{mid})} = -41.4$, which is equivalent to 32.3 dB. The lower 3-dB frequency therefore occurs where the gain is 29.3 dB, which turns out to

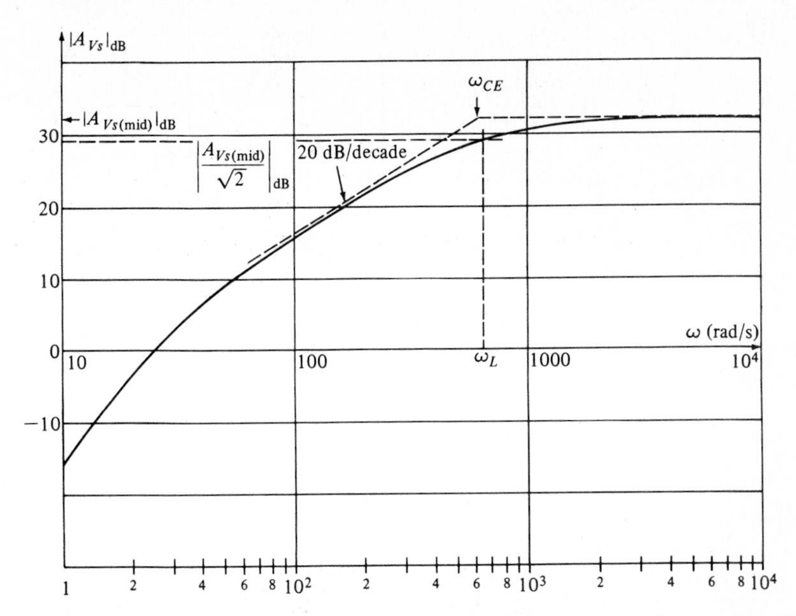

Fig. 7.25 The data from the computer analysis of the C-E example are plotted as a solid curve. From this, $\omega_L = 645$ rad/s, as compared with the approximate result, $\omega_L \doteq \omega_{CE} = 616$ rad/s. A portion of the asymptotic diagram is shown as a broken curve.

be at approximately 645 rad/s. The approximate value of 616 rad/s obtained above is thus quite close, showing approximately 5% error.

In determining ω_H by the open-circuit time-constant method, we added the time constants corresponding to each capacitance acting alone; the reciprocal of this sum represented a conservative estimate for ω_H. That is, our estimate was less than the true value. In an analogous way, we may establish a conservative estimate for ω_L by adding the three corner frequencies that appear in the denominator of the gain expression. Thus, the sum $\omega_{C1} + \omega_{C2} + \omega_{CE}$ is a more conservative estimate for ω_L; that is, this estimate is greater than the true value. For this example,

$$\omega_{C1} + \omega_{C2} + \omega_{CE} = 44.9 + 17.9 + 616 = 679 \text{ rad/s}$$

which is indeed greater than the true value of 645 rad/s.

The broken line on Fig. 7.25 is part of the asymptotic diagram, and it shows the corner at $\omega_{CE} = 616$ rad/s.

As an example of the low-frequency design of a common-emitter amplifier, let us continue with the mid-frequency design of the last chapter that is shown in Fig. 6.4*c*. The circuit is repeated in Fig. 7.26 for convenience, and some of the pertinent data are also listed near the circuit diagram. We shall design for a lower 3-dB frequency, $\omega_L = 1000$ rad/s. Our design pro-

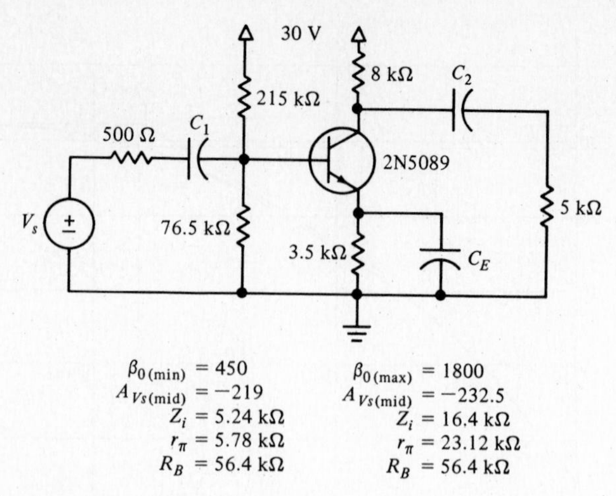

Fig. 7.26 The design of a common-emitter amplifier at mid-frequencies, taken from Fig. 6.4*c*.

cedure is a simple one. We let one of the three capacitors control the value of ω_L while both of the other two produce half-power frequencies less than one-tenth that value.

Suppose we select ω_{C1} as ω_L. Then

$$\omega_{C1} = \frac{1}{[R_s + (R_B \| r_\pi)]C_1} = \frac{10^{-3}}{[0.5 + (56.4 \| 5.78)]C_1}$$
$$= \omega_L = 1000$$

and $C_1 = 0.174\ \mu$F. Next, let us place ω_{C2} at $\omega_L/20$:

$$\omega_{C2} = \frac{1}{(R_L + R_C)C_2} = \frac{10^{-3}}{(8 + 5)C_2} = \frac{1000}{20}$$

so that $C_2 = 1.54\ \mu$F. Finally, we let $\omega_{CE} = \omega_L/10$, and

$$\omega_{CE} = \left[\frac{1}{R_E} + \frac{\beta_0 + 1}{r_\pi + (R_s \| R_B)}\right]\frac{1}{C_E}$$

$$= \left[\frac{1}{3.5} + \frac{451}{5.78 + (0.5 \| 56.4)}\right]\frac{10^{-3}}{C_E} = 100$$

and $C_E = 722\ \mu$F. This then places the zero break frequency at

$$\omega_z = \frac{1}{R_E C_E} = \frac{10^3}{3.5(722)} = 0.396 \text{ rad/s}$$

From these results, it is obvious that C_1 was not the best choice for the

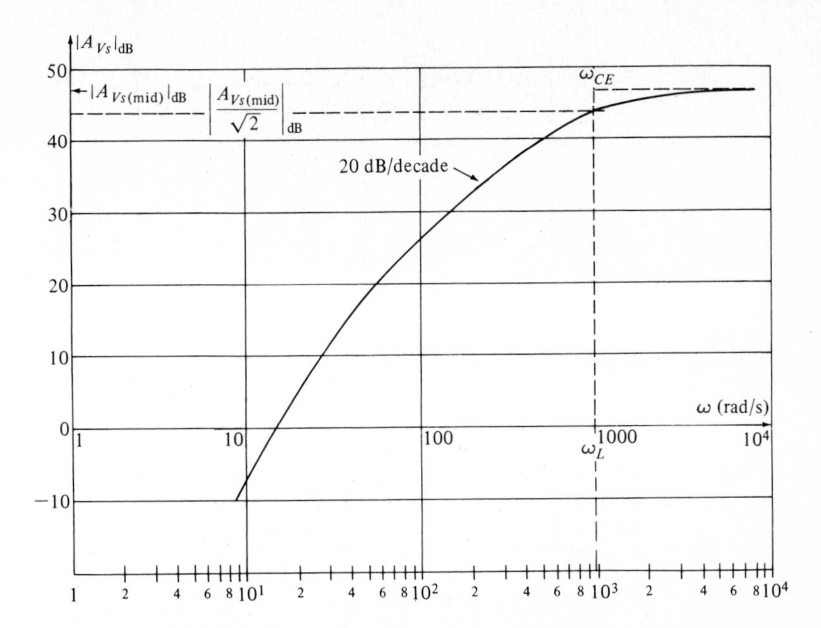

Fig. 7.27 The results of the computer simulation of the amplifier in Fig. 7.26 are shown in the solid line for $C_1 = C_2 = 2\ \mu F$, $C_E = 80\ \mu F$.

low-frequency control; C_E is inconveniently and uneconomically large. To reduce the size required for C_E, we should select it to control ω_L, a condition that was certainly suggested by the analysis example above, where R_{eq} was quite small. Letting $\omega_{CE} = 1000$, $\omega_{C1} = 100$, and $\omega_{C2} = 50$ rad/s, we find that $C_1 = 1.74\ \mu F$, $C_2 = 1.54\ \mu F$, $C_E = 72.2\ \mu F$, and $\omega_z = 3.96$ rad/s. If we increase each of the capacitor values to a commercially available size, say $C_1 = C_2 = 2\ \mu F$, and $C_E = 80\ \mu F$, then the response curve generated by computer simulation is that shown in Fig. 7.27. From the graph and the data, we find that $|A_{Vs(mid)}| = 218.7$ (or 46.8 dB), that $218.7/\sqrt{2} = 154.6$ (or 43.8 dB), and that $\omega_L = 982$ rad/s, a little bit better (lower) than the design value of 1000 rad/s.

Common-base

The common-base amplifier with its low-frequency equivalent circuit, shown in Fig. 7.28*a*, *b*, also contains three capacitors. We again avoid the difficult problem of deriving the gain expression; instead, we find the three corner frequencies that result from each capacitor controlling the response alone while the other two are replaced by short circuits. Using this method, we may approximate the lower 3-dB frequency ω_L quite closely by the highest corner frequency if that frequency is at least ten times as great as each of the lower two break frequencies.

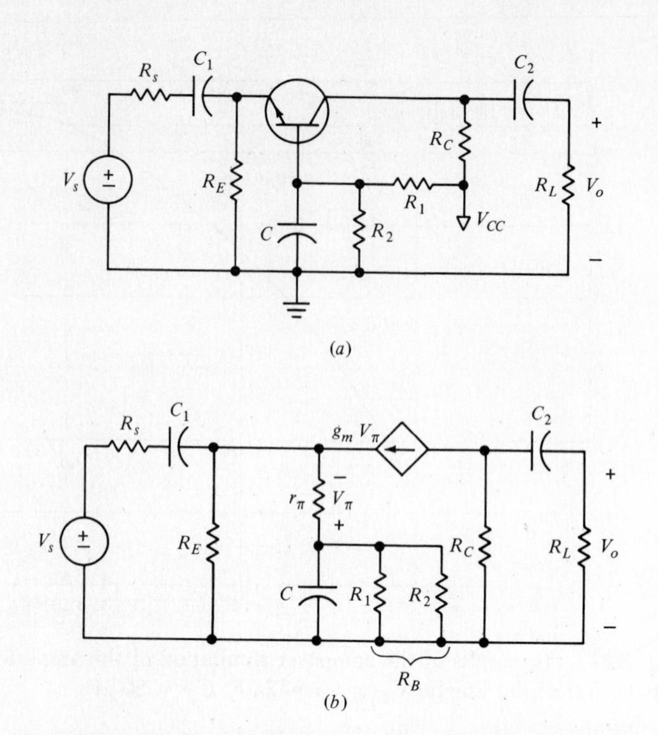

Fig. 7.28 (*a*) A common-base amplifier circuit, and (*b*) its low-frequency equivalent circuit.

We first replace C and C_2 by short circuits. Next we borrow the expression for the input impedance of a common-base amplifier, Eq. (10) in Section 6.3:

$$Z_i = R_E \| r_\pi \left\| \left(\frac{1}{g_m}\right) \right.$$

We identify the Thévenin resistance seen by C_1 as $R_s + Z_i$, yielding

$$\omega_{C1} = \frac{1}{[R_s + R_E \| r_\pi \| (1/g_m)]C_1} \tag{56}$$

The output capacitor C_2 by itself provides a corner at ω_{C2}:

$$\omega_{C2} = \frac{1}{(R_C + R_L)C_2} \tag{57}$$

The final case is not quite as simple. The impedance facing C in Fig. 7.28*b* with C_1, C_2, and V_s replaced by short circuits may be found by recalling our earlier discovery in Section 6.3 that any resistance connected to the emitter is increased by a factor of $\beta_0 + 1$ when it is viewed from the base.

Here, $R_s \| R_E$ is multiplied by $(\beta_0 + 1)$, and then combined in series with r_π. Including $R_B = R_1 \| R_2$, we have

$$R_{eq} = R_B \| [r_\pi + (\beta_0 + 1)(R_s \| R_E)] \tag{58}$$

and

$$\omega_C = \frac{1}{R_{eq} C}$$

It is informative to compare the values of the three resistances seen by C_1, C_2, and C. That presented to C_1 is apt to be low, since the input impedance of the common-base amplifier is quite low and operates best with a low-impedance source. Thus C_1 is probably associated with a few hundred ohms. The output capacitor C_2 sees R_C and R_L in series, or several thousand ohms. The base capacitor C faces R_B, generally quite large, in parallel with another resistance that is very large by virtue of the $\beta_0 + 1$ factor. Thus C looks at several tens of thousands of ohms. To provide identical corner frequencies, it would be necessary to make C_1 considerably larger than C_2, and C_2 much larger than C. The logical choice for ω_L is therefore ω_{C1}, which is usually the case in practice.

Common-collector

The common-collector amplifier in Fig. 7.29*a*, with its low-frequency equivalent circuit in Fig. 7.29*b*, contains only two capacitors. The corner frequency produced by C_1, with C_2 replaced by a short circuit, is given in terms of the input impedance to the common-collector circuit, Eq. (15) in Section 6.3, or by making another application of the $\beta_0 + 1$ factor:

$$Z_i = R_i = R_B \| [r_\pi + (\beta_0 + 1)(R_E \| R_L)] \tag{59}$$

so that

$$\omega_{C1} = \frac{1}{(R_s + R_i) C_1} \tag{60}$$

Since R_i is apt to be quite large, C_1 may be a relatively small value and yet permit good low-frequency performance.

The output coupling capacitor C_2 works in conjunction with R_L and the output impedance of the common-collector amplifier, Eq. (20) in Section 6.3:

$$Z_o = R_o = R_E \left\| \frac{r_\pi + (R_B \| R_s)}{(\beta_0 + 1)} \right. \tag{61}$$

We may also remember that, as seen by the emitter, any resistance connected to the base is divided by $(\beta_0 + 1)$. The corner is

$$\omega_{C2} = \frac{1}{(R_L + R_o)C_2} \tag{62}$$

Both R_o and R_L may be relatively low values of resistance, so it is therefore preferable to select a design in which $\omega_L \doteq \omega_{C2} \geq 10\omega_{C1}$, to avoid extremely large capacitor values.

However, we should also remember that the analysis of a common-base or a common-collector amplifier may be accomplished quite easily with a CAD program. This is particularly useful if one of the individual break frequencies is not considerably greater than the other two break frequencies. By all means, let us avoid calculating the exact expression for $A(j\omega)$, a task that only an eager graduate student would enjoy.

D7.14 The parameters of a common-emitter amplifier are $R_s = 300\ \Omega$, $R_B = 15\ \text{k}\Omega$, $r_\pi = 8\ \text{k}\Omega$, $R_E = 3\ \text{k}\Omega$, $g_m = 75\ \text{m}\mho$, $R_C = 6\ \text{k}\Omega$, $R_L = 3\ \text{k}\Omega$,

Fig. 7.29 (*a*) The common-collector or emitter-follower circuit. (*b*) Its low-frequency equivalent circuit, showing the points at which the input and output impedances are defined.

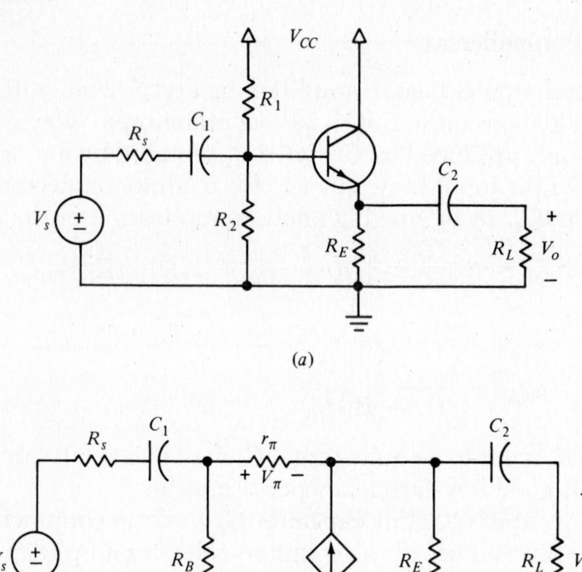

$C_1 = 4\ \mu F$, $C_2 = 2\ \mu F$, and $C_E = 80\ \mu F$. Determine (a) ω_{C1}, (b) ω_{C2}, (c) ω_{CE}, (d) ω_z. (e) Estimate ω_L.

Answers. 45.3; 55.6; 910; 4.2; $910 < \omega_L < 1011$ rad/s

D7.15 A common-emitter amplifier is to be designed for $R_s = 500\ \Omega$ and $R_L = 5\ k\Omega$. If $R_B = 20\ k\Omega$, $R_E = 2\ k\Omega$, $R_C = 4\ k\Omega$, $g_m = 50$ m℧, and $\beta_0 = 250$, let $\omega_L \doteq 400$ rad/s and select values for (a) C_1, (b) C_2, (c) C_E.

Answers. 5; 3; 125 μF

D7.16 A common-base amplifier operates between a 75-Ω source and a 2.5-kΩ load. If $\beta_0 = 125$, $g_m = 25$ m℧, $R_E = 1\ k\Omega$, $R_C = 5\ k\Omega$, $R_B = 20\ k\Omega$, $C_1 = 25\ \mu F$, $C_2 = 5\ \mu F$, and $C = 4\ \mu F$, find (a) ω_{C1}, (b) ω_{C2}, (c) ω_C. (d) Estimate ω_L.

Answers. 353; 26.7; 30.6; $353 < \omega_L < 411$ rad/s

D7.17 A common-collector amplifier operates between a 2.5-kΩ source and a 75-Ω load. If $\beta_0 = 125$, $g_m = 25$ m℧, $R_E = 1\ k\Omega$, $R_B = 20\ k\Omega$, $C_1 = 4\ \mu F$, and $C_2 = 25\ \mu F$, find (a) ω_{C1}, (b) ω_{C2}. (c) Estimate ω_L.

Answers. 23.4; 310; $310 < \omega_L < 333$ rad/s

7.10 A common-emitter example: analysis and design

Analysis

For an amplifier circuit such as the common-emitter stage shown in Fig. 7.30, we usually would like to know the several gains, the input impedance, and the bandwidth. The minimum value of the gain is of most concern, since some specified small input signal must be amplified to an acceptable level; larger values of gain make the system work even better in most applications. Therefore, we shall calculate the following six quantities for this amplifier:

1. Minimum mid-frequency voltage gain $V_o/V_s = A_{Vs}$
2. Input impedance $V_i/I_i = Z_i$
3. Current gain $I_o/I_i = A_I$
4. Power gain $(V_oI_o)/(V_iI_i) = A_P$
5. Upper half-power frequency ω_H
6. Lower half-power frequency ω_L

Before making a lot of unnecessary or unproductive calculations, we should give some consideration to problem identification and the plan of

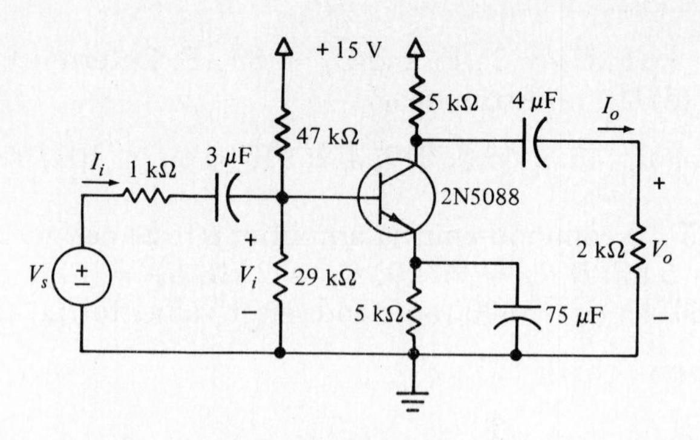

Fig. 7.30 A common-emitter amplifier circuit used as an example.

attack. First, we see that we require a rather complete analysis of this single-stage common-emitter amplifier. Thus a dc analysis must be performed first to provide parameter values for the hybrid-π model, which in turn is used to calculate gains and bandwidth. Next, we need the *minimum* voltage gain, so we must decide what combination of 2N5088 characteristics will result in this worst case. We certainly need to determine $g_{m(\min)}$; this occurs for $I_{C(\min)}$, which can be calculated by using $\beta_{dc(\min)}$. The gain also decreases with r_π, although the effect is less. That is, we should not reduce r_π by increasing g_m but rather let g_m be at its minimum value and then select $\beta_{0(\min)}$.

We let $V_{BE} = 0.65$ V, $T = 25°$C, and then proceed with the dc analysis:

$$V_{\text{Th}} = \frac{29}{29 + 47} 15 = 5.72 \text{ V}$$

$$R_{\text{Th}} = R_B = \frac{29(47)}{29 + 47} = 17.93 \text{ k}\Omega$$

From Fig. 9 of the 2N5088 data sheets in Appendix A,

$$\beta_{dc(\min)} = 300(1.2) = 360 \qquad (25°\text{C}, I_C = 1 \text{ mA}, V_{CE} = 5 \text{ V})$$

$$I_{C(\min)} = \frac{(360)(5.72 - 0.65)}{17.93 + 361(5)} = 1.002 \text{ mA}$$

$$V_{CE(\max)} = 15 - 5(1.00) - 5(1.00) = 5 \text{ V}$$

Since our calculated value of I_C agrees with the value we assumed in determining $\beta_{dc(\min)}$, we have a consistent set of data and can proceed to calculate values for the hybrid-π equivalent circuit:

$$g_{m(\min)} = 38.9(1.00) = 38.9 \text{ m℧}$$

$$\beta_{0(\min)} = 350 \qquad \text{(tabulated on data sheet)}$$

$$r_\pi = \frac{350}{38.9} = 9.00 \text{ k}\Omega$$

In the mid-frequency range, we neglect r_x and all the capacitors; by keeping our thoughts on the equivalent circuit, we can write

$$R_L' = 5\|2 = 1.429 \text{ k}\Omega$$

$$A_{Vi(\min)} = -38.9(1.429) = -55.6$$

$$Z_i = 9\|17.93 = 5.99 \text{ k}\Omega$$

$$A_{Vs(\min)} = -55.6\left(\frac{5.99}{1 + 5.99}\right) = -47.6 \text{ (33.6 dB in magnitude)}$$

$$A_I = -55.6 \times \frac{5.99}{2} = -166$$

$$A_P = 55.6(166) = 9240$$

To find the upper 3-dB frequency, we need to establish values for C_π and C_μ. A typical value for C_μ can be obtained from Fig. 12 on the 2N5088 data sheet with $V_{CB} = 5 - 0.65 = 4.35$ V. We find $C_\mu = C_{ob} = 2$ pF. For the assumed operating point, Fig. 11 shows $f_T = 300$ MHz, or $\omega_T = 1885$ Mrad/s, and $C_\pi = (38.9/1.885) - 2 = 18.6$ pF. We assume $r_x = 50\ \Omega$ and begin calculating:

$$R_\pi = R_s' = 9\|[0.05 + (1\|17.93)] = 0.898 \text{ k}\Omega$$

$$\tau_\pi = 0.898(18.6) = 16.7 \text{ ns}$$

$$R_\mu = 1.429 + 0.898(1 + 38.9 \times 1.429) = 52.2 \text{ k}\Omega$$

$$\tau_\mu = 52.2(2) = 104.5 \text{ ns}$$

Thus

$$\tau = 16.7 + 104.5 = 121.2 \text{ ns}$$

and

$$\omega_H = 8.25 \text{ Mrad/s}$$

Note how C_μ controls ω_H.

The lower half-power frequency is determined after evaluating ω_{C1},

ω_{C2}, and ω_{CE}:

$$\omega_{C1} = \frac{1000}{(1 + 5.99)3} = 47.7 \text{ rad/s}$$

$$\omega_{C2} = \frac{1000}{(2 + 5)4} = 35.7 \text{ rad/s}$$

$$\omega_{CE} = \left[\frac{1}{5} + \frac{351}{9 + (1 \| 17.93)}\right]\frac{1000}{75} = 473 \text{ rad/s}$$

Therefore, we know that ω_L is greater than 473 rad/s, and less than $\omega_{C1} + \omega_{C2} + \omega_{CE}$ or 556 rad/s. A reasonable estimate might be $\omega_L = 515$ rad/s.

This completes the analysis specified earlier, although we might also want to investigate the stability of the operating point with temperature or for higher-β transistors. The maximum gains could be found as well, but the above results are indicative of the procedures we need to follow.

Design

We now turn to a design problem. A common-emitter amplifier using a 2N5088 transistor is required to meet these specifications:

$$R_s = 1 \text{ k}\Omega$$

$$R_L = 2 \text{ k}\Omega$$

$$|A_{Vs}| \geq 25$$

$$f_L \leq 100 \text{ Hz}$$

$$f_H \geq 2 \text{ MHz}$$

$$|Z_i| \geq 2 \text{ k}\Omega \quad \text{(at mid-frequencies)}$$

Looking over these requirements and keeping the amplifier we just analyzed in mind, we should conclude that there will be no problem with ω_L, since we can always select larger capacitors; there will be no problem with the voltage gain, because we had almost twice that value before; the minimum allowable input impedance is a third of the earlier value, and we should be able to provide it easily. The real problem is with ω_H, which is about 50% larger than the earlier value (12.57 vs. 8.25 Mrad/s).

To increase ω_H, we should first consider reducing the Miller-effect capacitance. This is achieved by lowering R_L' and the voltage gain. We begin by letting the minimum gain be 25. At the same operating point used in the analysis problem,

$$25 \leq \frac{9 \| 17.93}{1 + (9 \| 17.93)} (38.9) R_L'$$

Therefore

$$R_L' \geq 0.75 \text{ k}\Omega$$

and it follows that

$$R_C \geq 1.2 \text{ k}\Omega$$

Since a nonzero value of r_x will reduce the gain slightly, let us try the circuit of Fig. 7.30 with R_C decreased to 1.25 kΩ ($R_L' = 0.769$ kΩ). The input impedance is still about 6 kΩ, and we need only check ω_H. The minimum value (worst case) of ω_H occurs when g_m is a maximum ($|A_{Vs(\text{mid})}|$ larger) and r_π is a maximum (R_s' larger). To find $g_{m(\max)}$, we have

$$\beta_{dc(\max)} = 900(1.2) = 1080$$

$$I_{C(\max)} = \frac{1080(5.72 - 0.65)}{17.93 + 1081(5)} = 1.011 \text{ mA}$$

$$g_{m(\max)} = 38.9(1.011) = 39.3 \text{ m℧}$$

Then we find r_π:

$$\beta_{0(\max)} = 1400$$

$$r_\pi = \frac{1400}{39.3} = 35.6 \text{ k}\Omega$$

Since V_{CB} has increased to $15 - 1.011(1.25 + 5) = 8.03$ V, C_μ decreases slightly to 1.6 pF, according to the data sheet, Fig. 12. Also, f_T increases to 320 MHz, or $\omega_H = 2011$ Mrad/s. Therefore

$$C_\pi = \frac{39.3}{2.011} - 1.6 = 17.95 \text{ pF}$$

$$R_\pi = R_s' = 35.6 \| [0.05 + (1 \| 17.93)] = 0.970 \text{ k}\Omega$$

$$\tau_\pi = 0.970(17.95) = 17.41 \text{ ns}$$

$$\tau_\mu = [0.769 + 0.970(1 + 39.3 \times 0.769)]1.6 = 49.7 \text{ ns}$$

$$\omega_H \doteq \frac{10^9}{17.41 + 49.7} = 14.9 \text{ Mrad/s}$$

$$f_H \doteq 2.37 \text{ MHz}$$

This is comfortably larger than the minimum required value, and all that remains of the design is the selection of C_1, C_2, and C_E. This is requested in Problem D7.18 and the answers appear there.

A computer analysis of the final design, similar to Fig. 7.30 but with R_C changed to 1.25 kΩ, may be made for the two conditions considered: $g_{m(\min)}$

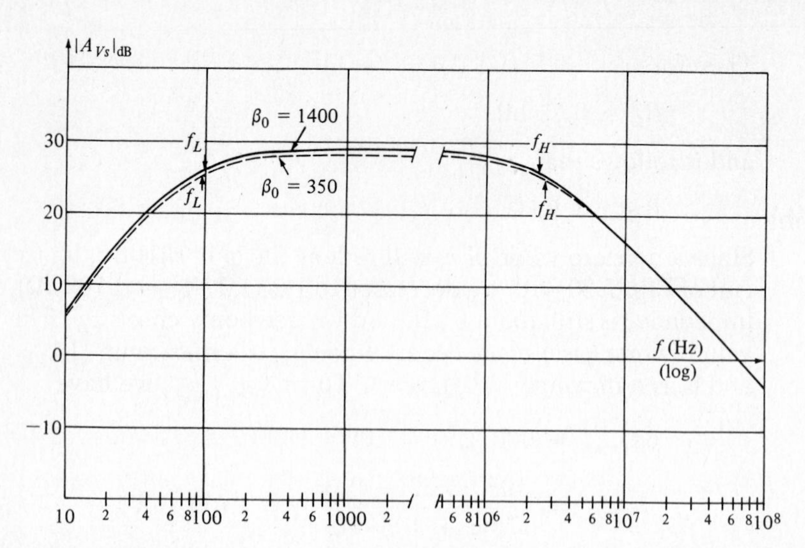

Fig. 7.31 A plot of the computer data for a common-emitter amplifier designed to provide a mid-frequency gain of 25 (27.96 dB) with $f_L \leq 100$ Hz and $f_H \geq 2$ MHz. Curves are shown for transistors with the minimum and maximum expected values of β_0.

$= 38.9$, $\beta_{0(\min)} = 350$; and $g_{m(\max)} = 39.3$, $\beta_{0(\max)} = 1400$. The values for C_1, C_2, and C_E are those given in the answers to Problem D7.18. The magnitude of A_{Vs} is shown as a function of frequency in Fig. 7.31 for the two cases, and the gains and half-power frequencies are tabulated in Table 7.2. Not only does our design meet the requirements, but the estimated (2.37 MHz) and computed (2.38 MHz) values of f_H show excellent agreement.

Suppose we had not met the ω_H requirement; how should we change our design to improve it? There are several possibilities. We might increase I_C to 1.5 or 2 mA, thus raising g_m and lowering r_π. This should then permit use of a smaller value for R_C while still permitting us to meet the voltage-gain requirement. We might also hunt up a better transistor, one having a smaller C_μ, a smaller r_x, or even a larger ω_T.

Table 7.2 CAD Analysis of a C-E Amplifier Design

Parameter	Specifications	$g_m = 38.9$ m℧ $\beta_0 = 350$	$g_m = 39.3$ m℧ $\beta_0 = 1400$
$\lvert A_{Vs(\mathrm{mid})}\rvert$	≥ 25	25.5	27.9
f_L	≤ 100 Hz	93.7 Hz	97.0 Hz
f_H	≥ 2 MHz	2.55 MHz	2.38 MHz

D7.18 Select values for C_1, C_2, and C_E to meet the design specifications of the common-emitter amplifier considered in Section 7.10.

Answers. 2.5; 5; 65 μF

Problems

1. The voltage gain of an amplifier is given as $A_V = j2\omega/(1 + j10^{-2}\omega)$. Calculate $|A_V|_{dB}$ if ω equals (a) 1, (b) 10, (c) 100, (d) 1000. (e) Sketch $|A_V|_{dB}$ vs. ω, $0.1 < \omega < 10{,}000$ rad/s, using a logarithmic frequency scale.
2. A resistor R_C is placed in parallel with capacitor C_S in the circuit of Fig. 7.4. Obtain expressions for (a) ω_L, (b) the high-frequency asymptotic value $|A_{V(\text{high})}|_{dB}$.
3. Determine the two straight-line asymptotes for the current gain $|A_I|_{dB}$, where $A_I = I_o/I_s$, for the circuit of Fig. 7.32. Prepare a sketch similar to Fig. 7.5.
4. A current source I_s, a 1-kΩ resistor, and the series combination of 250 Ω and 5 nF are in parallel. Let the gain function A be the ratio of the capacitor voltage to the current I_s in V/A. Plot $|A|$ for $10^4 < \omega < 10^7$ rad/s.
5. Given the gain function $A = 10{,}000/[100 + (j\omega)^2 + j101\omega]$, find A_{dB} when ω equals (a) 1, (b) 10, (c) 100, (d) 1000 rad/s. (e) Write A in a form that shows the presence of *two* break frequencies and determine the values predicted by the asymptotes at 1, 10, 100, and 1000 rad/s.
6. In Fig. 7.33, let $A_V = V_o/V_s$ and find (a) $|A_V|_{dB}$ at $\omega = 10^9$ rad/s; (b) the corner frequency.
7. If $A_V = 1/[(1 + j10^{-6}\omega)(1 + j0.2 \times 10^{-6}\omega)]$, (a) show that $|A_V| = 1/\sqrt{2}$ when $\omega = 0.964$ Mrad/s; (b) find the frequency at which $|A_V|_{dB} = -20$ dB.
8. The common-source amplifier shown in Fig. 6.10 has $g_m = 4$ m℧, $C_{gs} = 6$ pF, and $C_{gd} = 2.2$ pF. (a) Estimate ω_H. (b) Determine a new value for ω_H if R_L is reduced to 3 kΩ.

Fig. 7.32 See Problem 3.

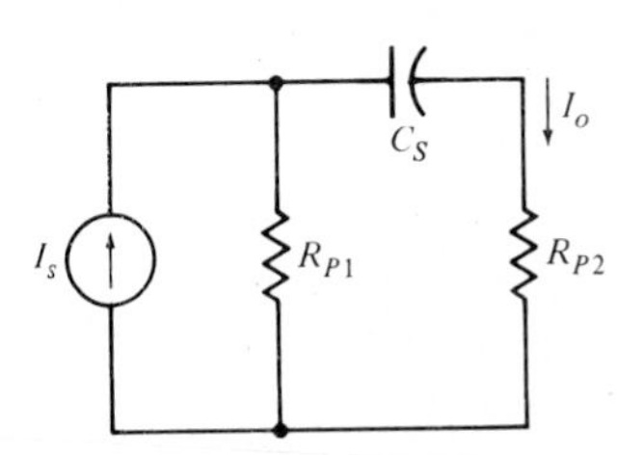

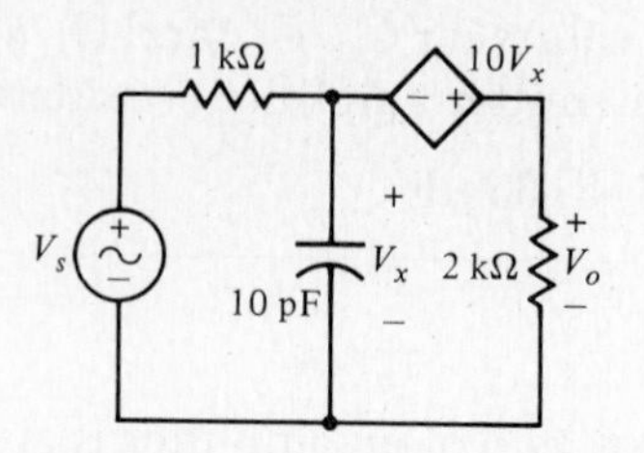

Fig. 7.33 See Problem 6.

9. Values for a C-S amplifier are $R_s = 500\ \Omega$, $R_1 = 900\ \text{k}\Omega$, $R_2 = 100\ \text{k}\Omega$, $R_{SS} = 250\ \Omega$, $R_D = 4\ \text{k}\Omega$, $R_L = 6\ \text{k}\Omega$, $g_m = 9\ \text{m℧}$, $C_{iss} = 5.7$ pF, and $C_{rss} = 1.3$ pF. (a) Find ω_H. (b) Find $A_{Vs(\text{mid})}$. (c) How do these values change if R_1 and R_2 are each halved?
10. Element values in a common-source amplifier include $R_s = 50\ \Omega$, $R_1 = 200\ \text{k}\Omega$, $R_2 = 50\ \text{k}\Omega$, $R_{SS} = 250\ \Omega$, $R_D = 2.5\ \text{k}\Omega$, $R_L = 1.5\ \text{k}\Omega$, and $V_{DD} = 17.5$ V. The *n*-channel JFET has $V_P = -1.5$ V and $I_{DSS} = 15$ mA. Let $C_{gs} = 4$ pF and $C_{gd} = 1.2$ pF and find (a) $A_{Vs(\text{mid})}$, (b) ω_H, (c) $|A_{Vs}|$ at $\omega = 10^9$ rad/s.
11. The high-frequency equivalent circuit of a C-S amplifier contains the elements $R_s = 300\ \Omega$, $R_B = 40\ \text{k}\Omega$, $R_D = 5\ \text{k}\Omega$, $R_L = 2.5\ \text{k}\Omega$, $C_{gs} = 4$ pF, and $C_{gd} = 1.3$ pF. Assuming ω_H is given by $1/\tau$, let g_m vary from 0 to 10 m℧ and plot a curve of (a) ω_H vs. g_m, (b) $|A_{Vs(\text{mid})}|$ vs. g_m, (c) $\omega_H |A_{Vs(\text{mid})}|$ vs. g_m.
12. If $R_1 = 300\ \text{k}\Omega$, $R_2 = 75\ \text{k}\Omega$, $R_D = 4\ \text{k}\Omega$, $R_L = 4\ \text{k}\Omega$, $C_{gd} = 1.5$ pF, $C_{gs} = 5$ pF, and $g_m = 8$ m℧ in a C-S amplifier, calculate the input admittance presented to the signal source (V_s and R_s) if ω equals (a) 10 Mrad/s, (b) 20 Mrad/s.
13. The box itself in Fig. 7.34 has the following characteristics: the input impedance is infinite, the output impedance is zero, and the ratio of the output voltage to the input voltage is A_V. Find Y_{in} at $\omega = 10^9$ rad/s

Fig. 7.34 See Problem 13.

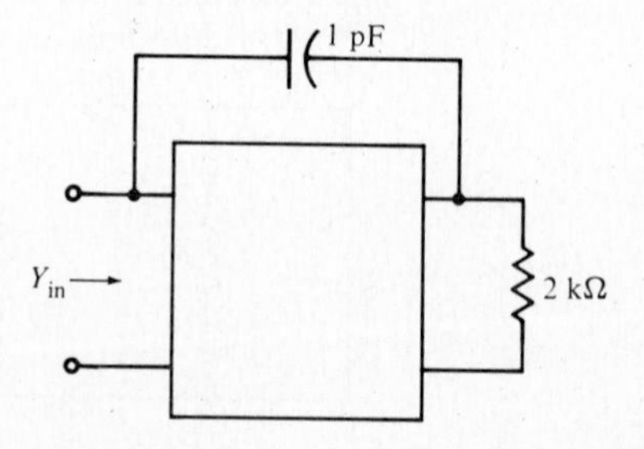

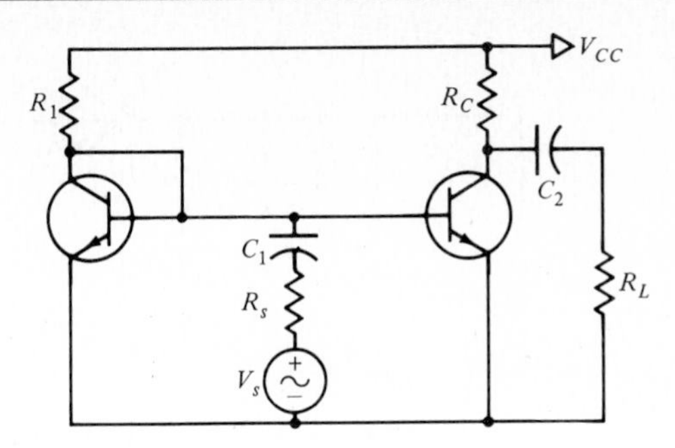

Fig. 7.35 See Problem 23.

if A_V equals (a) -10, (b) 10. (c) Repeat Parts (a) and (b) if a 1-kΩ resistor is inserted in series with the upper lead from the box at its output.

14. The amplifier of Problem D7.4 is reconnected as a common-gate amplifier with $R_s = 500\ \Omega$, $R_D = 4$ kΩ, $R_L = 6$ kΩ, $C_{gs} = 5.2$ pF, $C_{gd} = 1.2$ pF, and $g_m = 12$ m℧, as before. Also, let $R_{SS} = 1$ kΩ. Again, find (a) τ_{gs}, (b) τ_{gd}, (c) τ, (d) ω_H.
15. Let $V_P = -3$ V, $I_{DSS} = 9$ mA, $C_{iss} = 5$ pF, and $C_{rss} = 1.4$ pF for the amplifier of Fig. 6.15*a*. Find the upper half-power frequency.
16. The amplifier of Problem D7.4 is reconnected as a common-drain circuit with $R_s = 500\ \Omega$, $R_1 = 600$ kΩ, $R_2 = 200$ kΩ, $R_{SS} = 1$ kΩ, $R_L = 6$ kΩ, $C_{gs} = 5.2$ pF, $C_{gd} = 1.2$ pF, and $g_m = 12$ m℧. Find (a) τ_{gs}, (b) τ_{gd}, (c) τ, (d) ω_H.
17. Let $V_P = -3$ V, $I_{DSS} = 9$ mA, $C_{iss} = 5$ pF, and $C_{rss} = 1.4$ pF for the amplifier of Fig. 6.15*b*. Find (a) the mid-frequency gain $A_{Vs(\text{mid})}$; (b) the upper half-power frequency.
18. Parameter values in the high-frequency equivalent circuit of a common-emitter amplifier are $R_s = 0.5$ kΩ, $R_B = 40$ kΩ, $r_x = 30\ \Omega$, $r_\pi = 8$ kΩ, $R_C = 2.5$ kΩ, $R_L = 2.5$ kΩ, $C_\mu = 2$ pF, and $g_m = 50$ m℧. If $\omega_T = 2.5 \times 10^9$ rad/s, calculate (a) τ_π, (b) τ_μ, (c) τ, (d) ω_H, (e) C_{in}.
19. If $R_s' = 0.5$ kΩ, $R_C = 2$ kΩ, $g_m = 50$ m℧, $C_\pi = 15$ pF, and $C_\mu = 2.5$ pF, find the output impedance of the C-E high-frequency equivalent circuit at 2×10^7 rad/s.
20. Elements in the high-frequency equivalent circuit of a C-E amplifier include $R_s = 1$ kΩ, $r_x = 30\ \Omega$, $R_B = 20$ kΩ, $r_\pi = 5$ kΩ, $C_\pi = 12.5$ pF, $C_\mu = 2.5$ pF, $R_C = 4$ kΩ, $R_L = 6$ kΩ, and $g_m = 60$ m℧. Find (a) ω_T, (b) ω_H, (c) C_{in}.
21. A C-E amplifier equivalent circuit includes $r_x = 30\ \Omega$, $R_B = 20$ kΩ, $\beta_0 = 450$, $g_m = 30$ m℧, $R_C = 4$ kΩ, $C_\pi = 20$ pF, and $C_\mu = 2.5$ pF. Let $R_s = 1$ kΩ and $R_L = 6$ kΩ. (a) Calculate $A_{Vs(\text{mid})}$, ω_H, and the product $|A_{Vs(\text{mid})}|\,\omega_H$. (b) Let R_B, g_m, and R_C independently increase 10% in

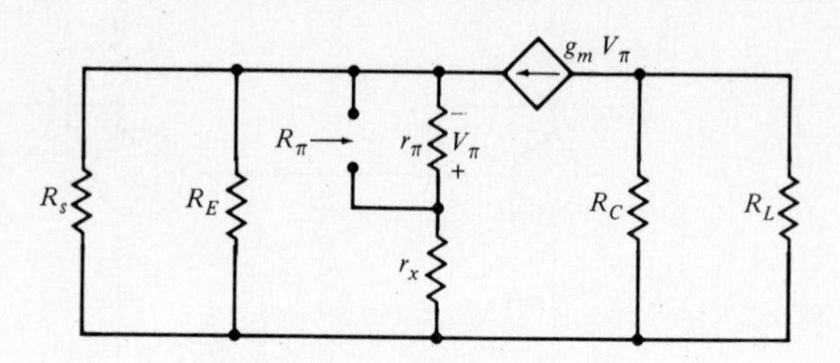

Fig. 7.36 See Problem 24.

value, and recalculate $A_{Vs(\text{mid})}$, ω_H, and $|A_{Vs(\text{mid})}|\omega_H$ each time. (c) Tabulate your results and identify the optimum changes to increase (independently) $A_{Vs(\text{mid})}$, ω_H, and $|A_{Vs(\text{mid})}|\omega_H$.

22. Using the data developed in Section 6.1 for the C-E amplifier of Fig. 6.3 with $\beta_{\text{dc}} = 1050$ and $\beta_0 = 1400$, assume $r_x = 20\ \Omega$, $C_\pi = 17.25$ pF, and $C_\mu = 1.8$ pF and find C_{in} and ω_H.
23. The integrated circuit amplifier of Fig. 7.35 contains a C-E stage with a current mirror stabilizing the operating point. Assuming two identical transistors, (a) draw the high-frequency equivalent circuit; (b) obtain expressions for τ_π and τ_μ for the C-E stage. (c) Let $R_1 = 10$ kΩ, $R_s = 1$ kΩ, $R_C = 5$ kΩ, $R_L = 10$ kΩ, $C_\mu = 1.2$ pF, $f_T = 100$ MHz, $V_{CC} = 15$ V, $V_{BE} = 0.65$ V, $\beta_{\text{dc}} = 100$, and $\beta_0 = 110$, and find $A_{Vs(\text{mid})}$ and ω_H.
24. If r_x is not neglected in the high-frequency equivalent circuit of the common-base amplifier, then the Thévenin resistance offered to C_π with C_μ open-circuited may be found by analyzing the circuit of Fig. 7.36. Do so, showing that R_π is given by the parallel combination of r_π and $(R_s' + r_x)/(1 + g_m R_s')$, where $R_s' = R_s \| R_E$.
25. To find R_μ for a C-B amplifier in which r_x is not negligible, it is necessary to find the Thévenin-equivalent resistance offered to C_μ with C_π open-circuited, as indicated by Fig. 7.37. Provide the derivation, showing that

$$R_\mu = r_x + R_L' + \frac{r_x(\beta_0 R_L' - r_x)}{r_x + r_\pi + (\beta_0 + 1)R_s'}$$

where $R_L' = R_C \| R_L$ and $R_s' = R_s \| R_E$.

26. Let $C_\pi = 17.25$ pF and $C_\mu = 1.8$ pF for the common-base amplifier shown in Fig. 6.6, where $I_C = 0.859$ mA and $\beta_0 = 350$. See Problems 24 and 25 for R_π and R_μ when r_x is not zero. Find ω_H if r_x equals (a) 0, (b) 20 Ω.
27. Elements in the circuit of a common-base amplifier (Fig. 6.5*a*, for example) are $R_s = 500\ \Omega$, $R_E = 5$ kΩ, $R_1 = 80$ kΩ, $R_2 = 40$ kΩ, $R_C =$

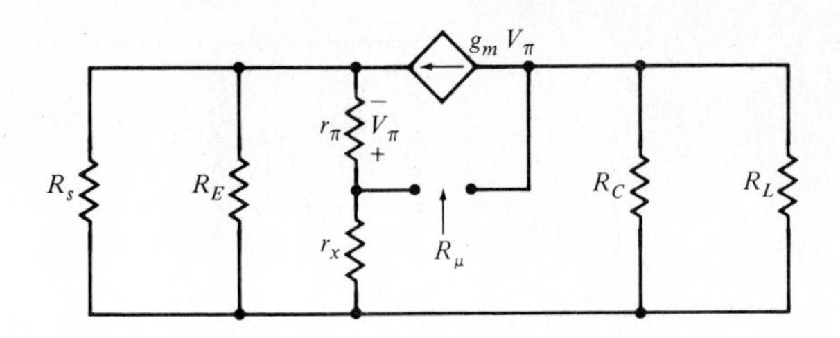

Fig. 7.37 See Problem 25.

6 kΩ, $R_L = 3$ kΩ, and $V_{CC} = 18$ V. Parameters of the transistor are $\beta_{dc} = \beta_0 = 200$, $n = 1$, $V_0 = 0.6$ V, $I_{CEO} \doteq 0$, $R_{BB} \doteq 0$, $\omega_T = 10^9$ rad/s, and $C_\mu = 3$ pF. Estimate ω_H if r_x may be neglected.

28. Let $C_\pi = 17.25$ pF and $C_\mu = 1.8$ pF for the common-collector amplifier of Fig. 6.8, where $I_C = 0.859$ mA and $\beta_0 = 350$. (a) Calculate ω_T. Estimate ω_H if r_x equals (b) 0, (c) 20 Ω.
29. Elements in the circuit of a common-collector amplifier (Fig. 6.7*a*, for example) are $R_s = 500$ Ω, $R_1 = 80$ kΩ, $R_2 = 40$ kΩ, $R_E = 5$ kΩ, $R_L = 3$ kΩ, and $V_{CC} = 15$ V. Parameters of the transistor are $\beta_{dc} = \beta_0 = 200$, $n = 1$, $V_0 = 0.6$ V, $I_{CEO} \doteq 0$, $R_{BB} \doteq 0$, $\omega_T = 1$ Grad/s, and $C_\mu = 3$ pF. Estimate ω_H if r_x equals (a) 0, (b) 50 Ω.
30. The transistor in the amplifier of Fig. 7.38 is characterized by $I_D = 20(0.3V_{GS} - 1)^2$ (mA) in the saturation region. (a) Find $A_{Vs(\text{mid})}$. (b) Specify values for C_1, C_2, and C_{SS} so that $\omega_L \doteq 100$ rad/s.
31. Element values in a low-frequency common-source equivalent circuit are $R_s = 1$ kΩ, $R_B = 99$ kΩ, $R_{SS} = 0.4$ kΩ, $R_D = 4$ kΩ, $R_L = 8$ kΩ, $g_m = 10$ m℧, $C_1 = 0.4$ μF, $C_2 = 10/3$ μF, and $C_{SS} = 250$ μF. (a) Find ω_{C1}, ω_{C2}, ω_{CSS}, and $\omega_z = 1/(R_{SS}C_{SS})$. (b) Use the exact expression $A_{Vs}/A_{Vs(\text{mid})} = [(j\omega)^2(j\omega + \omega_z)]/[(j\omega + \omega_{C1})(j\omega + \omega_{C2})(j\omega + \omega_{CSS})]$ to calculate $|A_{Vs}/A_{Vs(\text{mid})}|$ at $\omega = \omega_{CSS}$ and $\omega = 2\omega_{CSS}$.

Fig. 7.38 See Problems 30 and 33.

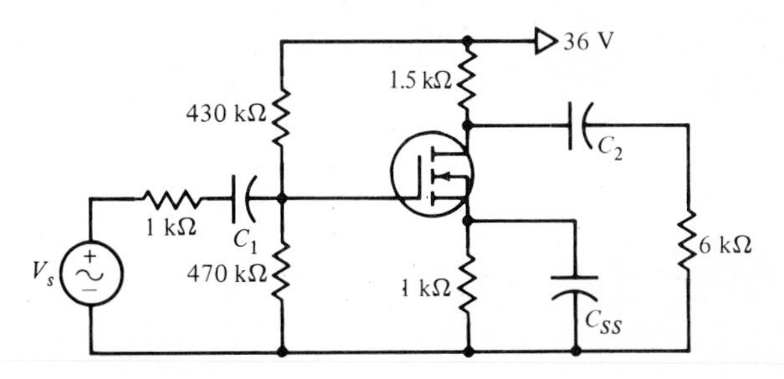

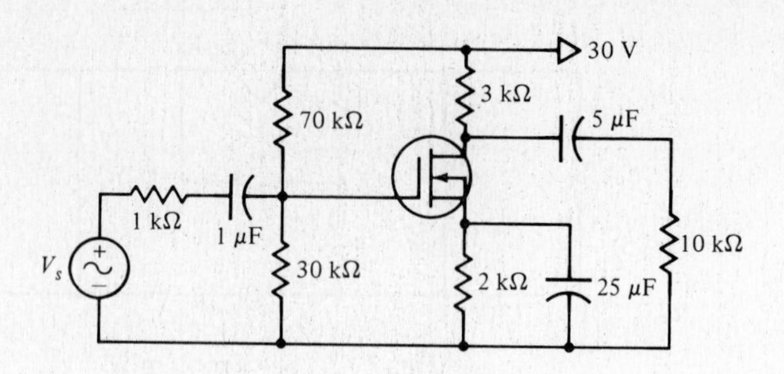

Fig. 7.39 See Problem 32.

32. Let $I_D = 5(V_{GS} + 2)^2$ (mA) for the MOSFET in Fig. 7.39. (a) Determine $A_{Vs(\text{mid})}$, ω_{C1}, ω_{C2}, and ω_{CSS}, and estimate ω_L. (b) Use a CAD program on the low-frequency equivalent circuit and obtain the necessary data to prepare a plot of $|A_{Vs}|$ vs. ω, $10 < \omega < 10{,}000$ rad/s, on a logarithmic frequency scale.
33. Let $C_1 = 5\ \mu$F, $C_2 = 2\ \mu$F, and $C_{SS} = 2\ \mu$F, and reassemble the elements in Fig. 7.38 as a common-gate amplifier. Use $I_D = 20(0.3V_{GS} - 1)^2$ (mA) for the MOSFET. (a) Find $A_{Vs(\text{mid})}$. (b) Estimate ω_L.
34. Let $I_D = 5(V_{GS} + 2)^2$ (mA) for the transistor used in the amplifier of Fig. 7.40. (a) Select standard values for C_1, C_2, and C_{SS} so that $80 < \omega_L < 100$ rad/s. (b) Check your design by a CAD analysis.
35. If the common-drain amplifier of Fig. 6.15*b* utilizes a transistor for which $I_D = (V_{GS} + 3)^2$ (mA), estimate the lower half-power frequency.
36. Element values in a common-drain circuit are $R_s = 1.5$ kΩ, $R_B = 125$ kΩ, $R_{SS} = 1.2$ kΩ, $R_L = 0.8$ kΩ, $C_1 = 0.04\ \mu$F, $C_2 = 40\ \mu$F, $g_m = 12.5$ m℧, $C_{iss} = 4$ pF, and $C_{rss} = 1$ pF. At mid-frequencies, calculate (a) A_{Vs}, (b) R_i, (c) R_o. Estimate (d) ω_L, (e) ω_H.
37. Values in the equivalent circuit of a common-emitter amplifier are $R_s = 1$ kΩ, $R_B = 20$ kΩ, $r_\pi = 10$ kΩ, $g_m = 50$ m℧, $R_E = 2.5$ kΩ, $R_C = 4$

Fig. 7.40 See Problem 34.

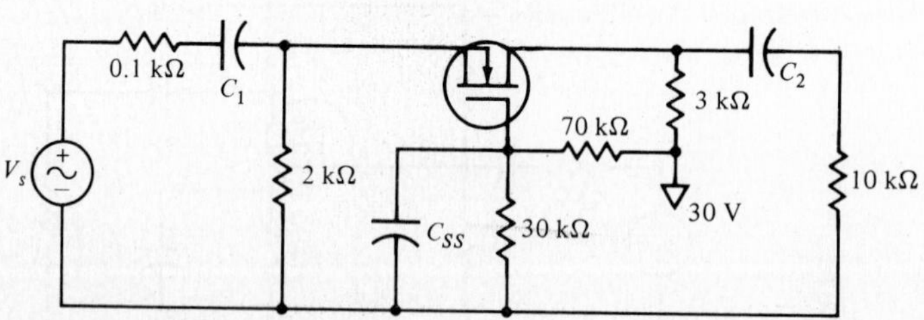

kΩ, R_L = 5 kΩ, C_1 = 5 μF, C_2 = 8 μF, C_E = 80 μF. (a) Find ω_{C1}, ω_{C2}, ω_{CE}, and ω_z. (b) Estimate ω_L.

38. For the transistor in Fig. 7.41, $\beta_{dc} = \beta_0 = 500$, $V_0 = 0.64$ V, and n = 1.945. (a) Calculate the mid-frequency gain $A_{Vs(mid)}$. (b) Select values of C_1, C_2, and C_E to provide a lower half-power frequency of 15 Hz. (c) Use a CAD program to obtain both low- and mid-frequency data for your design, and locate the lower half-power frequency accurately.
39. A C-E amplifier with R_s = 1 kΩ, R_1 = 30 kΩ, R_2 = 20 kΩ, R_E = 4 kΩ, R_C = 6 kΩ, R_L = 10 kΩ, C_1 = 5 μF, C_2 = 10 μF, and C_E = 50 μF is operating with β_0 = 120 and g_m = 40 m$\mho$. (a) Determine values for ω_{C1}, ω_{C2}, and ω_{CE}. (b) Estimate ω_L.
40. A standard C-E amplifier circuit includes R_s = 600 Ω, R_1 = 120 kΩ, R_2 = 30 kΩ, R_E = 2 kΩ, R_C = 8 kΩ, R_L = 6 kΩ, and V_{CC} = 15 V. Let $\beta_{dc} = \beta_0 = 250$, V_0 = ⅔ V, and n = 1.251 for the transistor. (a) Find $A_{Vs(mid)}$. (b) Specify values for C_1, C_2, and C_E so that $\omega_L \leq 100$ rad/s. Avoid using such large values that $\omega_L < 50$ rad/s. (c) Use a CAD program to find $A_{Vs(mid)}$ and ω_L.
41. A transistor for which $\beta_{dc} = \beta_0 = 200$, V_0 = ⅔, and n = 1 is used to design a C-E amplifier circuit with R_s = 500 Ω, R_1 = 50 kΩ, R_2 = 25 kΩ, R_E = 3 kΩ, R_C = 6 kΩ, R_L = 9 kΩ, V_{CC} = 12 V, C_1 = 6 μF, C_2 = 10 μF, and C_E = 50 μF. (a) Find ω_{C1}, ω_{C2}, ω_{CE}, and ω_z. (b) Estimate ω_L.
42. Element values in Fig. 7.22*a* are R_s = 100 Ω, R_1 = 90 kΩ, R_2 = 30 kΩ, R_E = 2.5 kΩ, R_C = 7.5 kΩ, and R_L = 10 kΩ. Let the transistor operate with β_0 = 300 and r_π = 6 kΩ. (a) Select values for C_1, C_2, and C_E such that $30 \leq \omega_L \leq 60$ rad/s. (b) Use data obtained with a CAD program to find an accurate value for ω_L.

Fig. 7.41 See Problem 38.

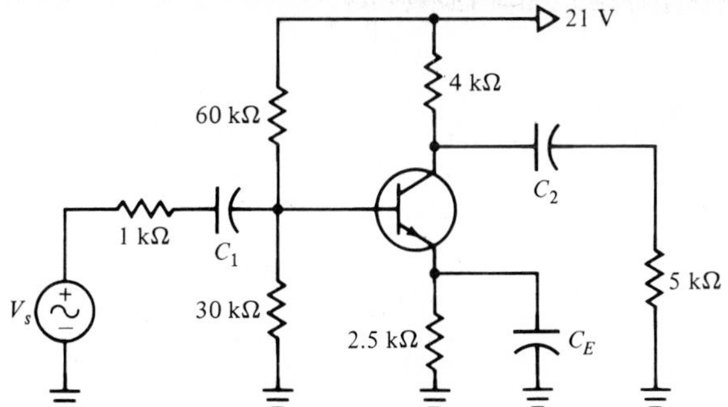

43. Values for a common-base amplifier circuit are R_s = 100 Ω, R_1 = 200 kΩ, R_2 = 50 kΩ, R_E = 5 kΩ, R_C = 20 kΩ, β_0 = 125, g_m = 50 m℧, and R_L = 10 kΩ. If f_L = 20 Hz, select suitable values for C_1, C_2, and C.
44. The low-frequency equivalent circuit of a common-base amplifier includes these elements: R_s = 100 Ω, R_1 = 200 kΩ, R_2 = 50 kΩ, R_E = 5 kΩ, r_π = 2.5 kΩ, R_C = 20 kΩ, g_m = 50 m℧, R_L = 10 kΩ, C_1 = 75 μF, C_2 = 3 μF, and C = 8 μF. Use a computer-analysis program to determine an accurate value for f_L.
45. Elements appearing in the low-frequency equivalent circuit of a common-base amplifier are R_s = 50 Ω, R_B = 25 kΩ, R_E = 4 kΩ, r_π = 5 kΩ, R_C = 12 kΩ, g_m = 80 m℧, R_L = 20 kΩ, C_1 = 200 μF, C_2 = 4 μF, and C = 10 μF. (a) Calculate values for ω_{C1}, ω_C, and ω_{C2}. (b) Estimate ω_L.
46. Elements in the low-frequency equivalent circuit of an emitter-follower are R_s = 1 kΩ, R_B = 20 kΩ, R_E = 5 kΩ, r_π = 7 kΩ, R_L = 100 Ω, and g_m = 50 m℧. (a) Select values for C_1 and C_2 so that $50 \le \omega_L \le 100$ rad/s. (b) If C_1 = 5 μF and C_2 = 250 μF, use a CAD program to determine ω_L.
47. A common-collector stage has R_s = 1 kΩ, R_1 = 100 kΩ, R_2 = 25 kΩ, R_E = 4 kΩ, and R_L = 250 Ω. The transistor is operating with r_π = 5 kΩ, β_0 = 200. (a) Select C_1 and C_2 so that $75 \le \omega_L \le 125$ rad/s. (b) Show that $75 \le \omega_L \le 125$ rad/s.
48. A C-E amplifier is to be designed to have $|A_{Vs(\text{mid})}| \ge 225$, $\omega_L \le 1800$ rad/s, and $\omega_H \ge 2$ Mrad/s for R_s = 0.5 kΩ and R_L = 5 kΩ. Use a 2N5089 transistor, assuming r_x = 50 Ω and a typical transistor is used. (a) Specify suitable values for R_C, V_{CC}, R_E, R_1, R_2, C_1, C_2, and C_E. (b) Check your design with a CAD program and prepare a plot of $|A_{Vs}(j\omega)|_{\text{dB}}$ vs. ω on a logarithmic frequency scale.

8

Multistage amplifiers

In the preceding two chapters we considered both bipolar and FET amplifiers having only a single stage. Three possible circuit configurations were studied in each case. Now we are ready to consider multistage amplifiers in which two or more stages are cascaded. Such an amplifier may be desirable for a number of reasons. Additional amplification can be required to provide a signal having some specified level. We may also need to furnish a high input impedance simultaneously with a large voltage gain, or perhaps a low output impedance and a large voltage gain. The first stage can be designed for input impedance, the last for output impedance, and one or more intermediate stages for voltage gain. We may also wish to meet a large bandwidth specification by supplying a number of stages, each having low gain and large bandwidth. With any of these conditions, the solution can be obtained by designing a multistage amplifier.

This chapter first considers gain calculations in the mid-frequency range, followed by approximate techniques of estimating ω_H and ω_L. Because of its important high-frequency applications, the cascode amplifier is the example we shall consider in the greatest detail.

8.1 The multistage amplifier at mid-frequencies

The analysis of a multistage amplifier in the mid-frequency range begins much like the familiar procedure for the single-stage circuit. First the dc operating point of each transistor is determined so that parameter values for a hybrid-π model can be calculated. Then we insert the hybrid-π model for the transistor into the external circuitry, and let the coupling and by-pass capacitors be short circuits while the interelectrode capacitances are open circuits. A two-stage common-emitter amplifier appears in Fig. 8.1*a*; its mid-frequency equivalent is shown in Fig. 8.1*b*.

To determine the voltage gain, it is best to avoid using long formulas, different for each configuration of course, and instead simply write down the proper expression factor by factor as we look at the equivalent circuit. This is always very easy to do for the hybrid-π model when there is no interaction between stages; that is, when the load impedance does not af-

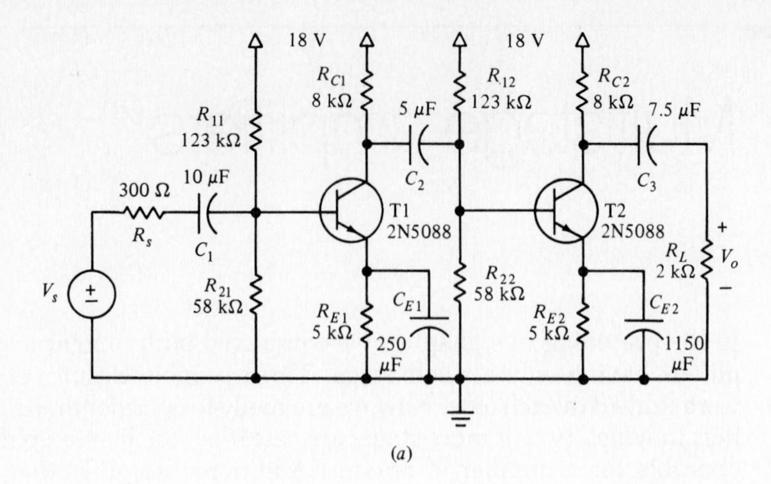

(*a*)

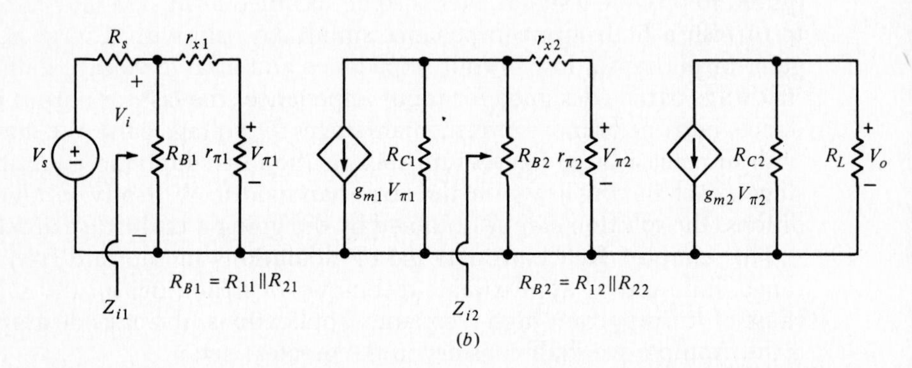

(*b*)

Fig. 8.1 (*a*) A two-stage CE-CE amplifier. (*b*) The mid-frequency model for the CE-CE amplifier.

fect the input resistance, which is the present case. For simplicity, let us take $r_{x1} = r_{x2} = 0$ for our first effort. We begin at the load resistor and work backwards:

$$V_o = (R_L \| R_{C2})(-g_{m2})V_{\pi 2}$$

and

$$V_{\pi 2} = (r_{\pi 2} \| R_{B2} \| R_{C1})(-g_{m1})V_{\pi 1}$$

so that

$$V_o = (R_L \| R_{C2})(-g_{m2})(r_{\pi 2} \| R_{B2} \| R_{C1})(-g_{m1})V_{\pi 1}$$

With $V_i = V_{\pi 1}$, then

$$A_{Vi} = \frac{V_o}{V_i} = (R_L \| R_{C2})(-g_{m2})(r_{\pi 2} \| R_{B2} \| R_{C1})(-g_{m1})$$

To obtain A_{Vs}, we also need the input impedance faced by the signal source:

$$Z_{i1} = R_{B1} \| r_{\pi 1}$$

Then

$$A_{Vs} = \frac{Z_{i1}}{Z_{i1} + R_s} A_{Vi}$$

and

$$A_{Vs} = \frac{V_o}{V_s} = (R_L \| R_{C2})(-g_{m2})(r_{\pi 2} \| R_{B2} \| R_{C1})$$

$$\times (-g_{m1}) \frac{(R_{B1} \| r_{\pi 1})}{(R_{B1} \| r_{\pi 1}) + R_s}$$

This last equation may be written directly; the intermediate steps have been included only to show the mental processes involved in approaching any multistage amplifier calculation.

Note that the effective load resistance of the first stage is the input resistance to the second stage, $Z_{i2} = R_{B2} \| r_{\pi 2}$. The larger we can make the value of Z_{i2}, the larger the gain of the first stage and hence the overall gain.

To calculate the minimum value of $A_{Vs(\text{mid})}$, we should use the minimum expected values of g_{m1} and g_{m2}, as well as the minimum values of $r_{\pi 1}$ and $r_{\pi 2}$, which occur for $\beta_{01(\text{min})}$ and $\beta_{02(\text{min})}$. Therefore, we seek the minimum values of I_C and β_0 for each transistor. Since the two stages have identical biasing circuitry, we have the single dc analysis:

$$R_{\text{Th}} = 123 \| 58 = 39.4 \text{ k}\Omega$$

$$V_{\text{Th}} = \frac{58}{58 + 123} 18 = 5.77 \text{ V}$$

The collector current can not be any larger than the value obtained by assuming that $V_{CE} = 0$, which is $18/(8 + 5)$, or 1.38 mA; a reasonable assumption for both β_{dc} and $\beta_{0(\text{min})}$ is therefore 350. Then

$$I_C = \frac{350(5.77 - 0.65)}{39.4 + 351(5)} = 0.998 \text{ mA} \doteq 1 \text{ mA}$$

$$V_{CE} = 18 - (8 + 5)(1) = 5 \text{ V}$$

$$V_{CB} = 5 - 0.65 = 4.35 \text{ V}$$

$$g_m = 38.9 \text{ m}\mho$$

$$r_\pi = \frac{350}{38.9} = 9 \text{ k}\Omega$$

Using these data, we have

$$A_{Vs(\text{mid})} = (2\|8)(-38.9)(9\|39.4\|8)(-38.9)\frac{(9\|39.4)}{0.3 + (9\|39.4)}$$

$$= 8890$$

This is a considerably larger value than any single-stage amplifier can provide. Note also that there is no phase reversal in this two-stage CE-CE amplifier.

Current and power gain can be calculated as we did in Chapter 6:

$$A_I = A_{Vi}\frac{Z_i}{R_L} \qquad \text{and} \qquad A_P = A_{Vi}A_I$$

If we had included r_{x1} and r_{x2} in our calculations, then the amplification would have been slightly less. Making use of voltage division twice, we have

$$A_{Vs(\text{mid})} = (R_L\|R_{C2})(-g_{m2})\frac{r_{\pi 2}}{r_{x2} + r_{\pi 2}}$$

$$\times [(r_{x2} + r_{\pi 2})\|R_{B2}\|R_{C1}](-g_{m1})$$

$$\times \left(\frac{r_{\pi 1}}{r_{x1} + r_{\pi 1}}\right)\left[\frac{(r_{\pi 1} + r_{x1})\|R_{B1}}{R_s + (r_{\pi 1} + r_{x1})\|R_{B1}}\right]$$

$$= 8820$$

for $r_{x1} = r_{x2} = 50\ \Omega$.

Note that the effect of including r_x in the model is less than a 1% reduction in gain, hardly worth the added complication. However, should the collector current be increased, r_π would decrease and the inclusion of r_x could be more significant.

Now let us look at three C-E stages in cascade, Fig. 8.2*a*, where each transistor has a different operating point. The small-signal model is shown in Fig. 8.2*b*. As element values, we have selected $R_{B1} = 30$ kΩ, $R_{B2} = 20$ kΩ, $R_{B3} = 10$ kΩ, $R_L = R_{C3} = 5$ kΩ, $R_{C2} = 10$ kΩ, $R_{C1} = 20$ kΩ, and $R_s = 1$ kΩ. The operating points furnish $I_{C1} = 1$ mA, $I_{C2} = 2$ mA, and $I_{C3} = 3$ mA, while $\beta_{01} = 300$, $\beta_{02} = 400$, and $\beta_{03} = 500$. This leads to $r_{\pi 1} = 7.71$ kΩ, $r_{\pi 2} = 5.14$ kΩ, and $r_{\pi 3} = 4.28$ kΩ. The voltage gain $A_{Vs} = V_o/V_s$ may be written by inspection, letting each r_x be zero:

$$A_{Vs} = (5\|5)(-38.9)(4.28\|10\|10)(-77.8)(5.14\|20\|20)$$

$$\times (-116.7)\frac{(7.71\|30)}{1 + (7.71\|30)}$$

$$= -5{,}940{,}000$$

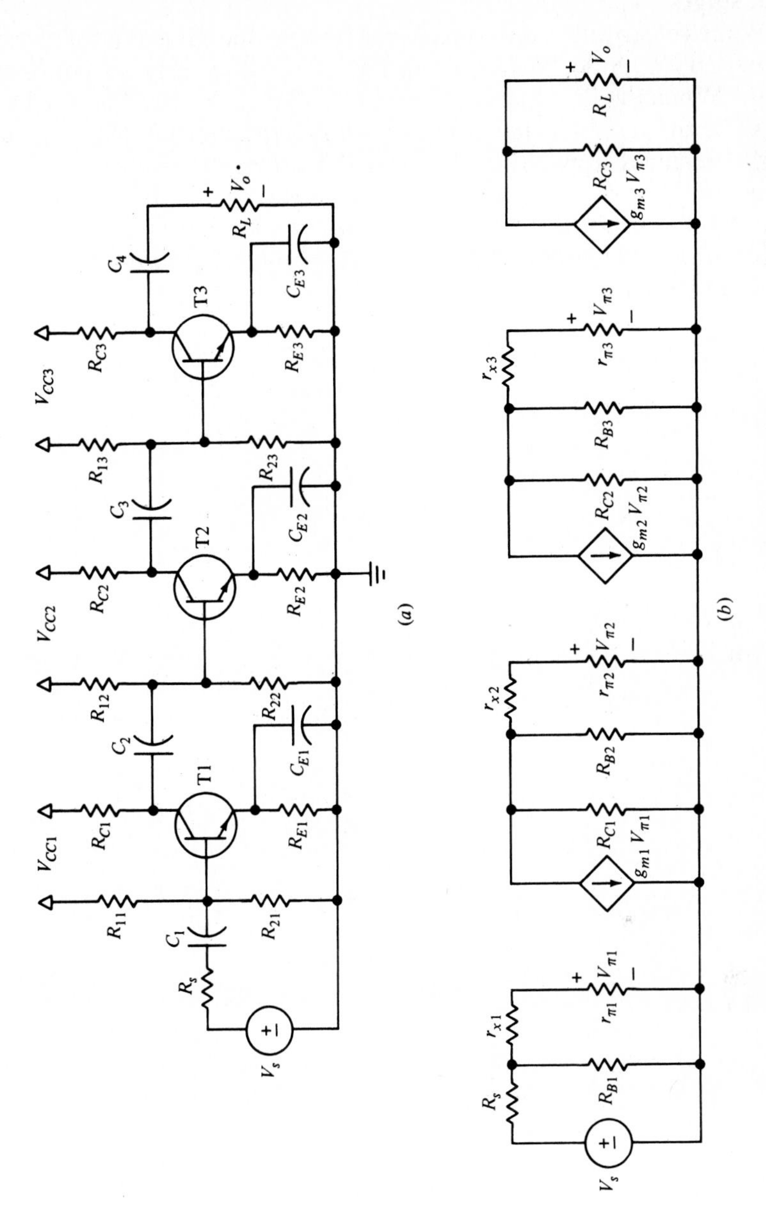

Fig. 8.2 (*a*) A CE-CE-CE three-stage amplifier. (*b*) The mid-frequency equivalent circuit.

This is greater than 135 dB, so we are now beginning to achieve some very large values of gain. Note that there is a net 180° phase reversal with three C-E stages.

We select a multistage amplifier including one FET as a third example, shown in Fig. 8.3*a*. Here, a common-source stage drives a common-base stage. When the first stage is connected C-S or C-E, and the second is C-G or C-B, the combination is called a *cascode* amplifier, used widely for high-frequency applications such as FM tuners because of its good gain, large ω_H, high Z_i, and good isolation between output and input. Here, we investigate its mid-frequency properties. Remembering that the input resistance to the common-base stage is $R_E \| r_\pi \| (1/g_{m2})$, we may again write an expression for $A_{Vs(\text{mid})}$ by inspection:

$$A_{Vs(\text{mid})} = (5\|2)(-33.5)(+4)\left(3\|5\|8.78\|\frac{1}{33.5}\right)\frac{255}{256}$$

$$= -5.58$$

Fig. 8.3 (*a*) A cascode amplifier composed of a common-source stage followed by a common-base stage. (*b*) The mid-frequency equivalent circuit for $I_D = 1$ mA, $I_C = 0.859$ mA.

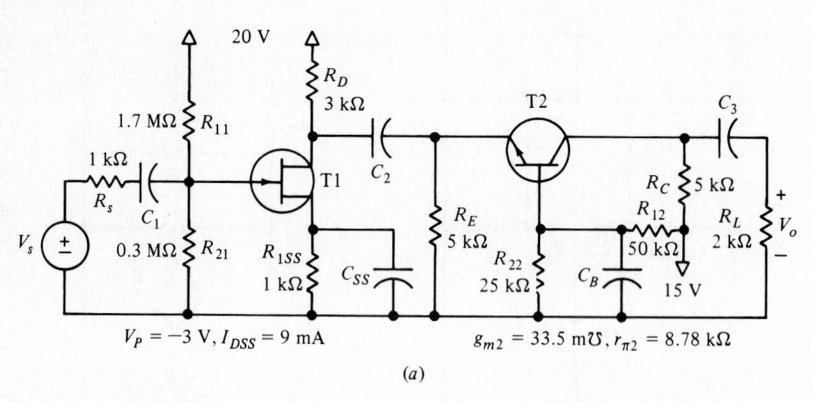

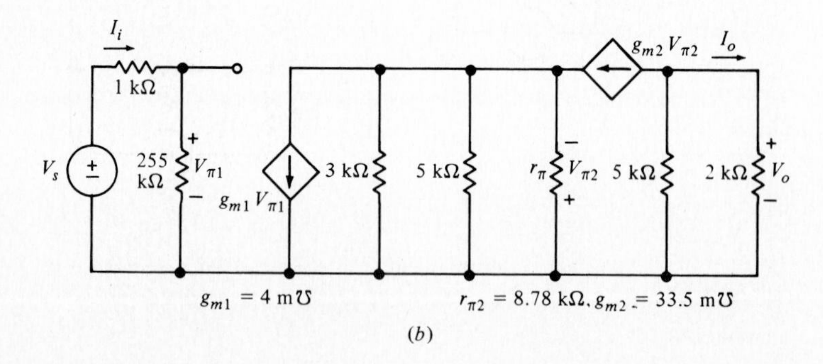

If necessary, the gain can be increased by using a bipolar transistor as a common-emitter first stage; however, the input resistance is considerably less.

There are many other combinations of bipolar transistors and FETs possible for multistage amplifiers. Several of these will appear as problems. In each case it is preferable to stare briefly at the equivalent circuit and then write the gain equation directly, rather than rely on a sheaf of special formulas. For example, we might want the current and power gain for the cascode amplifier above. First, we find A_{Vi}:

$$A_{Vi} = \frac{V_o}{V_i} = (5\|2)(-33.5)(4)\left(3\|5\|8.78\|\frac{1}{33.5}\right)$$

$$= -5.61 \text{ or } 15 \text{ dB}$$

The current gain then follows easily:

$$A_I = \frac{I_o}{I_i} = \frac{V_o}{R_L}\frac{Z_i}{V_i} = A_{Vi}\frac{Z_i}{R_L} = -5.61\,\frac{255}{2} = -715$$

The power gain is then

$$A_P = \frac{V_o I_o}{V_i I_i} = A_{Vi}A_I = 4010$$

D8.1 Increase V_{CC} to 20 V in the circuit of Fig. 8.1*a*, assume r_x is negligible, and calculate the minimum expected value of (a) $A_{Vi(\text{mid})}$, (b) $A_{Vs(\text{mid})}$, (c) $A_{I(\text{mid})}$.

Answers. 11,100; 10,600; 36,960

8.2 An approximation for ω_H

The high-frequency performance of a multistage amplifier can often be described satisfactorily by a single value, ω_H. This quantity is also closely equal to the bandwidth $\omega_H - \omega_L$, whenever $\omega_L \ll \omega_H$. In Chapter 7 we determined an approximate value of ω_H for a single-stage amplifier by considering the by-pass and coupling capacitors to be effectively short circuits when $\omega \doteq \omega_H$ and then finding the open-circuit time constant for each of the interelectrode capacitances individually. Here we extend that concept directly to multistage amplifiers.

We consider first the two-stage CE-CE amplifier of Fig. 8.1*a*, *b*. Its high-frequency equivalent circuit is shown in Fig. 8.4. Since this is a two-stage amplifier, we have four capacitors and we need to find the four time constants $\tau_{\pi 1}$, $\tau_{\pi 2}$, $\tau_{\mu 1}$, and $\tau_{\mu 2}$. Again following the procedure developed

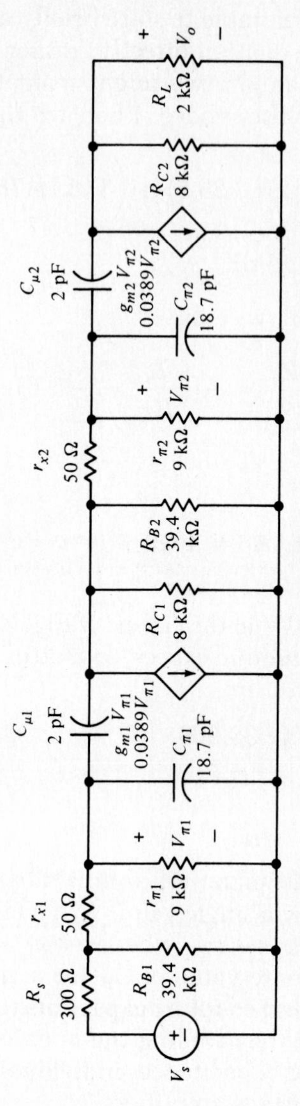

Fig. 8.4 A high-frequency model for the two-stage CE-CE amplifier of Fig. 8.1*a* with C_π = 18.7 pF, C_μ = 2 pF, and r_x = 50 Ω for both stages.

in the previous chapter, we shall sum the four time constants and use the reciprocal of this value as our estimate for ω_H:

$$\omega_H \doteq \frac{1}{\tau_{\pi 1} + \tau_{\pi 2} + \tau_{\mu 1} + \tau_{\mu 2}}$$

Let us begin by finding $\tau_{\pi 1} = R_{\pi 1} C_{\pi 1}$, where $R_{\pi 1}$ is the Thévenin-equivalent resistance presented to $C_{\pi 1}$ with $C_{\pi 2}$, $C_{\mu 1}$, and $C_{\mu 2}$ open-circuited. From Fig. 8.4,

$$R_{\pi 1} = r_{\pi 1} \| [r_{x1} + (R_s \| R_{B1})]$$

and

$$\tau_{\pi 1} = \{r_{\pi 1} \| [r_{x1} + (R_s \| R_{B1})]\} C_{\pi 1}$$

Similarly,

$$R_{\pi 2} = r_{\pi 2} \| [r_{x2} + (R_{C1} \| R_{B2})]$$

In order to find $R_{\mu 1}$ and $R_{\mu 2}$, we resurrect a very useful result from Chapter 7, Eq. (26):

$$R_\mu = R_L' + R_s'(1 + g_m R_L') \tag{1}$$

where R_L' is the equivalent load resistance of the C-E stage and R_s' is the equivalent source resistance. We make use of Eq. (1) in this multistage amplifer by looking at Fig. 8.4 and writing

$$R_{L1}' = R_{C1} \| R_{B2} \| (r_{x2} + r_{\pi 2})$$

$$R_{L2}' = R_{C2} \| R_L$$

$$R_{s1}' = r_{\pi 1} \| [r_{x1} + (R_s \| R_{B1})]$$

$$R_{s2}' = r_{\pi 2} \| [r_{x2} + (R_{C1} \| R_{B2})]$$

By combining these expressions with the appropriate value of g_m, we may obtain values for $R_{\mu 1}$ and $R_{\mu 2}$.

Let us see how this technique works by using the values given in Fig. 8.4. We let $r_{x1} = r_{x2} = 50\ \Omega$. Then

$$R_{L1}' = R_{C1} \| R_{B2} \| (r_{x2} + r_{\pi 2}) = 8 \| 39.4 \| (0.05 + 9) = 3.83\ \text{k}\Omega$$

$$R_{s1}' = r_{\pi 1} \| [r_{x1} + (R_s \| R_{B1})] = 9 \| [0.05 + (0.3 \| 39.4)]$$

$$= 0.335\ \text{k}\Omega$$

Therefore $R_{\pi 1} = R_{s1}' = 0.335\ \text{k}\Omega$, and

$$\tau_{\pi 1} = R_{\pi 1} C_{\pi 1} = 0.335 \times 18.7 = 6.3\ \text{ns}$$

Next,

$$R_{\mu 1} = R'_{L1} + R'_{s1}(1 + g_{m1}R'_{L1})$$
$$= 3.83 + 0.335(1 + 38.9 \times 3.83)$$
$$= 54.1 \text{ k}\Omega$$
$$\tau_{\mu 1} = R_{\mu 1}C_{\mu 1} = 54.1 \times 2 = 108.2 \text{ ns}$$

Continuing,

$$R'_{L2} = R_{C2} \| R_L = 8 \| 2 = 1.6 \text{ k}\Omega$$
$$R'_{s2} = r_{\pi 2} \| [r_{x2} + (R_{C1} \| R_{B2})] = 9 \| [0.05 + (8 \| 39.4)] = 3.84 \text{ k}\Omega$$

Therefore $R_{\pi 2} = R'_{s2} = 3.84 \text{ k}\Omega$, and

$$\tau_{\pi 2} = R_{\pi 2}C_{\pi 2} = 3.84 \times 18.7 = 71.8 \text{ ns}$$

Finally,

$$R_{\mu 2} = R'_{L2} + R'_{s2}(1 + g_{m2}R'_{L2}) = 1.6 + 3.84(1 + 38.9 \times 1.6)$$
$$= 244 \text{ k}\Omega$$
$$\tau_{\mu 2} = R_{\mu 2}C_{\mu 2} = 244 \times 2 = 489 \text{ ns}$$

Adding the four open-circuit time constants, we have

$$\tau = 6.3 + 108.2 + 71.8 + 489 = 675 \text{ ns}$$

and

$$\omega_H \doteq 1.481 \text{ Mrad/s}$$

A computer simulation of this circuit indicates that $\omega_H = 1.504$ Mrad/s, showing that our approximate method is quite accurate. Also, the error is on the conservative side, which is correct for design. The calculations above clearly point to $\tau_{\mu 2}$ as the chief factor in determining ω_H, while $\tau_{\mu 1}$ comes in second. Both of these include a large Miller effect.

To redesign for a larger ω_H, the $R_{\mu 2}C_{\mu 2}$ product must be made smaller. From the mid-frequency gain equation, we note that a reduction in either R'_{s2} or R'_{L2} yields a smaller gain for the amplifier. Again, we can trade gain for bandwidth. Of course, a reduction in C_μ for both transistors is the most effective method and causes no loss in gain. However, this probably means a more expensive transistor.

The method demonstrated above can be used to analyze other multistage amplifiers. In Section 8.5 we will examine a CS-CB cascode amplifier. Other useful multistage configurations include CC-CE, CS-CS-CS, CS-CG, and so forth. With the aid of the methods we have already formulated, these should be relatively straightforward problems.

As an example of an FET multistage amplifier, consider the CD-CS configuration of Fig. 8.5*a* and its high-frequency equivalent in Fig. 8.5*b*.

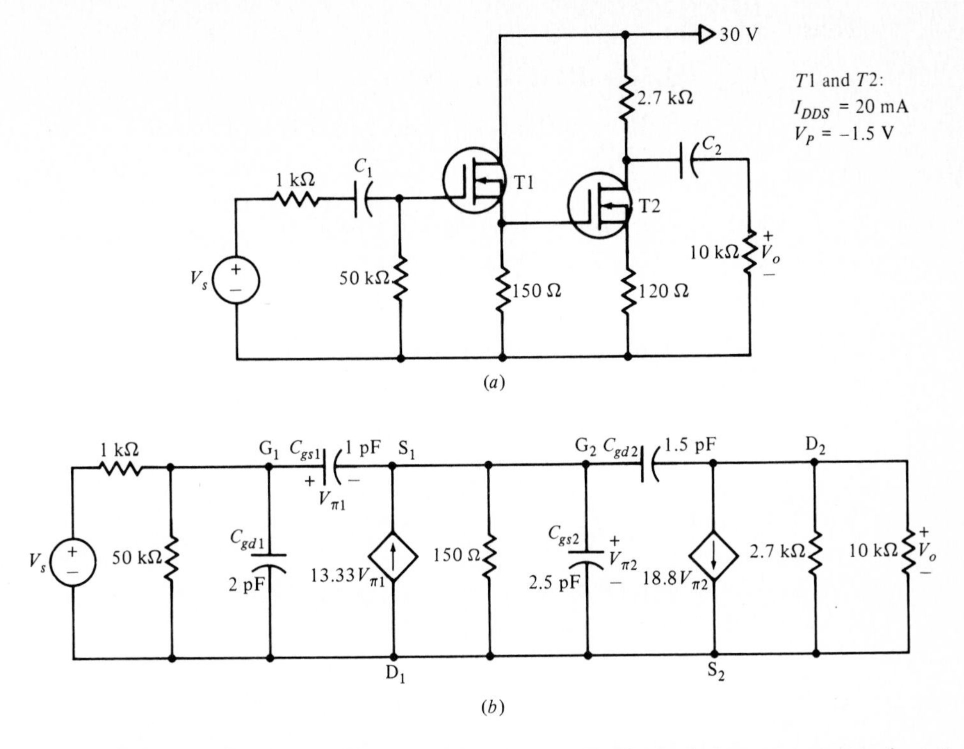

Fig. 8.5 (*a*) A CD-CS FET amplifier circuit. (b) The high-frequency equivalent circuit.

We might anticipate that $\tau_{gd2} = R_{gd2}C_{gd2}$ will be the largest time constant, since the C-D stage has no phase reversal and no Miller effect, leaving the Miller-effect capacitance of the C-S stage to dominate.

The dc analysis of the amplifier is quite uneventful, since the gate insulator of T2 isolates the two stages. This direct coupling also avoids one resistor and one capacitor.

We find that $I_{D1} = 5$ mA, $V_{GS1} = -0.75$ V, and $g_{m1} = 13.33$ m℧. The second stage has $V_{GS2} = -0.443$ V, $I_{D2} = 9.94$ mA, and $g_{m2} = 18.80$ m℧.

Interelectrode capacitance values are assigned as shown in Fig. 8.5*b*, $C_{gs1} = 2$ pF, $C_{gd1} = 1$ pF, $C_{gs2} = 2.5$ pF, and $C_{gd2} = 1.5$ pF. Beginning with the output stage, we see that C_{gs2} faces the output impedance of the C-D stage, $R_{SS1} \| (1/g_{m1})$. Therefore $R_{gs2} = 0.15 \| (1/13.33) = 0.05$ kΩ, and

$$\tau_{gs2} = R_{gs2}C_{gs2} = 0.05 \times 2.5 = 0.125 \text{ ns}$$

To find R_{gd2}, we apply Eq. (1) after finding R'_{L2} and R'_{s2}. We have

$$R'_{L2} = 2.7 \| 10 = 2.13 \text{ k}\Omega$$

$$R'_{s2} = R_{gs2} = 0.05 \text{ k}\Omega$$

Thus

$$R_{gd2} = R'_{L2} + R'_{s2}(1 + g_{m2}R'_{L2})$$
$$= 2.13 + 0.05(1 + 18.80 \times 2.13) = 4.17 \text{ k}\Omega$$

and

$$\tau_{gd2} = R_{gd2}C_{gd2} = 4.17 \times 1.5 = 6.26 \text{ ns}$$

Moving backwards to the first stage, we quickly calculate $R_{gd1} = 1 \| 50 = 0.980$ kΩ, and

$$\tau_{gd1} = R_{gd1}C_{gd1} = 0.980 \times 2 = 1.961 \text{ ns}$$

Finally, we require a value for R_{gs1}. The high-frequency performance of the common-drain circuit was analyzed in Section 7.5, where we obtained Eq. (23):

$$\tau_{gs} = R_{gs}C_{gs} = \frac{(R'_s + R'_L)C_{gs}}{1 + g_m R'_L}$$

We use that result now by interpreting R'_s as $R_{gd1} = 1 \| 50 = 0.980$ kΩ, and R'_L as 150 Ω. Thus

$$\tau_{gs1} = R_{gs1}C_{gs1} = \frac{(0.980 + 0.150)1}{1 + 13.33 \times 0.150} = 0.377 \text{ ns}$$

Summing these four values, we have

$$\tau = 0.125 + 6.26 + 1.961 + 0.377 = 8.72 \text{ ns}$$

and

$$\omega_H \doteq 114.6 \text{ Mrad/s}$$

This upper half-power frequency is associated with the mid-frequency gain:

$$A_{Vs(\text{mid})} = (2.7 \| 10)(-18.8)\frac{13.33 \times 0.15}{1 + 13.33 \times 0.15} \times \frac{50}{51} = -26.1$$

An exact analysis of the high-frequency equivalent circuit by computer simulation gives $\omega_H = 145.6$ Mrad/s, and we see again that our estimate is conservative.

D8.2 The elements in a CS-CS high-frequency equivalent circuit are $R_s = 0.5$ kΩ, $R_{B1} = 100$ kΩ, $g_{m1} = 8$ m℧, $C_{gs1} = 2.5$ pF, $C_{gd1} = 1$ pF, $R_{D1} = 5$ kΩ, $R_{B2} = 200$ kΩ, $g_{m2} = 10$ m℧, $C_{gs2} = 3$ pF, $C_{gd2} = 1.2$ pF, $R_{D2} = 4$ kΩ, and $R_L = 12$ kΩ. Determine (a) τ_{gs1}, (b) τ_{gd1}, (c) τ_{gs2}, (d) τ_{gd2}. (e) Estimate ω_H.

Answers. 1.24 ns; 24.8 ns; 14.6 ns; 185.1 ns; 4.43 Mrad/s

8.3 An approximation for ω_L

We now turn to the low-frequency end of multistage-amplifier performance and seek a value for ω_L. Again, we have to make approximations, but what we lose in accuracy we gain in savings of time and energy.

In the two-stage amplifier of Fig. 8.1*a*, there are five capacitors present in the low-frequency equivalent circuit: three coupling and two emitter–by-pass. A formal rendition of the transfer function A_{Vs} as a function of $j\omega$ must lead to a fifth-degree polynomial in the denominator. Thus the task of calculating ω_L would be an algebraic exercise suited only for the strong of will, the mighty of heart, and the misguided of purpose.

In analyzing the circuit, we might simulate it by a CAD routine and determine ω_L by interpolating the numerical output data. Such a use of the computer has been made several times before to check approximate analyses or verify designs. In the initial design of a multistage amplifier circuit, however, it is to our advantage if we have some indication of which capacitor controls ω_L and to what extent. This requires more information than the computer normally provides. The CAD program furnishes only answers; it offers no suggestions on what to change in order to improve the design. This is the mission of the design engineer. Most employers consider random guessing to be a technique embraced by the soon-to-be-unemployed; we shall therefore try an approximate procedure for finding ω_L as well as providing information for design and redesign.

In Chapter 7 we designed single-stage amplifiers for a given ω_L by selecting one capacitor to control ω_L while all the others were considered to be short circuits at that frequency. When the corner frequencies produced by each of the noncontrolling capacitors had magnitudes less than $0.1\omega_L$, we found that we had good accuracy when we let ω_L be equal to the largest corner frequency. The present method is an extension of that procedure. We again find a corner frequency ω_{Ci} for each capacitor acting alone, with the others treated as short circuits. The conservative approximation for ω_L is then taken as the sum of all the individual corners:

$$\begin{aligned}\omega_L &\doteq \omega_{C1} + \omega_{C2} + \cdots + \omega_{Cn} \\ &= \frac{1}{R_{\text{eq}1}C_1} + \frac{1}{R_{\text{eq}2}C_2} + \cdots + \frac{1}{R_{\text{eq}n}C_n}\end{aligned} \tag{2}$$

where $R_{\text{eq}i}$ is the Thévenin-equivalent resistance seen by C_i with all other capacitors short-circuited.

The lower 3-dB frequency calculated by this technique tends to be pessimistic; that is, it yields a value for ω_L that is greater than the actual value, so that a CAD analysis will show that we have actually designed a better amplifier than we thought we did.

As an illustration of the application and degree of accuracy of this procedure, let us return to the two-stage amplifier shown in Fig. 8.1*a*. Beginning with the coupling capacitors, and neglecting r_x, we have for C_1,

$$\omega_{C1} = \frac{1}{[R_s + (R_{B1} \| r_{\pi 1})]C_1} = \frac{1000}{[0.3 + (39.4 \| 9)]10}$$
$$= 13.11 \text{ rad/s}$$

For C_2,

$$\omega_{C2} = \frac{1}{[R_{C1} + (R_{B2} \| r_{\pi 2})]C_2} = \frac{1000}{[8 + (39.4 \| 9)]5}$$
$$= 13.05 \text{ rad/s}$$

Finally, for C_3,

$$\omega_{C3} = \frac{1}{(R_{C2} + R_L)C_3} = \frac{1000}{(8 + 2)7.5}$$
$$= 13.33 \text{ rad/s}$$

The emitter–by-pass capacitors see resistances that we designated as R_{eq} in Section 7.9:

$$R_{\text{eq}1} = R_{E1} \left\| \frac{r_{\pi 1} + (R_s \| R_{B1})}{1 + \beta_{01}} = 5 \right\| \frac{9 + (0.3 \| 39.4)}{351}$$
$$= 26.4\ \Omega$$

$$R_{\text{eq}2} = R_{E2} \left\| \frac{r_{\pi 2} + (Z_{o1} \| R_{B2})}{1 + \beta_{02}} = 5 \right\| \frac{9 + (8 \| 39.4)}{351}$$
$$= 44.2\ \Omega$$

Therefore

$$\omega_{CE1} = \frac{1}{R_{\text{eq}1} C_{E1}} = \frac{10^6}{26.4(250)} = 151.8 \text{ rad/s}$$

$$\omega_{CE2} = \frac{1}{R_{\text{eq}2} C_{E2}} = \frac{10^6}{44.2(1150)} = 19.68 \text{ rad/s}$$

Adding these five individual break frequencies, we obtain our estimate for ω_L:

$$\omega_L \doteq \omega_{C1} + \omega_{C2} + \omega_{C3} + \omega_{CE1} + \omega_{CE2}$$

or

$$\omega_L \doteq 13.11 + 13.05 + 13.33 + 151.8 + 19.68$$
$$= 211 \text{ rad/s}$$

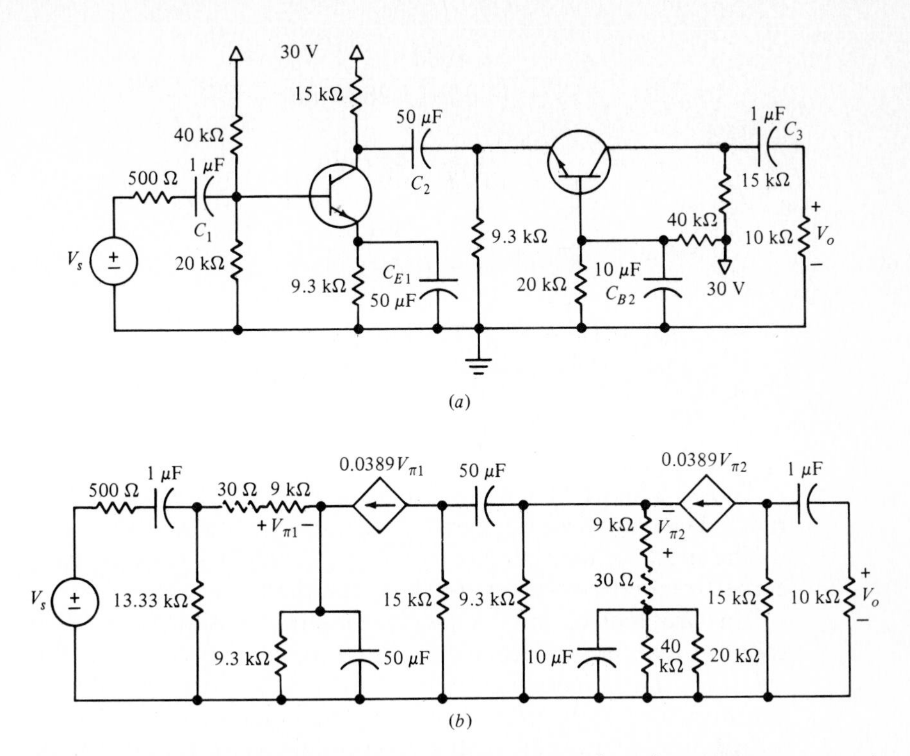

Fig. 8.6 (*a*) A bipolar cascode amplifier used as an example in determining an approximate value of ω_L. (*b*) The low-frequency equivalent circuit of the amplifier.

We expect ω_L actually to be slightly less than this value, and a computer analysis of this amplifier with $r_x = 0$ leads to 169 rad/s. Our estimate is therefore about 25% high. Note, however, that our approximate analysis shows us that C_{E1} is the major factor in determining ω_L. If we wished to lower ω_L by increasing some capacitance value, C_{E1} would be the one to increase. The frequency $\omega_{CE1} = 151.8$ rad/s is about eight times the next highest frequency, $\omega_{CE2} = 19.68$ rad/s. If they were separated by a factor of ten, we would be able to estimate ω_L as about 152 rad/s. Actually, the lower half-power frequency cannot be less than the highest individual break frequency, and it cannot be greater than the sum of all the individual break frequencies. We thus always have upper and lower bounds.

As a second example, we select the common-emitter–common-base cascode amplifier of Fig. 8.6*a*. The equivalent resistances seen by each capacitor can be found from the low-frequency equivalent circuit of Fig. 8.6*b*, with all but one of the capacitors replaced by short circuits. Letting $g_m = 38.9$ m℧, $r_\pi = 9$ kΩ, $\beta_o = 350$, and $r_x = 0$, we have

$$\omega_{C1} = \frac{1000}{[0.5 + (20 \| 40 \| 9)]1} = 170.3 \text{ rad/s}$$

$$\omega_{C2} = \frac{1000}{[15 + (9.3 \| 9 \| 1/38.9)]50} = 1.33 \text{ rad/s}$$

$$\omega_{C3} = \frac{1000}{(15 + 10)1} = 40 \text{ rad/s}$$

$$\omega_{CE1} = \left[\frac{1}{9.3} + \frac{351}{9 + (0.5 \| 20 \| 40)}\right]\frac{1000}{50} = 742.5 \text{ rad/s}$$

$$\omega_{CB2} = \left[\frac{1}{40} + \frac{1}{20} + \frac{1}{9 + 351(9.3 \| 15)}\right]\frac{1000}{10} = 7.55 \text{ rad/s}$$

Therefore

$$\omega_L \doteq 170.3 + 1.3 + 40 + 742.5 + 7.5 = 962 \text{ rad/s}$$

Thus the lower half-power frequency must lie between 742.5 and 962 rad/s. The computer divulges a more accurate value of 856 rad/s.

The most influential capacitor is C_{E1}, with C_1 ranking second; their corner frequencies are separated by less than a factor of five.

The problem of low-frequency amplifier design is that of selecting coupling and by-pass capacitors so that the sum of the individual frequencies is ω_L. The allocation of the values is at the discretion of the designer, but unreasonably large capacitance values should be avoided. If there are n capacitors and we let each one be equally effective in determining ω_L, we first determine the resistance offered to each capacitor alone, with all the others replaced by short circuits, and then calculate the necessary capacitance value $C = n/(R\omega_L)$. This provides information on the relative effect of the capacitors and the general size range required, as well as being a reasonable solution of the design problem. Adjustments to these values may be made to provide a more economical solution or to satisfy other criteria. Final values should be commercially available sizes. In any event, as a final check, the design should be simulated on the paper breadboard, a CAD program.

D8.3 The CS-CB cascode amplifier of Fig. 8.3 has the following capacitor values: $C_1 = 0.5\ \mu F$, $C_2 = 5\ \mu F$, $C_3 = 3\ \mu F$, $C_{SS} = 50\ \mu F$, and $C_B = 7.5\ \mu F$. Determine values for (a) ω_{C1}, (b) ω_{C2}, (c) ω_{C3}, (d) ω_{CSS}, (e) ω_{CB}. (f) Estimate ω_L.

Answers. 7.8; 66.0; 47.6; 100; 8.2; $100 < \omega_L < 230$ rad/s

8.4 An example of multistage amplifier design

Now that we have analyzed several amplifiers having two or three stages, let's consider the design of a typical multistage amplifier to meet a prescribed set of specifications. We require the mid-frequency voltage gain

$A_{Vs(\text{mid})}$ to have a magnitude greater than 10,000 under any conditions. The source has an internal resistance of 500 Ω, and the load is 5000 Ω. The lower half-power frequency must be less than 1000 rad/s, while the upper is greater than 1 Mrad/s. The operating temperature is guaranteed to be 25°C (somebody else's problem), and two 9-V batteries are available for the power supply.

As with most design problems, some of the biggest decisions have to be made before the actual design procedure is begun. In the present case, these preliminary choices involve deciding between bipolar transistors and FETs, the circuit configuration to be used, and the number of stages in the amplifier. These grandiose decisions are made most effectively from a solid base of experience, but even though we have only a limited background, we do have one that is sufficient for this case. After all, several different amplifiers have been analyzed in this chapter, and two of the C-E examples from Section 8.1 showed voltage gains of 8890 for two stages, and 5,940,000 for three. Since we do not require an extremely large or small input impedance, and 8890 is fairly close to 10,000 (it misses by only 1 dB), we presume that an improved two-stage C-E amplifier may work, while a three-stage C-E could be an overdesign. We tentatively select a CE-CE configuration similar to the amplifier of Fig. 8.1*a*. The proposed circuit is shown in Fig 8.7.

Fig. 8.7 A two-stage amplifier configuration selected to meet the requirements $|A_{Vs(\text{mid})}| \geq 10{,}000$, $\omega_L \leq 1000$ rad/s, and $\omega_H \geq 1$ Mrad/s. The final design values are given.

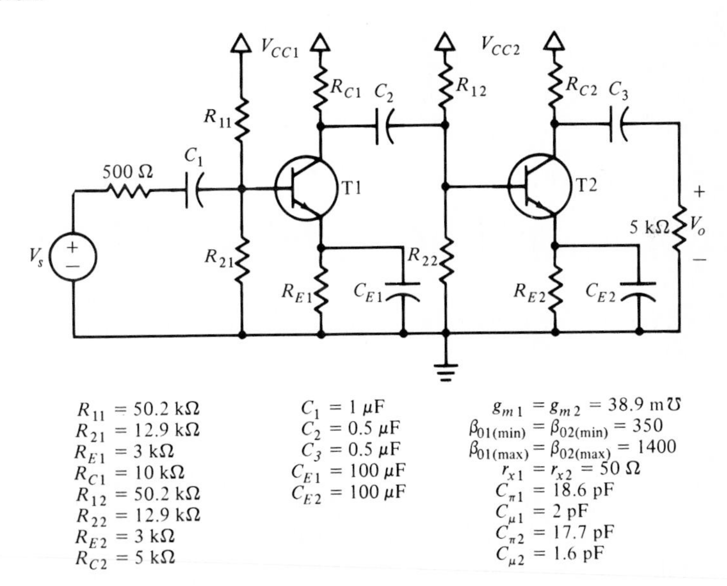

Some estimate of the gain required per stage should now be made so that operating points can be selected for each transistor. This exercise will also show us whether or not a two-stage design is possible. Neglecting r_x and R_B, we have

$$A_{Vs(\mathrm{mid})} \doteq (-g_{m2})(R_{C2} \| 5)(-g_{m1})(r_{\pi 2} \| R_{C1}) \frac{r_{\pi 1}}{r_{\pi 1} + 0.5}$$

Let us assume identical stages, although this is not always desirable for several reasons. One such reason is the signal level, which can be quite large in the last stage. Another reason may be the input or output impedance levels, or the frequency response, as we shall see shortly. We have, then,

$$A_{Vs(\mathrm{mid})} \doteq g_m{}^2 (R_C \| 5)(R_C \| r_\pi) \frac{r_\pi}{r_\pi + 0.5} \geq 10{,}000$$

When R_C and r_π are 10 kΩ, we find that g_m must be greater than 25.1 m℧. Allowing for the error introduced by our approximations above, let us try $g_m = 38.9$ m℧. Then $I_{C1} = I_{C2} = 1$ mA and $R_{C1} = R_{C2} = 10$ kΩ. If we let the voltage across R_E be 3 V, then $R_{E1} = R_{E2} = 3$ kΩ, and $V_{CE1} = V_{CE2} = 5$ V for $V_{CC1} = V_{CC2} = 18$ V. Continuing to use the same 2N5088 transistors that were used in the earlier two-stage amplifier, we must design with $\beta_{\mathrm{dc(min)}} = 350$ at this operating point. With $I_{R1} = 100 I_B$, we find

$$I_B = \frac{1000}{350} = 2.86\ \mu\mathrm{A}$$

$$R_1 = \frac{18 - 3 - 0.65}{286} = 50.2\ \mathrm{k\Omega}$$

$$R_2 = \frac{3 + 0.65}{286 - 2.86} = 12.90\ \mathrm{k\Omega}$$

$$R_{\mathrm{Th}} = 50.2 \| 12.90 = 10.3\ \mathrm{k\Omega}$$

$$V_{\mathrm{Th}} = \frac{12.90}{63.1}(18) = 3.68\ \mathrm{V}$$

As a check, we calculate I_C for $\beta_{\mathrm{dc(max)}} = 1050$:

$$I_{C(\mathrm{max})} = \frac{1050(3.68 - 0.65)}{10.3 + 1051(3)} = 1.005\ \mathrm{mA}$$

Our dc design is therefore relatively insensitive to the range of transistors we might expect.

A more accurate value for the mid-frequency gain can now be formulated. With $\beta_{0(\mathrm{min})} = 350$, we have $g_{m1} = g_{m2} = 38.9$ m℧, and $r_{\pi 1} = r_{\pi 2}$

$= 350/38.9 = 9.00$ kΩ. Therefore

$$A_{Vs(\text{mid})} = 38.9(5 \| 10)38.9(9 \| 10.3 \| 10)\frac{(9 \| 10.3)}{0.5 + (9 \| 10.3)}$$

$$= 14{,}820$$

We see that we have plenty of gain, almost 50% more than necessary.

The low-frequency requirements can always be met by selecting sufficiently large capacitors, and we therefore turn to what may be the real problem in this design, ω_H.

The lowest value of ω_H will occur when the gain is a maximum, because the Miller effect predominates in common-emitter amplifiers. We use $\beta_{0(\max)} = 1400$, $g_{m(\max)} = 38.9$ m℧, and $r_\pi = 1400/38.9 = 36.0$ kΩ. The high-frequency equivalent circuit requires values for r_x, C_μ, and C_π. We select $r_x = 50\ \Omega$ for both stages, and turn to the 2N5088 data sheets in Appendix A for typical values of C_μ and C_π. With $V_{CB} = 4.35$ V, we find $C_\mu \doteq C_{ob} = 2$ pF. Also, at $I_C = 1$ mA, $V_{CE} = 5$ V, the value given for f_T is 300 MHz, so that

$$C_\pi = \frac{38.9 \times 10^3}{2\pi(300)} - 2 = 18.6 \text{ pF}$$

The calculations upon which ω_H depends may now be carried out. For the first stage, we again make use of Eq. (1),

$$R_\mu = R'_L + R'_s(1 + g_m R'_L)$$

by calculating

$$R'_{L1} = 10 \| 10.3 \| (36 + 0.05) = 4.45 \text{ k}\Omega$$

$$R'_{s1} = 36 \| [0.05 + (0.5 \| 10.3)] = 0.519 \text{ k}\Omega$$

$$R_{\mu 1} = 4.45 + 0.519(1 + 38.9 \times 4.45) = 94.8 \text{ k}\Omega$$

so that

$$\tau_{\mu 1} = R_{\mu 1} C_{\mu 1} = 94.8 \times 2 = 189.6 \text{ ns}$$

Next,

$$\tau_{\pi 1} = R_{\pi 1} C_{\pi 1} = R'_{s1} C_{\pi 1} = 0.519 \times 18.6 = 9.66 \text{ ns}$$

The second-stage calculations are

$$R'_{L2} = 10 \| 5 = 3.33 \text{ k}\Omega$$

$$R'_{s2} = 36 \| [0.05 + (10 \| 10.3)] = 4.49 \text{ k}\Omega$$

$$R_{\mu 2} = 3.33 + 4.49(1 + 38.9 \times 3.33) = 589 \text{ k}\Omega$$

$$\tau_{\mu 2} = R_{\mu 2} C_{\mu 2} = 589 \times 2 = 1179 \text{ ns}$$

$$\tau_{\pi 2} = R_{\pi 2} C_{\pi 2} = R'_{s2} C_{\pi 2} = 4.49 \times 18.6 = 83.4 \text{ ns}$$

Therefore

$$\tau = 189.6 + 9.66 + 1179 + 83.4 = 1462 \text{ ns}$$

and

$$\omega_H \doteq \frac{1}{\tau} = 0.684 \text{ Mrad/s}$$

We can see that we have failed in meeting the specification for ω_H.

Some redesign is required, and the calculations above show clearly that the culprit is the Miller effect in the second stage, $\tau_{\mu 2}$. To increase ω_H to 1 Mrad/s, we need to reduce τ to 1000 ns, about ⅔ of its previous value. However, we saw above that we had approximately 3/2 times the necessary gain; therefore a trade-off between gain and bandwidth is indicated. A simple way to do this is to decrease R_{C2}. We try a reduction from 10 to 5 kΩ:

$$R_{C2} = 5 \text{ k}\Omega$$

$$R'_{L2} = 2.5 \text{ k}\Omega$$

$$R'_{s2} = 4.49 \text{ k}\Omega$$

A slight change in $C_{\mu 2}$ occurs since V_{CB2} is now the larger value, 18 − (5 + 3)(1) − 0.65 = 9.35 V. We find

$$C_{\mu 2} = 1.6 \text{ pF}$$

This decrease helps. Next, we find that f_T has increased slightly to 320 MHz, and therefore

$$C_{\pi 2} = \frac{38.9 \times 10^3}{2\pi(320)} - 1.6 = 17.7 \text{ pF}$$

$$R_{\mu 2} = 2.5 + 4.49(1 + 38.9 \times 2.5) = 443 \text{ k}\Omega$$

$$\tau_{\mu 2} = 443(1.6) = 709.1 \text{ ns}$$

$$\tau = 189.6 + 9.66 + 709.1 + 83.4 = 992 \text{ ns}$$

Thus

$$\omega_H \doteq \frac{1}{\tau} = 1.008 \text{ Mrad/s}$$

and we come in just under the wire.

The mid-frequency gain must now be recalculated:

$$A_{Vs(\text{mid})} = 38.9(5 \| 5)38.9(9 \| 10.3 \| 10)\frac{(9 \| 10.3)}{0.5 + (9 \| 10.3)}$$

$$= 11{,}100$$

This is also within the specifications. Letting R_{C2} be one-half its previous value has resulted in $A_{Vs(\text{mid})}$ dropping 25% below its former value, while the bandwidth increased 48%. The relatively large improvement in bandwidth has resulted from our clever decision to change R_{C2}, because it not only reduced the gain but $C_{\mu 2}$ as well.

Approximating ω_H from the values of $\tau_{\mu 1}$, $\tau_{\mu 2}$, $\tau_{\pi 1}$, and $\tau_{\pi 2}$ in the case of a CE-CE amplifier is again seen to be a very useful technique, for it gives considerable insight into the necessary redesign process.

The design is completed with the selection of values for the five large capacitors. We may place the corner frequency for one of these capacitors, typically C_{E1} or C_{E2}, just below $\omega_L = 1000$ rad/s, and then let the remaining corners be near $\omega = 100$. Let us try instead to minimize the size of C_{E1} and C_{E2} by selecting $\omega_{CE1} = \omega_{CE2} = 400$ rad/s, and then letting $\omega_{C1} = \omega_{C2} = \omega_{C3} = 200$ rad/s. This should keep ω_L well below 1000 rad/s, even though the sum of the corners is 1400 rad/s. We can then choose the first commercial-sized capacitor values larger than our design values. We have

$$\omega_{C1} = 200 = \frac{1000}{C_1[0.5 + (10.3\|9)]} \qquad C_1 = 0.944\ \mu\text{F}$$

$$\omega_{C2} = 200 = \frac{1000}{C_2[10 + (10.3\|9)]} \qquad C_2 = 0.338\ \mu\text{F}$$

$$\omega_{C3} = 200 = \frac{1000}{C_3(5 + 5)} \qquad C_3 = 0.5\ \mu\text{F}$$

The equivalent resistance $R_{\text{eq}1}$ seen by C_{E1} depends on the value of β_0. For $\beta_{0(\max)}$ we have

$$R_{\text{eq}1} = 3 \left\| \frac{36 + (0.5\|10.3)}{1401} \right. = 25.8\ \Omega \qquad (\beta_{0(\max)})$$

while $\beta_{0(\min)}$ gives

$$R_{\text{eq}1} = 3 \left\| \frac{9 + (0.5\|10.3)}{351} \right. = 26.8\ \Omega \qquad (\beta_{0(\min)})$$

The largest value (worst case) of C_{E1} needed is therefore

$$C_{E1} = \frac{1}{25.8(400)} = 96.9\ \mu\text{F}$$

Similarly, we calculate $R_{\text{eq}2}$ for $\beta_{0(\max)}$:

$$R_{\text{eq}2} = 3 \left\| \frac{36 + (10.3\|10)}{1401} \right. = 29.0\ \Omega \qquad (\beta_{0(\max)})$$

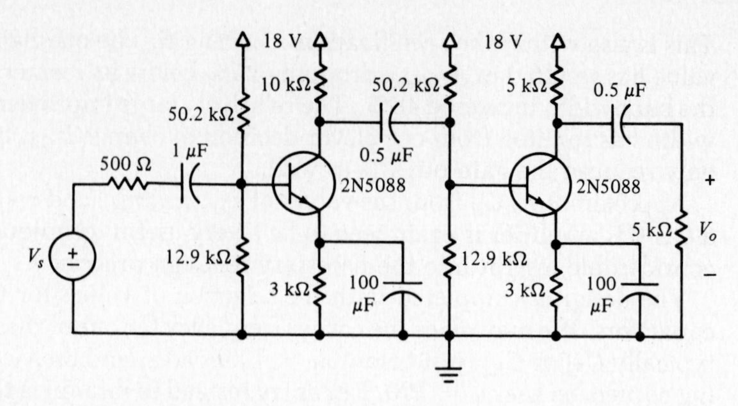

Fig. 8.8 The final design of an amplifier that meets the requirements $|A_{Vs(\text{mid})}| \geq 10{,}000$, $\omega_L \leq 1000$ rad/s, and $\omega_H \geq 1$ Mrad/s.

which is the worst-case condition, and then

$$C_{E2} = \frac{1}{29.0(400)} = 86.1\ \mu\text{F}$$

We therefore select the commercial sizes:

$$C_1 = 1\ \mu\text{F}$$
$$C_2 = 0.5\ \mu\text{F}$$
$$C_3 = 0.5\ \mu\text{F}$$
$$C_{E1} = 100\ \mu\text{F}$$
$$C_{E2} = 100\ \mu\text{F}$$

The final design is shown in Fig. 8.8, and data from a CAD simulation of this circuit appear in Table 8.1. The "calculated" results are either those

Table 8.1 CAD Analysis of CE-CE Amplifier Design

		CAD		Calculated	
Parameter specification		$\beta_0 = 350$	$\beta_0 = 1400$	$\beta_0 = 350$	$\beta_0 = 1400$
$A_{VS(\text{mid})}$	$\geq 10{,}000$	11,101	15,805	11,101	15,805
ω_L	$\leq 10^3$ rad/s	700 rad/s	681 rad/s	1,149 rad/s	1,162 rad/s
ω_H	≥ 1 Mrad/s	1,374 krad/s	1,020 krad/s	1,393 krad/s	1,008 krad/s

found in the design procedure above or those requested in Problem D8.4 and Problem 22. At any rate, the calculated values for ω_L and ω_H are approximate.

D8.4 (a) For the capacitor values shown in the circuit of Fig. 8.8, what are the five half-power frequencies obtained when each of the five large capacitors acts alone with the others being short-circuited, assuming $\beta_0 = 1400$? (b) How does their sum compare with the CAD value for ω_L?

Answers. 111; 118; 200; 345; 388; 1162 vs. 681 rad/s

8.5 The design of a broadband amplifier

Many applications in the high-frequency band (HF band, 3 to 30 MHz) and very-high-frequency band (VHF band, 30 to 300 MHz) call for amplifiers that have an upper half-power frequency in that region of the frequency spectrum, as well as a large bandwidth. These amplifiers are often components in a system that also requires them to have a high input impedance. In our final multistage design example, we shall therefore try to meet the following set of specifications:

$$|A_{Vs(\mathrm{mid})}| \geq 50$$

$$R_{\mathrm{in}} \geq 120\ \mathrm{k}\Omega \qquad \text{(at mid-frequencies)}$$

$$\omega_H \geq 100\ \mathrm{Mrad/s}$$

$$R_s = 300\ \Omega$$

$$R_L = 5\ \mathrm{k}\Omega$$

The large value of input resistance that is required tells us that the input stage should be a bipolar transistor in the common-collector mode, or an FET that is operated common-drain or common-source. However, 120 kΩ is larger than typical values of R_B needed for good operating-point stability of the bipolar transistor, and it therefore appears that the FET is preferable. We shall try a CS-CB cascode arrangement. The FET provides high R_{in} and the cascode arrangement avoids a large Miller-effect capacitance, while the C-B stage should deliver the necessary mid-frequency gain. The proposed circuit is shown in Fig. 8.9*a*. Note that we are planning on a direct-coupled arrangement in order to avoid using two elements, a large coupling capacitor between stages, and the emitter-resistor. For the FET, we assume that there is an *n*-channel JFET available with these characteristics:

$$I_{DSS} = 20 \text{ mA}$$
$$V_P = -2 \text{ V}$$
$$C_{gs} = 5 \text{ pF}$$
$$C_{gd} = 2 \text{ pF}$$

while the bipolar transistor is a 2N5088.

We first make a rough estimate of the gain required in each stage in order to select suitable values of g_m, specify the operating points, and then design the bias circuitry. The mid-frequency equivalent circuit appears in Fig. 8.9*b*. The input impedance to the C-B stage is approximately $1/g_{m2}$, so we have the gain

$$|A_{Vs(\text{mid})}| \doteq g_{m2}(R'_{L2})g_{m1}\left(R_D \,\middle\|\, \frac{1}{g_{m2}}\right)$$
$$\doteq g_{m2}(R'_{L2})g_{m1}\left(\frac{1}{g_{m2}}\right)$$

Fig. 8.9 (*a*) A proposed circuit configuration for a high-frequency broadband amplifier. (*b*) The mid-frequency equivalent circuit.

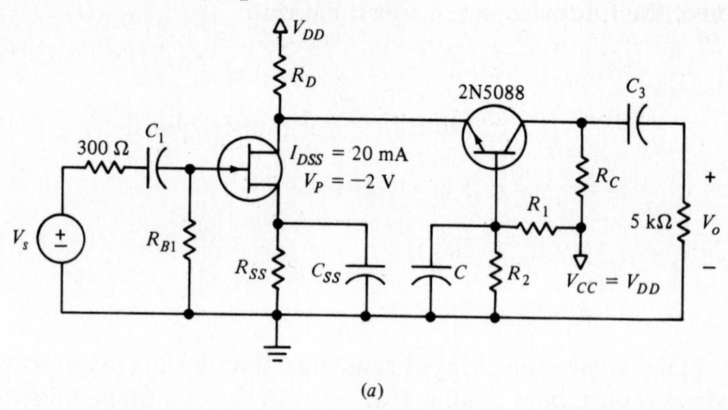

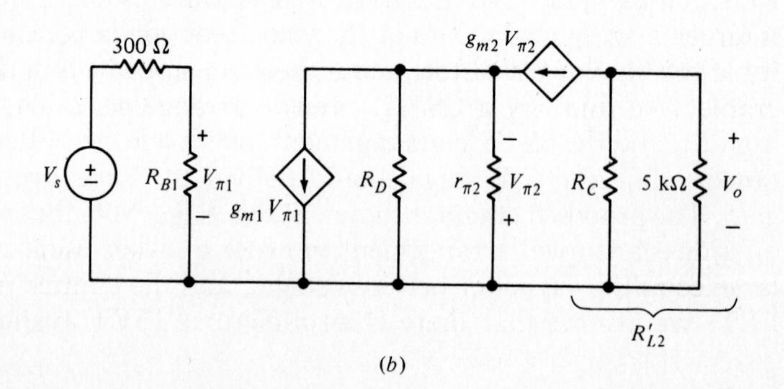

or

$$|A_{Vs(\text{mid})}| \doteq g_{m1} R'_{L2}$$

The effective load $R'_{L2} = R_{C2} \| 5$ cannot be greater than 5 kΩ, g_{m1} cannot be greater than $2I_{DSS}/|V_P| = 20$ m℧, and their product must be greater than the minimum gain, 50. Let us try $g_{m1} = 18$ m℧, just below the 20-m℧ maximum value. The JFET operating point is then found:

$$I_D = \frac{20}{4}(V_{GS} + 2)^2 \qquad \text{(mA)}$$

$$g_{m1} = \frac{2(20)}{4}(V_{GS} + 2) = 18 \text{ m℧}$$

$$V_{GS} = -0.2 \text{ V}$$

$$I_D = 16.2 \text{ mA}$$

To bias the gate at -0.2 V with respect to the source, we select a self-biasing scheme using a source resistor whose value is

$$R_{SS} = \frac{0.2}{16.2} = 12.3 \ \Omega$$

This low value portends poor operating-point stability for the JFET if there is a wide variation among transistors. We assume that we can pay for uniformity in I_{DSS} and V_P, at least for this input stage.

Selecting an 18-V power supply, $V_{DD} = V_{CC} = 18$ V, we now refer to the dc-equivalent circuit of Fig. 8.10*a*. We assume a low value of V_{DS} in order to keep the emitter of the second stage at a reasonably low potential. We also need $V_{DS} \geq -0.2 + 2 = 1.8$ V for saturated operation. Suppose $V_{DS} = 2$ V, and therefore the voltage from drain to ground is 2.2 V.

A convenient value for g_{m2} is 38.9 m℧, and then $I_C = 1$ mA. We now see that

$$R_D = \frac{18 - 2.2}{16.2 - 1} = 1.04 \text{ k}\Omega$$

Since the input resistance must be equal to or greater than 120 kΩ, we may select $R_B = 150$ kΩ, thus completing the design of the first stage.

To bias the second stage at $I_C = 1$ mA and, say, $V_{CE} = 5$ V, we use $\beta_{dc} = 350$, $V_{BE} = 0.65$ V, and have

$$I_{R1} = 50 I_B = 50\left(\frac{1}{350}\right) = 143 \ \mu\text{A}$$

$$I_{R2} = 143 - \frac{1000}{350} = 140 \ \mu\text{A}$$

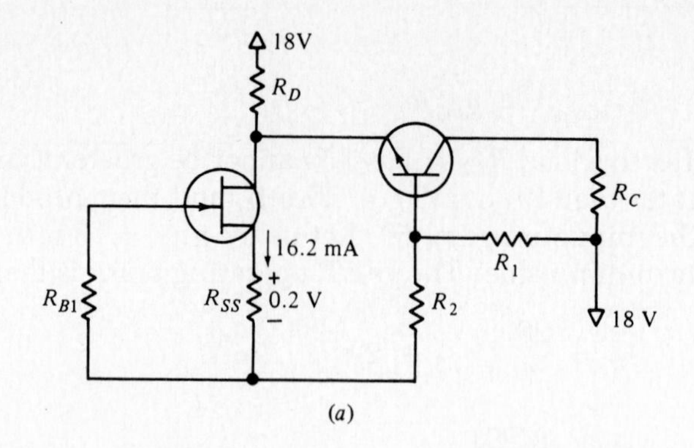

(a)

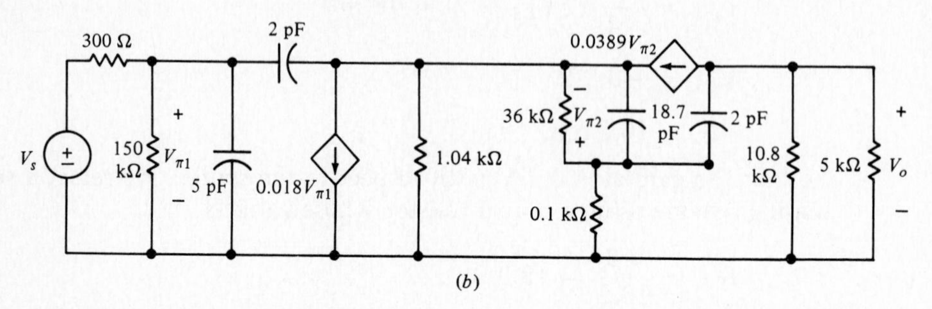

(b)

Fig. 8.10 (*a*) The dc equivalent of the cascode amplifier circuit of Fig. 8.9*a*. (*b*) The high-frequency equivalent circuit.

Since V_{BE} is 0.65 V,

$$R_2 = \frac{2.2 + 0.65}{0.140} = 20.4\ \text{k}\Omega$$

$$R_1 = \frac{18 - 2.2 - 0.65}{0.143} = 105.9\ \text{k}\Omega$$

$$R_C = \frac{18 - 5 - 2.2}{1} = 10.8\ \text{k}\Omega$$

Let us now obtain an accurate value for $|A_{Vs(\text{mid})}|_{(\text{min})}$. Using $r_{\pi 2} = 350/38.9 = 9\ \text{k}\Omega$ and neglecting r_x, we find the input resistance to the second stage, which is also the load for the first stage:

$$R_{L1} = R_{\text{in}2} = r_{\pi 2} \left\| \frac{1}{g_{m2}} \right. = 9 \| 0.0257 = 25.6\ \Omega$$

and then

$$A_{Vs(\text{mid})} = -38.9(10.8 \| 5)(18)(1.04 \| 0.0256)\frac{150}{150.3}$$

$$= -59.7 \qquad \text{or} \qquad |A_{Vs(\text{mid})}| = 59.7$$

This is satisfactory. Since R_{in} is 150 kΩ, we have also met that requirement.

We now turn to the high-frequency response. The equivalent circuit is shown in Fig. 8.10*b*. We choose $\beta_0 = \beta_{0(\max)} = 1400$, since this will lead to the minimum ω_H. We find $r_{\pi 2}$ is 36 kΩ and use typical values for $C_{\pi 2}$ (18.7 pF) and $C_{\mu 2}$ (2 pF) that we have used in previous examples; r_{x2} is assumed to be 100 Ω. For the JFET C-S stage,

$$R'_{s1} = 0.3 \| 150 = 0.299 \text{ k}\Omega$$

$$R'_{L1} \doteq 1.04 \| 0.0256 = 0.0250 \text{ k}\Omega \qquad (\text{neglecting } r_{x2})$$

$$\tau_{gs} = R_{gs}C_{gs} = R'_{s1}C_{gs} = 0.299 \times 5 = 1.497 \text{ ns}$$

$$R_{gd} = R'_{L1} + R'_{s1}(1 + g_{m1}R'_{L1})$$

$$= 0.0250 + 0.299(1 + 18 \times 0.0250) = 0.459 \text{ k}\Omega$$

$$\tau_{gd} = R_{gd}C_{gd} = 0.459 \times 2 = 0.918 \text{ ns}$$

Calculations for the C-B stage should include the effect of r_{x2}; these expressions are available as Eqs. (35) and (36) in Chapter 7:

$$R_{\pi 2} = r_{\pi 2} \left\| \frac{R'_{s2} + r_{x2}}{1 + g_{m2}R'_{s2}} \right.$$

and

$$R_{\mu 2} = r_{x2} + R'_{L2} + \frac{r_{x2}(\beta_{02}R'_{L2} - r_{x2})}{r_{x2} + r_{\pi 2} + (\beta_{02} + 1)R'_{s2}}$$

We find

$$R'_{s2} = R_{C1} = 1.04 \text{ k}\Omega$$

$$R'_{L2} = R_{C2} \| R_{L2} = 10.8 \| 5 = 3.42 \text{ k}\Omega$$

Therefore

$$R_{\pi 2} = 36 \left\| \frac{1.04 + 0.1}{1 + 38.9 \times 1.04} \right. = 0.0275 \text{ k}\Omega$$

$$\tau_{\pi 2} = R_{\pi 2}C_{\pi 2} = 0.0275 \times 18.7 = 0.514 \text{ ns}$$

$$R_{\mu 2} = 0.1 + 3.42 + \frac{0.1(1400 \times 3.42 - 0.1)}{0.1 + 36 + 1401 \times 1.04} = 3.84\ \text{k}\Omega$$

$$\tau_{\mu 2} = R_{\mu 2} C_{\mu 2} = 3.84 \times 2 = 7.676\ \text{ns}$$

Adding, we have

$$\tau = 1.497 + 0.918 + 0.514 + 7.676 = 10.606\ \text{ns}$$

and

$$\omega_H \doteq \frac{1}{\tau} = 94.3\ \text{Mrad/s}$$

Although this estimate is below the required value of 100 Mrad/s, our analyses are always conservative, and we are probably within requirements. Our alternatives are to do a slight redesign, or to make an accurate check by CAD. In this case, an analysis by SPICE2 leads to the values $|A_{Vs(\text{mid})}| = 59.9$ and $\omega_H = 128.6$ Mrad/s. Both these values are obtained with $\beta_0 = 1400$ and $r_{x2} = 100\ \Omega$.

Before leaving the high-frequency range, we might also find the input capacitance of our cascode amplifier. From Section 7.4 we have

$$C_{\text{in}1} = C_{gs1} + C_{\text{Miller}(1)} = C_{gs1} + C_{gd1}(1 + g_{m1}R'_{L1})$$

or

$$C_{\text{in}1} = 5 + 2(1 + 18 \times 0.0250) = 7.90\ \text{pF}$$

This is a low value, although it is not insignificant. It should be kept in mind as a part of the load that the amplifier offers to the source.

We have met all the high- and mid-frequency specifications, and only the low-frequency design remains. This is saved to make an exciting Problem 23 at the end of the chapter.

D8.5 At the operating point $I_C = 1$ mA, $V_{CE} = 5$ V, assume that the particular transistor used for the second stage in the amplifier of Fig. 8.9*a* has an unusually low value of β_0, 100. Find τ_{gs1}, τ_{gd1}, $\tau_{\pi 2}$, and $\tau_{\mu 2}$ and estimate ω_H. Assume all other values are as designed.

Answers. 1.497 ns; 0.915 ns; 0.509 ns; 7.675 ns; 94.3 Mrad/s

Problems

1. The amplifier in Fig. 8.1*a* is modified by increasing the positive supply voltage from 18 to 24 V, changing R_{21} from 58 to 39 kΩ, changing R_{22} from 58 to 100 kΩ, and reducing R_{C2} to 5 kΩ. Assume $r_x = 0$ and calculate (a) $A_{Vs(\text{mid})}$, (b) Z_i, (c) Z_o.

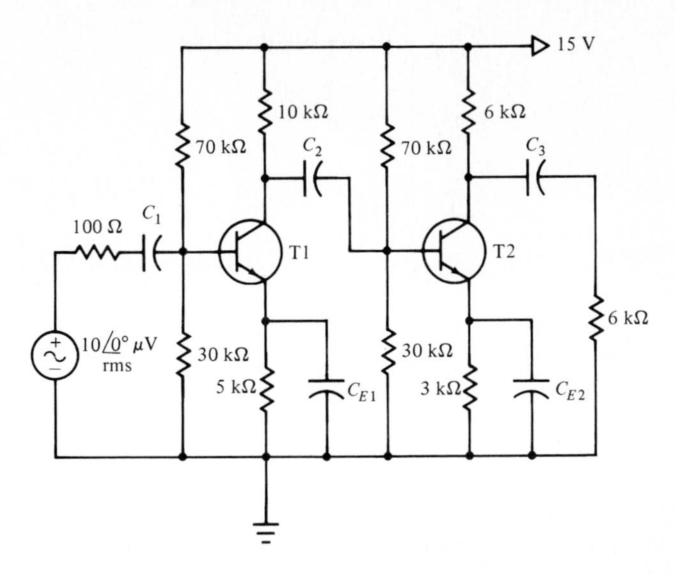

Fig. 8.11 See Problems 2, 7, and 15.

2. For T1 in Fig. 8.11, let $\beta_{dc1} = \beta_{01} = 200$. For T2, $\beta_{dc2} = \beta_{02} = 250$. For both, $n = 1$, $V_0 = 0.65$ V. At mid-frequencies, find (a) the signal-source current, (b) the load current, (c) the load voltage, (d) the dc battery current.
3. (a) Calculate the mid-frequency voltage gain $A_{Vs(\text{mid})}$ for the two-stage CS-CS amplifier shown in Fig. 8.12. Both transistors have $V_T = 2$ V and $I_D = 6$ mA when $V_{GS} = 3$ V. (b) Increase R_{D1} and R_{D2} until T1 and T2 are both at the edge of the pinch-off region and recalculate $A_{Vs(\text{mid})}$.

Fig. 8.12 See Problems 3, 8, and 18.

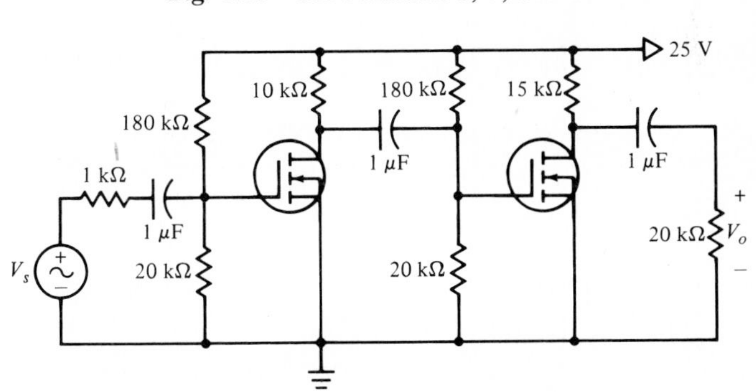

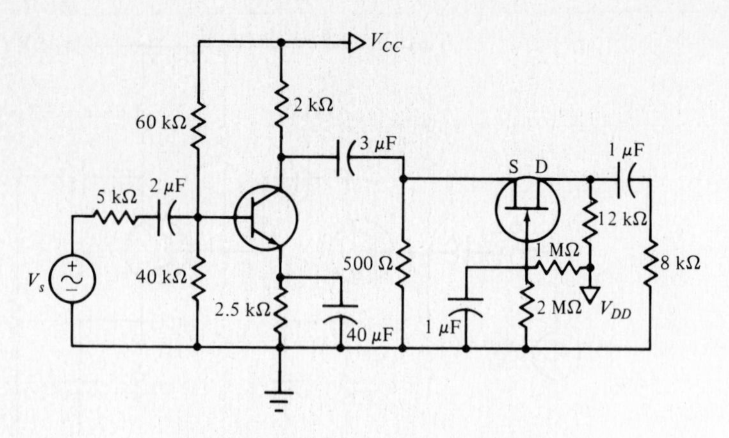

Fig. 8.13 See Problems 4, 9, and 19.

4. In the multistage amplifier of Fig. 8.13, let $\beta_0 = 240$ and $r_\pi = 8$ kΩ for the bipolar transistor, while $g_m = 6$ m℧ for the JFET. Find the mid-frequency values of (a) A_{Vi}, (b) A_{Vs}, (c) A_I, (d) A_P.
5. Parameter values for a CE-CE-CC three-stage amplifier are $R_s = 1$ kΩ, $R_{B1} = 25$ kΩ, $r_{\pi 1} = 4$ kΩ, $g_{m1} = 60$ m℧, $R_{C1} = 5$ kΩ, $R_{B2} = 20$ kΩ, $r_{\pi 2} = 6$ kΩ, $g_{m2} = 40$ m℧, $R_{C2} = 4$ kΩ, $R_{B3} = 40$ kΩ, $r_{\pi 3} = 3$ kΩ, $g_{m3} = 50$ m℧, $R_{E3} = 2$ kΩ, and $R_L = 3$ kΩ. (a) Find $A_{Vs(\text{mid})}$. (b) Find Z_{o3}.
6. The combination of two transistors coupled together as shown in Fig. 8.14 is called a *Darlington pair* or a *Darlington transistor*. It is used here as a C-E stage. (a) Draw the mid-frequency equivalent circuit. (b) Show that $Z_i = R_B \| [r_{\pi 1} + r_{\pi 2}(\beta_{01} + 1)]$, a large value. (c) Show that $V_o/V_i = -\{[\beta_{02}(\beta_{01} + 1) + \beta_{01}]/[r_{\pi 1} + r_{\pi 2}(\beta_{01} + 1)]\}(R_C \| R_L)$, also a large value.

Fig. 8.14 See Problem 6.

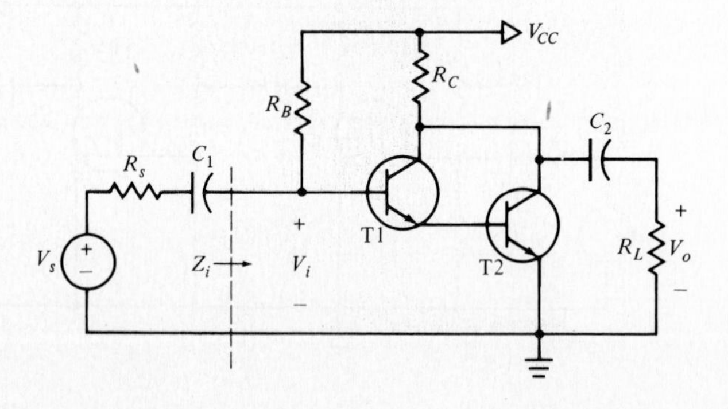

7. For the two-stage CE-CE amplifier of Fig. 8.11, select $\omega_{T1} = 600$ Mrad/s, $C_{\mu 1} = 2$ pF, $r_{x1} = 50\ \Omega$, $\omega_{T2} = 700$ Mrad/s, $C_{\mu 2} = 1.8$ pF, and $r_{x2} = 40\ \Omega$. Estimate ω_H.
8. The two-stage CS-CS amplifier of Fig. 8.12 uses transistors for which $V_T = 2$ V and $I_D = 6$ mA when $V_{GS} = 3$ V. Estimate ω_H if $C_{gd} = 1.2$ pF and $C_{gs} = 4.5$ pF.
9. The CE-CG amplifier of Fig. 8.13 has $r_x = 80\ \Omega$, $\beta_0 = 240$, $r_\pi = 8$ kΩ, $C_\pi = 60$ pF, and $C_\mu = 4$ pF for the bipolar transistor, while $g_m = 6$ m℧, $C_{gs} = 4.5$ pF, and $C_{gd} = 0.4$ pF for the FET. Determine a value for ω_H.
10. Figure 8.15 shows the *low-frequency* equivalent circuit of a certain CS-CG cascode amplifier. Assume interelectrode capacitances of $C_{gs1} = 5$ pF, $C_{gd1} = 1.5$ pF, $C_{gs2} = 6$ pF, and $C_{gd2} = 2$ pF. (a) Draw the high-frequency equivalent circuit. (b) Estimate ω_H. (c) Find $A_{Vs(\text{mid})}$.
11. Element values in the high-frequency equivalent circuit of a CS-CD two-stage amplifier are $R_s = 1$ kΩ, $R_{B1} = 49$ kΩ, $C_{gs1} = 3$ pF, $C_{gd1} = 0.8$ pF, $g_{m1} = 10$ m℧, $R_{D1} = 5$ kΩ, $R_{B2} = 49$ kΩ, $C_{gs2} = 3$ pF, $C_{gd2} = 0.8$ pF, $g_{m2} = 10$ m℧, $R_{SS2} = 2$ kΩ, and $R_L = 4$ kΩ. (a) Find $A_{Vs(\text{mid})}$. (b) Estimate ω_H by analytical methods. (c) Find ω_H by a CAD program.
12. Interelectrode capacitance values for the three-stage CE-CE-CC amplifier of Problem 5 are $C_{\pi 1} = 60$ pF, $C_{\mu 1} = 3$ pF, $C_{\pi 2} = 45$ pF, $C_{\mu 2} = 2$ pF, $C_{\pi 3} = 35$ pF, and $C_{\mu 3} = 2.5$ pF. Estimate ω_H.
13. Use the following parameters for the three-stage amplifier of Fig. 8.2*a* (resistances in kΩ, capacitances in pF): $R_s = 1$, $R_{B1} = 30$, $r_{x1} = 0.05$, $r_{\pi 1} = 7.71$, $C_{\pi 1} = 25$, $C_{\mu 1} = 3$, $\beta_{01} = 300$, $R_{C1} = R_{B2} = 20$, $r_{x2} = 0.08$, $r_{\pi 2} = 5.14$, $C_{\pi 2} = 35$, $C_{\mu 2} = 2.5$, $\beta_{02} = 400$, $R_{C2} = R_{B3} = 10$, $r_{x3} = 0.1$, $r_{\pi 3} = 4.28$, $C_{\pi 3} = 50$, $C_{\mu 3} = 1.5$, $\beta_{03} = 500$, and $R_{C3} = R_L = 5$. (a) Estimate ω_H. (b) Calculate $A_{Vs(\text{mid})}$, including r_x.
14. Estimate ω_H for the two-stage amplifier shown in Fig. 8.4 if β_0 is assumed to be 1400, $r_\pi = 36$ kΩ, and all other parameters of the

Fig. 8.15 See Problems 10 and 20.

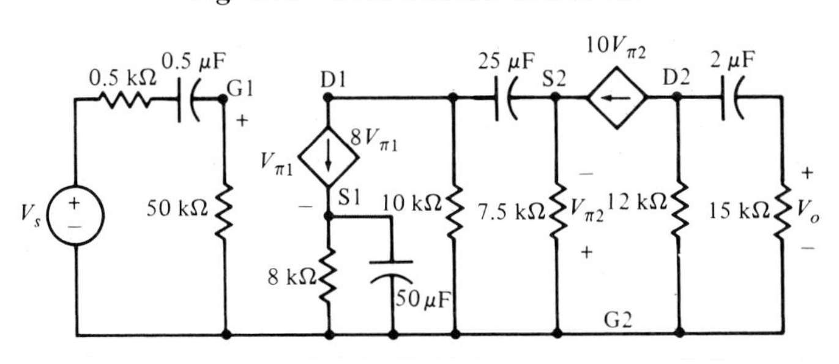

high-frequency equivalent circuit are unchanged. How does this increase in β_0 affect ω_H?

15. Select reasonable values (not too big and not too small) for the five capacitors shown in the CE-CE amplifier of Fig. 8.11 to provide a lower half-power frequency of 20 Hz.
16. In Fig. 8.16, let $V_0 = 0.7$ V and $n = 1$ for each transistor, while $\beta_{dc1} = 250$, $\beta_{01} = 220$, $\beta_{dc2} = 350$, and $\beta_{02} = 320$. (a) Find V_o at midfrequencies. (b) Select standard values for C_1, C_2, C_3, C_{E1}, and C_{E2} so that $15 < f_L < 30$ Hz. (c) Check Parts (a) and (b) by CAD.
17. The CE-CE-CE amplifier of Fig. 8.2*a* contains the following elements in its low-frequency equivalent circuit (resistance in kΩ, capacitance in μF): $R_s = 1$, $R_{B1} = 30$, $C_1 = 4$, $r_{\pi 1} = 7.71$, $\beta_{01} = 300$, $C_{E1} = C_{E2} = C_{E3} = 80$, $R_{E1} = 4$, $R_{C1} = R_{B2} = 20$, $C_2 = 2.5$, $r_{\pi 2} = 5.14$, $\beta_{02} = 400$, $R_{E2} = 2$, $R_{C2} = R_{B3} = 10$, $C_3 = 2.5$, $r_{\pi 3} = 4.28$, $\beta_{03} = 500$, $R_{E3} = 1$, $R_{C3} = R_L = 5$, and $C_4 = 5$. Calculate ω_{C1}, ω_{C2}, ω_{C3}, ω_{C4}, ω_{CE1}, ω_{CE2}, and ω_{CE3} and estimate ω_L.
18. (a) Estimate the lower half-power frequency for the CS-CS amplifier shown in Fig. 8.12. (b) Obtain frequency data by computer to determine an accurate value for f_L.
19. (a) Estimate ω_L for the CE-CG two-stage amplifier shown in Fig. 8.13. Use $g_{m1} = 30$ m℧, $r_{\pi 1} = 8$ kΩ, and $g_{m2} = 6$ m℧. (b) Select new capacitor values so that $\omega_L \doteq 500$ rad/s and each capacitor is equally effective in controlling ω_L.
20. (a) Estimate ω_L for the amplifier shown in Fig. 8.15 and give upper and lower bounds for your estimate. (b) Find Z_i and Z_o at midfrequencies.
21. Choose values for the five capacitors appearing in the CE-CE amplifier of Fig. 8.1*a* to provide $\omega_L \doteq 700$ rad/s. Use $r_\pi = 9$ kΩ, $g_m = 38.9$ m℧, and $\beta_0 = 350$. Set the corner frequency corresponding to

Fig. 8.16 See Problem 16.

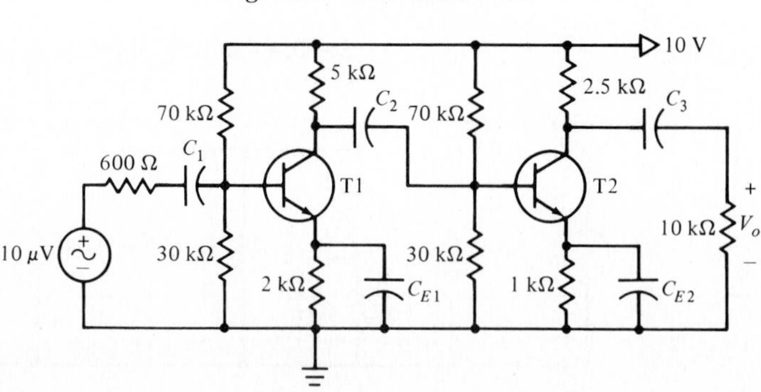

the smallest R_{eq} at 400 rad/s, the next larger at 150 rad/s, and the remaining three all at the same frequency.

22. Show that the following calculated values in Table 8.1 at the end of Section 8.4 are correct for $\beta_0 = 350$: (a) $\omega_L = 1149$ rad/s, (b) $\omega_H = 1.393$ Mrad/s.
23. Select values for C_1, C_{SS}, C, and C_3 in the circuit shown in Fig. 8.9*a* if ω_L is to be approximately 10^4 rad/s. Use $R_{B1} = 150$ kΩ, $R_{SS} = 12.3$ Ω, $R_D = 1.04$ kΩ, $R_1 = 105.9$ kΩ, $R_2 = 20.4$ kΩ, $R_C = 10.8$ kΩ, $g_{m1} = 18$ m℧, $g_{m2} = 38.9$ m℧, $r_{\pi 2} = 9$ kΩ, and $r_{x2} = 100$ Ω.
24. Design an amplifier to meet the following specifications using 2N5089 transistors and check your design with a CAD program: $|A_{Vs(\text{mid})}| > 25{,}000$, $R_s = 100$ Ω, $R_L = 8$ kΩ, $\omega_L < 500$ rad/s, and $\omega_H > 2 \times 10^5$ rad/s. Let $r_x = 50$ Ω and use "typical" values for C_π and C_μ.
25. Design a two-stage CE-CE amplifier to provide $|A_{Vs(\text{mid})}| > 200$ and $f_H > 3$ MHz using 2N5088 transistors operating with $I_C = 1$ mA, $V_{CE} = 5$ V, $\beta_0 = 350$, $C_\pi = 18$ pF, and $C_\mu = 2$ pF. Let $R_s = 300$ Ω and $R_L = 5$ kΩ and neglect r_x.
26. Design an amplifier to have $|A_{Vs}| \geq 100$ and $R_{in} \geq 10$ kΩ at mid-frequencies, and $f_H \geq 10$ MHz for average transistors. The internal resistance of the source is 300 Ω and $R_L = 5$ kΩ.
27. Design an output stage for the amplifier of Problem 26 above to replace the 5-kΩ load. It must have an input resistance equal to 5 kΩ and an output resistance equal to or less than 100 Ω. The upper 3-dB frequency should be at least 5 MHz.

9

The operational amplifier

In the past several hundred pages we have studied diodes, bipolar transistors, field-effect transistors, and the design and analysis of basic amplifier circuits, including input and output impedance and frequency effects. We have also looked briefly at several special circuits, such as the basic rectifiers and regulators, the current mirror, and the phase splitter. We are now ready to widen our field of view and study the external behavior of a *system* of amplifiers without becoming too involved with the individual units that compose it. The particular system we shall consider is the *operational amplifier*.

The term *operational amplifier* or *op amp* was originally used to describe amplifiers that performed various mathematical operations in analog computers. We shall show that the application of negative feedback around a high-gain dc amplifier leads to circuits that can add, subtract, average, integrate, or differentiate, all of which are useful mathematical operations. However, today's applications of operational amplifiers go far beyond simple mathematical operations. Op amps are used in many control and instrumentation systems to perform a myriad of tasks as voltage regulators, oscillators, logarithmic amplifiers, peak detectors, voltage comparators, and preamplifiers with special frequency characteristics for use in record-playing equipment.

The op amp has a very high gain, ranging from 10,000 to more than a million; a high input resistance, from 10^3 to 10^{15} Ω; a low output resistance, 1 to 1000 Ω; and a bandwidth that extends from dc to an upper half-power frequency that ranges from 100 kHz to several hundred megahertz.

There are a number of practical reasons for the appearance of the op amp in so many new circuit designs. The main one is the significant improvement in performance over that obtainable with vacuum tubes in early analog computers. Second, operational amplifiers containing about 30 transistors, 10 resistors, and a few diodes cost as little as 20 cents, even in small quantities. Their small size and low power consumption allow any number of them to be incorporated easily into complex systems. Their good reliability and stability make them versatile and predictable building blocks.

9.1 The ideal operational amplifier

Figure 9.1*a* shows the standard symbol for an operational amplifier. There are two input terminals, one output terminal, and a ground terminal, which is usually not shown explicitly. The terminal marked with the minus sign is called the *inverting input*, because a signal applied between that terminal and ground appears at the output with a 180° phase shift. The terminal with the plus sign is the *noninverting input*; a signal applied between that terminal and ground appears between the output terminal and ground with 0° phase shift. Figure 9.1*b* shows a signal v_1 applied between the inverting input and ground, a signal v_2 between the noninverting input and ground, and the output v_o. We use the lower-case v to indicate instantaneous voltage, either ac or dc.

Many applications of operational amplifiers require the use of only one input. Such an amplifier is termed *single-ended*. Either the inverting or noninverting input can be used with the other having no signal applied. This second (unused) input might have other circuit elements connected to it or it might be grounded. Other applications require that signals be applied to both inputs so that the amplified difference of the two signals appears at the output. This configuration is called a *differential amplifier* or a *double-ended amplifier*.

Figure 9.2*a* illustrates a simplified low-frequency equivalent circuit for an operational amplifier. The voltage between the inverting and noninverting input terminals is marked v_i, where $v_i = v_1 - v_2$; v_1 is the voltage between the inverting input and ground, while v_2 is that present between the noninverting input and ground. The input resistance is represented by R_i and the output resistance by R_o. The amplification is supplied by the dependent voltage source $A(v_1 - v_2) = Av_i$, where A is the *open-loop gain*. It is real and positive. Note that the positive terminal of the dependent source is connected to ground.

Fig. 9.1 (*a*) The standard symbol for an operational amplifier. (*b*) Two input voltages, v_1 and v_2, and the output voltage v_o are defined with respect to ground.

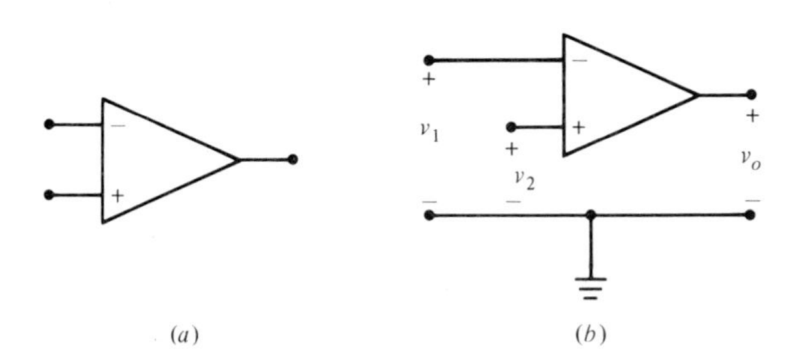

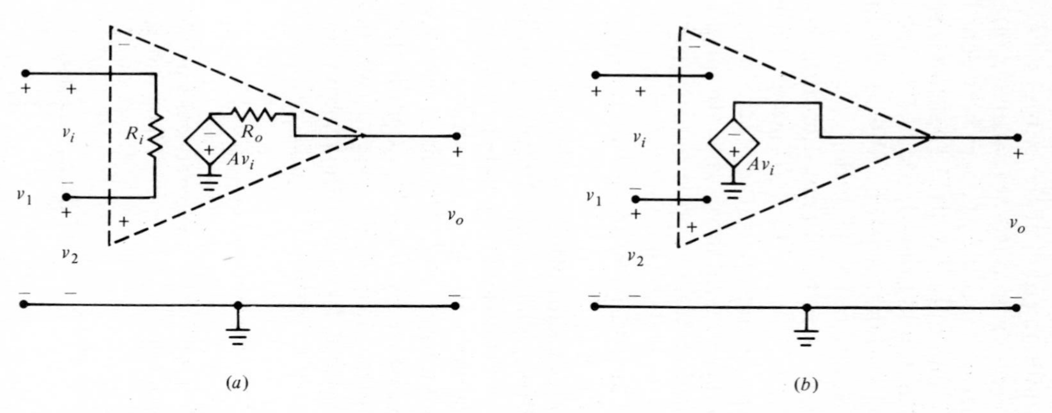

Fig. 9.2 (*a*) A low-frequency equivalent circuit for an op amp includes the input resistance R_i, the output resistance R_o, and a dependent voltage source supplying the open-loop gain A. (*b*) The equivalent circuit of an ideal op amp; $A \to \infty$.

In the next few sections we shall study several simple applications of the op amp by simplifying the equivalent circuit of Fig. 9.2*a* to the so-called *ideal operational amplifier* shown in Fig. 9.2*b*. Here we have let R_i become infinite and set R_o equal to zero; we shall also let A approach infinity.

Let us try out the equivalent circuit of the ideal op amp on the *inverting amplifier* shown in Fig. 9.3*a*. The signal v_s is applied through R_1 to the inverting input, while the noninverting input is grounded. A feedback resistor R_f is connected between input and output. The op amp is replaced with its ideal equivalent circuit in Fig. 9.3*b*, and we see that $v_i = v_1$ since $v_2 = 0$. Then, with $R_o = 0$, we must have $v_o = -Av_i$.

At the input,

$$v_s = R_1 i_1 + v_i$$

and therefore

$$v_s = R_1 i_1 - \frac{v_o}{A} \tag{1}$$

Applying Kirchhoff's voltage law through R_f and the outside loop,

$$-v_s + R_1 i_1 + R_f i_1 + v_o = 0 \tag{2}$$

Solving Eq. (1) for i_1, we have

$$i_1 = \left(v_s + \frac{v_o}{A}\right) \bigg/ R_1$$

Using this result to eliminate i_1 in Eq. (2), we obtain

$$-v_s + v_o + (R_1 + R_f)\left(v_s + \frac{v_o}{A}\right) \bigg/ R_1 = 0$$

Fig. 9.3 (*a*) An op amp is used as an inverting amplifier. (*b*) The op amp is replaced by an equivalent circuit in which $R_i = \infty$ and $R_o = 0$.

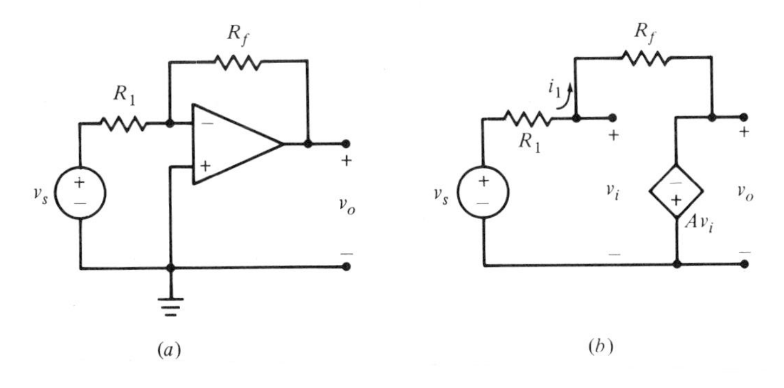

This is easily solved for the *closed-loop gain* v_o/v_s:

$$\frac{v_o}{v_s} = \frac{-(R_f/R_1)}{1 + [1 + (R_f/R_1)]/A} \tag{3}$$

As the gain A becomes very large, the fraction in the denominator approaches zero, and we have

$$\frac{v_o}{v_s} \doteq -\frac{R_f}{R_1} \tag{4}$$

Equation (4) implies that the gain of the circuit with feedback is fixed by the ratio of R_f to R_1. This is an important result. First, it shows that the *closed-loop gain* is independent of the op amp's *open-loop gain* and any variations it may have; that is, the gain depends only on two fixed resistors. Second, any desired gain may be obtained; the accuracy is limited only by the accuracy of the resistance ratio. This means that we can pick an op amp "off the shelf" and fix its gain to our specific needs.

Let us summarize the characteristics of an ideal operational amplifier:

1. $A \to \infty$
2. $R_i \to \infty$
3. $R_o \to 0$
4. $\omega_L = 0$ and $\omega_H \to \infty$
5. Characteristics are constant with time and temperature.
6. $v_o = 0$ when $v_1 = v_2$

For the ideal op amp the relationship given in Eq. (4) becomes exact:

$$\frac{v_o}{v_s} = -\frac{R_f}{R_1} \qquad \text{(ideal op amp)} \tag{5}$$

This result may be obtained much more easily by using the concept of a "virtual ground" as it applies to an operational amplifier. As $A \to \infty$, v_i in Fig. 9.4 must approach zero in order to have a finite output voltage v_o. Thus, both inputs are at the same potential; since the noninverting input is grounded, we find a "virtual ground" at the inverting input. For an ideal op amp with R_i and A infinite, we see that not only is there no voltage between the input terminals ($v_i = 0$), but also that no current flows into either input terminal ($i_i = 0$).

Let us now use this virtual-ground concept to simplify the derivation of Eq. (5). In Fig. 9.4, $i_i = 0$, and thus i_1 flows through both R_1 and R_f. Also, $v_i = 0$, so

$$v_s = R_1 i_1$$

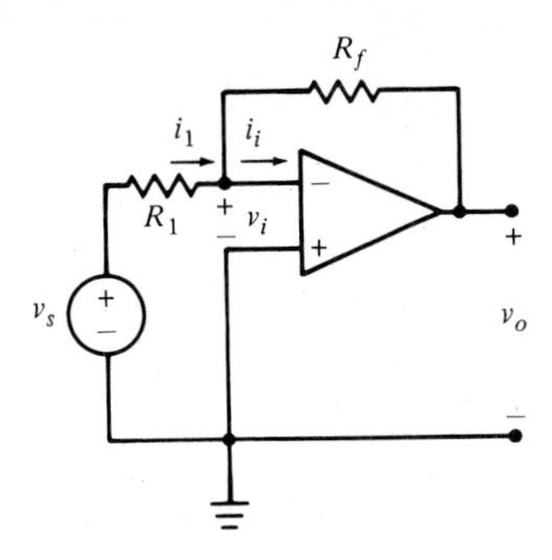

Fig. 9.4 As $A \to \infty$, a virtual ground is found at the input and $v_i \to 0$ and $i_i \to 0$.

and

$$v_o = -R_f i_1$$

By dividing,

$$\frac{v_o}{v_s} = -\frac{R_f}{R_1} \qquad \text{(ideal op amp)}$$

which is the same result we obtained before.

Even though this result was derived for the ideal op amp, the errors incurred in applying it to a real device are only a few tenths of a percent in most cases. For example, the effect of a finite value for A may be appreciated by using both Eqs. (3) and (5) to calculate the closed-loop gain for an inverting amplifier with $R_1 = 10\ \text{k}\Omega$, $R_f = 100\ \text{k}\Omega$, and $A = 10{,}000$. From Eq. (3),

$$\frac{v_o}{v_s} = \frac{-(10^5/10^4)}{1 + \dfrac{1 + (10^5/10^4)}{10{,}000}} = -9.989$$

while Eq. (5) gives $v_o/v_s = -10$. The error in this latter result is only 0.11%.

We might also check to see how good our virtual ground really is. If $v_o = 1$ V with $A = 10{,}000$, then $v_i = -10^{-4}$ V, or $-100\ \mu$V. This is small compared with both the output and with $v_s = -0.1$ V.

D9.1 Let $R_1 = 20\ \text{k}\Omega$ and $R_f = 100\ \text{k}\Omega$ for an inverting amplifier having $R_i = \infty$ and $R_o = 0$. Find an accurate value for the closed-loop gain if A equals (a) 10^2, (b) 10^4, (c) 10^6.

Answers. -4.717; -4.9970; -4.999970

D9.2 Let $R_1 = 10\ \text{k}\Omega$ and $R_f = 80\ \text{k}\Omega$ for an inverting amplifier having $R_i = \infty$, $R_o = 0$, $A = 10^5$, and $v_s = 1$ V. Find (a) v_o, (b) v_i, (c) i_1 (Fig. 9.4).

Answers. -8 V; 80 μV; 100 μA

9.2 The noninverting amplifier

In instrumentation and control circuits where a high input impedance and a gain without phase reversal are required, the *noninverting operational amplifier* is appropriate. This circuit is shown in Fig. 9.5.

Let us first assume a virtual ground at the input terminals. Therefore $v_i = 0$ and $i_i = 0$. It follows that $v_1 = v_2 = v_s$; i_1 is the total current in both R_1 and R_f. Thus these two resistors form a voltage divider and

$$v_1 = v_2 = v_s = \frac{R_1}{R_1 + R_f} v_o$$

so that

$$\frac{v_o}{v_s} = \frac{R_1 + R_f}{R_1} = 1 + \frac{R_f}{R_1} \qquad \text{(ideal op amp)} \tag{6}$$

We see that we can obtain any value of gain greater than unity with no phase reversal by selecting suitable values of R_1 and R_f.

The source v_s sees a very high input impedance, since i_i is so small; a load placed across the output terminals sees zero output impedance if we assume that R_o is zero. We shall obtain more accurate values for both R_{in} and R_{out} later in the chapter.

Now let us consider the case in which A is finite. Again we let $i_i = 0$, and we have

$$v_1 = \frac{R_1}{R_1 + R_f} v_o$$

Also,

$$v_o = -Av_i = -A(v_1 - v_s)$$

Therefore

$$v_o = -A\left(\frac{R_1}{R_1 + R_f} v_o - v_s\right)$$

Solving for the closed-loop gain,

$$\frac{v_o}{v_s} = \frac{A}{1 + \dfrac{AR_1}{R_1 + R_f}} = \frac{R_1 + R_f}{R_1 + \dfrac{R_1 + R_f}{A}} \tag{7}$$

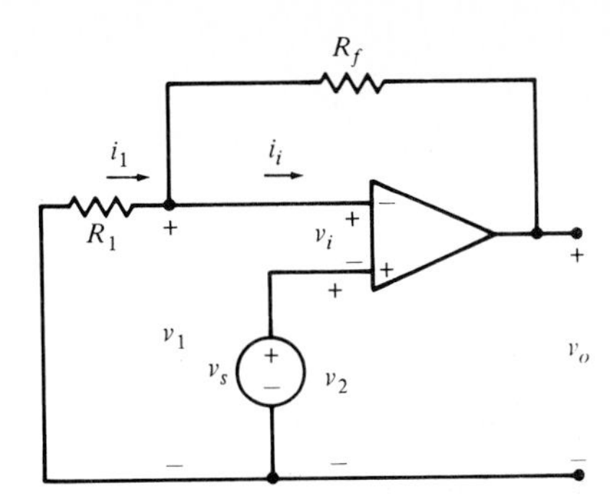

Fig. 9.5 The basic noninverting amplifier has the gain $v_o/v_s = 1 + (R_f/R_1)$.

As A approaches infinity, it is obvious that this gain approaches that given by Eq. (6).

The *voltage follower* is an important form of the noninverting operational amplifier that is obtained by letting $R_f = 0$. When this substitution is made in Eq. (7), we have a result that is independent of R_1:

$$\frac{v_o}{v_s} = \frac{1}{1 + (1/A)} \tag{8}$$

Equation (7) also yields Eq. (8) when $R_1 \to \infty$. As $A \to \infty$ in Eq. (8),

$$\frac{v_o}{v_s} = 1 \qquad \text{(ideal op amp)} \tag{9}$$

Figure 9.6 shows a voltage follower in its barest form. Although R_1 appears to be infinite, we shall see later that a dc path to ground is required at every input.

We now have a circuit with unity gain and no phase shift (hence the name *voltage follower*), very high input impedance, and very low output impedance. Such an arrangement is useful for isolating two circuits so that they do not interact with each other. This circuit is also known as a *buffer amplifier* or *isolation amplifier*.

D9.3 Let $R_i = \infty$ and $R_o = 0$ for a noninverting amplifier. (a) Calculate v_o/v_s if $R_1 = 100$ kΩ, $R_f = 300$ kΩ, and $A = \infty$. (b) Let $R_1 = 20$ kΩ, $A = \infty$, and $v_o/v_s = 10$, and determine R_f. Find v_o/v_s if $R_1 = 10$ kΩ, $R_f = 200$ kΩ, and (c) $A = 1000$, (d) $A = 10^5$.

Answers. 4; 180 kΩ; 20.57; 20.96

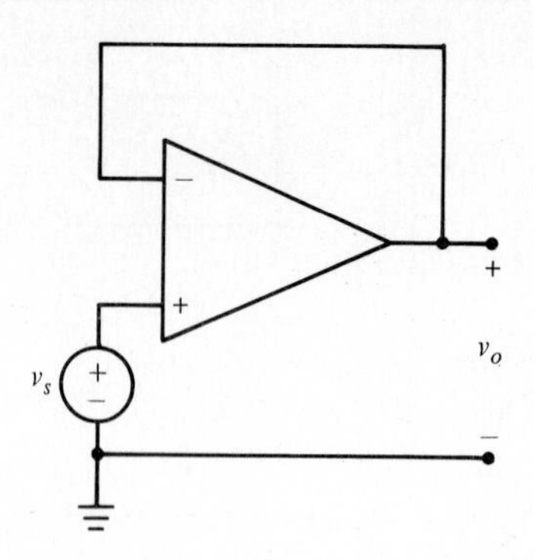

Fig. 9.6 The voltage follower has unity gain, $v_o/v_s = 1$, very high input impedance, and very low output impedance.

D9.4 A voltage follower is designed to provide $v_o = 1$ V. Determine v_1, v_i, and v_2 if $A =$ (a) 10^2, (b) 10^4.

Answers. 1, −0.01, 1.01; 1, −0.0001, 1.0001 V

9.3 The inverting amplifier

The inverting operational amplifier was first shown in Fig. 9.3*a*, and it is drawn again as Fig. 9.7. If we assume that R_i and A are infinite while R_o is zero, then once again

$$\frac{v_o}{v_s} = -\frac{R_f}{R_1} \qquad \text{(ideal op amp)} \tag{10}$$

Now let us consider the input impedance offered to the ideal voltage source v_s. With the virtual ground at the input, $v_i = 0$, and it follows that

$$R_{\text{in}} = R_1 \qquad \text{(ideal op amp)} \tag{11}$$

Since $R_o = 0$, the Thévenin impedance as seen at the output terminals is

$$R_{\text{out}} = 0 \qquad \text{(ideal op amp)} \tag{12}$$

These results indicate why this circuit is such a popular and versatile amplifier. Any values of closed-loop gain and input impedance can be obtained by the appropriate selection of R_1 and R_f.

The basic inverting amplifier of Fig. 9.7 may also be used as a constant-current source. If we again consider $v_i = 0$, then the current to the right in R_1 is v_s/R_1; this current also flows to the right in R_f, since $i_i = 0$. Thus

$$i = \frac{v_s}{R_1} \qquad \text{(ideal op amp)} \tag{13}$$

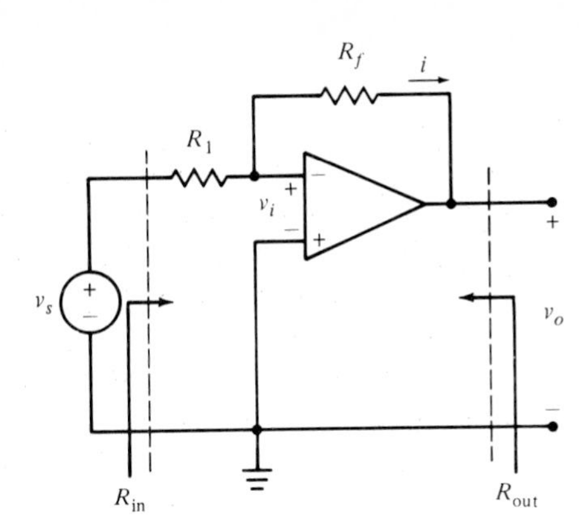

Fig. 9.7 Using an ideal op amp, the inverting amplifier is characterized by $v_o/v_s = -R_f/R_1$, $R_{in} = R_1$, and $R_o = 0$.

This is completely independent of R_f, and it therefore acts like a constant-current source. The value of the current is proportional to v_s, and the circuit might therefore be used as a linear amplifier to cause a meter current to be directly proportional to v_s, or as the source supplying the deflection-coil current for the sweep in a television set.

Many applications of the inverting op amp involve the summing of several signals. One circuit that accomplishes this is the *summer* or *summing amplifier* shown in Fig. 9.8 for the case of three input signals. If we again let $v_i = 0$, then we see that the currents in R_1, R_2, and R_3 are v_{s1}/R_1, v_{s2}/R_2, and v_{s3}/R_3 respectively, while $i = -v_o/R_f$. With $i_i = 0$, we may apply Kirchhoff's current law at the input and obtain

$$i = -\frac{v_o}{R_f} = \frac{v_{s1}}{R_1} + \frac{v_{s2}}{R_2} + \frac{v_{s3}}{R_3}$$

Thus

$$v_o = -R_f\left(\frac{v_{s1}}{R_1} + \frac{v_{s2}}{R_2} + \frac{v_{s3}}{R_3}\right) \qquad \text{(ideal op amp)} \qquad (14)$$

If $R_1 = R_2 = R_3$, then the summing is accomplished without scaling the individual signals:

$$v_o = -\frac{R_f}{R_1}(v_{s1} + v_{s2} + v_{s3}) \qquad \text{(ideal op amp)} \qquad (15)$$

Note that the input impedance to each source is given by the resistance in that branch.

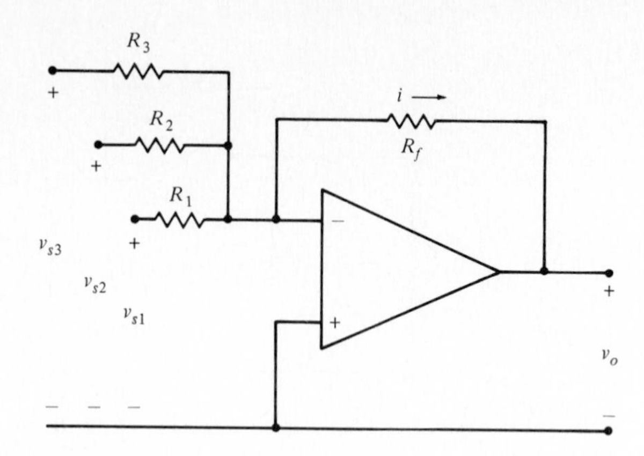

Fig. 9.8 This *summing amplifier* or *summer* has an output $v_o = -R_f[(v_{s1}/R_1) + (v_{s2}/R_2) + (v_{s3}/R_3)]$.

Subtraction and addition may be accomplished simultaneously through the use of an inverting amplifier with unity gain for each signal to be subtracted.

D9.5 Let $R_i = \infty$, $R_o = 0$, $R_1 = 10$ kΩ, and $R_f = 200$ kΩ in the circuit of Fig. 9.7. If $A = \infty$ and $v_s = 1$ V, find (a) R_{in}, (b) i. (c) Find i if $A = 100$.

Answers. 10 kΩ; 0.1 mA; 0.0835 mA

D9.6 Design an amplifier to carry out the following operations: (a) $v_o = -10v_{s1} - 100v_{s2} - 5v_{s3}$; (b) $v_o = v_{s1} + 2v_{s2} + 4v_{s3}$.

Answers. $R_f = 100$ kΩ: 10 kΩ, 1 kΩ, 20 kΩ; $R_f = 100$ kΩ: 100 kΩ, 50 kΩ, 25 kΩ, and unity-gain inverting

9.4 The differential amplifier

There are numerous instrumentation applications in which we wish to measure the difference between two signals. For example, in a heat-flow problem we may wish to measure the temperature difference between two points and not the specific temperature of each point. This could be done with two thermocouples and a differential amplifier. Or perhaps we might want to compare a temperature with a standard, often done in precision temperature measurements.

Figure 9.9 shows a differential amplifier that is easily analyzed by superposition. Let us first set $v_{s2} = 0$ and call the partial output that v_{s1} produces v_{o1}. The parallel resistors to ground at the lower input have no effect since $v_i = 0$, and we have an inverting amplifier for which

$$v_{o1} = -\frac{R_{f1}}{R_1}v_{s1}$$

With v_{s2} turned back on and $v_{s1} = 0$, we see a voltage divider at the noninverting input, and

$$v_2 = \frac{R_{f2}}{R_{f2} + R_2}v_{s2}$$

Therefore

$$v_{o2} = \left(1 + \frac{R_{f2}}{R_2}\right)\left(\frac{R_{f2}}{R_{f2} + R_2}v_{s2}\right)$$

Combining these results,

$$v_o = v_{o1} + v_{o2} = -\frac{R_{f1}}{R_1}v_{s1} + \frac{R_{f2} + R_2}{R_2} \times \frac{R_{f2}}{R_{f2} + R_2}v_{s2}$$

or

$$v_o = \frac{R_{f2}}{R_2}v_{s2} - \frac{R_{f1}}{R_1}v_{s1} \tag{16}$$

In order to achieve $v_o = 0$ with $v_{s1} = v_{s2}$, we put

$$\frac{R_{f1}}{R_1} = \frac{R_{f2}}{R_2} \tag{17}$$

Fig. 9.9 The output of the differential amplifier is $v_o = (R_{f1}/R_1)(v_{s2} - v_{s1}) = (R_{f2}/R_2)(v_{s2} - v_{s1})$ for $R_{f1}/R_1 = R_{f2}/R_2$.

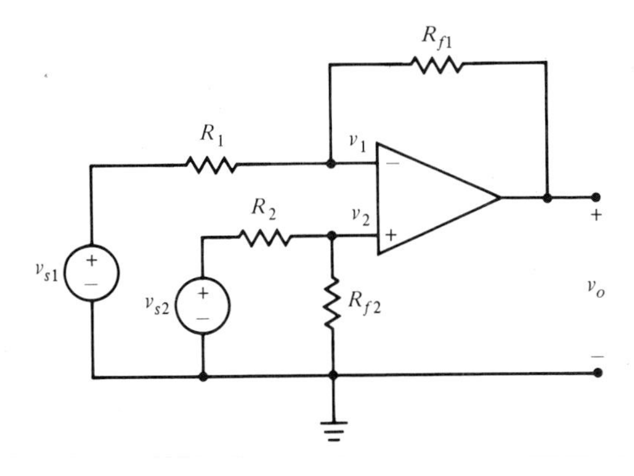

Then,

$$v_o = \frac{R_{f1}}{R_1}(v_{s2} - v_{s1}) = \frac{R_{f2}}{R_2}(v_{s2} - v_{s1}) \qquad \text{(ideal op amp)} \quad (18)$$

The ratios R_{f1}/R_1 and R_{f2}/R_2 should be exactly the same; otherwise, two equal input signals will cause a nonzero output.

The input impedances offered to the two sources in Fig. 9.9 are not the same. At the noninverting input, we must have

$$R_{\text{in}2} = R_2 + R_{f2} \qquad (19)$$

since the op amp draws no input current. To find $R_{\text{in}1}$, we again set $v_{s2} = 0$, therefore causing v_2 to be zero. Thus, the virtual ground at the op amp input forces the condition $v_1 = 0$, and

$$R_{\text{in}1} = R_1 \qquad (20)$$

D9.7 A differential amplifier is to give an output $v_o = 3(v_{s2} - v_{s1})$. The input impedance at each input is to be 10 kΩ. If the op amp has a large open-loop gain, specify the value of (a) R_1, (b) R_{f1}, (c) R_2, (d) R_{f2}.

Answers. 10; 30; 2.5; 7.5 kΩ

9.5 Characteristics of a real operational amplifier

Up to this point, we have mainly considered the ideal operational amplifier with an open-loop gain approaching infinity, zero output impedance, and infinite input impedance. These assumptions led us to the concept of the virtual ground and a fairly simple analysis of several op-amp circuits. Throughout the remainder of this chapter, we shall consider a real operational amplifier, beginning with the inspection of a typical manufacturer's data sheets.

There are many types of op amp commercially available to the circuit designer, covering a wide range of performance characteristics (and costs). Some have a very specialized nature, while others are designed to be used in many different applications. Figure 9.10 is the first page of data for the Harris Semiconductor HA-2107 operational amplifier. The complete data sheet is reproduced in Appendix A. Note from Fig. 9.10 that this is a "general purpose operational amplifier" that can operate over a temperature range from −55°C to 125°C. The pin connections are shown for a metal can encapsulation (TO-99) in the lower right corner of Fig. 9.10. A square metal tab identifies Pin No. 8, and the pins are numbered in a counterclockwise direction, as viewed from the top. The inverting input is Pin 2, the noninverting input is Pin 3, the positive power supply connection is Pin 7, and the negative supply connection is Pin 4. This information

HA-2107/2207/2307

Operational Amplifiers

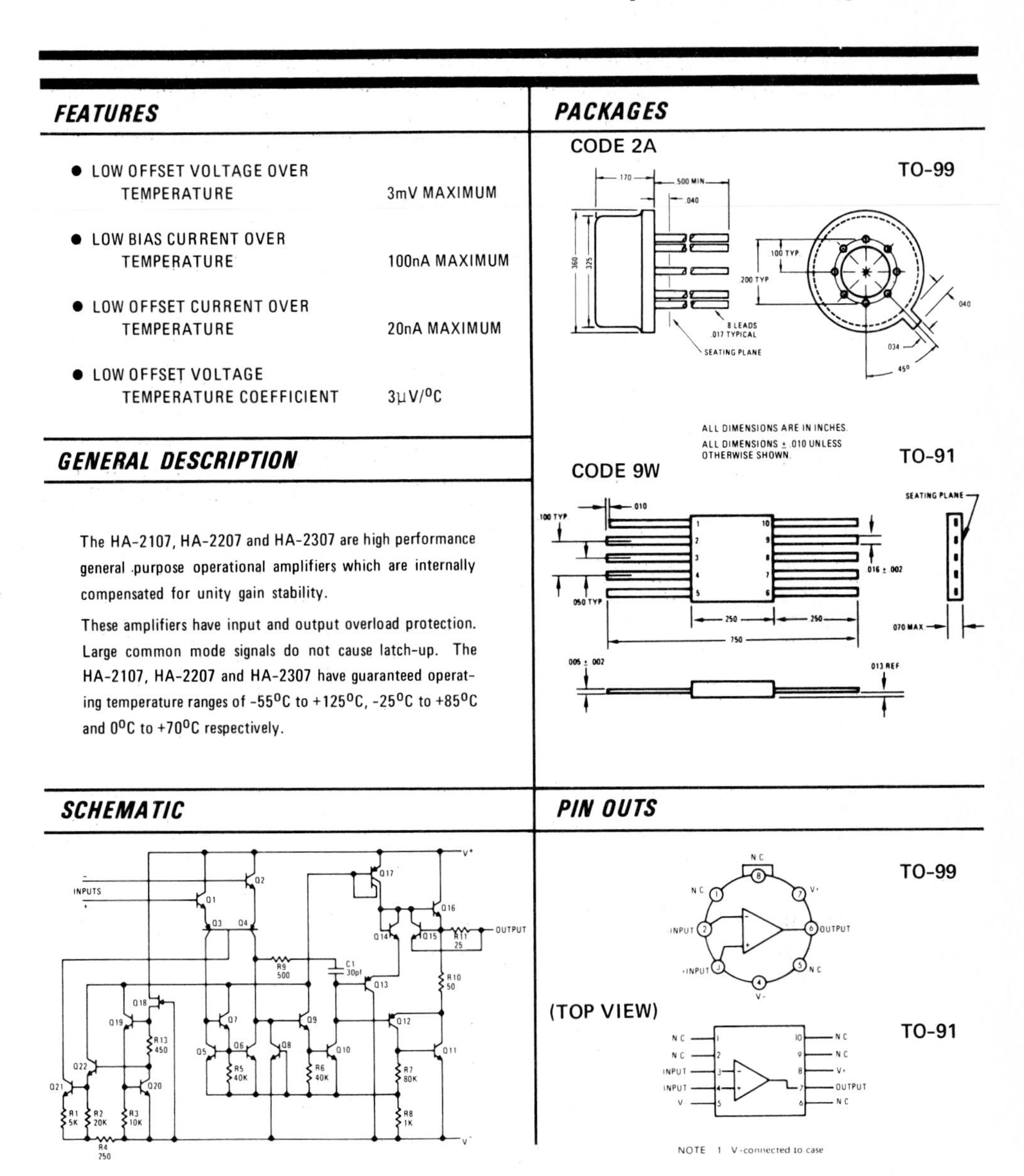

FEATURES

- LOW OFFSET VOLTAGE OVER TEMPERATURE — 3mV MAXIMUM
- LOW BIAS CURRENT OVER TEMPERATURE — 100nA MAXIMUM
- LOW OFFSET CURRENT OVER TEMPERATURE — 20nA MAXIMUM
- LOW OFFSET VOLTAGE TEMPERATURE COEFFICIENT — 3μV/°C

GENERAL DESCRIPTION

The HA-2107, HA-2207 and HA-2307 are high performance general purpose operational amplifiers which are internally compensated for unity gain stability.

These amplifiers have input and output overload protection. Large common mode signals do not cause latch-up. The HA-2107, HA-2207 and HA-2307 have guaranteed operating temperature ranges of -55°C to +125°C, -25°C to +85°C and 0°C to +70°C respectively.

PACKAGES

SCHEMATIC

PIN OUTS

Fig. 9.10 The first sheet of specifications for the HA-2107 operational amplifier, courtesy of Harris Semiconductor. See Appendix A for complete data sheets.

enables us to make connections to the external circuit. The leads are also identified for the mini-DIP (dual in-line package) arrangement. A dot, a notch, or a shorter lead indicates Pin 1; the pins again are numbered in a counterclockwise direction as viewed from the top.

Figure 9.10 also gives a schematic diagram for the 2107 op amp, showing the interior circuitry. Although this is of great interest to operational-amplifier and linear integrated-circuit designers, it need not be mastered by the likes of us. However, there are some points of interest in this schematic. For example, we see that the op amp contains 22 transistors, 12 resistors, and one capacitor. Also, the noninverting and inverting inputs are connected directly to the bases of $Q1$ and $Q2$, both of which are *npn* bipolar transistors. This indicates that some dc biasing current is necessary. Furthermore, since the inputs are direct-coupled, the op amp is capable of amplifying dc voltages and currents.

Figure 9.11 lists all the absolute maximum ratings and gives some electrical characteristics of the 2107 op amp for various temperature conditions. If the absolute maximum ratings are exceeded, damage to the integrated circuit may be catastrophic. At the least, some degradation in performance will likely occur. Notice that there is a limit to the temperature range and the input voltage.

As a comparison of the ideal and real operational amplifier, consider the *input resistance*, the sixth listing under "Electrical Characteristics" in Fig. 9.11. At 25°C, the 2107 op amp has a minimum R_{in} of 1.5 MΩ and a typical value of 4 MΩ. This is not infinite, but in comparison to a l-kΩ resistance in the external circuit, it is very large. Also notice that an input *bias current* no greater than 75 nA at 25°C is required to bias the bipolar transistors $Q1$ and $Q2$. The minimum bias current increases to 100 nA at some point within the temperature range from −55°C to 125°C.

The large signal voltage gain at 25°C with $R_L = 2$ kΩ and the supply voltage $V_S = \pm 15$ V is listed as 50,000 minimum and 160,000 typical. Thus A is not infinite, but it is sufficiently high to make the closed-loop gain essentially independent of any changes that might occur in the op amp.

We see from Note 3 at the bottom of Fig. 9.11 that the given data are valid for a supply voltage in the range from ±5 V to ±20 V. Thus, we might connect −15 V to Pin 4 and +15 V to Pin 7.

There are many unfamiliar terms in these data sheets, but most of them will be discussed later in the text.

As an example comparing an ideal, a fairly low-gain, and a real inverting operational amplifier, consider Table 9.1. The low-gain example has $A = 10^4$, $R_i = 50$ kΩ, and $R_o = 100$ Ω. Corresponding values for the HA-2107 op amp are $A = 5 \times 10^4$, $R_i = 1.5$ MΩ, and $R_o = 100$ Ω. The exact values of v_o/v_s have been calculated by expressions that are developed in the next section, and we see that the gain differs from the ideal value by

ABSOLUTE MAXIMUM RATINGS

SPECIFICATIONS

	HA-2107 HA-2207	HA-2307
Supply Voltage	±22.0V	±18.0V
Power Dissipation (Note 1)	500mW	
Differential Input Voltage	±30.0V	
Input Voltage (Note 2)	±15.0V	
Output Short Circuit Duration	Indefinite	

Operating Temperature Range	
HA-2107	-55°C to +125°C
HA-2207	-25°C to +85°C
HA-2307	0°C to +70°C
Storage Temperature Range	-65°C to +150°C
Lead Temperature (Soldering 60 seconds)	300°C

ELECTRICAL CHARACTERISTICS

PARAMETER	TEMPERATURE (Note 7)	HA-2107 HA-2207 MIN.	TYP.	MAX.	HA-2307 MIN.	TYP.	MAX.	UNITS
INPUT CHARACTERISTICS								
Offset Voltage	+25°C		0.7	2.0		2.0	7.5	mV
	Full			3.0			10	mV
Average Offset Voltage Temperature Coefficient	Full		3.0	15		6.0	30	μV/°C
Bias Current	+25°C		30	75		70	250	nA
	Full			100			300	nA
Offset Current	+25°C		1.5	10		3	50	nA
	Full			20			70	nA
Offset Current Temperature Coefficient	T_{Low} to +25°C		0.02	0.2		0.02	0.6	nA/°C
	+25°C to T_{High}		0.01	0.1		0.01	0.3	nA/°C
Input Resistance	+25°C	1.5	4		0.5	2		MΩ
Input Voltage Range (Note 4)	Full	±15.0			±12			V
TRANSFER CHARACTERISTICS								
Large Signal Voltage Gain (Notes 5 & 6)	+25°C	50K	160K		25K	160K		V/V
	Full	25K			15K			V/V
Common Mode Rejection Ratio	Full	80	96		70	90		dB
Output Voltage Swing								
R_L = 10KΩ	Full	±12.0	±14.0		±12.0	±14.0		V
R_L = 2KΩ (Note 6)	Full	±10.0	±13.0		±10.0	±13.0		V
POWER SUPPLY CHARACTERISTICS								
Supply Current (Note 4)	+25°C		1.8	3.0		1.8	3.0	mA
	125°C		1.2	2.5				mA
Supply Voltage Rejection Ratio	Full	80	96		70	96		dB

NOTES:

1. Derate TO-99 package at 6.8mW/°C for operation ambient temperature above 75°C and 4.9mW/°C above 50°C for the TO-91 package.
2. For supply voltages less than +15.0V, the absolute maximum input voltage is equal to supply voltage.
3. These specifications apply for $\pm 5.0V \leq V_S \leq +20.0V$ unless otherwise specified.

(4) $V_S = \pm 20.0V$

(5) $R_L = 2K\Omega$, $V_{OUT} = \pm 10.0V$

(6) $V_S = \pm 15.0V$

(7) Full = T_{Low} to T_{High} = -55°C to +125°C (HA-2107)
-25°C to +85°C (HA-2207)
0°C to +70°C (HA-2307)

Fig. 9.11 Page 2 of the HA-2107 specifications gives maximum ratings and certain electrical characteristics.

Table 9.1 A Comparison of the Closed-Loop Gain for Three Inverting Op Amps with Various R_1 and R_f if $R_L = 10\ k\Omega$

		v_o/v_s		
R_1	R_f	Ideal $R_i = \infty$ $R_o = 0$ $A = \infty$	Example $R_i = 50\ k\Omega$ $R_o = 100\ \Omega$ $A = 10^4$	HA-2107 $R_i = 1.5\ M\Omega$ $R_o = 100\ \Omega$ $A = 5 \times 10^4$
10 kΩ	10 kΩ	−1	−0.999776	−0.999959
10 kΩ	100 kΩ	−10	−9.98687	−9.99776
10 kΩ	1 MΩ	−100	−98.793	−99.79503
10 kΩ	10 MΩ	−1000	−891.82	−980.051
1 kΩ	1 kΩ	−1	−0.999776	−0.999956
1 kΩ	100 kΩ	−100	−98.9694	−99.796
1 kΩ	10 MΩ	-10^4	−4925.4	−8318.38

the largest amount when the closed-loop gain is greatest. Still larger values of A and R_i for the HA-2107 op amp cause the error to be even less.

The values in Table 9.2 are presented to compare the real HA-2107 and an ideal op amp in an inverting amplifier using a smaller load resistance $R_L = 2\ k\Omega$. "Typical" values of $R_i = 4\ M\Omega$ and $A = 160{,}000$ are used. With R_o assumed to be 100 Ω, the real and ideal gains differ by less than 1%, even at the highest closed-loop gain. The righthand column

Table 9.2 A Comparison of v_o/v_s Values for an Inverting Amplifier Using an Ideal Op Amp and the HA-2107 Op Amp; $R_L = 2\ k\Omega$ and $R_1 = 10\ k\Omega$

	v_o/v_s		
		HA-2107 $R_i = 4\ M\Omega$ $A = 160{,}000$	
R_f	Ideal $R_i = \infty$ $R_o = 0$ $A = \infty$	$R_o = 100\ \Omega$	$R_o = 1\ k\Omega$
10 kΩ	−1	−0.999987	−0.999980
100 kΩ	−10	−9.99928	−9.99896
1 MΩ	−100	−99.934	−99.905
10 MΩ	−1000	−993.458	−990.679

indicates slightly more error when the assumed value of R_o is increased to 1 kΩ.

In summary, Tables 9.1 and 9.2 show that for a typical set of values in an inverting amplifier, the ideal op-amp assumption is really quite good. In fact, the error, as compared to the ideal case, could easily be less than the calibration error; it certainly could be cancelled by a slight adjustment of the feedback resistor R_f.

D9.8 If the HA-2107 op amp is operated at 100°C ambient temperature, find the maximum permissible power dissipation if the unit is encapsulated in (a) a TO-99 package, (b) a TO-91 package.

Answers. 330; 255 mW

9.6 The real inverting amplifier

In order to calculate the accurate values of closed-loop gain that appear in the tables above, it is necessary to make an exact analysis of the complete inverting-amplifier circuit. Figure 9.12*a* shows such an amplifier with a load resistor R_L that must be included in the analysis if R_o is not zero. The op amp is replaced with its equivalent circuit containing R_i, R_o, and the dependent voltage source in Fig. 9.12*b*. We could now write three mesh equations and then solve for v_o/v_s, but we can also save ourselves some work by transforming both voltage sources into current sources, as shown in Fig. 9.12*c*. Only two nodal equations are required with this approach.

At the v_i node, we equate the sum of the currents leaving the node to zero:

$$-\frac{v_s}{R_1} + \frac{v_i}{R_1} + \frac{v_i}{R_i} + \frac{v_i - v_o}{R_f} = 0$$

or

$$v_i\left(\frac{1}{R_1} + \frac{1}{R_i} + \frac{1}{R_f}\right) - v_o\left(\frac{1}{R_f}\right) = v_s\left(\frac{1}{R_1}\right) \tag{21}$$

At the output node,

$$\frac{v_o - v_i}{R_f} + \frac{Av_i}{R_o} + \frac{v_o}{R_o} + \frac{v_o}{R_L} = 0$$

and

$$v_i\left(\frac{A}{R_o} - \frac{1}{R_f}\right) + v_o\left(\frac{1}{R_f} + \frac{1}{R_o} + \frac{1}{R_L}\right) = 0 \tag{22}$$

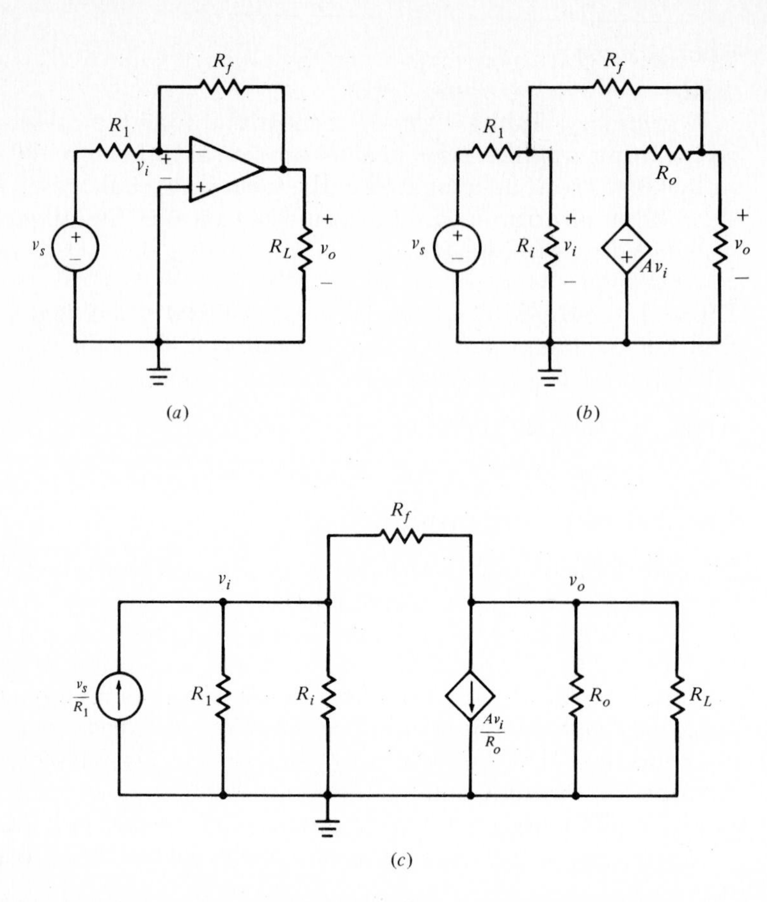

Fig. 9.12 (*a*) The inverting amplifier, including load R_L. (*b*) The op amp is replaced by an equivalent circuit that includes R_i and R_o. (*c*) The two voltage sources are replaced by current sources.

We can solve Eq. (22) for v_i, substitute into Eq. (21), and then solve for the gain v_o/v_s. Or we can simply use Cramer's rule and determinants to find v_o/v_s directly. Either way, the result is

$$\frac{v_o}{v_s} = \frac{-\dfrac{R_f}{R_1}}{1 + \dfrac{\left(1 + \dfrac{R_f}{R_1} + \dfrac{R_f}{R_i}\right)\left(1 + \dfrac{R_o}{R_f} + \dfrac{R_o}{R_L}\right)}{A - \dfrac{R_o}{R_f}}} \tag{23}$$

When the result is written in this form, it is apparent that the closed-loop gain v_o/v_s is equal to $-R_f/R_1$ in the limit as A becomes infinite, regardless of the values of R_i, R_o, and R_L (we exclude $R_i = 0$, $R_o = \infty$, and $R_L = 0$ as choices that lead obviously to zero closed-loop gain).

If we set $R_o = 0$, then v_o/v_s is not a function of the load resistance R_L. As the ratio R_o/R_L increases, the closed-loop gain decreases and becomes more dependent on the open-loop gain.

While we are obtaining exact expressions for the inverting amplifier of Fig. 9.12*a*, we should also find its input and output impedance. The input impedance, here the resistance R_{in}, is seen to the right of the source v_s in Fig. 9.12*a* or *b*. It must be given by the ratio of v_s to the current in R_1:

$$R_{\text{in}} = \frac{v_s}{(v_s - v_i)/R_1} = \frac{R_1}{1 - (v_i/v_s)}$$

We use Eqs. (22) and (23) first to find

$$\frac{v_i}{v_s} = \frac{\dfrac{R_f}{R_1}\left(1 + \dfrac{R_o}{R_f} + \dfrac{R_o}{R_L}\right)}{\left(1 + \dfrac{R_f}{R_1} + \dfrac{R_f}{R_i}\right)\left(1 + \dfrac{R_o}{R_f} + \dfrac{R_o}{R_L}\right) - \dfrac{R_o}{R_f} + A}$$

We then obtain

$$R_{\text{in}} = R_1 \frac{\left(1 + \dfrac{R_f}{R_1} + \dfrac{R_f}{R_i}\right)\left(1 + \dfrac{R_o}{R_f} + \dfrac{R_o}{R_L}\right) + A - \dfrac{R_o}{R_f}}{\left(1 + \dfrac{R_f}{R_i}\right)\left(1 + \dfrac{R_o}{R_f} + \dfrac{R_o}{R_L}\right) + A - \dfrac{R_o}{R_f}} \tag{24}$$

As A becomes large, R_{in} approaches R_1, the value we obtained with our initial approximate analysis.

The procedure for finding the output impedance requires us to set $v_s = 0$ and replace R_L with a 1-A current source, as shown in Fig. 9.13. At the left node,

$$\frac{v_i}{R_1} + \frac{v_i}{R_i} + \frac{v_i - v_o}{R_f} = 0$$

or

$$v_i\left(\frac{1}{R_1} + \frac{1}{R_i} + \frac{1}{R_f}\right) - v_o\left(\frac{1}{R_f}\right) = 0$$

At the right,

$$\frac{v_o - v_i}{R_f} + \frac{Av_i}{R_o} + \frac{v_o}{R_o} = 1$$

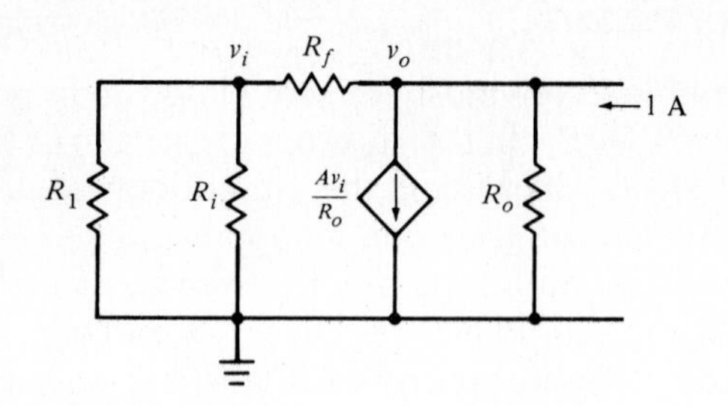

Fig. 9.13 The output resistance of the inverting amplifier is $R_{out} = v_o/1 = v_o$.

Thus

$$v_i\left(\frac{A}{R_o} - \frac{1}{R_f}\right) + v_o\left(\frac{1}{R_f} + \frac{1}{R_o}\right) = 1$$

Solving for v_o, we find that

$$R_{out} = \frac{v_o}{1} = \frac{R_o\left(1 + \frac{R_f}{R_1} + \frac{R_f}{R_i}\right)}{\left(1 + \frac{R_f}{R_1} + \frac{R_f}{R_i}\right)\left(1 + \frac{R_o}{R_f}\right) + A - \frac{R_o}{R_f}} \tag{25}$$

Note that $R_{out} \to 0$ as $R_o \to 0$, regardless of the value of A, and also that $R_{out} \to 0$ as $A \to \infty$, regardless of the value of R_o.

For an inverting amplifier with $R_1 = 10$ kΩ, $R_f = 100$ kΩ, $R_i = 50$ kΩ, $R_o = 100$ Ω, $R_L = 10$ kΩ, and $A = 10^4$ (all given in the second line of Table 9.1), we find that $v_o/v_s = -9.98687$, as given in the table, and also that $R_{in} = 10.01011$ kΩ and $R_{out} = 0.12983$ Ω. Thus, $R_{in} \doteq R_1$ and $R_{out} \doteq 0$.

D9.9 An inverting amplifier is constructed with $R_1 = 20$ kΩ, $R_f = 200$ kΩ, $R_L = 5$ kΩ, $A = 5000$, $R_i = 25$ kΩ, and $R_o = 500$ Ω. Find (a) v_o/v_s, (b) R_{in}, (c) R_{out}.

Answers. -9.96; 9.86 kΩ; 1.893 Ω

9.7 The real voltage follower

The voltage follower was first shown in Fig. 9.6 and discussed briefly at that time. We found that it has a very high input impedance so that it does not load the circuit that drives it, a very low output impedance so that it appears as a nearly ideal voltage source to the circuit following it, and a gain of unity. For these reasons, we also called it a buffer or an isolation amplifier.

Let us now consider a voltage follower with an external load resistance R_L that is not infinite and a feedback resistor R_f that need not be zero. The circuit is shown in Fig. 9.14*a*. A nonzero R_f is sometimes desirable for balancing the effects of the input biasing currents, a point we consider in the next section. In Fig. 9.14*b* the op amp is replaced by its equivalent circuit containing R_i, R_o, and the dependent voltage source Av_i. Any resistance R_s in series with the source v_s is usually much smaller than R_i and is therefore neglected.

Our object is to analyze this circuit and obtain accurate expressions for the voltage gain v_o/v_s, the input resistance R_{in} seen by v_s, and the Thévenin output resistance R_{out} presented to R_L. Having the complete expressions, we may then consider the effects of certain conditions, such as $R_f = 0$, $R_o = 0$, and so forth.

We write two mesh equations in terms of the mesh currents i_i and i_o:

$$-v_s + R_i i_i + R_f i_i + R_o(i_i - i_o) - A(-R_i i_i) = 0$$

or

$$(R_i + R_f + R_o + AR_i)i_i - R_o i_o = v_s \tag{26}$$

and

$$A(-R_i i_i) + R_o(i_o - i_i) + R_L i_o = 0$$

or

$$-(AR_i + R_o)i_i + (R_o + R_L)i_o = 0 \tag{27}$$

If we solve Eq. (27) for i_i and then eliminate i_i in Eq. (26), we have

$$i_o = \frac{v_s(AR_i + R_o)}{(R_i + R_f + R_o + AR_i)(R_o + R_L) - R_o(AR_i + R_o)}$$

Fig. 9.14 (*a*) A voltage follower may include $R_f \neq 0$ and $R_L \neq \infty$. (*b*) The op amp is replaced by its equivalent circuit.

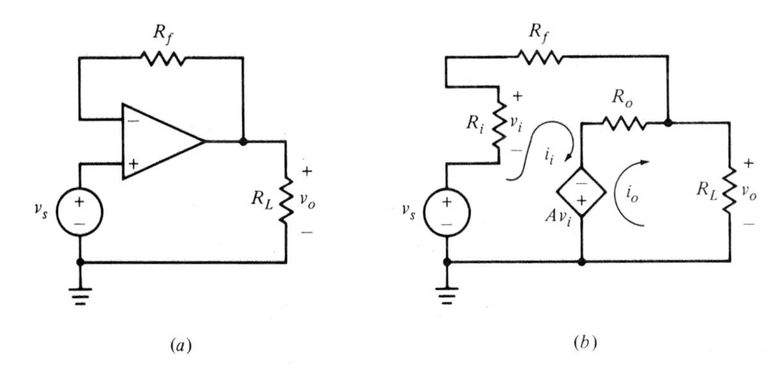

Since $v_o = R_L i_o$, the gain becomes

$$\frac{v_o}{v_s} = \frac{R_L(AR_i + R_o)}{(R_i + R_f + R_o + AR_i)(R_o + R_L) - R_o(AR_i + R_o)} \tag{28}$$

Dividing numerator and denominator by $R_L(AR_i + R_o)$, we obtain a somewhat neater expression:

$$\frac{v_o}{v_s} = \frac{1}{1 + \left(1 + \dfrac{R_o}{R_L}\right)\left(\dfrac{R_f + R_i}{R_o + AR_i}\right)} \tag{29}$$

Note that $v_o/v_s \to 1$ as $A \to \infty$.

Let us apply this result to two examples. In each, we shall let $R_L = 2\ \text{k}\Omega$ and $R_f = 10\ \text{k}\Omega$. The first represents a fairly poor op amp with $A = 10^4$, $R_i = 100\ \text{k}\Omega$, and $R_o = 1\ \text{k}\Omega$. Using Eq. (29), we find $v_o/v_s = 0.99984$, which is unity on almost anyone's voltmeter. For the second example, we use some data for the HA-2107 op amp: $A = 160{,}000$, $R_i = 4\ \text{M}\Omega$, and $R_o = 100\ \Omega$. We now obtain $v_o/v_s = 0.9999934$, even closer to our mythical ideal.

The input resistance is given by

$$R_{\text{in}} = \frac{v_s}{i_i}$$

This is readily obtained from Eqs. (26) and (27) by solving the latter for i_o, substituting this result into the former, and finding

$$R_{\text{in}} = \frac{(R_i + R_f + R_o + AR_i)(R_o + R_L) - R_o(AR_i + R_o)}{R_o + R_L} \tag{30}$$

This may be rearranged to give the slightly simpler result:

$$R_{\text{in}} = (AR_i + R_o)\frac{R_L}{R_o + R_L} + R_i + R_f \tag{31}$$

As either A or R_i becomes infinite, so does R_{in}.

For the first of our examples, we find $R_{\text{in}} = 667\ \text{M}\Omega$, while the HA-2107 op amp gives $R_{\text{in}} = 609{,}500\ \text{M}\Omega$, both essentially open circuits compared to almost anything.

We now consider the output resistance. Figure 9.15 shows the follower with $v_s = 0$ and R_L replaced by a 1-A current source. Around the left mesh,

$$R_i i_i + R_f i_i + R_o(i_i + 1) - A(-R_i i_i) = 0$$

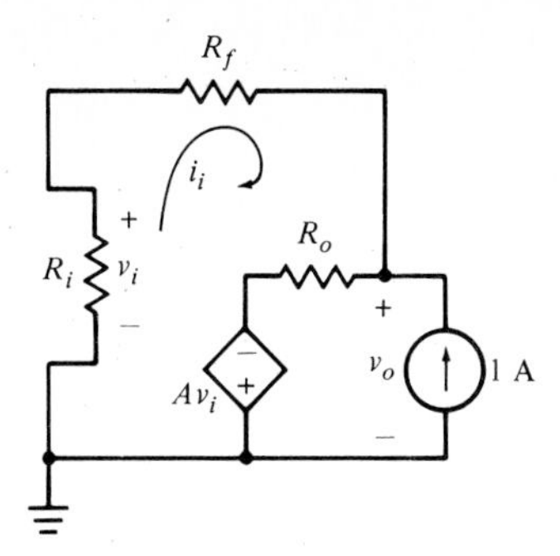

Fig. 9.15 The output resistance of the voltage follower of Fig. 9.14*a* is $R_{out} = v_o/1 = v_o$.

or

$$(R_i + R_f + R_o + AR_i)i_i = -R_o$$

so that

$$i_i = \frac{-R_o}{R_i(A + 1) + R_f + R_o} \tag{32}$$

Now, R_{out} is the effective resistance offered to the 1-A source, and

$$R_{out} = \frac{v_o}{1} = v_o = -i_i(R_i + R_f)$$

Using i_i from Eq. (32), we have

$$R_{out} = \frac{R_o(R_i + R_f)}{R_i(A + 1) + R_f + R_o} \tag{33}$$

or

$$R_{out} = \frac{R_o\left(1 + \dfrac{R_f}{R_i}\right)}{A + 1 + \dfrac{R_f}{R_i} + \dfrac{R_o}{R_i}} \tag{34}$$

We note that $R_{out} \to 0$ as $A \to \infty$ or $R_o \to 0$. As $R_i \to \infty$, $R_{out} \to R_o/(A + 1)$.

For the example with $A = 10^4$, we find $R_{out} = 0.110\ \Omega$; the HA-2107 op amp gives an even smaller value, $R_{out} = 0.00063\ \Omega$.

We therefore may conclude that R_{in} is indeed very large, R_{out} is negligibly small, and the gain is unity for the voltage follower.

D9.10 A voltage follower uses $R_f = 20\ \text{k}\Omega$, $R_L = 1\ \text{k}\Omega$, and an op amp with $R_i = 50\ \text{k}\Omega$, $A = 1000$, and $R_o = 2\ \text{k}\Omega$. Calculate (a) v_o/v_s, (b) R_{in}, (c) R_{out}.

Answers. 0.996; 16.74 MΩ; 2.80 Ω

9.8 Offset and drift

In this section we shall consider some of those characteristics of an operational amplifier that cause it to be less than ideal. We look first at discrepancies in the dc behavior, saving frequency response, transient response, and stability for the following chapter.

The op amp is almost never used in an open-loop situation because the open-loop gain A is so large that a very small input signal will produce an output that is saturated and no longer a function of the input. This is shown as a plot of output voltage v_o versus input voltage $v_i = v_1 - v_2$ in Fig. 9.16*a* for a certain op amp saturating at $v_o = \pm 13$ V whenever the input exceeds 0.13 mV in magnitude. The open-loop gain is given by the negative of the slope of the line passing through the origin; here, $A = 13/(0.13 \times 10^{-3}) = 100{,}000$. This op amp only amplifies when the input voltage is in the range $|v_i| < 0.13$ mV. The value at which v_o saturates depends on the supply voltage and is typically 1 or 2 V less than the supply voltage. Here, the supply is probably ± 15 V.

A real operational-amplifier open-loop transfer curve seldom passes through the origin; Fig. 9.16*b* is more typical. Note that zero output occurs when $v_i = 0.7$ mV. We define the *input offset voltage* V_{os} as that value of v_i required to cause v_o to equal zero. For the particular op amp illustrated in Fig. 9.16*b*, $V_{os} = 0.7$ mV. If $A = 100{,}000$, then amplification occurs only when v_i lies in the range $0.7 - 0.13 < v_i < 0.7 + 0.13$ mV, or $0.57 < v_i < 0.83$ mV. As the curve shows, when $v_i = 0$ or $v_1 = v_2$, the output is saturated. Although this sketch applies to a particular unit

Fig. 9.16 (*a*) The output voltage of an op amp saturates when the magnitude of the input is too large. (*b*) This op amp has an input offset voltage of +0.7 mV.

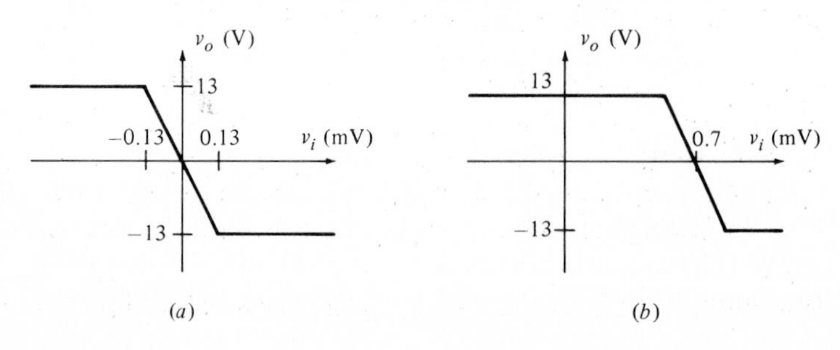

for which $V_{os} > 0$, negative values are equally likely. Data on the HA-2107 op amp indicate a typical value for $|V_{os}|$ of 0.7 mV, and a maximum value of 2 mV at 25°C; over the complete temperature range, the maximum value increases to 3 mV.

It is obvious that some provisions must be made to cancel the effects of the input offset voltage. That is, we need to have $v_o = 0$ when $v_i = 0$. Most op amps have two terminals marked "balance" or "offset null" to which a simple external circuit may be connected to set $v_o = 0$ when the inputs are tied together. For example, the second page of data on the Analog Devices AD547, given in Appendix A, includes the pin configuration shown in Fig. 9.17. The ends of a 10-kΩ potentiometer are connected to the "offset null" pins, Nos. 1 and 5, and the movable arm is connected to the negative supply voltage at pin No. 4.

There are no provisions for nulling the output on the HA-2107 op amp, and suitable circuitry must be provided at the input terminals. One method of doing this is illustrated in Fig. 9.18. Here, the input is balanced by applying an adjustable voltage to the inverting input through a resistance R_x. This is a high resistance in order to avoid affecting the ratio R_f/R_1. The two ends of the potentiometer are connected to the positive and negative supply voltages. A similar circuit may be used for a noninverting amplifier.

The offset voltage may change with temperature, with time, and with a changing supply voltage. Thus an adjustment setting $v_o = 0$ when $v_1 = v_2$ may not be valid as conditions change. The drift of the input offset voltage or the *input offset voltage drift* with temperature is often given in the data

Fig. 9.17 Many op amps have two terminals at which a simple external circuit may be connected to set $v_o = 0$ when $v_1 = v_2$.

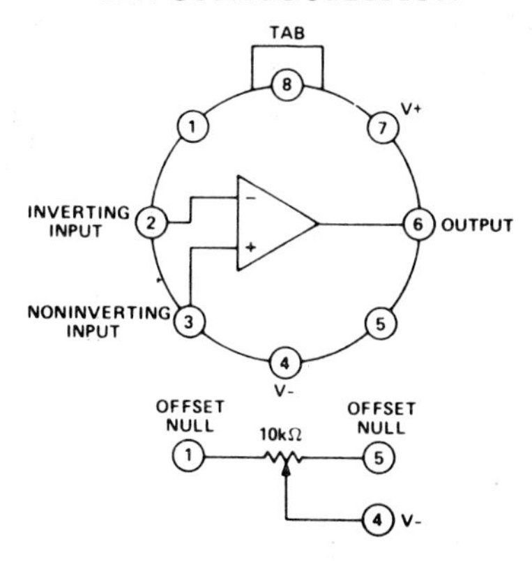

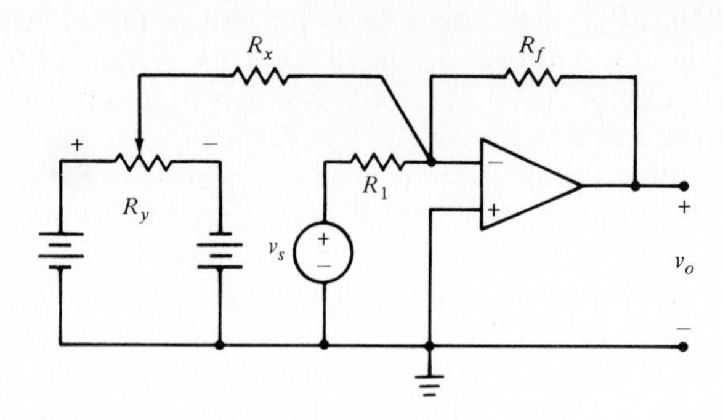

Fig. 9.18 The effect of the input offset voltage may be nullified by providing an adjustable voltage at one input.

sheets as a temperature coefficient. For the HA-2107 op amp, the typical value is 3 μV/°C; a maximum value of 15 μV/°C is listed in Fig. 9.11. These are reasonably low values. Op amps that do not list a temperature coefficient will usually have values that are larger. Special units may be obtained with temperature coefficients as low as 0.1 μV/°C.

Now let us look at the bias currents. We define the *input bias current* as the average of the dc bias currents required at the two inputs. This is given for the HA-2107 op amp in Fig. 9.11 as 30 nA typical and 75 nA maximum at 25°C. Over the complete temperature range, the maximum value increases to 100 nA. Data are also provided for the *input offset current*, the difference in the two bias currents necessary to produce zero output voltage. Figure 9.11 gives a typical value of 1.5 nA at 25°C. The *input offset current drift* is given as a temperature coefficient, here 0.02 nA/°C below 25°C and 0.01 nA/°C above 25°C.

Even if the offset current is zero, the bias currents at the two inputs may produce a nonzero input voltage if they flow through unequal resistances in the external circuit. For example, Fig. 9.19 shows a circuit in which all signal sources have been set equal to zero (superposition). At balance, $v_o = 0$, and the output is effectively grounded. Thus, I_{B1} flows through R_1 and R_f in parallel, while I_{B2} flows through R_2. If we have zero input offset current, or $I_{B1} = I_{B2}$, then equal voltages will be produced at the two inputs if

$$R_2 = R_1 \| R_f \tag{35}$$

This condition minimizes the unbalancing effect of the input bias currents. Smaller dc voltages are also produced by the bias currents if the resistor values are kept as small as is consistent with input impedance requirements.

In order to put some of these concepts on a more quantitative basis, let us consider Fig. 9.20, in which we account for an offset voltage V_{os} and

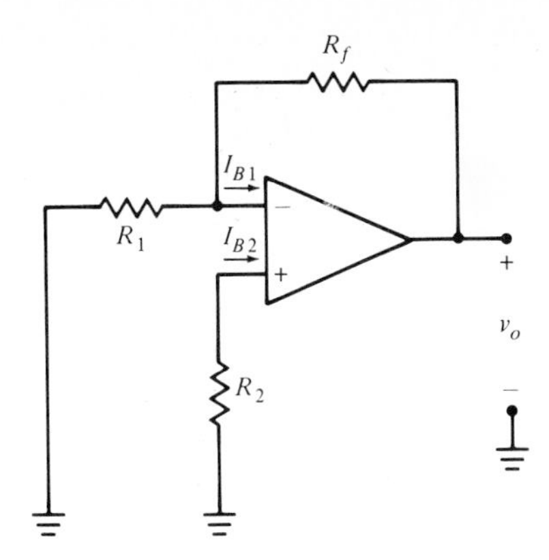

Fig. 9.19 The voltage imbalance caused by the bias currents may be minimized by setting $R_2 = R_1 \| R_f$.

bias currents I_{B1} and I_{B2} by sources external to an *ideal* op amp. No signal sources are shown, and the output is due entirely to V_{os}, I_{B1}, and I_{B2}. There are no restrictions on the value of R_2 at this time. We shall use superposition, considering the effect of the sources one at a time.

First, we set I_{B1} and I_{B2} equal to zero. Since the op amp is ideal, $v_i = v_1 - v_2 = 0$. Also, there is no input current and therefore $v_2 = 0$. Thus the circuit may be drawn as Fig. 9.21. Also, the same current, $v_{oV}/(R_1 + R_f)$,

Fig. 9.20 The effects of an input offset voltage v_{os} and input bias currents I_{B1} and I_{B2} are provided by three ideal sources.

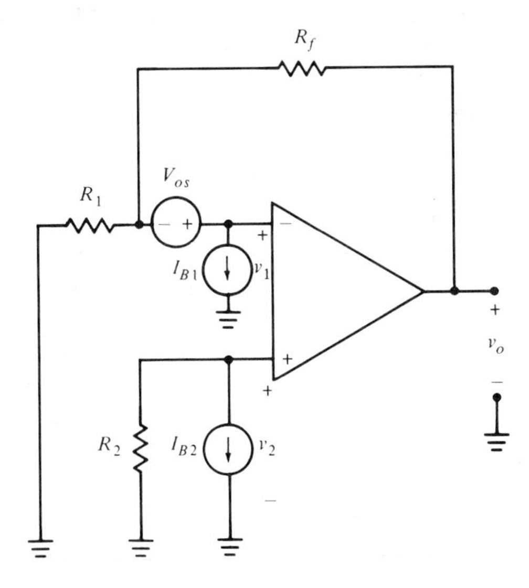

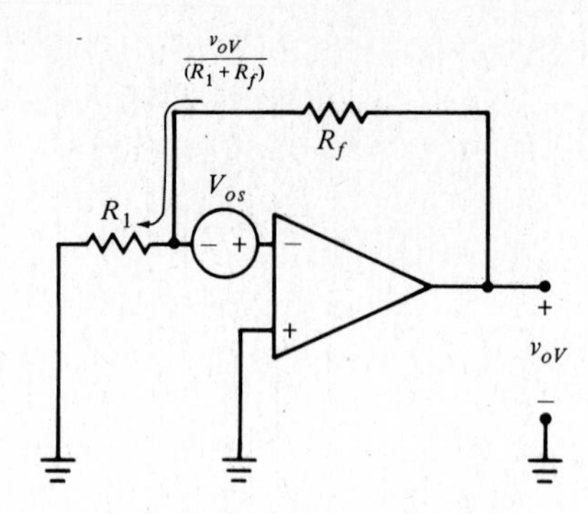

Fig. 9.21 With $I_{B1} = I_{B2} = 0$, the output voltage is $v_{oV} = -V_{os}(R_1 + R_f)/R_1$.

flows through both R_1 and R_f, and therefore we may sum voltages around the input mesh:

$$V_{os} + R_1 \frac{v_{oV}}{R_1 + R_f} + 0 = 0$$

and

$$v_{oV} = -\frac{R_1 + R_f}{R_1} V_{os} \tag{36}$$

Next, we set V_{os} and I_{B2} equal to zero. It is then helpful to replace the parallel combination of I_{B1} and R_1 with its Thévenin equivalent, as shown in Fig. 9.22. Once again, $v_2 = 0$ and therefore R_2 may be ignored. We now see an inverting amplifier, and

$$v_{o1} = -\frac{R_f}{R_1}(-R_1 I_{B1}) = R_f I_{B1} \tag{37}$$

Fig. 9.22 If $V_{os} = 0$ and $I_{B2} = 0$, then $v_{o1} = R_f I_{B1}$.

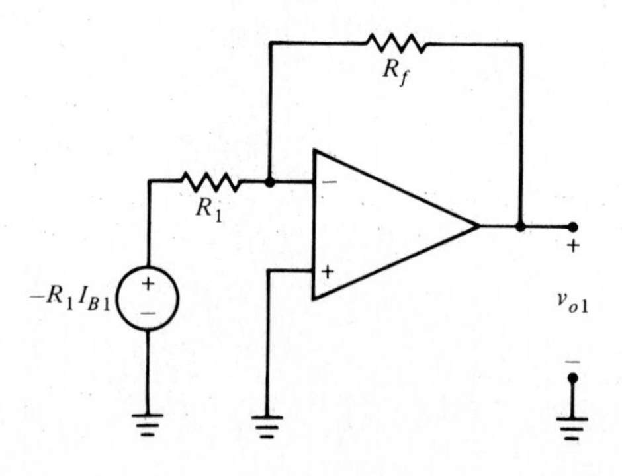

Finally, we set $V_{os} = 0$ and $I_{B1} = 0$, replace the parallel combination of R_2 and I_{B2} by its Thévenin equivalent, and obtain the circuit shown in Fig. 9.23. This is a noninverting amplifier, so we have

$$v_{o2} = \frac{R_1 + R_f}{R_1}(-R_2 I_{B2}) \tag{38}$$

Combining these three results gives

$$v_o = -\frac{R_1 + R_f}{R_1}(V_{os} + R_2 I_{B2}) + R_f I_{B1} \tag{39}$$

Let us use this result first to check a conclusion we reached earlier. We consider only the effects of I_{B1} and I_{B2}. If $V_{os} = 0$ and $I_{B1} = I_{B2} = I_B$, then

$$v_o = I_B\left[R_f - \frac{R_2}{R_1}(R_1 + R_f)\right]$$

To obtain zero output, we must have

$$R_f = \frac{R_2}{R_1}(R_1 + R_f)$$

or

$$R_2 = \frac{R_1 R_f}{R_1 + R_f} = R_1 \| R_f \tag{40}$$

as we found before.

Now let us see what Eq. (39) implies for an HA-2107 op amp with a maximum $|V_{os}|$ of 3 mV and a maximum input offset current $|I_{B1} - I_{B2}|$

Fig. 9.23 Letting $V_{os} = 0$ and $I_{B1} = 0$, we find $v_{o2} = [(R_1 + R_f)/R_1](-R_2 I_{B2})$.

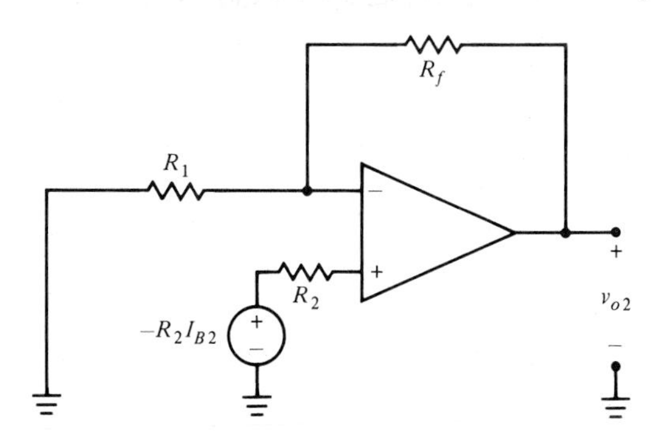

of 20 nA. We shall select circuit values of $R_1 = 100$ kΩ, $R_f = 1$ MΩ, and $R_2 = R_1 \| R_f = 90.9$ kΩ. From Eq. (39),

$$v_o = -\frac{100 + 1000}{100}(V_{os} + 90.9I_{B2}) + 1000I_{B1}$$

or

$$v_o = -11V_{os} + 1000(I_{B1} - I_{B2})$$

We thus see that V_{os} may contribute an output voltage with a magnitude of 33 mV, while the offset current can give a magnitude of 20 mV. Although these voltages might subtract, Murphy's law says they'll probably add. Smaller resistance values will reduce the 20-mV component.

D9.11 Use data sheets for the LM308 operational amplifier in Appendix A to determine (a) the typical value of V_{os} at 25°C; (b) the maximum value of V_{os} at 25°C; (c) the maximum value of V_{os} for 0°C to 70°C; (d) the maximum offset bias current at any temperature. (e) If $R_1 = 300$ kΩ, $R_f = 2700$ kΩ, and $R_2 = R_1 \| R_f$, what is the maximum possible value of v_o at any temperature that might be caused by offset voltage and current?

Answers. 2 mV; 7.5 mV; 10 mV; 1.5 nA; 104 mV

Problems

1. The inverting operational amplifier of Fig. 9.3*a* has $R_1 = 5$ kΩ, $R_f = 40$ kΩ, $R_i = \infty$, $R_o = 0$, and $A = 10{,}000$. If $V_s = 1$ V rms, determine the signal power that is (a) dissipated in R_1, (b) dissipated in R_f, (c) supplied by V_s, (d) supplied by the op amp.
2. Element values in the inverting amplifier of Fig. 9.3*a* are $R_1 = 10$ kΩ and $R_f = 100$ kΩ. The op amp is ideal. (a) Calculate v_o/v_s. (b) A 40-kΩ resistor is now connected from the inverting input to ground. Find v_o/v_s.
3. (a) If the op amp in Fig. 9.24 is ideal, calculate i_o/i_s. (b) If $R_i = \infty$, $R_o = 0$, $A = 1000$, $R_1 = 10$ kΩ, and $R_f = 100$ kΩ, find i_o/i_s.
4. Let $R_i = \infty$ and $R_o = 0$ for the op amp of Fig. 9.25. Find v_o/v_s if $A = \infty$.
5. Element values in the noninverting amplifier circuit of Fig. 9.5 are $R_1 = 4.7$ kΩ and $R_f = 47$ kΩ. Let $v_s = 1$ V and $A = 1000$, and determine (a) v_o, (b) v_i, (c) v_2, (d) v_1, (e) i_1.
6. Let the voltage follower of Fig. 9.6 have $R_i = \infty$, $R_o = 0$, and $A = 10^4$. Determine v_o/v_s if a 10-kΩ resistor is placed (a) in series with v_s, (b) in parallel with v_s, (c) from the inverting input to ground, (d)

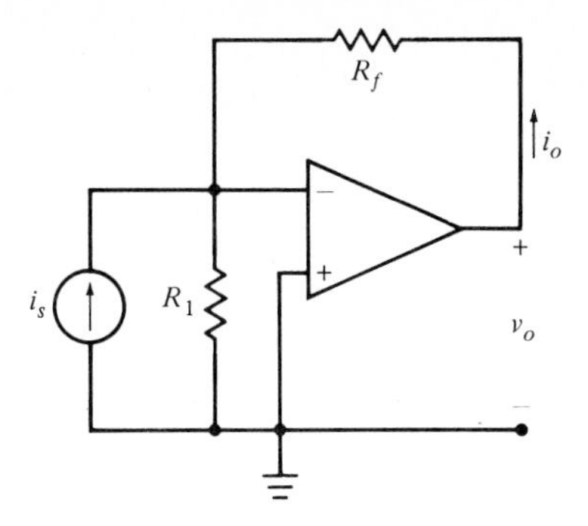

Fig. 9.24 See Problem 3.

from output to ground, (e) between output and inverting input (in place of the short circuit).

7. Find v_o/v_s, R_{in}, and R_{out} for the network shown in Fig. 9.26 if both op amps are considered to be ideal.
8. Find v_o/v_s for the circuit shown in Fig. 9.27 if the op amp is ideal and Switch S is (a) closed, (b) open.
9. If the op amp in Fig. 9.28 is ideal, express v_o as a function of the four input voltages.
10. (a) Find v_o/v_s in terms of R_1, R_A, R_B, and R_C for the circuit shown in Fig. 9.29. Assume the op amp is ideal. (b) If R_1 is set equal to 100 kΩ in a standard inverting amplifier to obtain a high value for R_{in}, what

Fig. 9.25 See Problem 4.

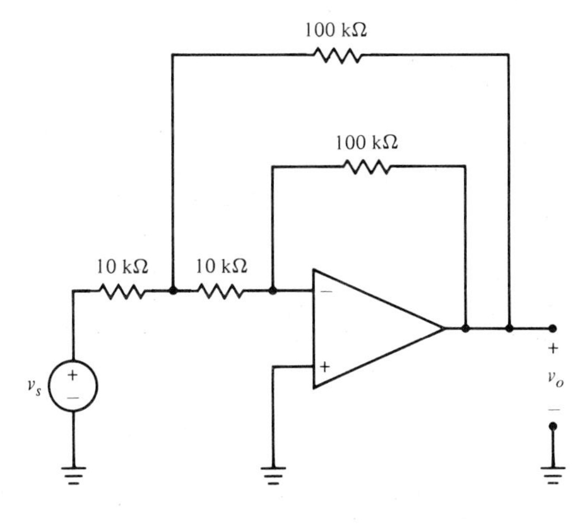

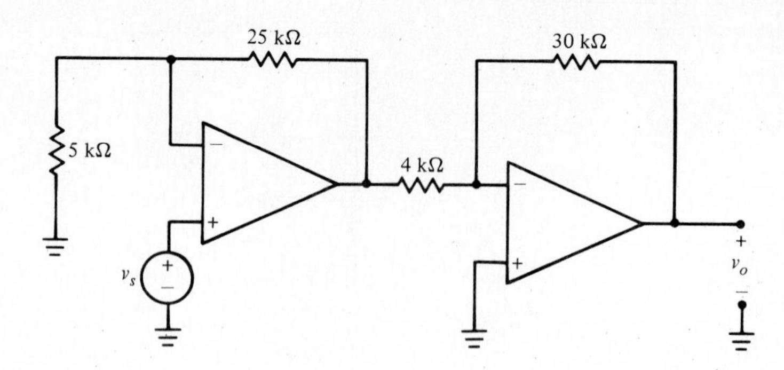

Fig. 9.26 See Problem 7.

value would a single feedback resistor R_f have to be in order that $v_o/v_s = -50$? (c) Now select values for R_A, R_B, and R_C, none greater than 100 kΩ, to achieve $v_o/v_s = -50$ for the network shown.

11. The differential amplifier shown in Fig. 9.9 has $R_1 = 3.2$ kΩ, $R_{f1} = 80$ kΩ, $R_2 = 1$ kΩ, and $R_{f2} = 25$ kΩ. Assume the op amp is ideal, let $v_{s1} = 0.2$ V, and (a) sketch v_o vs. v_{s2}, (b) find $R_{\text{in}1}$, (c) find $R_{\text{in}2}$.
12. A differential amplifier includes an op amp for which we may assume $R_i = \infty$ and $R_o = 0$. (a) What relationship must exist between R_1, R_{f1}, R_2, and R_{f2} to give $v_o = 0$ with $v_{s1} = v_{s2}$ if the open-loop gain of the op amp is A? (b) Find v_o as a function of v_{s1} and v_{s2} if $R_1 = R_2 = 10$ kΩ, $R_{f1} = 100$ kΩ, $A = 1000$, and $v_o = 0$ when $v_{s1} = v_{s2}$.
13. A differential amplifier similar to that shown in Fig. 9.9 gives $v_o = 0$ when $v_{s1} = v_{s2}$. If $R_1 = 10$ kΩ, $R_{f1} = 50$ kΩ, $R_2 = 4$ kΩ, $v_{s1} = 2$ V, v_{s2}

Fig. 9.27 See Problem 8.

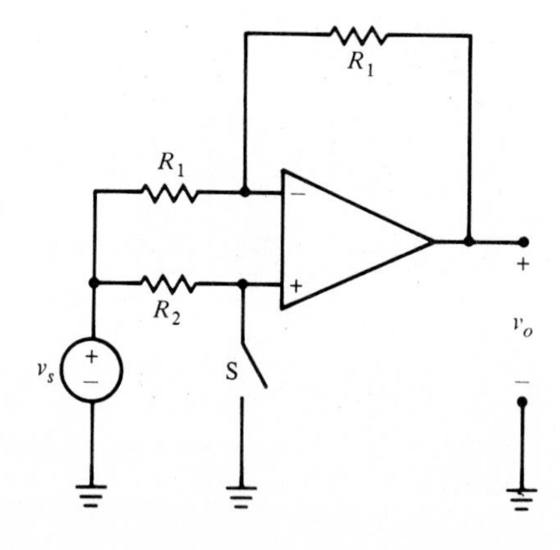

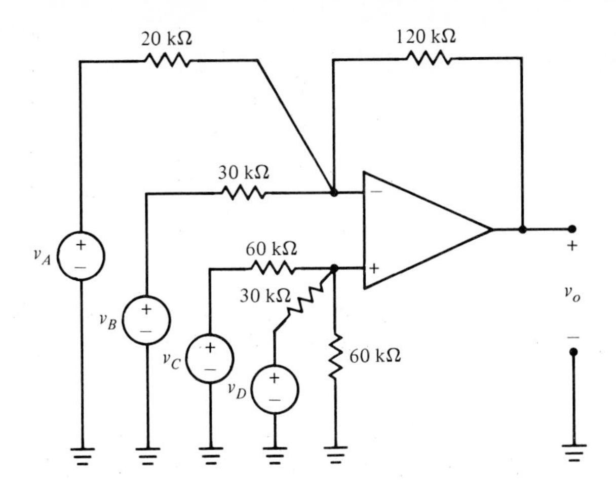

Fig. 9.28 See Problem 9.

= 3 V, and the op amp is ideal, find the current magnitude in each of the four resistors.

14. Refer to the performance curves in the data sheets for the HA-2107 op amp in Appendix A for the supply voltage of $V_S = \pm 15$ V at 25°C and use the current-limiting graph to prepare a curve of output current swing as a function of load resistance R_L.
15. Use data on the LM308 op amp in Appendix A to determine (a) A_{min} at 25°C, (b) A_{min} for $0° \leq T \leq 70°$C, (c) A_{typical} at 25°C, (d) the factor by which A increases at 25°C as the supply voltage goes from ± 5 V to ± 18 V.

Fig. 9.29 See Problem 10.

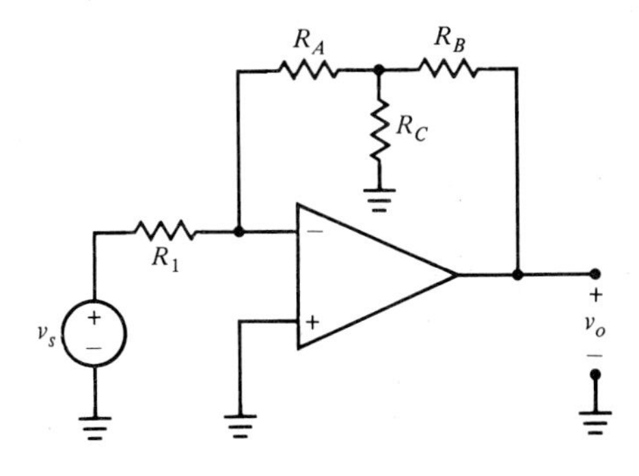

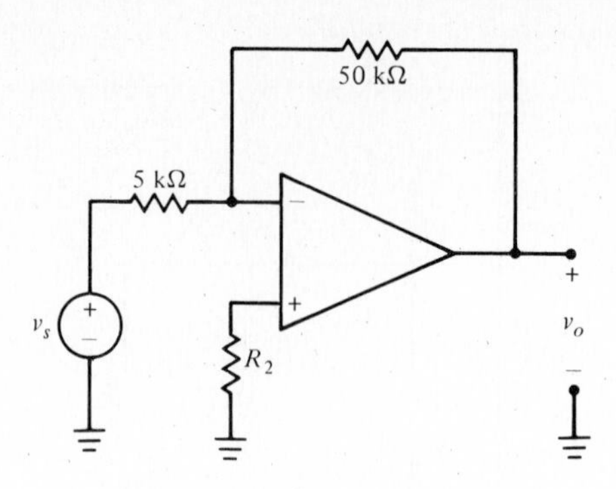

Fig. 9.30 See Problem 18.

16. The inverting amplifier of Fig. 9.12*a* has $R_1 = 10$ kΩ, $R_f = 200$ kΩ, $R_L = 5$ kΩ, $R_i = 100$ kΩ, and $A = 10^4$. Calculate v_o/v_s, R_{in}, and R_{out} if R_o equals (a) 100 Ω, (b) 2000 Ω.
17. An op amp with $R_i = 200$ kΩ and $R_o = 1$ kΩ is used in an inverting amplifier with $R_1 = 5$ kΩ, $R_f = 100$ kΩ, and $R_L = 2.5$ kΩ. What value of A is required to give (a) $v_o/v_s = -19.9$; (b) $R_{\text{in}} = 5.1$ kΩ; (c) $R_{\text{out}} = 1$ Ω?
18. The op amp of Fig. 9.30 is an ancient vacuum-tube contraption having $R_i = 200$ kΩ, $A = 100$, and $R_o = 500$ Ω. Find v_o/v_s if R_2 equals (a) 0, (b) 20 kΩ.
19. A voltage follower (Fig. 9.14*a*) operates with $R_f = 20$ kΩ and $R_L = 4$ kΩ. If $R_i = 50$ kΩ and $R_o = 1$ kΩ, what range of values for the open-loop gain will ensure (a) $0.999 \leq v_o/v_s \leq 1$? (b) $R_{\text{in}} \geq 100$ MΩ?
20. Find v_o for the follower shown in Fig. 9.31 if (a) the op amp is ideal. Repeat if $R_o = 1$ kΩ, $A = 1000$, and R_i equals (b) ∞, (c) 100 kΩ.

Fig. 9.31 See Problem 20.

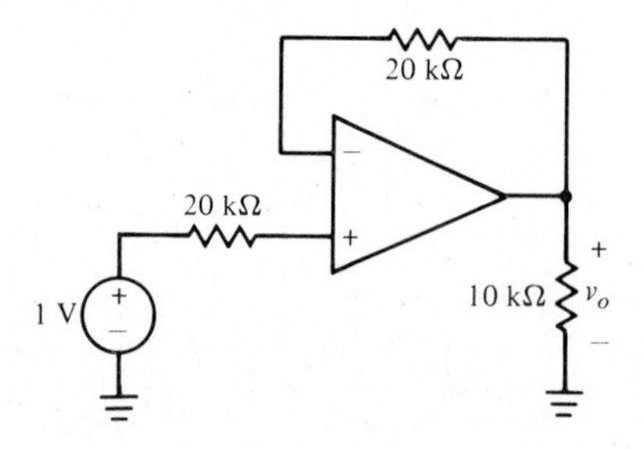

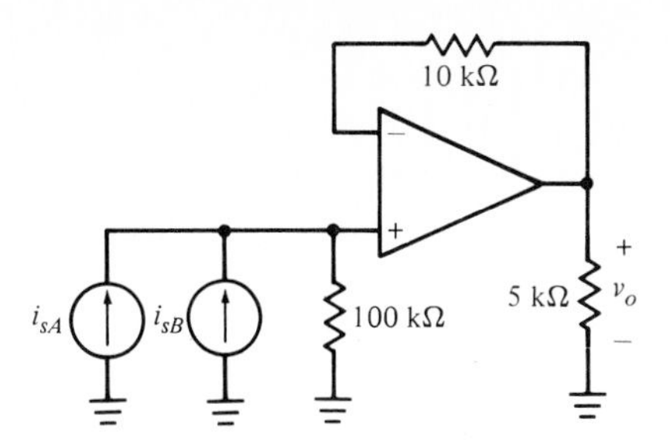

Fig. 9.32 See Problem 21.

21. If $R_i = \infty$, $R_o = 500\ \Omega$, and $A = 10^4$ for the op amp in Fig. 9.32, find v_o in terms of i_{sA} and i_{sB}.
22. The circuit of Fig. 9.19 contains $R_1 = 20\ \text{k}\Omega$ and $R_f = 180\ \text{k}\Omega$. (a) What is the optimum value for R_2? If the op amp is an LM308, what is the maximum output voltage magnitude that might occur because of (b) only voltage offset? (c) only bias current offset? (d) both?
23. The AD547J op amp, for which data may be found in Appendix A, is advertised as an "ultra low drift" unit. If this op amp is used in a circuit for which $R_1 = 30\ \text{k}\Omega$ and $R_f = 120\ \text{k}\Omega$, (a) what should be the value of R_2? What is the maximum output voltage magnitude that might occur because of (b) voltage offset alone? (c) bias current offset alone? (d) both? (e) What change might occur in v_o if the temperature increased 50°C?

10

Applications of operational amplifiers

In this chapter we shall continue our study of operational amplifiers by looking at several useful circuits built around the op amp. The first several are said to be dc applications, such as voltage reference supplies, or bridge amplifiers. We then conclude our investigation by looking at op-amp behavior with frequency and for signals that may change rapidly with time.

10.1 Reference voltage sources

Many measurement and instrumentation problems require the establishment of a fixed voltage reference with an accuracy of a fraction of a percent, particularly in making precision measurements. The precision voltage source is often a *standard cell*, designed to provide a reference voltage with an accuracy as high as one part in a million under strict laboratory conditions where the temperature is maintained constant and the current through the cell is essentially zero. Even a lower quality standard cell can take hours to recover when an ordinary voltmeter is used to check its terminal voltage. The current drawn by the meter causes the cell voltage to change, and it is necessary to isolate the cell from the voltage measurement.

One circuit that accomplishes this is shown in Fig. 10.1. The voltage follower has an input resistance ranging from 10^9 or 10^{10} Ω for bipolar input circuits to 10^{15} Ω for MOSFETs. This circuit also has a gain very close to unity and an output resistance that can be less than a milliohm. The circuit can easily supply currents of 1 mA or more to a load. As an example, Eqs. (29), (31), and (34) of Chapter 9 can be used to show that a circuit using an op amp for which $R_i = 4$ MΩ, $A = 10^5$, and $R_o = 100$ Ω will enable a 1-V standard cell to supply a voltage only 12 μV less than 1 V and a current only 12 nA less than 1 mA to a 1-kΩ load. Since $R_{in} = 3.64 \times 10^{11}$ Ω, the current through the cell is limited to 1 V divided by R_{in}, or 2.75 pA, plus the necessary bias current, usually 50 to 100 nA for bipolar transistors. Thus the cell current is essentially equal to the bias current. Note also that the output voltage is independent of R_L if $R_{out} = 0$. For this example, $R_{out} = 1$ mΩ.

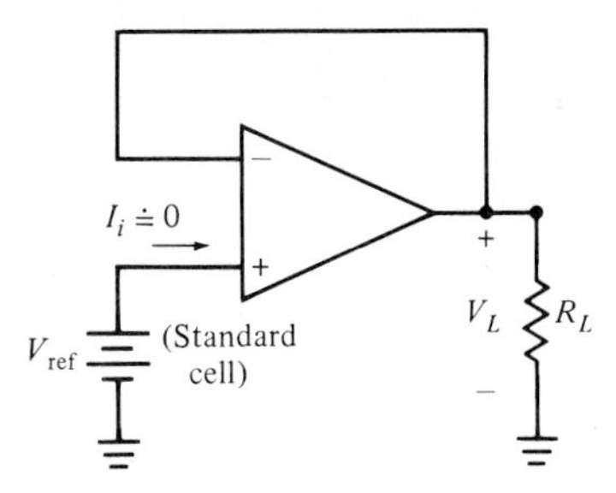

Fig. 10.1 A voltage follower may be used to isolate a standard cell from the load R_L and yet provide $V_L \doteq V_{ref}$ within a small fraction of a percent.

Figure 10.2 shows another circuit used as a voltage reference. This non-inverting operational amplifier provides an output voltage,

$$V_o = \left(1 + \frac{R_f}{R_1}\right) V_{ref} \tag{1}$$

We see that R_f or R_1 could be a variable resistance or a decade resistance box to make possible a range of output voltages. Again, the cell current is essentially the input bias current, say 100 nA for bipolar inputs and negligible for FET input circuits.

A more common source of reference voltage is the Zener diode described in Chapter 1. This is a reliable and inexpensive device that finds wide use. In the simple circuit shown in Fig. 10.3*a*, the voltage supply V_S and resistance R_S are selected to place the Zener diode at an operating point safely in the breakdown section of its characteristic. The current I_Z is given closely by

$$I_Z = \frac{V_S - V_{BR}}{R_S}$$

Fig. 10.2 A voltage-reference circuit using a noninverting amplifier provides an output voltage $V_o = V_{ref}(1 + R_f/R_1)$.

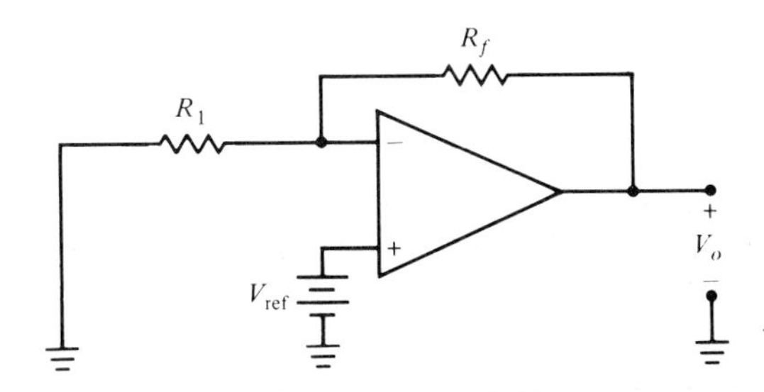

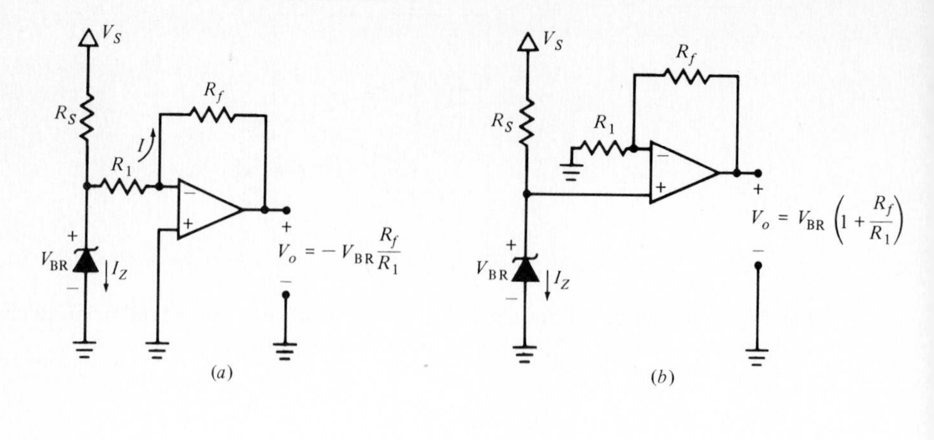

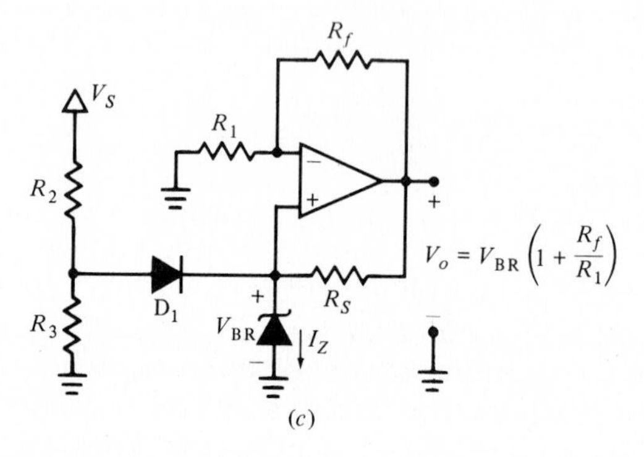

Fig. 10.3 Three voltage-reference circuits using Zener diodes. (*a*) The sign of V_o is reversed with respect to V_{BR}. (*b*) No sign reversal occurs. (*c*) Once the Zener conducts, its supply voltage is V_o rather than V_S.

since little current is drawn through R_1 and R_f. The current I is

$$I = \frac{V_{BR} - 0}{R_1} = \frac{0 - V_o}{R_f}$$

and therefore

$$V_o = -\frac{R_f}{R_1} V_{BR} \tag{2}$$

Note the reversal of polarity between V_{BR} and V_o. This circuit may supply V_o to a wide range of loads at currents ranging from zero to the maximum permissible output current for the op amp.

Figure 10.3*b* provides an output voltage with the same sign as V_{BR},

$$V_o = \left(1 + \frac{R_f}{R_1}\right) V_{BR} \tag{3}$$

In either of these circuits, changes in V_S cause a shift in the operating point of the diode, and therefore a slight change in V_{BR}. This is minimized in the circuit of Fig. 10.3*c*. Here, V_S, R_2, R_3, and the diode D1 put the Zener diode into breakdown initially, thus establishing a value of V_o given by Eq. (3). The resistor R_S is now used to drive the Zener diode from the output V_o. If V_o is slightly greater than the start-up value, $V_S R_3/(R_2 + R_3)$, then D1 becomes reverse-biased, the Zener diode is driven from a more constant source, and its operating point is thus more fixed.

D10.1 In Fig. 10.1 let $R_i = 1$ MΩ, $R_o = 1$ kΩ, $V_{ref} = 1.5$ V, $A = 1000$, and $R_L = 750$ Ω. Calculate (a) R_{in}, (b) R_{out}, (c) V_L.

Answers. 430 MΩ; 0.999 Ω; 1.4965 V

D10.2 Element values for the circuit of Fig. 10.3*b* are $R_S = 2$ kΩ, $R_1 = 10$ kΩ, and $R_f = 40$ kΩ. Assume an ideal op amp. If the Zener diode has a dynamic resistance $R_Z = 100$ Ω and it is modeled by $V_{BR} = 6 + 100I_Z$, find V_o if V_S equals (a) 10 V, (b) 12 V, (c) 10 V, but $V_{BR} = 6 + 10I_Z$, (d) 12 V with $V_{BR} = 6 + 10I_Z$.

Answers. 30.95; 31.43; 30.10; 30.15 V

10.2 Voltage detectors and comparators

There are many applications for a device that can sense whether an input voltage signal is larger or smaller than a specific reference voltage and then provide an output voltage that essentially says *yes* or *no*. The input voltage is an *analog* signal, in that it is the analog of some continuously varying function, such as the signal provided by a thermocouple or a microphone. The output voltage has only two states, and is a *digital* signal. We therefore are considering an analog-to-digital device.

A simple circuit that accomplishes this operation is the *comparator*, shown in Fig. 10.4. There is no feedback around the op amp and the large open-loop gain A causes the output to be at $\pm V_{sat}$ whenever the input voltage magnitude is greater than a fraction of a millivolt. If we neglect the small input voltage range for which the output is not saturated (shown in Fig. 9.16) and connect the signal to the inverting input and the reference voltage to the noninverting input, then $v_o = -V_{sat}$ when $v_s > V_{ref}$, and $v_o = V_{sat}$ when $v_s < V_{ref}$. The signal and reference voltage sources could be interchanged so that $v_o = +V_{sat}$ for $v_s > V_{ref}$ if desired. The reference voltage is often obtained with a Zener diode.

Note that several of our rules of thumb for an op amp do not hold for a comparator. In particular, the input voltage v_i is not necessarily small,

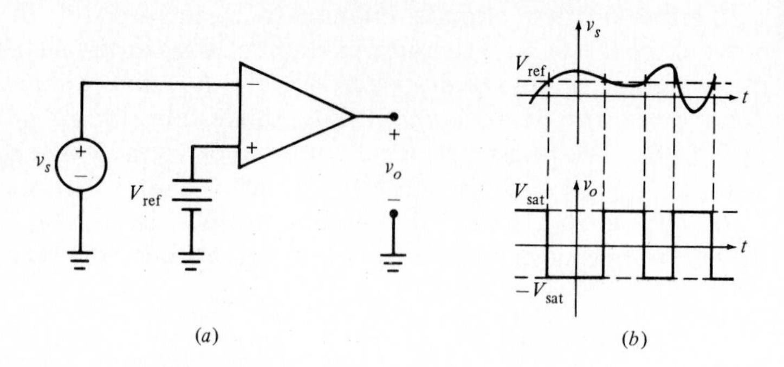

Fig. 10.4 The comparator gives an output $v_o = -V_{sat}$ when $v_s > V_{ref}$; $v_o = +V_{sat}$ when $v_s < V_{ref}$.

and the input current may switch suddenly from near zero to an input bias current of 50 to 100 nA, and vice versa.

A different type of comparator is shown in Fig. 10.5. If we apply the superposition theorem to this circuit and neglect any input current, then the input voltage is

$$v_i = v_s \frac{R_2}{R_1 + R_2} + V_{ref} \frac{R_1}{R_1 + R_2} \tag{4}$$

When v_i is positive, $v_o = -V_{sat}$, and when v_i is negative, $v_o = +V_{sat}$. The threshold voltage $V_{threshold}$ is that value of the signal voltage v_s that causes v_i to equal 0. From Eq. (4), it is

$$V_{threshold} = -V_{ref} \frac{R_1}{R_2} \tag{5}$$

This circuit allows us to scale the reference voltage by changing R_1 or R_2. Note also that the input impedance is now determined by these two resistors.

Fig. 10.5 A scaled voltage comparator for which $v_o = -V_{sat}$ when $v_s > V_{threshold}$, and $v_o = +V_{sat}$ when $v_s < V_{threshold}$; $V_{threshold} = -V_{ref}(R_1/R_2)$.

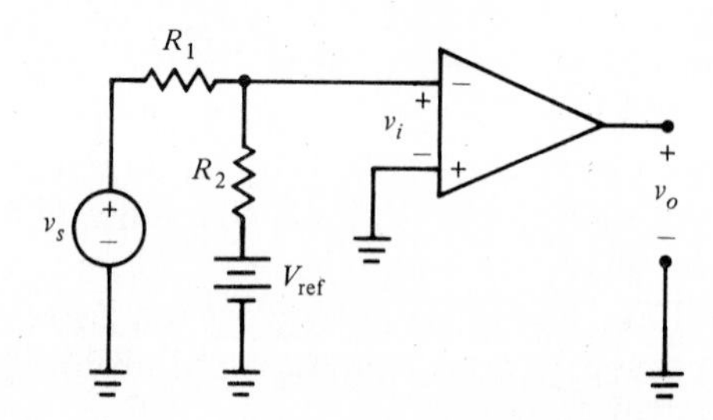

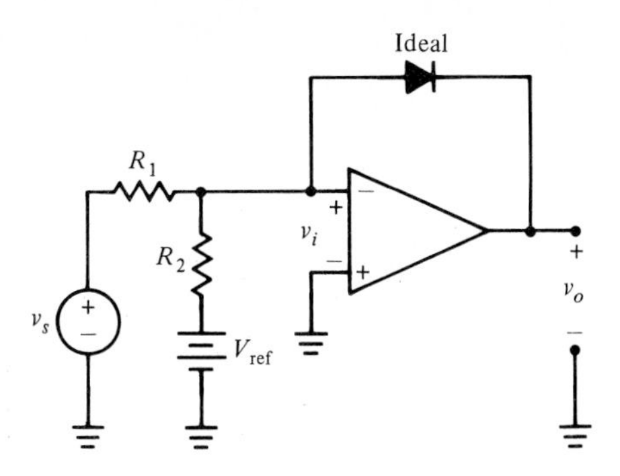

Fig. 10.6 This scaled comparator bottom clamps at $v_o = 0$. With $v_s < V_{\text{threshold}}$, $v_o = +V_{\text{sat}}$; for $v_s > V_{\text{threshold}}$, $v_o = 0$; $V_{\text{threshold}} = -V_{\text{ref}}(R_1/R_2)$.

As an example, if we let $R_1 = 10\text{ k}\Omega$, $R_2 = 100\text{ k}\Omega$, $V_{\text{ref}} = -20$ V, and $V_{\text{sat}} = \pm 10$ V, then $v_o = -10$ V when $v_s > 2$ V, and $v_o = +10$ V for $v_s < 2$ V.

Figure 10.6 is similar to Fig. 10.5 except that an ideal diode is connected between the output and the inverting input. When $v_s < V_{\text{threshold}}$, $v_o = +V_{\text{sat}}$, the diode is reverse-biased, and the circuit behaves exactly like that of Fig. 10.5. When $v_s = V_{\text{threshold}}$, the output heads toward $-V_{\text{sat}}$; in doing so, however, the diode becomes forward-biased, since $v_i = 0$. With the diode forward-biased, $v_o = v_i = 0$, and we see that the output cannot go negative. We say that it is *clamped* to zero with the ideal diode. Not only does the forward-biased diode serve to connect the output to the input ($v_o = v_i = 0$), we also have the equivalent of a typical negative feedback network with $R_f = 0$. Hence there is no gain.

D10.3 Let $R_1 = 0.1\text{ M}\Omega$, $R_2 = 1\text{ M}\Omega$, $V_{\text{ref}} = -15$ V, and $V_{\text{sat}} = \pm 12$ V for the circuit of Fig. 10.5. Find v_o if v_s equals (a) 1 V, (b) 2 V, (c) -1 V.

Answers. 12; -12; 12 V

D10.4 In Fig. 10.6, let $R_1 = 0.1\text{ M}\Omega$, $R_2 = 1\text{ M}\Omega$, $V_{\text{ref}} = 15$ V, and $V_{\text{sat}} = \pm 12$ V. The diode is reversed. Find v_o if v_s equals (a) -2 V, (b) 0 V, (c) 2 V.

Answers. 0; -12; -12 V

10.3 Differential amplifiers

We took a preliminary look at the differential amplifier in Section 9.4, and we now consider some of its imperfections and their consequences.

The basic circuit is repeated as Fig. 10.7. We concluded before that

$$v_o = \frac{R_{f1}}{R_1}(v_{s2} - v_{s1}) = \frac{R_{f2}}{R_2}(v_{s2} - v_{s1}) \tag{6}$$

when the resistor ratios were chosen such that $R_{f1}/R_1 = R_{f2}/R_2$. Under these conditions there should be no output when $v_{s1} = v_{s2}$, regardless of the magnitude of these two voltages (assuming they fall within the limits imposed by the specifications). In practice, however, we have seen that there generally is an output voltage when $v_1 = v_2 = 0$, and that the input offset voltage is the input voltage magnitude required to obtain $v_o = 0$.

Let us now define several new terms that deal with this same problem for the differential amplifier. We call the average of v_1 and v_2 the *common-mode signal* v_{CM}:

$$v_{CM} = \tfrac{1}{2}(v_1 + v_2) \tag{7}$$

For the special case where $v_1 = v_2$,

$$v_{CM} = v_1 = v_2$$

There should be no output for such an input, but all practical amplifiers have some small lack of symmetry in their transistors and resistors, not to mention the presence of nonlinearities, and we find that v_{CM} does generate an output $v_{o(CM)}$. The ratio of the common-mode output to the common-mode input is called the *common-mode gain*:

$$A_{CM} = \left| \frac{v_{o(CM)}}{v_{CM}} \right| \tag{8}$$

For a good op amp, it is a very small number, although it may be greater than unity. Note that this result implies that when $v_1 = v_2 = 1$ mV, there will be an output, say $v_o = 2$ mV if $A_{CM} = 2$. Moreover, if $v_1 = v_2$ is increased to 3 mV, the output increases to 6 mV.

Fig. 10.7 The differential amplifier. Both v_1 and v_2 are referenced to ground.

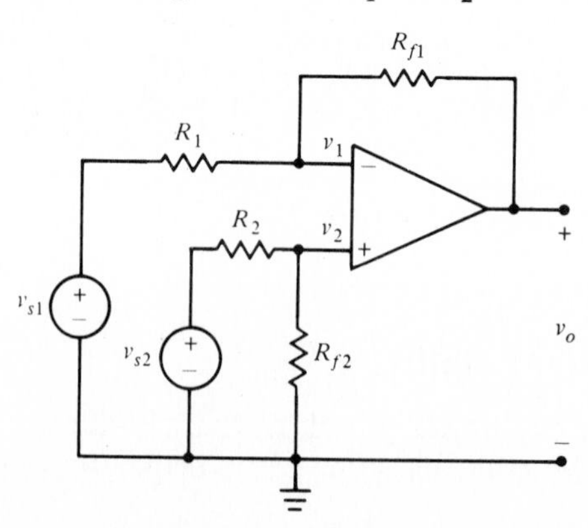

It is apparent in all our previous discussion of op amps that the input $v_i = v_1 - v_2$, the output v_o, and the open-loop gain A are really *differential-mode* concepts. It is customary to define the ratio of the differential-mode gain to the common-mode gain as the *common-mode rejection ratio* CMRR:

$$\text{CMRR} = \left|\frac{A}{A_{\text{CM}}}\right| \tag{9}$$

This is often given in dB:

$$\text{CMRR}_{\text{dB}} = 20 \log |A/A_{\text{CM}}| \tag{10}$$

For example, page 2 of the HA-2107 data sheets (Fig. 9.11) gives typical and minimum values for the CMRR_{dB} at any temperature as 96 and 80 dB respectively. The typical value of 96 dB should be associated with the typical value for A of 160,000. Therefore

$$A_{\text{CM}} = \frac{160{,}000}{10^{96/20}} = \frac{160{,}000}{63{,}096} = 2.54$$

To see the effect of the common-mode signal, let us express the output in terms of the differential input $v_1 - v_2$ and the common-mode input $(v_1 + v_2)/2$. Each input is subject to a different gain, so we have

$$v_o = A(v_2 - v_1) \pm \frac{A_{\text{CM}}(v_1 + v_2)}{2} \tag{11}$$

Since $|A/A_{\text{CM}}| = \text{CMRR}$, we may write

$$v_o = A(v_2 - v_1) \pm \frac{A(v_1 + v_2)}{2(\text{CMRR})}$$

or

$$v_o = A(v_2 - v_1)\left[1 \pm \frac{1}{2(\text{CMRR})}\,\frac{v_2 + v_1}{v_2 - v_1}\right] \tag{12}$$

As $\text{CMRR} \to \infty$, the output approaches the desired value, $A(v_2 - v_1)$. As an example of the errors involved, let us consider several simple examples. We first let $v_1 = -1$ mV, $v_2 = 1$ mV, $A = 1000$, and $\text{CMRR} = 100$. Thus

$$v_o = 1000(1 + 1)(10^{-3})\left[1 \pm \frac{1}{2 \times 100}\,\frac{0}{1 + 1}\right] = 2 \text{ V}$$

This is the correct output, and there is no error. We now obtain the same difference input with a nonzero common-mode signal: $v_1 = 9$ mV and $v_2 = 11$ mV. Thus, if we use the + sign,

$$v_o = 1000(11 - 9)(10^{-3})\left[1 + \frac{1}{2 \times 100}\,\frac{11 + 9}{11 - 9}\right] = 2.1 \text{ V}$$

and we see a 5% error. If the CMRR increases to 1000, the error decreases to 0.5%. Of course, percentage errors can be misleading. If $v_1 = v_2 = 1$ mV, with $A = 1000$ and CMRR $= 100$, then $A_{CM} = A/\text{CMRR} = 10$. Equation (11) then gives

$$v_o = 1000(1 - 1)(10^{-3}) \pm 10(1 + 1)(10^{-3})/2 = \pm 10 \text{ mV}$$

whereas the correct output is zero. This would be an infinite percentage error if we cared to calculate it.

D10.5 A differential amplifier has a differential gain of 86 dB and a CMRR_{dB} of 95 dB. Find the magnitude of the differential-mode output $v_{o(DM)}$ and the common-mode output $v_{o(CM)}$ if (a) $v_1 = 1.6\ \mu$V, $v_2 = 2\ \mu$V; (b) $v_1 = -1.6\ \mu$V, $v_2 = 2\ \mu$V; (c) $v_1 = 10.6\ \mu$V, $v_2 = 11\ \mu$V.

Answers. 7981, 0.639; 71,800, 0.071; 7981, 3.83 μV

10.4 Bridge amplifiers

The most common application of the dc-coupled differential amplifier is in amplifying the output of a transducer bridge, in which one of the four resistive elements in the bridge is a transducer such as a strain gauge in which R changes with elongation, or a thermistor in which R changes with temperature. In either of these applications there is a large common-mode signal, and the small difference signal must be detected in its presence. A typical strain gauge may provide a maximum differential output of 25 mV for a common-mode voltage of 5 V.

Figure 10.8 shows a bridge circuit containing a differential amplifier and three equal resistors of R ohms, as well as one resistor $R + \Delta R$ where ΔR is proportional to the strain or elongation in the case of a strain gauge, or to temperature in the case of a thermistor in a temperature bridge. The

Fig. 10.8 A differential amplifier is used as a bridge amplifier.

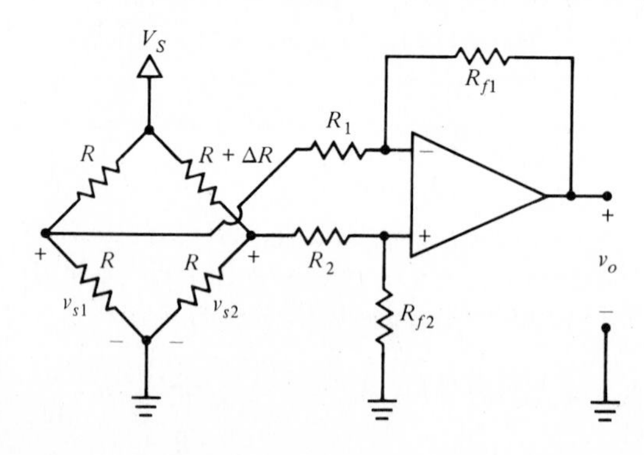

value of R is usually several hundred ohms, whereas the differential amplifier inputs have much higher input impedances, R_1 for the inverting and $R_2 + R_{f2}$ for the noninverting input. If we assume that the currents in R_1 and R_2 are negligible in comparison with the bridge currents, then voltage division leads to

$$v_{s1} = V_S \frac{R}{R + R} = \frac{V_S}{2}$$

and

$$v_{s2} = V_S \frac{R}{R + (R + \Delta R)}$$

Therefore

$$v_{s2} - v_{s1} = V_S \left[\frac{R}{2R + \Delta R} - \frac{1}{2} \right] = \frac{1}{2} V_S \left[\frac{1}{1 + (\Delta R/2R)} - 1 \right]$$

We now let

$$\delta = \frac{\Delta R}{R} \tag{13}$$

and therefore

$$v_{s2} - v_{s1} = \frac{1}{2} V_S \left[\frac{1}{1 + (\delta/2)} - \frac{1 + (\delta/2)}{1 + (\delta/2)} \right]$$

or

$$v_{s2} - v_{s1} = -\frac{1}{4} V_S \frac{\delta}{1 + (\delta/2)}$$

We may take the differential amplifier gain as K:

$$K = \frac{v_o}{v_{s2} - v_{s1}} = \frac{R_{f1}}{R_1} = \frac{R_{f2}}{R_2} \tag{14}$$

and then

$$v_o = -\frac{1}{4} K V_S \frac{\delta}{1 + (\delta/2)} \tag{15}$$

If $\delta \ll 1$, we may use the simpler but approximate result

$$v_o \doteq -\tfrac{1}{4} K V_S \delta \tag{16}$$

Under these conditions the op-amp output is a linear function of the change of resistance and thus of the change in elongation or temperature. If larger fractional changes are involved, Eq. (15) must be used.

Another circuit that is sometimes useful is the half-bridge amplifier shown in Fig. 10.9. Only one fixed resistor R is used with the transducer $R + \Delta R$, and the voltage supply uses both $\pm V_S$. The half-bridge output is connected directly to the inverting terminal of the op amp, while the non-inverting input is grounded. Since $v_i \doteq 0$, the half-bridge output is effectively grounded and it is the current i that reflects the change in resistance. The current i is obtained by adding the currents flowing through R and $R + \Delta R$:

$$i = \frac{V_S}{R} - \frac{V_S}{R + \Delta R} = \frac{V_S}{R}\left(1 - \frac{1}{1 + \delta}\right)$$

Thus

$$i = \frac{V_S}{R}\left(\frac{\delta}{1 + \delta}\right) \tag{17}$$

where δ is again $\Delta R/R$. If $\delta \ll 1$, then Eq. (17) simplifies to

$$i \doteq \frac{V_S}{R}\,\delta$$

This current also flows through R_f, and therefore

$$v_o = -iR_f$$

since $v_i \doteq 0$. We can now express v_o in terms of the fractional change in resistance:

$$v_o = -V_S \frac{R_f}{R}\frac{\delta}{1 + \delta} \doteq -V_S \frac{R_f}{R}\,\delta \tag{18}$$

Again we have an output that is approximately a linear function of δ.

Fig. 10.9 A half-bridge current amplifier has an output

$$v_o \doteq -\frac{R_f}{R}\frac{\Delta R}{R}V_S$$

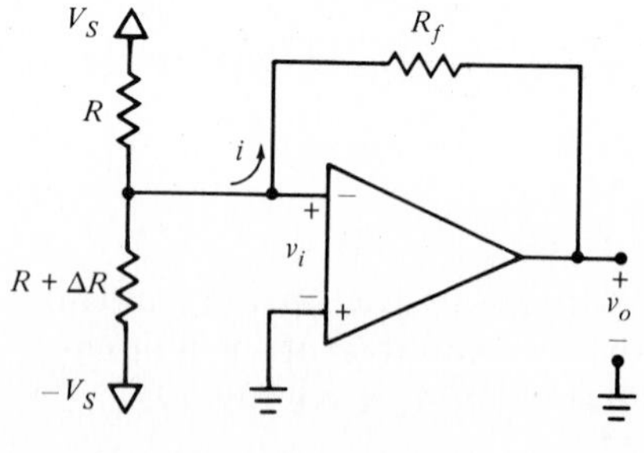

Compared with the full-bridge circuit and differential amplifier, this circuit has advantages and disadvantages. One advantage is that the voltage V_S applied to the half bridge is not limited by the common-mode voltage limit of the op amp, since the half-bridge output is at ground potential. Larger supply voltages lead to larger signals and increased sensitivity. Thus smaller changes in resistance can be detected. The major disadvantage of the half-bridge circuit is that any variations or noise in the supply voltage cannot be differentiated from the desired signal. In the full-bridge circuit, these noise signals are just a common-mode signal for which the differential amplifier should have good rejection. Thus the half-bridge circuit must be carefully constructed to reduce noise and power supply ripple. All wiring should be kept short and it should be well shielded.

As a last example of bridge amplifiers, let us consider the wide deviation bridge shown in Fig. 10.10. We shall show that this circuit provides an output that is a linear function of δ even when δ is large.

The circuit is similar to a differential amplifier itself, with the transducer replacing one of the feedback resistors. Again we assume that $v_i = 0$ and the amplifier draws no input current. We therefore have $v_1 = v_2$, and we obtain v_2 by voltage division:

$$v_2 = V_S \frac{R_f}{R_1 + R_f} = v_1$$

The voltage across the upper R_1 is $V_S - v_1$, so that

$$i = \frac{V_S - v_1}{R_1} = \frac{V_S}{R_1}\left(1 - \frac{R_f}{R_1 + R_f}\right) = \frac{V_S}{R_1 + R_f}$$

Fig. 10.10 The output of a wide deviation bridge

$$v_o = -V_S\delta \frac{R_f}{R_1 + R_f}$$

is a linear function of δ, even for large δ.

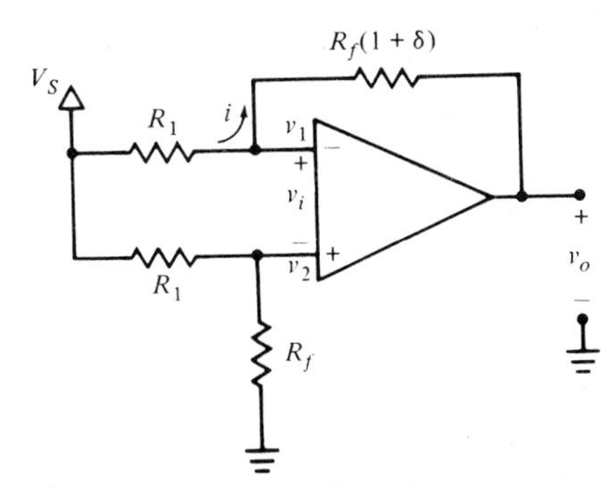

The voltage across the transducer is $v_1 - v_o = iR_f(1 + \delta)$, and we may solve for v_o:

$$v_o = v_1 - iR_f(1 + \delta) = V_S \frac{R_f}{R_1 + R_f} - \frac{V_S}{R_1 + R_f} R_f(1 + \delta)$$

or

$$v_o = -\frac{V_S R_f \delta}{R_1 + R_f} \tag{19}$$

This is an exact expression, assuming an ideal op amp, and not an approximate result, as Eqs. (16) and (18) were. Thus the op-amp output is a linear function of the change in resistance, even for large changes. This is useful for strain gauges incorporating semiconductor elements that have high sensitivity, and for thermistors.

It is necessary that the two R_1 resistors be well matched, and also that the resistance of the transducer be exactly R_f when the bridge is balanced.

D10.6 Element values in Fig. 10.8 are $R_1 = 25\ \text{k}\Omega$, $R_{f1} = 250\ \text{k}\Omega$, $R_2 = 10\ \text{k}\Omega$, $R_{f2} = 100\ \text{k}\Omega$, $R = 250\ \Omega$, and $V_S = 10$ V. Find v_o if $\Delta R/R$ equals (a) 10^{-3}, (b) 10^{-2}, (c) 10^{-1}.

Answers. -25.0 mV; -249 mV; -2.38 V

10.5 Frequency dependence of open-loop gain

Many pages ago in Chapters 7 and 8 we looked at the frequency response of single-stage and multistage amplifiers. In the process we learned how to calculate or estimate upper and lower half-power frequencies. We also discovered the significance of corner frequencies in establishing the frequency behavior. Now we shall recall some of those results as we investigate the frequency response of an op amp. Only the high-frequency performance is necessary, for these amplifiers are dc-coupled, and $\omega_L = 0$.

An op amp contains several stages, usually two or three; for bipolar transistors these usually consist of common-emitter and emitter-follower circuits, although an occasional common-base stage is found in a cascode arrangement. Some of these stages are connected as differential amplifiers, having two inputs, the inverting and noninverting. For example, the input stage for the HA-2107 op amp is a common-collector–common-base differential amplifier. The common-collector stage provides a large R_{in}. Each stage has a different high-frequency response, contributing several corner or break frequencies to the gain function. The corner frequencies of greatest interest appear as factors in the denominator (poles), although numerator factors (zeros) also appear, usually with break frequencies well above the frequency range of interest.

Each one of the poles starts a decrease of 20 dB per decade in the magnitude of the open-loop gain, and each zero causes an increase of 20 dB per decade. If the pole and zero frequencies were all well separated, it would be possible to see these slopes and changes in slope, and to identify all the corner frequencies from a plot of $|A(j\omega)|_{dB}$ vs. ω. Often, however, the poles are close enough together that they interact, and a curve of open-loop gain vs. frequency is obtained that is similar to one of those appearing in Fig. 10.11. Curve (*a*) apparently has all its poles at frequencies greater than 10^5 rad/s. At the frequency for which $|A|_{dB} = 0$, or the magnitude of the open-loop gain is unity, the slope of the curve is approaching -60 dB per decade in this case.

Curve (*b*) in Fig. 10.11 has one pole at a much lower frequency, about 100 rad/s. At this frequency the amplitude begins to decrease at 20 dB per decade, staying at a fairly constant slope up to about 10^6 rad/s. At 0 dB the slope is apparently between -40 and -60 dB per decade.

The corner at 100 rad/s is, of course, a surprising characteristic of the *high*-frequency response, and it is not typical of any CE stage that we have looked at before, even one with a severe Miller effect. This pole is actually introduced internally into the op amp, usually by artificially increasing C_μ in a high-gain common-emitter stage. We shall see in the following two sections that most op amps have a tendency to oscillate unless certain specific preventive measures are taken. One such method is to introduce a low-frequency pole in the open-loop gain. This pole is either provided externally by the user or internally by the manufacturer. The process of preventing oscillations by the addition of capacitance is called *compensation*. Curve (*a*) in Fig. 10.11 is for an uncompensated op amp, and we can be sure that it will have to have capacitance added externally to secure an open-loop gain curve with a much lower upper half-power frequency, although not necessarily as low as 100 rad/s. Curve (*b*) applies to an

Fig. 10.11 Open-loop gain $|A|_{dB}$ is plotted in decibels against ω on a logarithmic frequency scale for a typical (*a*) uncompensated op amp; (*b*) internally compensated op amp.

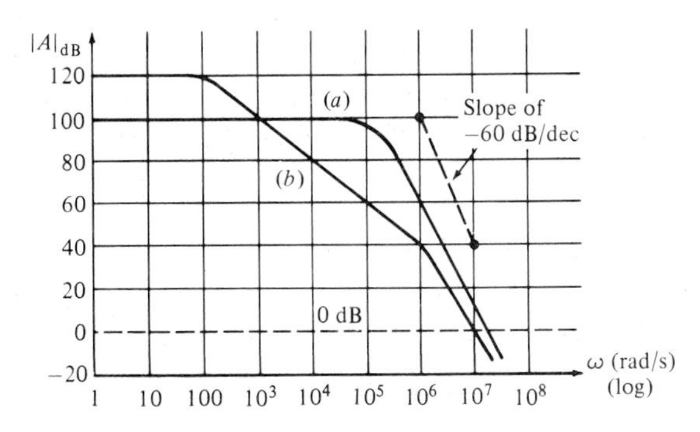

internally compensated op amp. Of the three op amps whose data sheets appear in Appendix A (LM308, HA-2107, and AD547), two are internally compensated and one requires compensation to be provided externally.

We shall see in the next section that a much wider mid-frequency region, a much greater upper half-power frequency, and a much greater bandwidth can be obtained for the *closed*-loop gain after negative feedback is applied around the amplifier.

The open-loop phase characteristic must also be considered whenever feedback is to be applied and the involuntary construction of an oscillator becomes possible. We recall from Chapter 7 that each pole, or corner frequency in the denominator of the gain function, introduces a factor of the form $[1 + j(\omega/\omega_1)]$, where ω_1 is the frequency of the pole. As ω proceeds from zero to infinity, the phase angle thus *decreases* 90°. A zero contributes a total phase angle *increase* of 90°.

A typical op amp shows a phase angle change of −270° from very low to very high frequencies; the open-loop phase curves for the two op amps of Fig. 10.11 are shown in Fig. 10.12. Curve (*a*) of Fig. 10.12 applies to the uncompensated amplifier, and Curve (*b*) is for the internally compensated unit. Note that the angle scale is linear, although the frequency scale remains logarithmic. Also, the angle is that of the open-loop gain A, and 180° must be added to it (or subtracted from it, take your choice) when the signal is applied to the inverting input.

D10.7 Refer to Appendix A and answer the following questions: (a) Which of the three op amps described must be compensated externally? (b) What is f_H for the HA-2107? (c) At what frequency is $|A|_{\text{dB}} = 0$ for the AD547?

Answers. LM308; 8 Hz; 0.9 MHz (curves) or 1 MHz (specs)

Fig. 10.12 The open-loop phase angle of A is plotted against ω on a logarithmic frequency scale for the (*a*) uncompensated and (*b*) internally compensated amplifier of Fig. 10.11.

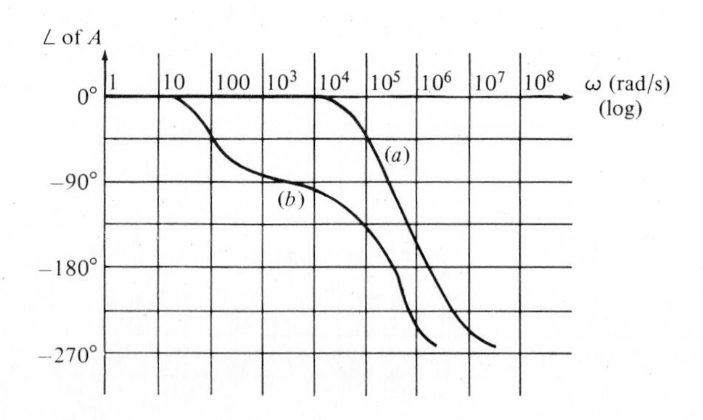

10.6 Closed-loop frequency response

In Section 10.5 we discussed the open-loop frequency response of an operational amplifier and showed that it reflected the presence of several RC networks within the op amp and the corner frequencies they produced. These RC combinations cause the output to fall, in the limit, at a rate of 20, 40, or 60 dB per decade, or perhaps at an even greater rate. This section deals with the *closed-loop* frequency response.

Most operational amplifiers are used in circuits that provide some negative feedback around the amplifier. In contrast, most oscillators depend on positive feedback to establish and maintain the oscillations. The distinction between negative and positive feedback, however, becomes a little fuzzy as frequency increases and phase shift is present between the input and output of the operational amplifier, or sometimes in the feedback path between output and input. Thus a 180° phase shift in the op amp at some frequency can turn negative feedback into positive feedback, and an inverting operational amplifier into an oscillator.

Other values of phase shift in the open-loop gain of the op amp are also possible, of course. For example, let us assume a basic inverting amplifier with the usual resistance R_f between the output and the inverting input, Fig. 10.13. We assume ideal conditions in that $R_i = \infty$ and $R_o = 0$, but we let $A = |A|\underline{/\phi}$. The voltage gain V_o/V_s is then the ratio of two sinusoidal steady-state signal quantities.

At first glance, we might feel that there must be 0° phase shift through a resistive feedback path. However, this would cause a paradox when we closed the loop. That is, there certainly is a phase shift $\phi + 180°$ (180° since $V_o = -AV_i$) between V_i and V_o through the op amp. With R_f in place, a part of V_o is returned to the input and this must produce the same V_i we started with. Thus a phase shift of $-\phi - 180°$ must occur somewhere in the circuit. The answer, of course, is that the signal fed back combines with V_s to produce V_i, and the addition or subtraction of two

Fig. 10.13 A basic inverting amplifier for which $A = V_o/V_i = |A|\underline{/\phi}$. The currents and voltages are sinusoidal steady-state quantities.

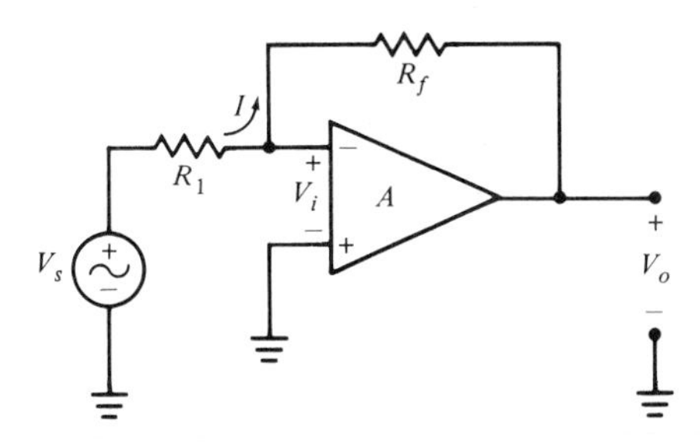

phasors with different amplitudes and different phase angles can produce a wide range of resultant phasor amplitudes and angles.

In Chapter 9 we obtained Eq. (3) for the inverting amplifier,

$$\frac{v_o}{v_s} = \frac{-R_f/R_1}{1 + \dfrac{1 + (R_f/R_1)}{A}}$$

We now write it in terms of sinusoidal steady-state quantities,

$$\frac{V_o}{V_s} = \frac{-R_f/R_1}{1 + \dfrac{1 + (R_f/R_1)}{|A|\underline{/\phi}}} \tag{20}$$

Let us now consider a simple example in which the open-loop gain A has a phase shift of $-90°$. Specifically, we take $A = 10\underline{/-90°}$, a very small gain magnitude for an op amp, but one that enables us to see the effects of the phase shift more easily. If $R_1 = 10\ \mathrm{k\Omega}$, $R_f = 50\ \mathrm{k\Omega}$, and $V_s = 1\underline{/0°}$ V, then Eq. (20) gives

$$V_o = \frac{-5(1\underline{/0°})}{1 + \dfrac{1+5}{10\underline{/-90°}}} = \frac{-5}{1 + j0.6} = -4.29\underline{/-31.0°}\ \mathrm{V}$$

and

$$V_i = -\frac{V_o}{A} = \frac{4.29\underline{/-31.0°}}{10\underline{/-90°}} = 0.429\underline{/59.0°}\ \mathrm{V}$$

Note the required 90° phase shift and magnitude ratio of 10 between output and input. If A had a larger magnitude, say $A = 10^4\underline{/-90°}$, then we should obtain $V_o = -4.999\,999\underline{/-0.0344°}$ V, very close to the ideal value, $-5\underline{/0°}$ V.

One additional informative result from the example above (where $A = 10\underline{/-90°}$) is the value of the current I in both R_1 and the feedback resistor R_f. It is

$$I = \frac{V_i - V_o}{R_f} = \frac{V_s - V_i}{R_1} = \frac{1 - 0.429\underline{/59.0°}}{10{,}000} = 86.2\underline{/-25.3°}\ \mu\mathrm{A}$$

We note that this current is not in phase with either V_o, V_i, or V_s, even though there are apparently no circuit elements external to the op amp that can cause a phase shift. The phase shift is once again seen to result from the addition and subtraction of out-of-phase quantities.

Let us now inspect the frequency behavior of the closed-loop gain for an inverting amplifier. We use Eq. (20). At low or mid-frequencies where $|A|$ is large, we have $V_o/V_s = -R_f/R_1$. This is the value of the closed-loop gain at dc, and we now let G represent the closed-loop gain V_o/V_s, so that

$$G|_{dc} = -R_f/R_1 = -G_0$$

where G_0 is the magnitude of the closed-loop gain at dc when A is large. Equation (20) therefore becomes

$$G = \frac{-G_0}{1 + \dfrac{1 + G_0}{A}} \tag{21}$$

The open-loop gain A is a complicated function of ω, since it might contain as many as six poles and several zeros. In order to make our point as simply as possible, let us consider an op amp that is internally compensated like that shown as Curve (b) in Fig. 10.11. There is one dominant pole at a relatively low frequency, and the remaining poles (and zeros) are at much higher frequencies. We thus approximate A by using only the lowest corner frequency ω_1,

$$A = \frac{A_0}{1 + j(\omega/\omega_1)} = |A|\underline{/\phi}$$

where A_0 is the value of A at dc. If we let $x = \omega/\omega_1$, we have

$$A = \frac{A_0}{1 + jx}$$

This may be used for A in Eq. (21):

$$G = \frac{-G_0}{1 + \dfrac{(1 + G_0)(1 + jx)}{A_0}}$$

We now multiply throughout by A_0 and separate the real and imaginary parts of the denominator:

$$G = \frac{-G_0 A_0}{(1 + G_0 + A_0) + (1 + G_0)jx}$$

The upper half-power frequency occurs when the real and imaginary parts of the denominator are equal. Then,

$$x_H = \frac{\omega_H}{\omega_1} = \frac{1 + G_0 + A_0}{1 + G_0}$$

or

$$\omega_H = \omega_1 \frac{1 + G_0 + A_0}{1 + G_0} \tag{22}$$

This is the accurate result for our assumed form of A, but a simpler approximate form is obtained by keeping only the largest terms of the numerator and denominator:

$$\omega_H \doteq \frac{A_0}{G_0} \omega_1 \tag{23}$$

The magic number is the ratio of open-loop to closed-loop gain, A_0/G_0. This is the factor by which negative feedback has *reduced* the gain. Now we see that it is also the factor by which negative feedback has *increased* ω_H. We have a result similar to that we obtained several times in Chapters 7 and 8: The gain-bandwidth product tends to remain constant.

This result stands out clearly when we plot curves of gain vs. frequency. Figure 10.14 shows open-loop gain $|A|_{dB}$ and closed-loop gain $|G|_{dB}$ vs. ω for the internally compensated op amp of Fig. 10.11. We see an open-loop gain at dc $A_{0(dB)} = 120$ dB, $A_0 = 10^6$, and an upper half-power frequency $\omega_{H(A)} = 100$ rad/s. Negative feedback is then added with $R_f/R_1 = 200$. Thus the closed-loop gain magnitude at low frequencies is $G_0 = 200$, $G_{0(dB)} = 46$ dB; the new upper half-power frequency is $\omega_{H(G)} = (A_0/G_0)\omega_{H(A)} = (10^6/200)100 = 5 \times 10^5$ rad/s. If G_0 is made smaller, $\omega_{H(G)}$ becomes correspondingly larger.

D10.8 The open-loop gain of an inverting operational amplifier is $A = 100/[1 + j(\omega/10)]$, $R_i = \infty$, and $R_o = 0$. Let $R_f/R_1 = 5$ and find $|A|$ and $|G|$ if ω equals (a) 0, (b) 10, (c) 200 rad/s.

Answers. 100, 4.72; 70.7, 4.71; 4.99, 3.12

Fig. 10.14 The open-loop gain-bandwidth product, $10^6 \times 100$, is equal to the closed-loop gain-bandwidth product, $200(5 \times 10^5)$.

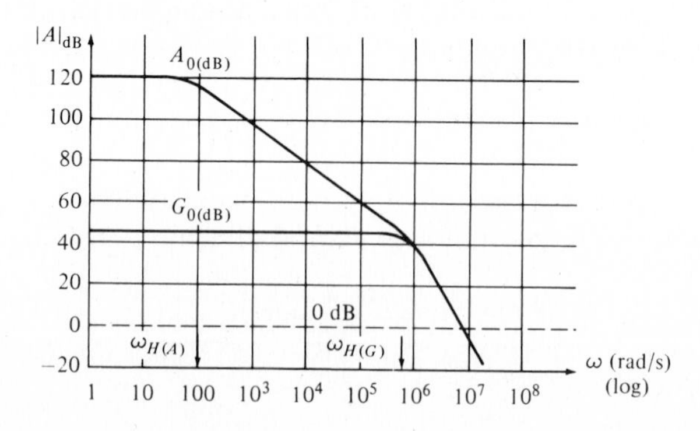

10.7 Stability and compensation

The possibility of oscillations occurring when negative feedback is applied to an operational amplifier has been mentioned several times before. The condition under which oscillations arise is termed *instability*, while *stability* is the desirable situation for an inverting amplifier or a voltage follower.

There are numerous analysis techniques by which the stability of a circuit can be investigated. We shall not go into details on any of them because a thorough study requires more writing for the authors and more reading for the students than is really required for a happy and productive life. Instead, let us just outline several methods so that we shall know what to look for in a later course in system theory, automatic control, or feedback theory. The third technique we look at is easy to apply, and we shall use it as a predictor of instability.

One method uses plots of gain magnitude and angle vs. frequency. The gain, however, is neither the open-loop nor the closed-loop gain, but a quantity called the *loop gain*. For most practical op amp circuits, the loop gain is closely equal to the ratio of the open-loop to the closed-loop gain. In decibels it would be the difference, $|A|_{\mathrm{dB}} - |G|_{\mathrm{dB}}$. Thus, for the inverting amplifier discussed in Fig. 10.14, the open-loop gain at low frequencies is 120 dB, the closed-loop gain is 46 dB, and the loop gain is 74 dB. The phase angle of the loop gain is the difference of the phase angles of A and G. In a study of feedback, the loop gain T is found by writing the closed-loop gain in the form

$$G = \frac{-A}{1 - T} \tag{24}$$

At low frequencies for an inverting amplifier, G is real and negative, A is real and positive, and the loop gain T is real and negative.

Instability threatens as the denominator of Eq. (24) approaches zero. This is investigated by preparing plots of $|T|$ and $\angle T$ vs. ω and paying particular attention to the point at which $\angle T = 0°$. Here the magnitude of T should be safely less than unity, or $|T|_{\mathrm{dB}} < 0$. For example, if $|T| = 2$ when $\angle T = 0°$, then an oscillatory unstable condition would be reached when the amplifier was turned on and $|T|$ passed through unity in its initial increase from zero.

A second technique for investigating stability also requires use of the loop gain T. Here we visualize a phasor diagram of the complex phasor T for every value of ω from 0 to ∞. The actual plot made is only of the locus of the tip of T for $0 < \omega < \infty$. The conjugate plot for $-\infty < \omega < 0$ is then added to the sketch, and instability is indicated if the locus encircles the point $1 + j0$. The locus is known as a *Nyquist plot* and the stability condition is called the *Nyquist criterion*.

We shall use a simpler approximate technique that does not involve the loop gain T. It requires only a plot of open-loop gain magnitude $|A|_{dB}$ vs. ω and a knowledge of the low-frequency closed-loop gain magnitude $|G_0|_{dB}$. Figure 10.15 shows two sketches of the open-loop gain vs. frequency, one for an uncompensated op amp and one for a compensated unit. The value of $|G_0|_{dB}$ is indicated. Stability is indicated by the difference in slope of the horizontal line marked $|G_0|_{dB}$ and the open-loop gain curve at their intersection. A difference in slope of more than 40 dB per decade predicts an unstable system, while a slope difference less than 40 dB per decade indicates a stable system. For added safety from oscillation, the 40-dB-per-decade number should be reduced to 34 dB per decade. As the slope difference approaches 40 dB per decade, it is found that the closed-loop gain begins to show a peak in the gain vs. frequency characteristic in the neighborhood of the frequency at which the intersection occurs. A small peak may be helpful in increasing the bandwidth by a few percent, but a large peak produces distortion in pulses and other rapidly changing waveforms.

We note from Fig. 10.15 that the compensated op amp gives a stable system, while the uncompensated amplifier is unstable. Both open-loop gain curves decrease at 60 dB per decade for sufficiently high frequencies, and the reason for compensation is now evident. It is necessary to introduce a low-frequency corner frequency so that the gain drops at a 20 dB per decade rate up to the intersection at A.

A few moments of serious thought should also show us that the instability problem is severest for a voltage follower where $|G_0|_{dB} = 0$. As a matter of fact, if the $|G_0|_{dB}$ curve in Fig. 10.15 were lowered to 0 dB, it looks like it would intersect the compensated amplifier curve at a point where the difference in slope is approximately 60 dB per decade; hence instability would result. In that case, it would become necessary to reduce the

Fig. 10.15 At the intersection marked A, the difference in slopes is less than 40 dB per decade and the system is stable. At B, the difference is greater than 40 dB per decade and the amplifier is unstable.

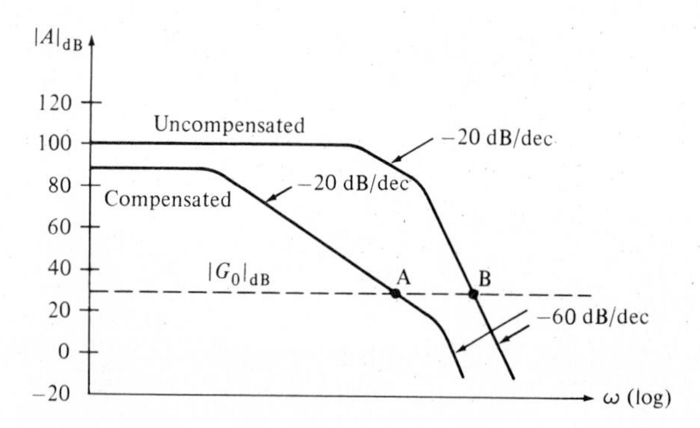

frequency of the lowest-frequency pole by using a larger capacitor for compensation. Internally compensated op amps are available that are guaranteed to be stable only down to a specified minimum closed-loop gain, usually 5 or 10 (14 or 20 dB). They are not intended for use as voltage followers unless additional external compensation is added.

An op amp such as the LM308 that requires external compensation often gives specifications for two different values of added capacitance. For example, the performance curves for the LM308 on the third page of those data sheets in Appendix A show the addition of a capacitance C_f between Pins 1 and 8. If $C_f = 3$ pF, $f_H = 20$ Hz; if $C_f = 30$ pF, then $f_H \doteq 1$ Hz. The curves also show that the slope is greater than 20 db per decade at 0 dB when $C_f = 3$ pF. Instability might therefore result if the unit is used as a voltage follower.

The procedure of compensating an operational amplifier by adding capacitance and shifting one pole to a very low frequency is called *dominant-pole compensation*. Other methods are also used, such as *lag compensation*, *lead compensation*, and *pole-zero compensation*. Each requires more elements and a more careful selection of element values than does the dominant-pole method. Each also provides a better high-frequency response for the open-loop gain.

D10.9 For the LM308 op amp with a compensating capacitance $C_f = 3$ pF, (a) find $|A|_{dB}$ and $\angle A$ at $f = 1$ MHz. (b) At what f is $|A|_{dB} = 0$? (c) At what f is $\angle A = 180°$?

Answers. 14 dB, 140°; 3 MHz; 3 MHz

10.8 Differentiators and integrators

Specific applications of the op amps that we have investigated include summers, subtractors, comparators, and buffers, not to mention inverting and noninverting amplifiers. In this section we shall look at differentiators and integrators, two circuits that have important applications in analog computers, wave-shaping circuits, and instrumentation systems. Other important circuits that we shall not have time to consider are clamps, triggers, oscillators, rectifiers, voltage regulators, and logarithmic amplifiers and multipliers.

Both the differentiator and the integrator represent more traditional uses for an operational amplifier—performing the mathematical operations of differentiation and integration. Their circuits are obtained by generalizing the basic inverting amplifier circuit. Consider the differentiator shown in Fig. 10.16. Here the resistor R_1 is replaced by a capacitor C_1. Let us assume an op amp with $R_i = \infty$, $R_o = 0$, and open-loop gain A.

There is no input current to the op amp, so

$$i = C_1 \frac{d}{dt}(v_s - v_i) = \frac{1}{R_f}(v_i - v_o)$$

and

$$v_o = -Av_i \qquad \text{or} \qquad v_i = -\frac{1}{A}v_o$$

Eliminating v_i, we have

$$C_1 \frac{dv_s}{dt} + C_1\left(\frac{1}{A}\right)\frac{dv_o}{dt} = -\frac{1}{R_f}\left(\frac{v_o}{A} + v_o\right)$$

or

$$R_f C_1 \frac{dv_s}{dt} = -\frac{v_o}{A} - v_o - R_f C_1\left(\frac{1}{A}\right)\frac{dv_o}{dt} \tag{25}$$

Now let us simplify the right side of Eq. (25) by assuming that $|A| \gg 1$. We can therefore discard v_o/A, since it is much smaller than v_o in magnitude. We next assume that $|A|$ is sufficiently large so that

$$\left|R_f C_1\left(\frac{1}{A}\right)\frac{dv_o}{dt}\right| \ll |v_o| \tag{26}$$

This produces the desired result:

$$v_o = -R_f C_1 \frac{dv_s}{dt} \tag{27}$$

We should investigate the validity of the inequality marked Eq. (26). One way to do this is to assume that all the voltages (and currents) are sinusoids. If $v_o = V_m \sin \omega t$, then $dv_o/dt = \omega V_m \cos \omega t$, and the maximum amplitudes are V_m and ωV_m respectively. Our inequality is now

$$R_f C_1 \omega V_m \ll |A| V_m$$

Fig. 10.16 The basic op amp differentiator for which $v_o \doteq -R_f C_1\, dv_s/dt$.

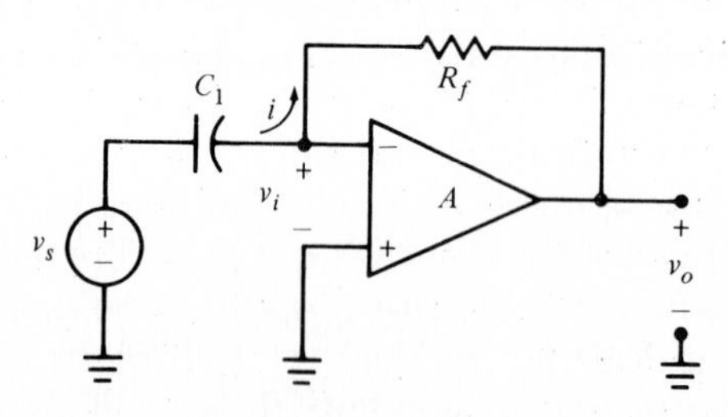

or

$$|A| \gg R_f C_1 \omega \tag{28}$$

If we are using large element values, such as $R_f = 1\ \mathrm{M}\Omega$ and $C_1 = 1\ \mu\mathrm{F}$, with $|A| = 10^5$, then we should probably limit the highest-frequency component of our signal to 10^4 rad/s, an audio frequency.

The differentiator may also be analyzed by treating it as an inverting amplifier operating in the sinusoidal steady state. Let us resurrect Eq. (3) from Chapter 9,

$$\frac{v_o}{v_s} = \frac{-R_f/R_1}{1 + \dfrac{1 + (R_f/R_1)}{A}} \qquad (R_i = \infty,\ R_o = 0)$$

which applies when $R_i = \infty$ and $R_o = 0$. We now extend this result to the sinusoidal steady state by taking v_o and v_s as the phasors V_o and V_s, and replacing R_f and R_1 by the impedances Z_f and Z_1,

$$\frac{V_o}{V_s} = G = \frac{-Z_f/Z_1}{1 + \dfrac{1 + (Z_f/Z_1)}{A}} \qquad (Z_i = \infty,\ Z_o = 0) \tag{29}$$

It follows that the open-loop gain A and the closed-loop gain G are both complex functions of frequency.

For our differentiator, $Z_f = R_f$ and $Z_1 = 1/(j\omega C_1)$. Therefore $Z_f/Z_1 = j\omega C_1 R_f$ and

$$G = \frac{-j\omega C_1 R_f}{1 + \dfrac{1 + j\omega C_1 R_f}{A}} \tag{30}$$

Let us first apply Eq. (30) to the region in which $|A| \gg 1$. Thus

$$G \doteq -\frac{Z_f}{Z_1} = -j\omega C_1 R_f \tag{31}$$

We might have inferred this result from Eq. (27), since differentiation in the time domain is equivalent to multiplication by $j\omega$ in the frequency domain.

Equation (31) indicates one of the troubles that may arise using differentiators in practice. The gain magnitude is proportional to frequency. Therefore relatively small high-frequency noise components, such as television snow or hi-fi hiss, may represent objectionably large components of the output. Under severe conditions, they might even mask the output completely.

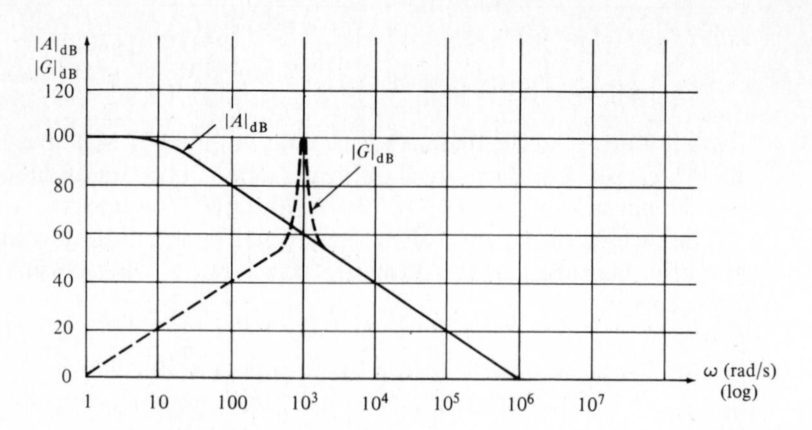

Fig. 10.17 $|A|_{dB}$ (solid line) and $|G|_{dB}$ (dashed line) are shown for a differentiator having $R_fC_1 = 1$ and $\omega_H = 10$ for A.

Another problem that can arise with a differentiator may be illustrated by an example. Let us take $R_i = \infty$, $R_o = 0$, $R_fC_1 = 1$, and an internally compensated op amp with an upper half-power frequency $\omega_H = 10$ and a low-frequency (dc) gain of 10^5, $A = 10^5/(1 + j0.1\omega)$. Using Eq. (30), the closed-loop gain is

$$G = \frac{-j\omega}{1 + \dfrac{1 + j\omega}{10^5/(1 + j0.1\omega)}} = \frac{-j10^5\omega}{10^5 + (1 + j\omega)(1 + j0.1\omega)}$$

or

$$G = \frac{-j10^5\omega}{(100{,}001 - 0.1\omega^2) + j1.1\omega} \tag{32}$$

We now use this equation to find $|G|$ or $|G|_{dB}$ at various values of ω. The results are plotted in Fig. 10.17. The solid curve shows $|A|_{dB}$ vs. ω, with a corner at $\omega = 10$ and a slope of -20 dB per decade at higher frequencies. The dashed curve gives $|G|_{dB}$ vs. ω. When $|A| \gg 1$, we see the linear increase with ω, predicted by Eq. (31). Of course, the plot is $|G|_{dB}$ vs. $\log \omega$, but this is also a straight line having a slope of $+20$ dB per decade. As $|A|$ or $|A|_{dB}$ begin to decrease with frequency, Eq. (31) ceases to be valid and we must use results obtained from Eq. (32) above.

Two characteristics of this dashed curve should be noted. First, for sufficiently large ω (here $\omega \gg 10^3$), $|G| \doteq |A|$. This occurs because the inequality of Eq. (28) is no longer satisfied. Treating this expression as an equality, then, we would expect the two curves to merge somewhere beyond $\omega = |A|/(R_fC_1)$. In this case, $R_fC_1 = 1$ and $|A| =$

$10^5/\sqrt{1 + 0.01\omega^2}$, which leads to

$$\omega \doteq \frac{10^5}{\sqrt{1 + 0.01\omega^2}}$$

for which the solution is $\omega \doteq 1000$. The curves appear to join around $\omega = 3000$ rad/s.

The second important characteristic is the large peak that occurs at $\omega = 1000$. Not only does this accentuate any desired signals or undesired noise in this frequency range, it also is a warning that instability lurks nearby.

There are several methods by which the differentiator circuit may be modified to improve its stability. One such improvement is shown in Fig. 10.18*a*. A resistor R_1 has been added in series with C_1 so that when $|A|$ is large,

$$G = \frac{V_o}{V_s} = -\frac{Z_f}{Z_1} = -\frac{R_f}{R_1 + (1/j\omega C_1)} = -\frac{j\omega C_1 R_f}{1 + j\omega C_1 R_1}$$

At low frequencies where $\omega \ll 1/C_1 R_1$,

$$G \doteq -j\omega C_1 R_f \qquad (\omega \ll 1/C_1 R_1)$$

This circuit gives the same closed-loop gain as does our original differentiator. At high frequencies with $\omega \gg 1/R_1 C_1$,

$$G \doteq -\frac{R_f}{R_1} \qquad (\omega \gg 1/C_1 R_1)$$

which is a constant value. Thus the curve of $|G|_{dB}$ vs. ω appears as the dashed line in Fig. 10.18*b*, drawn for $R_f C_1 = 1$, $R_1 C_1 = 0.01$, and A

Fig. 10.18 (*a*) The addition of R_1 to the basic differentiator circuit ensures a stable circuit. (*b*) Differentiation occurs only for $\omega < 100$ rad/s, where the slope of $|G|_{dB}$ vs. ω is 20 dB per decade.

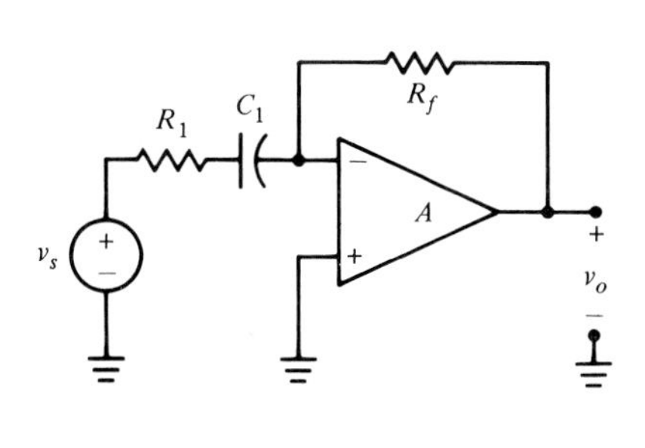

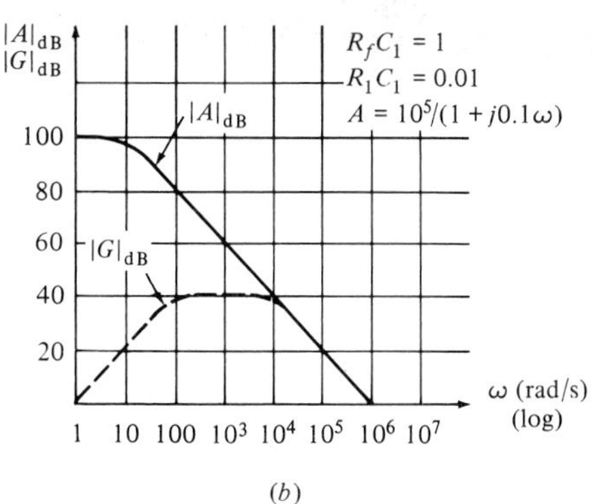

$= 10^5/(1 + j0.1\omega)$. There is no longer any peak or possibility of instability. Differentiator action and a slope of + 20 dB per decade are found only up to about 100 rad/s, however. For large ω, $|G|_{dB} \doteq |A|_{dB}$.

Figure 10.19*a* shows a second stable differentiator circuit with R_f paralleled by C_f. At low frequencies ($\omega \ll 1/R_fC_f$), this circuit behaves like the basic differentiator of Fig. 10.16, since C_f has little effect. At high frequencies, C_f causes a reduction in Z_f and the closed-loop gain. The net result is a curve of $|G|_{dB}$ that shows a good range for satisfactory differentiation and no instabilities. The curves in Fig. 10.19*b* are drawn for $R_fC_f = 0.001$, $R_fC_1 = 1$, and $A = 10^5/(1 + j0.1\omega)$.

It is possible to design a differentiator with all four external elements, R_1, C_1, R_f, and C_f, and to obtain a response curve for the closed-loop gain that is a combination of those appearing in Figs. 10.18*b* and 10.19*b*.

The last circuit that we shall study is the integrator, shown in Fig. 10.20. In terms of instantaneous voltages and currents for an op amp with $R_i = \infty$ and $R_o = 0$, we have

$$v_i - v_o = \frac{1}{C_f}\int_{-\infty}^{t} i\,dt = \frac{1}{C_f}\int_{-\infty}^{t} \frac{1}{R_1}(v_s - v_i)\,dt$$

Looking at the left member of this equation, we now neglect v_i in comparison with v_o. Also, in the integrand on the right, we neglect v_i as being much smaller than v_s. With these approximations,

$$v_o = -\frac{1}{R_1C_f}\int_{-\infty}^{t} v_s\,dt \tag{33}$$

Note that the output is the negative integral of the input voltage. If the range of integration is from t_0 to t, then an initial voltage $v_o(t_0)$ must be

Fig. 10.19 (*a*) A capacitor C_f in parallel with the feedback resistor R_f results in a stable differentiator. (*b*) The open- and closed-loop gains are shown for $R_fC_1 = 1$, $R_fC_f = 0.001$, and $A = 10^5/(1 + j0.1\omega)$.

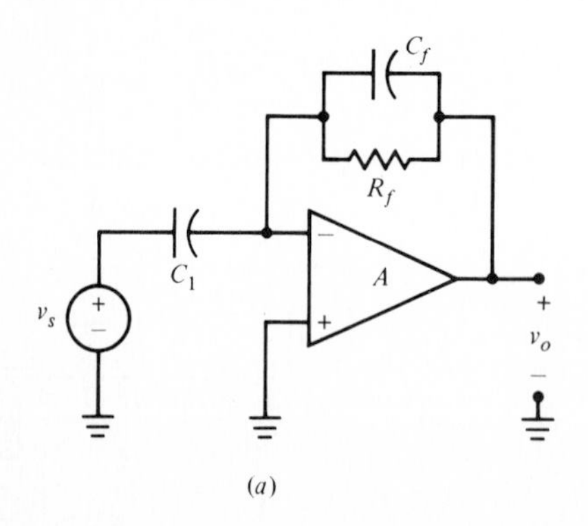

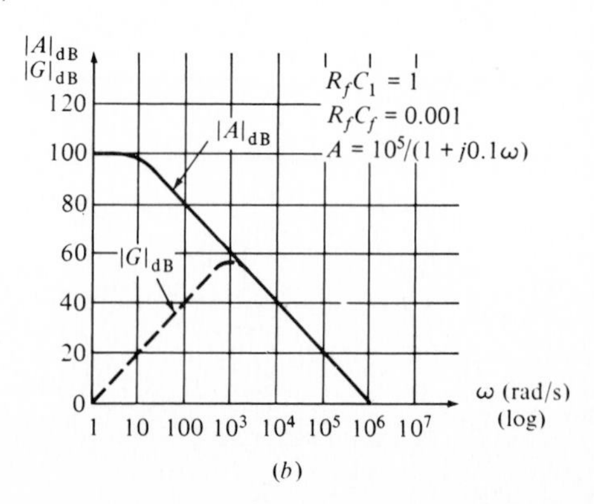

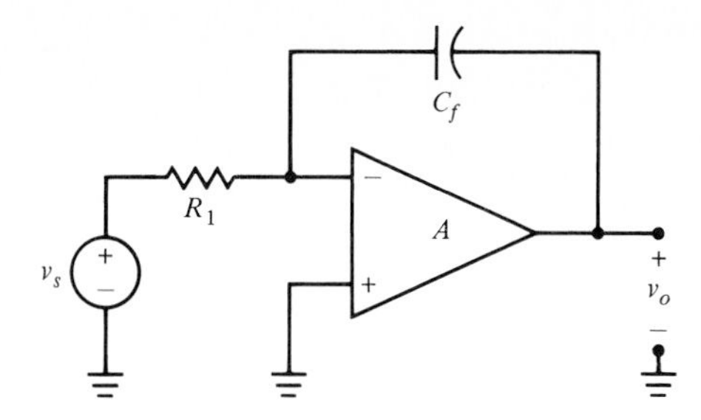

Fig. 10.20 For this op-amp integrator,

$$v_o = -\frac{1}{R_f C_1}\int_{-\infty}^{t} v_s\,dt$$

for sufficiently large $|A|$.

added to the integral. This may also be done in practice by using additional circuit elements to discharge C_f and then to charge it to the desired value.

We can obtain additional information from the sinusoidal steady state. Starting with Eq. (29),

$$G = \frac{V_o}{V_s} = \frac{-Z_f/Z_1}{1 + \dfrac{1 + (Z_f/Z_1)}{A}}$$

we let $Z_1 = R_1$ and $Z_f = 1/(j\omega C_f)$, so that $Z_f/Z_1 = 1/(j\omega C_f R_1)$. Then

$$G = \frac{-1/(j\omega C_f R_1)}{1 + \dfrac{1 + 1/(j\omega C_f R_1)}{A}} = \frac{-1}{j\omega C_f R_1 + \dfrac{1 + j\omega C_f R_1}{A}} \tag{34}$$

Now we must satisfy the condition,

$$|A| \gg |1 + j\omega C_f R_1| \tag{35}$$

in order to have

$$G = \frac{-1}{j\omega C_f R_1}$$

or

$$|G| = \frac{1}{\omega C_f R_1}$$

Since integration in the time domain corresponds to division by $j\omega$ in the frequency domain, we again see an integration if $|A|$ is large. Inequality (35) becomes harder to satisfy as ω increases.

Let us illustrate this by one example. We let $R_1C_f = 10^{-4}$ and $A = 10^4/(1 + j0.1\omega)$. Equation (34) becomes

$$G = \frac{-1}{j10^{-4}\omega + (1 + j10^{-4}\omega)(1 + j0.1\omega)10^{-4}}$$

or

$$G = \frac{-10^4}{1 - 10^{-5}\omega^2 + j1.1001\omega}$$

Plots of $|A|_{dB}$ and $|G|_{dB}$ are shown in Fig. 10.21; we see from the slope of $|G|_{dB}$ that the integration is accurate for frequency components from about 10 to 10,000 rad/s. The low-frequency performance improves as the product of C_fR_1 and the dc open-loop gain of the op amp increases. The high-frequency limit is extended as the upper half-power frequency and dc gain of the op-amp increase.

One problem in a practical integrator is caused by the input offset voltage and input bias current, both of which provide a constant input signal. When integrated, this produces an increasing output voltage magnitude that will certainly cause errors, and may lead to saturation. The cure is usually found by integrating over only a finite interval. That is, C_f is initially discharged, then set to a specified initial voltage if desired, and finally allowed to integrate and charge up for a period of time during which the offset voltage and bias current cannot introduce any appreciable error.

Fig. 10.21 Open-loop gain $|A|_{dB}$ and closed-loop gain $|G|_{dB}$ are plotted against ω for an integrator with $R_fC_1 = 10^{-4}$ and $A = 10^4/(1 + j0.1\omega)$.

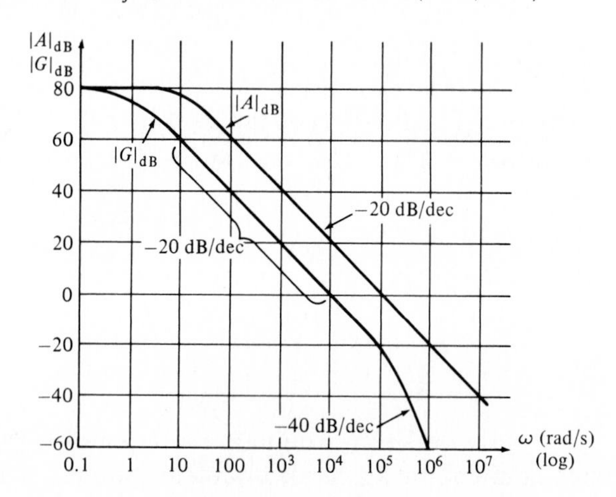

D10.10 For the differentiator of Fig. 10.16, let $C_1 = 100$ pF, $R_f = 20$ kΩ, and $V_s = 2\underline{/0°}$ V. If $A = 1000/(1 + j\omega)$, find $|V_o|$ at ω equals (a) 1 rad/s, (b) 1 krad/s, (c) 1 Mrad/s.

Answers. 3.996 μV; 2.83 mV; 1.790 mV

D10.11 An integrator is constructed with $R_1 = 20$ kΩ, $C_f = 100$ pF, and $A = 1000/(1 + j\omega)$. Find G if ω equals (a) 1 rad/s, (b) 100 rad/s, (c) 10 krad/s.

Answers. $-706\underline{/-45.1°}$; $-9.98\underline{/-89.4°}$; $-0.0998\underline{/-91.1°}$

Problems

1. The circuit of Fig. 10.1 contains a standard cell with $V_{ref} = 1.01830$ V. The op amp has $A = 5 \times 10^5$, $R_i = 2$ MΩ, and $R_o = 200$ Ω. Find the voltage across R_L if R_L equals (a) 10 kΩ, (b) 100 Ω. (c) What is the voltage across the terminals of the standard cell if it has an internal resistance of 5 Ω and the bias current for the op amp is 30 nA?
2. The voltage-reference circuit of Fig. 10.3*a* uses an ideal op amp and a Zener diode that may be modeled by $V_{BR} = 15$ V in series with $R_Z = 10$ Ω and an ideal diode. If $V_S = 30$ V, (a) select R_S so that $I_Z = 3$ mA, (b) let $R_1 = R_f = 10$ kΩ and determine V_o, (c) find V_o if V_S decreases to 20 V.
3. The circuit of Fig. 10.3*c* contains an ideal op amp and resistances $R_1 = 5$ kΩ and $R_f = 8$ kΩ. Let $V_S = 10$ V and let diode D1 have a voltage drop of 0.7 V in the forward direction. Diode D1 and the Zener diode have negligible resistance. If $V_{BR} = 5$ V, (a) neglect R_S and select R_2 and R_3 so that the initial current through the Zener diode is 5 mA and that supplied by V_S is 6 mA. (b) Select R_S so that the Zener current is 2 mA during normal operations. (c) Using these values, determine V_o.
4. Calculate V_o for the scaled voltage comparator shown in Fig. 10.22 if the op amp saturates at ± 12 V but is otherwise ideal, and V_{s1} equals (a) 1 V, (b) 2 V, (c) 3 V.
5. In the comparator of Fig. 10.4, let $v_s = 2 \sin 10t$ V and let V_{ref} be replaced by the sinusoidal voltage $v_{ref} = 2 \sin 20t$ V. Sketch $v_o(t)$, $0 \le t \le 0.2\pi$ s, if $V_{sat} = \pm 15$ V.
6. A simple voltage comparator, such as that shown in Fig. 10.4, is operating with $V_{ref} = 3$ V and with an op amp having $R_i = 100$ kΩ, $R_o = 1$ kΩ, $A = 1000$, and $R_L = 4$ kΩ. Plot a curve of V_o vs. V_s if $V_{sat} = \pm 10$ V.
7. A sinusoidal voltage $v_s = 2 \sin 500t$ V is applied at the input of the

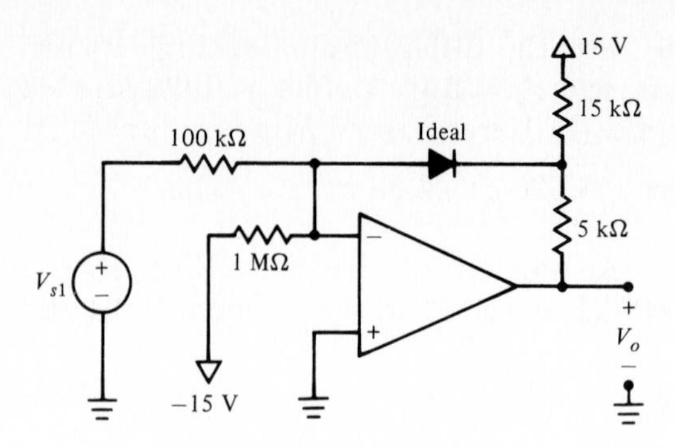

Fig. 10.22 See Problem 4.

comparator shown in Fig. 10.4. Assume an ideal op amp with $V_{sat} = \pm 12$ V, and plot v_o vs. t if the reference voltage is (a) $+1$ V, (b) -1 V, (c) $v_s/2$, (d) v_o.

8. The common-mode and differential-mode inputs to a differential amplifier are $v_{CM} = 200\ \mu$V and $v_{DM} = 50\ \mu$V respectively. Let $A = 10^4$ and $A_{CM} = 3$. Assume $v_{o(CM)}$ and $v_{o(DM)}$ add. Find (a) CMRR_{dB}, (b) v_1, (c) v_2, (d) $v_{o(CM)}$, (e) $v_{o(DM)}$, (f) v_o.
9. A differential amplifier has $A = 5000$ and a CMRR_{dB} of 60 dB. Let $v_o = v_{o(CM)} + v_{o(DM)} = 2$ V. Construct a graph of v_2 vs. v_1 showing the locus of all possible inputs that provide this output. Assume that $v_2 \geq v_1$ and the outputs add, and maintain $|v_2| \leq 5$ V.
10. A simple voltage follower, such as that shown in Fig. 9.6, uses an op amp for which $A = 1000$, $R_i = \infty$, and $R_o = 0$. Find v_o/v_s if (a) $\text{CMRR} = \infty$, (b) $A_{CM} = 2$ (find both possible answers).
11. The gain of the op amp used in a differential amplifier is $A = 2 \times 10^4$ and $\text{CMRR}_{dB} = 80$ dB. Assume that $v_o = v_{o(CM)} + v_{o(DM)}$ and find $v_{o(DM)}$, $v_{o(CM)}$, and v_o if (a) $v_1 = v_2 = 10\ \mu$V, (b) $v_1 = -v_2 = 10\ \mu$V, (c) $v_1 = 10\ \mu$V and $v_2 = 0$, (d) $v_1 = 0$ and $v_2 = 10\ \mu$V.
12. In the half-bridge amplifier shown in Fig. 10.9, $R = 400\ \Omega$, $R_f = 20$ kΩ, and $V_S = \pm 12$ V. The op amp is ideal. Find v_o if (a) $\Delta R/R = 10^{-3}$, (b) $\Delta R/R = 10^{-3}$ and $V_S = \pm 12.1$ V, (c) $\Delta R/R = 10^{-3}$ and the upper voltage supply increases to 12.1 V while the lower remains at -12 V.
13. Values for the bridge circuit of Fig. 10.10 are $R_1 = 200\ \Omega$, $R_f = 600\ \Omega$, and $V_S = 20$ V. The op amp is ideal. Find v_o if (a) $\delta = 10^{-2}$, (b) $\delta = 10^{-2}$ and V_S increases to 20.1 V.
14. Find v_o for a bridge amplifier similar to that of Fig. 10.8 except that the transducer appears as the lower right element in the bridge. Assume an ideal op amp.

15. Let $R_1 = R_2 = 10\ \text{k}\Omega$, $R_{f1} = R_{f2} = 50\ \text{k}\Omega$, $R = 250\ \Omega$, and $V_S = 12$ V in the bridge amplifier of Fig. 10.8. Assume that the op amp is ideal. Find v_{s1}, v_{s2}, v_o, and the voltage across R_{f2} if (a) $\delta = 0$, (b) $\delta = 10^{-3}$, (c) $\delta = -10^{-3}$.
16. The open-loop gain of an uncompensated op amp is given by $A = 10^5(1 + j10^{-7}\omega)/[(1 + j10^{-4}\omega)(1 + j10^{-5}\omega)(1 + j10^{-6}\omega)^2]$. Plot curves of $|A|_{\text{dB}}$ and $\angle A$ vs. ω, $10^3 \leq \omega \leq 10^8$ rad/s, using a logarithmic frequency scale.
17. If a gain magnitude is $|A| = (1 + \omega^2)^{-1}$, what is the slope in decibels per decade at $\omega = 2$? [Hint: Slope in decibels per decade equals $d|A|_{\text{dB}}/d\ (\log \omega)$.]
18. A curve of $|A|_{\text{dB}}$ vs. ω on a logarithmic frequency scale is represented by its asymptotes (straight lines). It is a constant 120 dB up to $\omega = 10^3$, where it begins to drop at 20 dB per decade. At $\omega = 10^6$, the slope magnitude increases to 40 dB per decade. At $\omega = 10^7$, it increases to 60 dB per decade. (a) At what frequency does the asymptotic curve pass through 0 dB? (b) What is the true value of $|A|_{\text{dB}}$ at $\omega = 10^7$? (c) What is the gain angle at $\omega = 10^5$?
19. An inverting amplifier is constructed with $R_1 = 5\ \text{k}\Omega$ and $R_f = 45\ \text{k}\Omega$. Assume an open-loop gain expression for the op amp (an inferior type used only for problems) of $A = 100/(1 + j0.1\omega)$, and let R_i be infinite and R_o be zero. (a) Find G at $\omega = 10$. (b) Find ω_H for the closed-loop system.
20. A voltage follower (with $R_f = 0$) is constructed with the op amp of Problem 19. Find (a) A for $\omega = 10$, (b) G for $\omega = 10$, (c) ω_H for the follower.
21. A noninverting amplifier is constructed with $R_1 = 5\ \text{k}\Omega$, $R_f = 45\ \text{k}\Omega$, and the op amp of Problem 19. Find (a) A at $\omega = 10$, (b) G at $\omega = 10$, (c) ω_H for the noninverting amplifier.
22. Sketch curves of $|A|_{\text{dB}}$ and $|G|_{\text{dB}}$ vs. ω, using a logarithmic frequency scale, for an inverting amplifier with $R_1 = 10\ \text{k}\Omega$, $R_f = 100\ \text{k}\Omega$, and an op amp modeled by $A = 1000/(1 + j0.02\omega)$, $R_i = \infty$, and $R_o = 0$.
23. Assume that an op amp has an open-loop response of $A = 10^5/(1 + j10^{-3}\omega)^3$ with $R_i = \infty$ and $R_o = 0$. (a) At what frequency is $|A| = 1$? (b) At what frequency is the angle of A $= -180°$? (c) If this op amp is used in a voltage follower, at what frequency is $\angle G = 90°$?
24. If $T = -250/[(1 + j\omega)(1 + j0.01\omega)^2]$ is the loop gain of an inverting amplifier, (a) plot $|T|_{\text{dB}}$ vs. ω, $50 \leq \omega \leq 150$, using a linear frequency scale. (b) Using the same frequency scale, plot $\angle T$ vs. ω. (c) By an inspection of these two curves, state whether the amplifier is stable or not.

25. The loop gain of an inverting amplifier is given by the function $T = -5/(1 + j0.01\omega)^3$. (a) Plot a Nyquist diagram for T, showing the locus of the complex quantity $T = T_{real} + jT_{imag}$, using T_{real} as the abscissa and T_{imag} as the ordinate. (b) Is the amplifier stable? (c) What value of the constant in the numerator of T defines the borderline between stability and instability?
26. Use the method illustrated in Fig. 10.15 to investigate the stability of an inverting amplifier having a low-frequency gain of 20 dB and an open-loop gain specified by the following asymptotes: 100 dB constant, $0 \le \omega \le 100$; -20 dB per decade, $100 \le \omega \le 10^5$; -60 dB per decade, $\omega > 10^5$. What value of G_0 would provide reasonable gain and ensure stability?
27. An op amp has an open-loop gain function

$$A = \frac{10^5}{\left(1 + \dfrac{j\omega}{100}\right)\left(1 + \dfrac{j\omega}{10^5}\right)^2}$$

It is used in an inverting amplifier with $R_1 = 1\ \text{k}\Omega$ and $R_f = 125\ \text{k}\Omega$. Plot $|G|_{dB}$ vs. ω on a logarithmic frequency scale and try to discover a peak in the response.
28. Elements used in the differentiator of Fig. 10.16 are $C_1 = 1\ \mu\text{F}$, $R_f = 1\ \text{M}\Omega$, $A = 10^4$, $R_i = \infty$, and $R_o = 0$. (a) Show that the voltages $v_s = v_o = 0$ for $t < 0$ and $v_s = -1.0001t$, $v_o = 1 - e^{-10{,}001t}$ for $t > 0$ satisfy the differential equation in Eq. (25). (b) Sketch v_o and $-R_fC_1\, dv_s/dt$ on the same time axis. How accurate is the differentiation?
29. A differentiator circuit similar to Fig. 10.16 uses $C_1 = 100$ pF, $R_f = 10\ \text{k}\Omega$, and an op amp whose gain is approximated by $A = 10^5/(1 + j0.01\omega)$. If $R_i = \infty$ and $R_o = 0$, (a) in what frequency range is differentiation accurate? (b) What is the frequency of the peak differentiator output?

Fig. 10.23 See Problem 33.

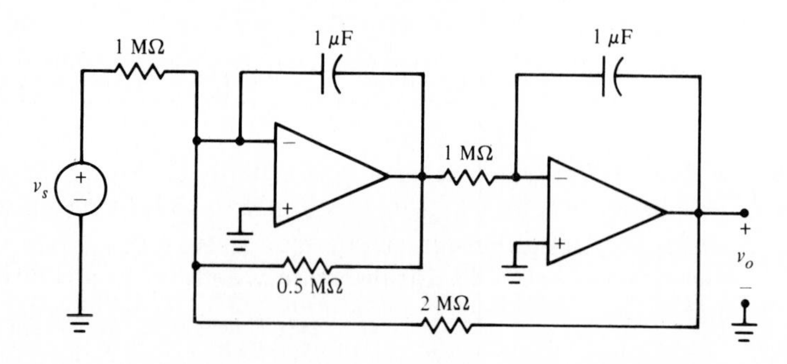

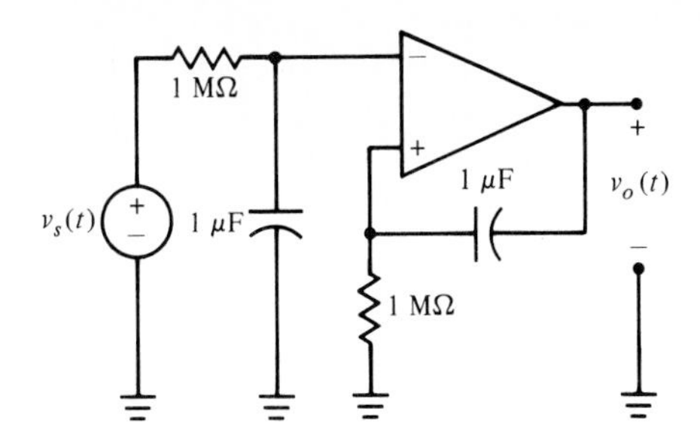

Fig. 10.24 See Problem 34.

30. The improved differentiator circuit of Fig. 10.19 uses passive elements $C_1 = 100$ pF, $R_f = 10$ kΩ, and $C_f = 100$ pF. If the op amp is represented by $A = 10^5/(1 + j0.01\omega)$, (a) sketch a curve of $|G|_{\text{dB}}$ vs. ω, $10^3 \leq \omega \leq 10^8$. (b) Identify the range of satisfactory differentiation.
31. Element values in the integrator circuit of Fig. 10.20 are $C_f = 1$ nF, $R_1 = 10$ kΩ, and $A = 10^4/(1 + j0.01\omega)$ for the op amp. (a) Sketch $|G|_{\text{dB}}$ vs. ω, $1 \leq \omega \leq 10^7$ rad/s. (b) Specify the range over which integration is reasonably accurate.
32. In the example for which $|A|_{\text{dB}}$ and $|G|_{\text{dB}}$ vs. ω are shown in Fig. 10.21, make a single change in the conditions by letting $A = 10^6/(1 + j0.1\omega)$ and plot new curves of $|A|_{\text{dB}}$ and $|G|_{\text{dB}}$.
33. If both op amps in Fig. 10.23 are ideal, show that v_o and v_s are related by

$$\frac{d^2v_o}{dt^2} + 2\frac{dv_o}{dt} - \frac{v_o}{2} = v_s$$

34. Assume an ideal op amp in the circuit of Fig. 10.24 and determine $v_o(t)$ as a function of $v_s(t)$.

Appendixes

LM308 operational amplifier

general description

The LM308 is a precision operational amplifier featuring input currents nearly a thousand times lower than industry standards like the LM709C. In fact, its performance approaches that of high quality FET amplifiers. The circuit is directly interchangeable with the LM301A in low frequency circuits and incorporates the same protective features which make its application nearly foolproof.

The device operates with supply voltages from ±2V to ±15V and has sufficient supply rejection to use unregulated supplies. Although the circuit is designed to work with the standard compensation for the LM301A, an alternate compensation scheme can be used to make it particularly insensitive to power supply noise and to make supply bypass capacitors unnecessary. Power consumption is extremely low, so the amplifiers are ideally suited for battery powered applications.

features

- Maximum input bias current of 7.0 nA
- Offset current less than 1.0 nA
- Supply current of only 300 μA, even in saturation
- Guaranteed drift characteristics

The low current error of the LM308 makes possible many designs that are not practical with conventional amplifiers. In fact, it operates from 10 MΩ source resistances, introducing less error than devices like the 709C with 10 kΩ sources. Integrators with worst case drifts less than 1 mV/sec and analog time delays in excess of one hour can be made using capacitors no larger than 1 μF. The device is well suited for use with piezoelectric, electrostatic or other capacitive transducers, in addition to low frequency active filters with small capacitor values.

schematic diagram and compensation circuits

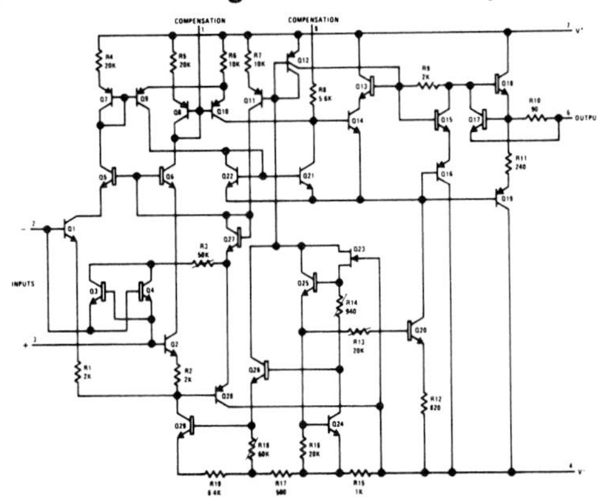

Standard Compensation Circuit

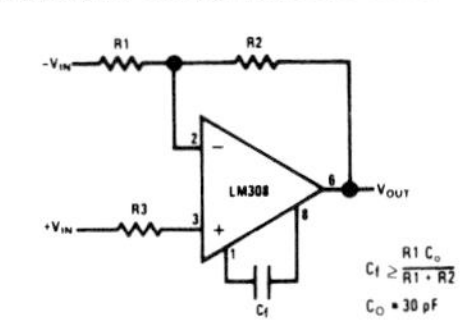

Alternate* Frequency Compensation

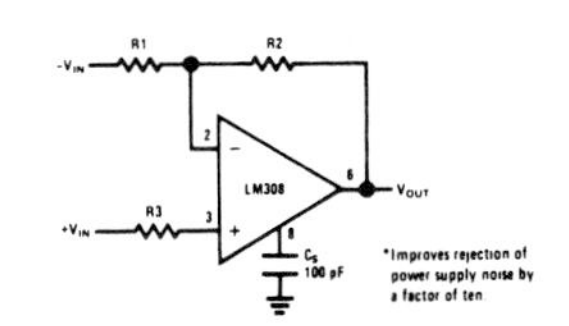

typical applications

Temperature Compensated Logarithmic Converter

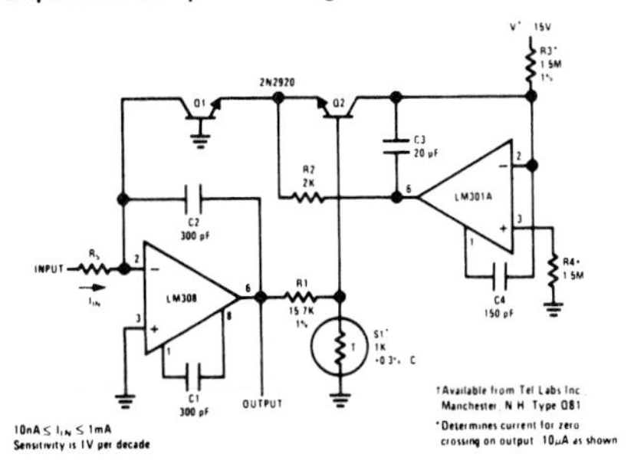

Fast† Summing Amplifier

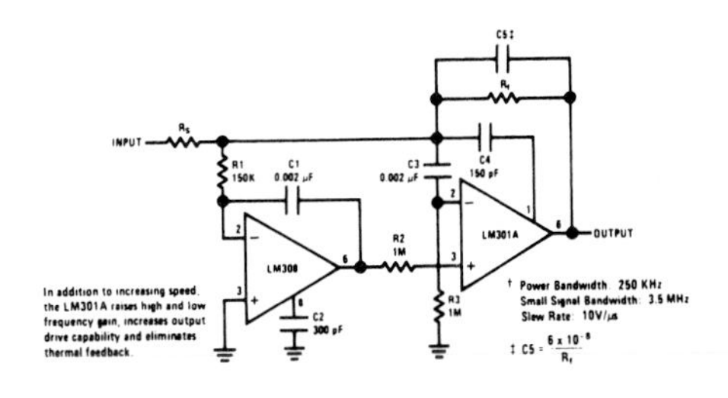

absolute maximum ratings

Supply Voltage	±18V
Power Dissipation (Note 1)	500 mW
Differential Input Current (Note 2)	±10 mA
Input Voltage (Note 3)	±15V
Output Short-Circuit Duration	Indefinite
Operating Temperature Range	0°C to 70°C
Storage Temperature Range	−65°C to 150°C
Lead Temperature (Soldering, 60 sec)	300°C

electrical characteristics (Note 4)

PARAMETER	CONDITIONS	MIN	TYP	MAX	UNITS
Input Offset Voltage	$T_A = 25°C$		2.0	7.5	mV
Input Offset Current	$T_A = 25°C$		0.2	1	nA
Input Bias Current	$T_A = 25°C$		1.5	7	nA
Input Resistance	$T_A = 25°C$	10	40		MΩ
Supply Current	$T_A = 25°C$, $V_S = \pm 15V$		0.3	0.8	mA
Large Signal Voltage Gain	$T_A = 25°C$, $V_S = \pm 15V$ $V_{OUT} = \pm 10V$, $R_L \geq 10\ k\Omega$	25	300		V/mV
Input Offset Voltage				10	mV
Average Temperature Coefficient of Input Offset Voltage			6.0	30	μV/°C
Input Offset Current				1.5	nA
Average Temperature Coefficient of Input Offset Current			2.0	10	pA/°C
Input Bias Current				10	nA
Large Signal Voltage Gain	$V_S = \pm 15V$, $V_{OUT} = \pm 10V$ $R_L \geq 10\ k\Omega$	15			V/mV
Output Voltage Swing	$V_S = \pm 15V$, $R_L = 10\ k\Omega$	±13	±14		V
Input Voltage Range	$V_S = \pm 15V$	±14			V
Common Mode Rejection Ratio		80	100		dB
Supply Voltage Rejection Ratio		80	96		dB

Note 1: The maximum junction temperature of the LM308 is 85°C. For operating at elevated temperatures, devices in the TO-5 package must be derated based on a thermal resistance of 150°C/W, junction to ambient, or 45°C/W, junction to case.

Note 2: The inputs are shunted with back-to-back diodes for overvoltage protection. Therefore, excessive current will flow if a differential input voltage in excess of 1V is applied between the inputs unless some limiting resistance is used.

Note 3: For supply voltages less than ±15V, the absolute maximum input voltage is equal to the supply voltage.

Note 4: These specifications apply for $\pm 5V \leq V_S \leq \pm 15V$ and $0°C \leq T_A \leq 70°C$, unless otherwise specified.

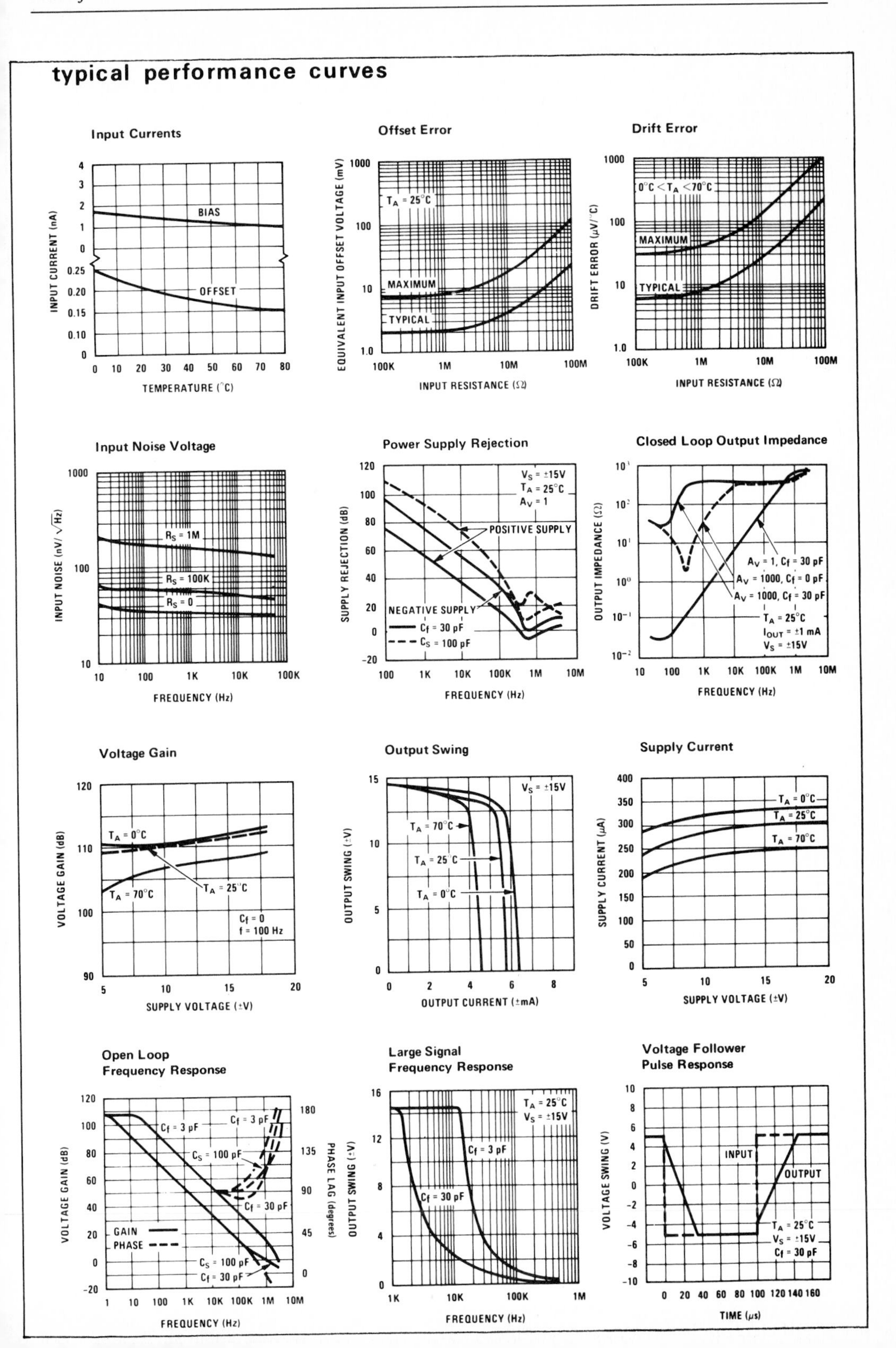
typical performance curves
Input Currents
BIAS
OFFSET
INPUT CURRENT (nA)
TEMPERATURE (°C)
Offset Error
T_A = 25°C
MAXIMUM
TYPICAL
EQUIVALENT INPUT OFFSET VOLTAGE (mV)
INPUT RESISTANCE (Ω)
Drift Error
0°C < T_A < 70°C
MAXIMUM
TYPICAL
DRIFT ERROR (μV/°C)
INPUT RESISTANCE (Ω)
Input Noise Voltage
R_S = 1M
R_S = 100K
R_S = 0
INPUT NOISE (nV/√Hz)
FREQUENCY (Hz)
Power Supply Rejection
V_S = ±15V
T_A = 25°C
A_V = 1
POSITIVE SUPPLY
NEGATIVE SUPPLY
C_f = 30 pF
C_S = 100 pF
SUPPLY REJECTION (dB)
FREQUENCY (Hz)
Closed Loop Output Impedance
A_V = 1, C_f = 30 pF
A_V = 1000, C_f = 0 pF
A_V = 1000, C_f = 30 pF
T_A = 25°C
I_{OUT} = ±1 mA
V_S = ±15V
OUTPUT IMPEDANCE (Ω)
FREQUENCY (Hz)
Voltage Gain
T_A = 0°C
T_A = 70°C
T_A = 25°C
C_f = 0
f = 100 Hz
VOLTAGE GAIN (dB)
SUPPLY VOLTAGE (±V)
Output Swing
V_S = ±15V
T_A = 70°C
T_A = 25°C
T_A = 0°C
OUTPUT SWING (±V)
OUTPUT CURRENT (±mA)
Supply Current
T_A = 0°C
T_A = 25°C
T_A = 70°C
SUPPLY CURRENT (μA)
SUPPLY VOLTAGE (±V)
Open Loop Frequency Response
C_f = 3 pF
C_f = 3 pF
C_S = 100 pF
C_f = 30 pF
GAIN
PHASE
C_S = 100 pF
C_f = 30 pF
VOLTAGE GAIN (dB)
PHASE LAG (degrees)
FREQUENCY (Hz)
Large Signal Frequency Response
T_A = 25°C
V_S = ±15V
C_f = 3 pF
C_f = 30 pF
OUTPUT SWING (±V)
FREQUENCY (Hz)
Voltage Follower Pulse Response
INPUT
OUTPUT
T_A = 25°C
V_S = ±15V
C_f = 30 pF
VOLTAGE SWING (V)
TIME (μs)

definition of terms

Input Offset Voltage: That voltage which must be applied between the input terminals through two equal resistances to obtain zero output voltage.

Input Offset Current: The difference in the currents into the two input terminals when the output is at zero.

Input Voltage Range: The range of voltages on the input terminals for which the offset specifications apply.

Input Bias Current: The average of the two input currents.

Common Mode Rejection Ratio: The ratio of the input voltage range to the peak-to-peak change in input offset voltage over this range.

Input Resistance: The ratio of the change in input voltage to the change in input current on either input with the other grounded.

Supply Current: The current required from the power supply to operate the amplifier with no load and the output at zero.

Output Voltage Swing: The peak output voltage swing, referred to zero, that can be obtained without clipping.

Large-Signal Voltage Gain: The ratio of the output voltage swing to the change in input voltage required to drive the output from zero to this voltage.

Power Supply Rejection: The ratio of the change in input offset voltage to the change in power supply voltages producing it.

connection diagrams

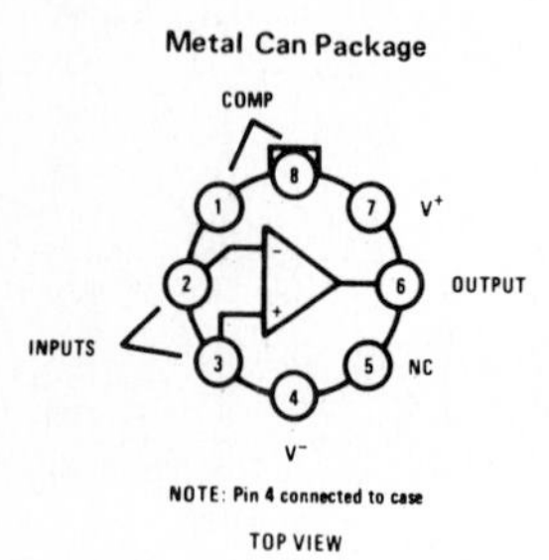

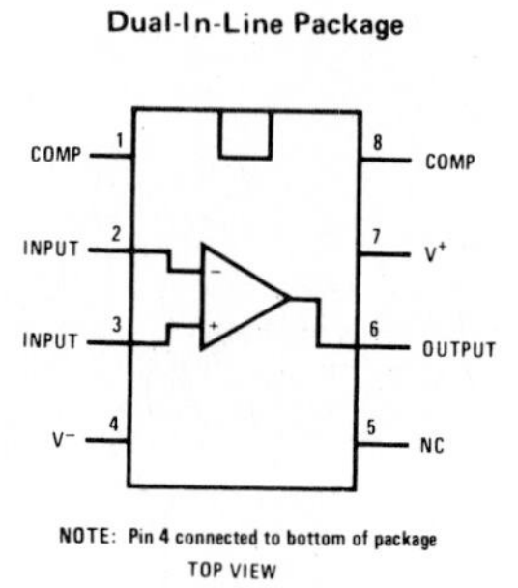

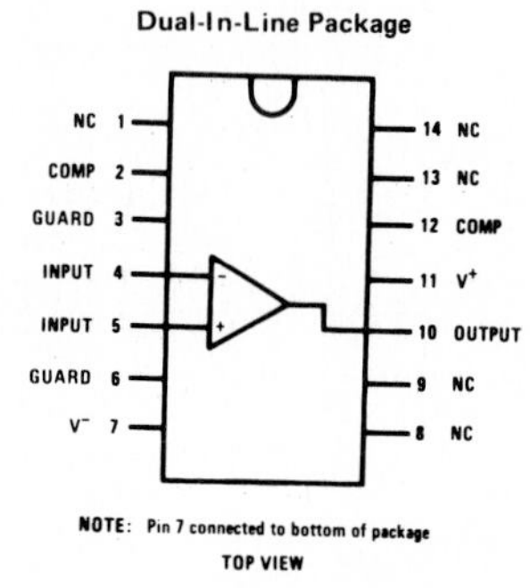

physical dimensions

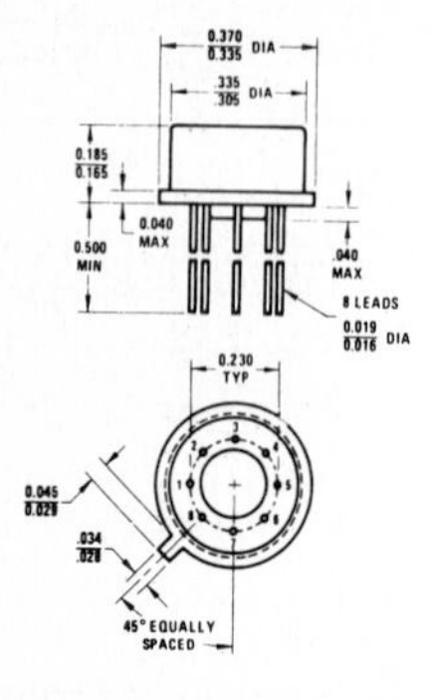

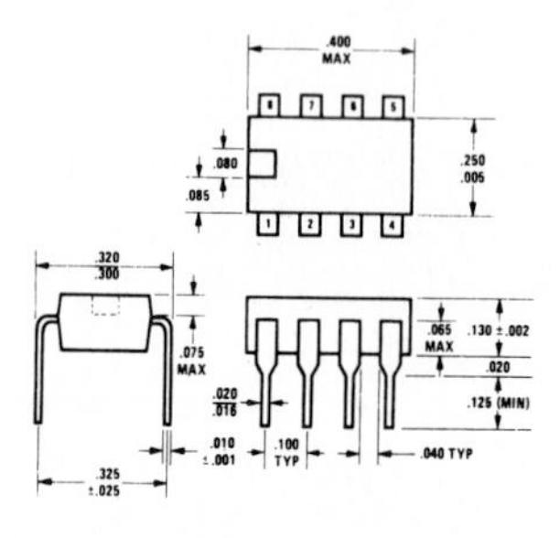

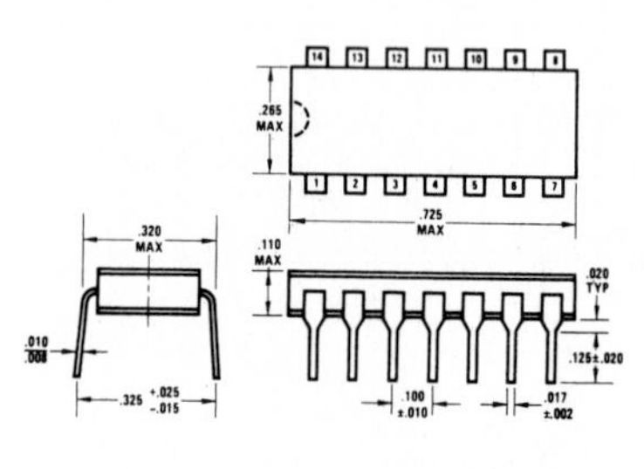

Order Number LM308(H)

Order Number LM308(N)

Order Number LM308(D)

Manufactured under one or more of the following U.S. patents: 3083262, 3189758, 3231797, 3303356, 3317671, 3323071, 3381071, 3408542, 3421025, 3426423, 3440498, 3518750, 3519897, 3557431, 3560765, 3566218, 3571630, 3575609, 3579059, 3593069, 3597640, 3607469, 3617859, 3631312, 3633052, 3638131, 3648071, 3651565, 3693248.

National Semiconductor Corporation
2900 Semiconductor Drive, Santa Clara, California 95051, (408) 732-5000/TWX (910) 339-9240

National Semiconductor GmbH
8080 Fuerstenfeldbruck, Industriestrasse 10, West Germany, Tele. (08141) 1371, 1372, 1373, 1374,1484/Telex 527-649

National Semiconductor (UK) Ltd.
Larkfield Industrial Estate, Greenock, Scotland, Tele. 33251/Telex 778-632

Ultra Low Drift BIFET Operational Amplifier

AD547

ATURES

ra Low Drift (1μV/°C–AD547L)
w Offset Voltage (0.25mV–AD547L)
w Input Bias Currents (25pA–AD547L, K)
w Quiescent Current (1.5mA)
w Noise (2μV p-p)
h Open Loop Gain (108dB–AD547K, L, S)

DUCT DESCRIPTION

AD547 is a monolithic, FET input operational amplifier ining the very low input bias current advantages of a T op amp with offset and drift performance previous- ailable only from high quality bipolar amplifiers.

exclusive Analog Devices laser wafer trim process trims the input offset voltage and offset voltage drift to levels wer than any competing BIFET amplifier (1mV, 5μV/°C– 47JH, 0.25mV, 1μV/°C–AD547LH).

dition to superior low drift performance, the AD547 s the lowest guaranteed input bias currents of any BIFET ifier with 50pA max for the J grade and 25pA max for grade. Since Analog Devices, unlike most other manu- rers, specifies input bias current with the amplifiers ed-up, our BIFET amplifiers are specified under actual ting conditions.

AD547 is especially designed for use in applications, as instrumentation signal conditioning and analog utation, that require a high degree of precision at low

AD547 is offered in three commercial versions, J, K and cified from 0 to +70°C and one military version, the S, fied from –55°C to +125°C. All grades are packaged in etically sealed TO-99 cans. The S grade is available ned to MIL-STD-883, Level B.

PRODUCT HIGHLIGHTS

1. Advanced laser wafer trimming techniques reduce offset voltage drift to 1μV/°C max and reduce offset voltage to only 0.25mV max on the AD547L.
2. Analog Devices BIFET processing provides 25pA max (10pA typical) bias currents specified after 5 minutes of warm-up.
3. Low voltage noise, high open loop gain and outstanding offset performance make the AD547 a true precision BIFET amplifier.
4. The low quiescent supply current, typically 1.1mA, enables the AD547 to bring a new level of precision to applications where low power consumption is essential.
5. A further benefit on the AD547's low power consumption and low offset voltage drift is a minimal warm-up drift after power is applied (typically 7μV shift for the AD547L).

Route 1 Industrial Park; P.O. Box 280; Norwood, Mass. 02062
Tel: 617/329-4700 **TWX: 710/394-6577**
West Coast 714/842-1717 **Mid-West 312/653-5000** **Texas 214/231-5094**

SPECIFICATIONS (typical @ +25°C and V_S = ±15V dc unless otherwise specified)

MODEL	AD547J	AD547K	AD547L	AD547SH (AD547SH/883B
OPEN LOOP GAIN[2]				
V_{out} = ±10V, $R_L \geqslant 2k\Omega$	100,000 min	250,000 min	**	**
T_A = min to max	100,000 min	250,000 min	**	**
OUTPUT CHARACTERISTICS				
Voltage @ R_L = 2kΩ, T_A = min to max	±10V min (±12V typ)	*	*	*
Voltage @ R_L = 10kΩ, T_A = min to max	±12V min (±13V typ)	*	*	*
Short Circuit Current	25mA	*	*	*
FREQUENCY RESPONSE				
Unity Gain, Small Signal	1.0MHz	*	*	*
Full Power Response	50kHz	*	*	*
Slew Rate, Unity Gain	3.0V/µs	*	*	*
INPUT OFFSET VOLTAGE[3]	1.0mV max	0.5mV max	0.25mV max	**
vs. Temperature[4]	5µV/°C max	2µV/°C max	1µV/°C max	5µV/°C max
vs. Supply, T_A = min to max	200µV/V max	100µV/V max	**	**
INPUT BIAS CURRENT				
Either Input[5]	50pA max (10pA typ)	25pA max (10pA typ)		**
Input Offset Current	5pA	2pA		**
INPUT IMPEDANCE				
Differential	$10^{12}\Omega\|\|6pF$	*	*	*
Common Mode	$10^{12}\Omega\|\|6pF$	*	*	*
INPUT VOLTAGE RANGE				
Differential[6]	±20V	*	*	*
Common Mode	±10V min (±12V typ)	*	*	*
Common Mode Rejection, V_{IN} = ±10V	76dB min	80dB min	**	**
POWER SUPPLY				
Rated Performance	±15V	*	*	*
Operating	±(5 to 18)V	*	*	*
Quiescent Current	1.5mA max (1.1mA typ)	*	*	*
VOLTAGE NOISE				
0.1-10Hz	2µV p-p typ	4µV p-p max	**	**
10Hz	$70nV/\sqrt{Hz}$	*	*	*
100Hz	$45nV/\sqrt{Hz}$	*	*	*
1kHz	$30nV/\sqrt{Hz}$	*	*	*
10kHz	$25nV/\sqrt{Hz}$	*	*	*
TEMPERATURE RANGE				
Operating, Rated Performance	0 to +70°C	*	*	-55°C to +125°C
Storage	-65°C to +150°C	*	*	*

NOTES

[1] The AD547SH is offered screened to MIL-STD-883, Level B.

[2] Open Loop Gain is specified with V_{OS} both nulled and unnulled.

[3] Input Offset Voltage specifications are guaranteed after 5 minutes of operation at T_A = +25°C.

[4] Input Offset Voltage Drift is specified with the offset voltage unnulled. Nulling will induce an additional 3µV/°C/mV of nulled offset.

[5] Bias Current specifications are guaranteed maximum at either input after 5 minutes of operation at T_A = +25°C. For higher temperatures, the current doubles every 10°C.

[6] Defined as the maximum safe voltage between inputs, such that neither exceeds ±10V from ground.

*Specifications same as AD547J.
**Specifications same as AD547K.

Specifications subject to change without notice.

OUTLINE DIMENSIONS

Dimensions shown in inches and (mm).

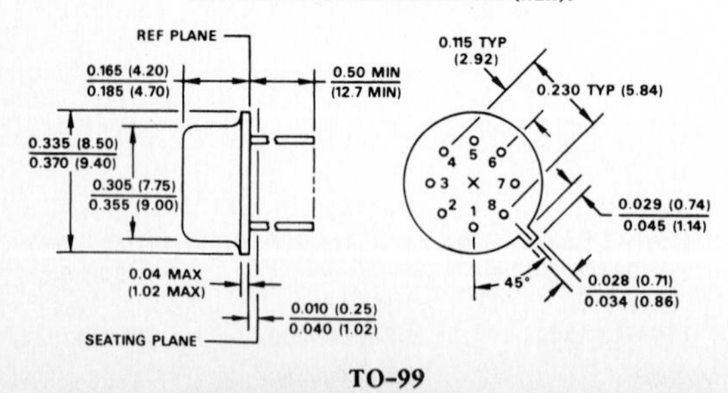

TO-99

PIN CONFIGURATION

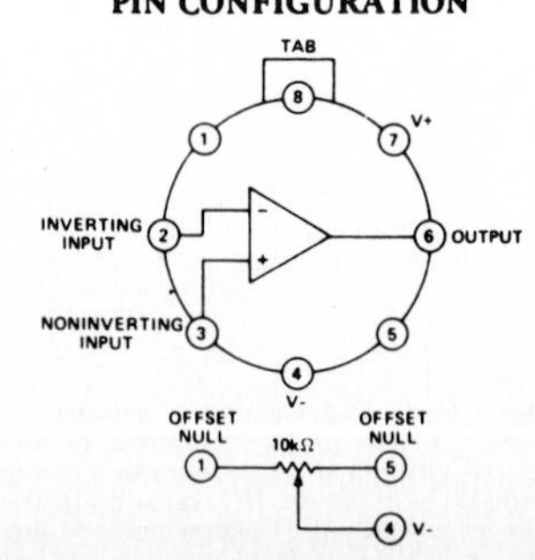

TOP VIEW

Typical Characteristics

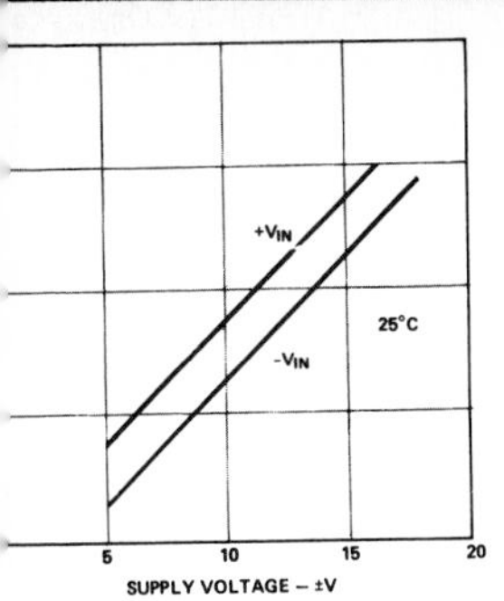

ure 1. Input Voltage Range vs. ·ply Voltage

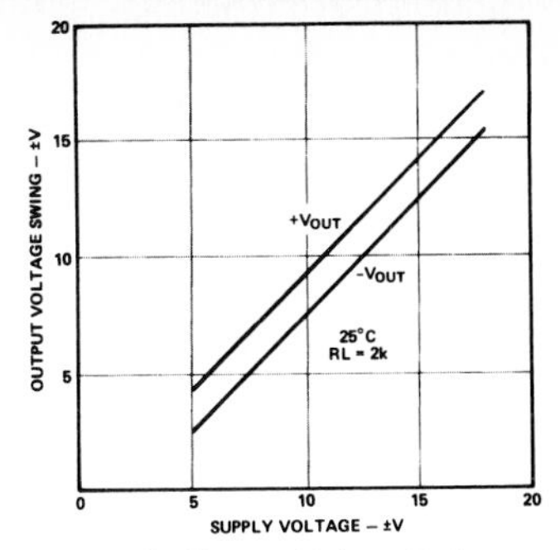

Figure 2. Output Voltage Swing vs. Supply Voltage

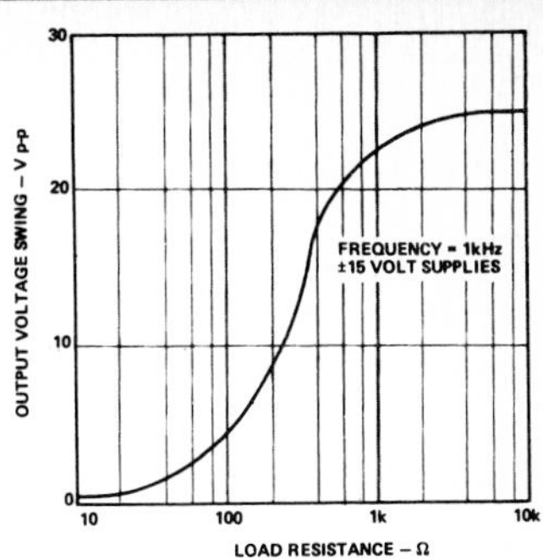

Figure 3. Output Voltage Swing vs. Resistive Load

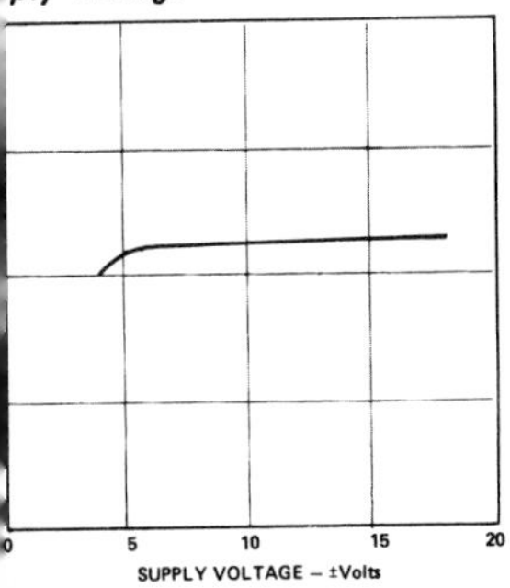

ure 4. Quiescent Current vs. ·ply Voltage

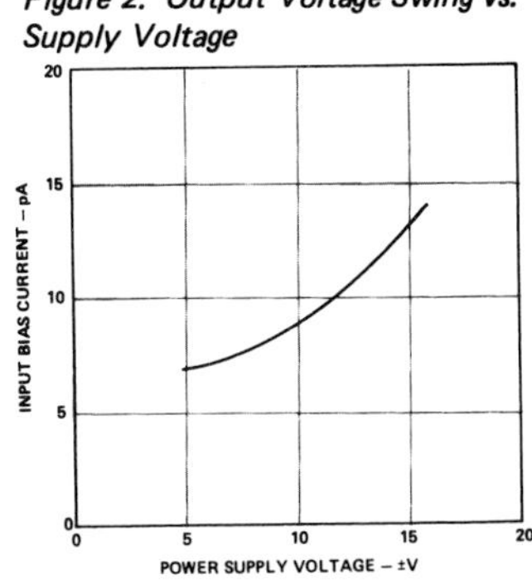

Figure 5. Input Bias Current vs. Supply Voltage

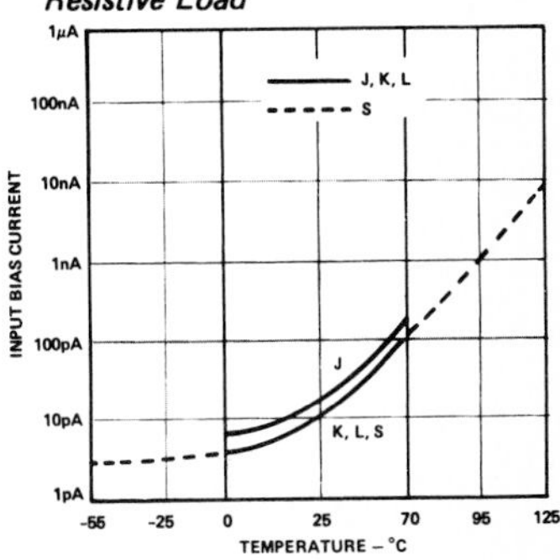

Figure 6. Input Bias Current vs. Temperature

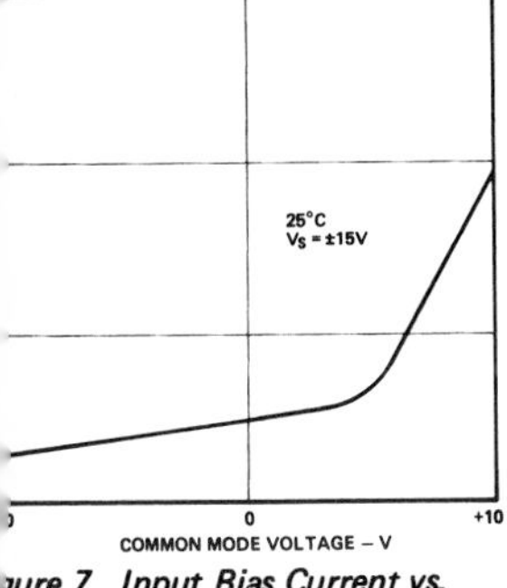

ure 7. Input Bias Current vs. ·V

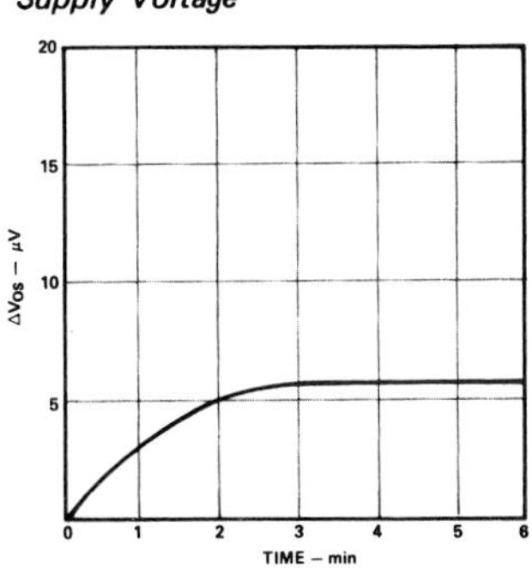

Figure 8. Input Offset Voltage Turn On Drift vs. Time

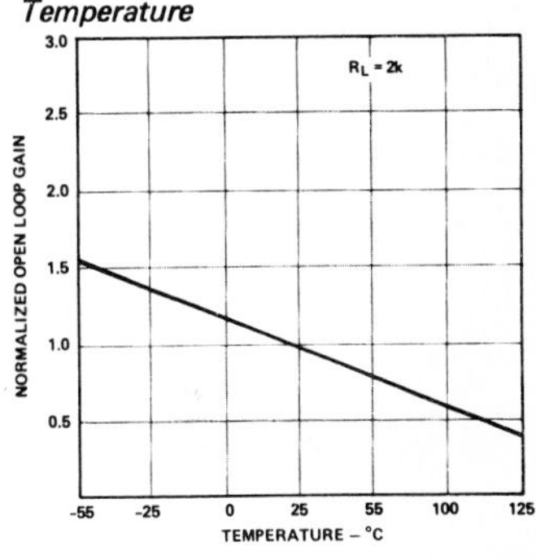

Figure 9. Open Loop Gain vs. Temperature

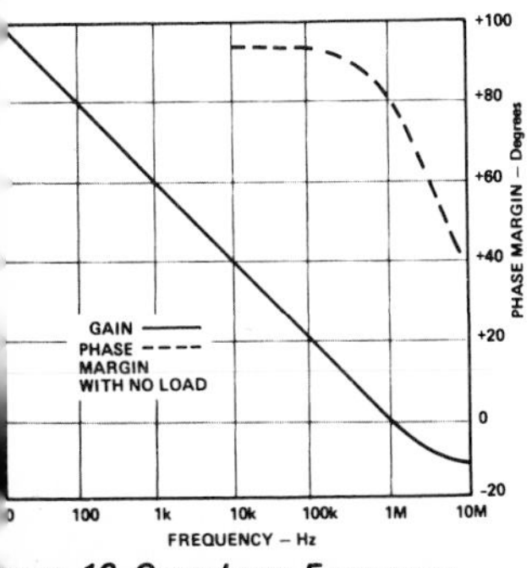

ure. 10 Open Loop Frequency ·sponse

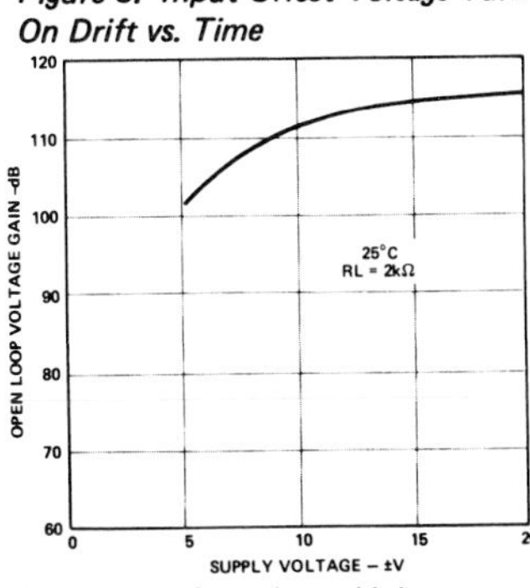

Figure 11. Open Loop Voltage Gain vs. Supply Voltage

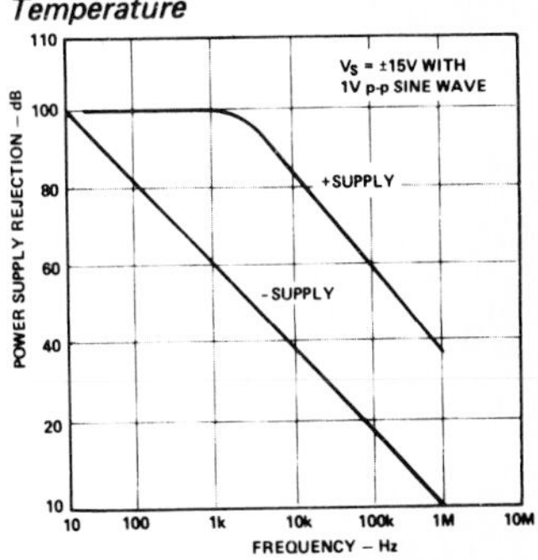

Figure 12. Power Supply Rejection vs. Frequency

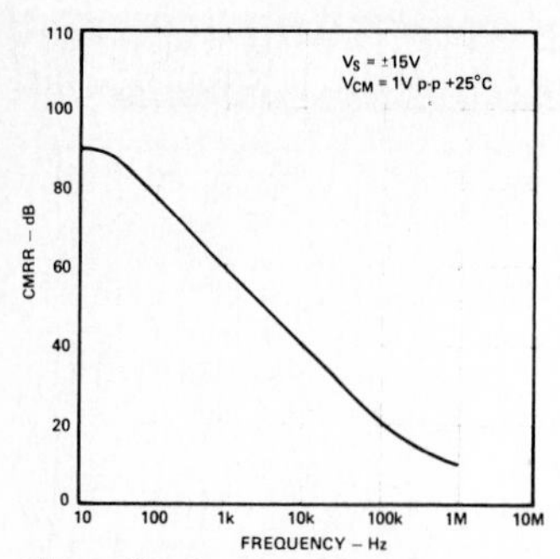

Figure 13. Common Mode Rejection vs. Frequency

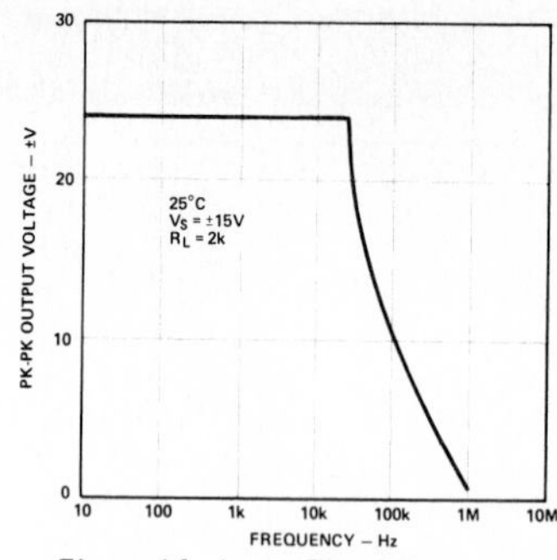

Figure 14. Large Signal Frequency Response

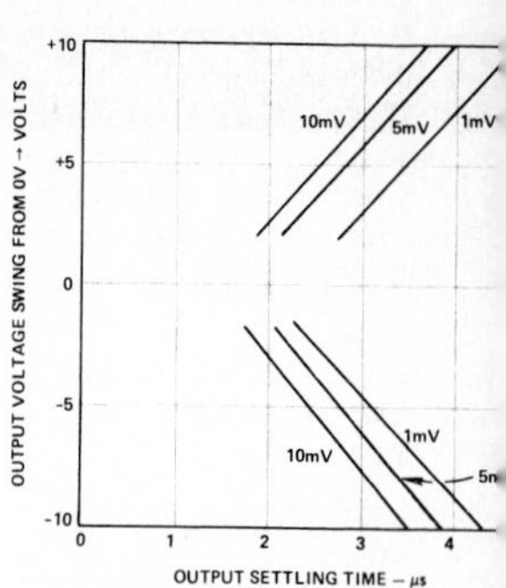

Figure 15. Output Settling Time vs. Output Swing and Error (Circuit of Figure 20)

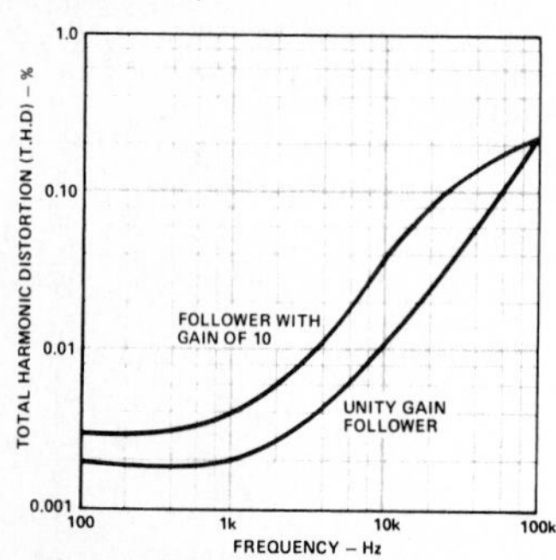

Figure 16. Total Harmonic Distortion vs. Frequency

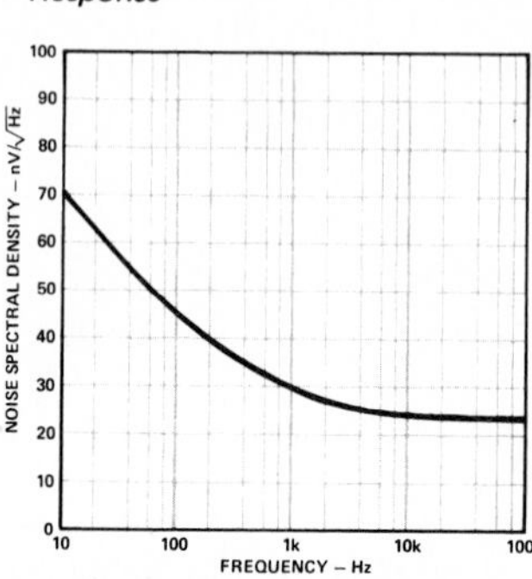

Figure 17. Input Noise Voltage Spectral Density

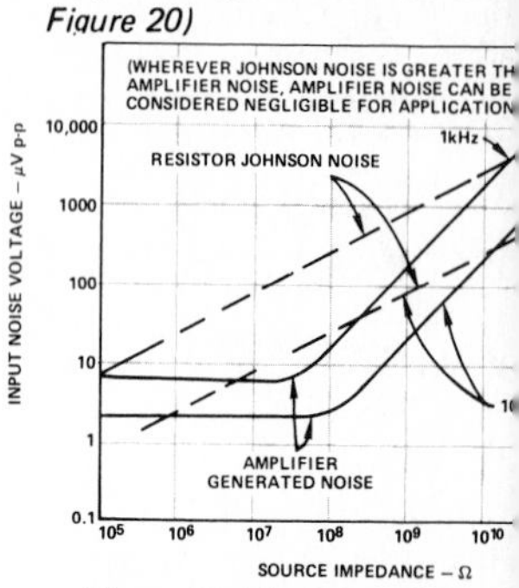

Figure 18. Total Input Noise vs. Source Resistance

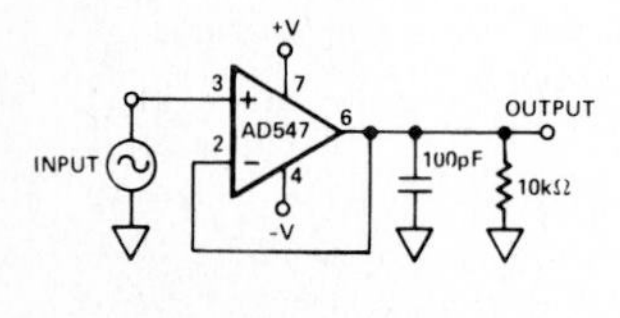

a. Unity Gain Follower

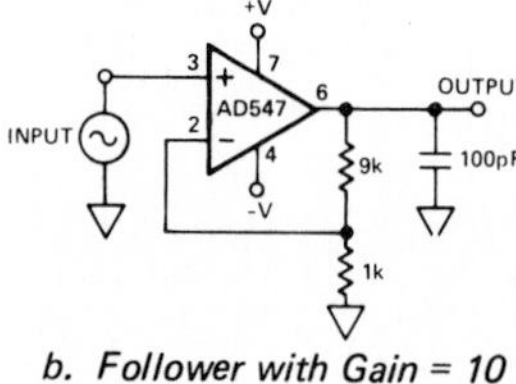

b. Follower with Gain = 10

Figure 19. T.H.D. Test Circuits

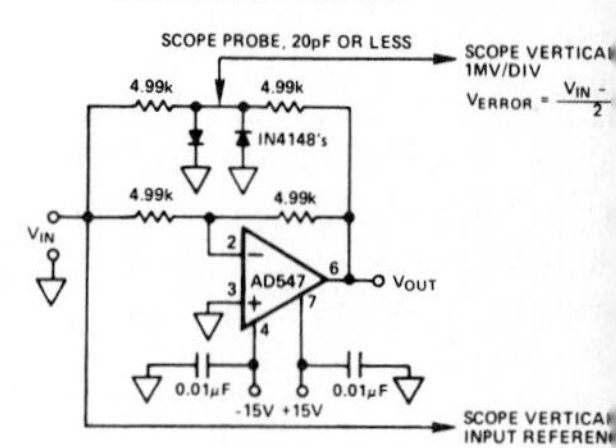

Figure 20. Settling Time Test Circuit

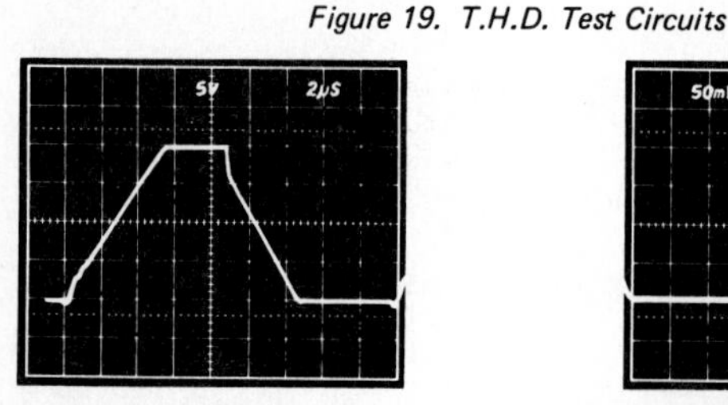

Figure 21a. Unity Gain Follower Pulse Response (Large Signal)

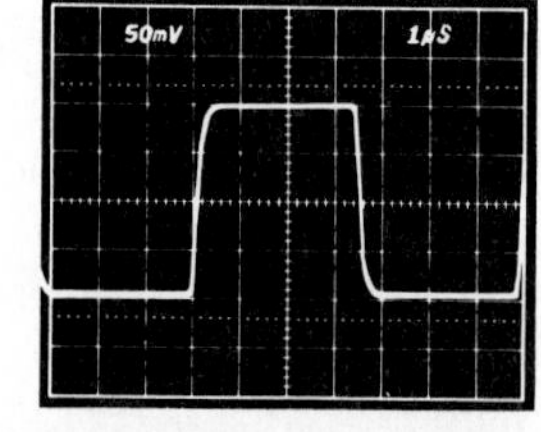

Figure 21b. Unity Gain Follower Pulse Response (Small Signal)

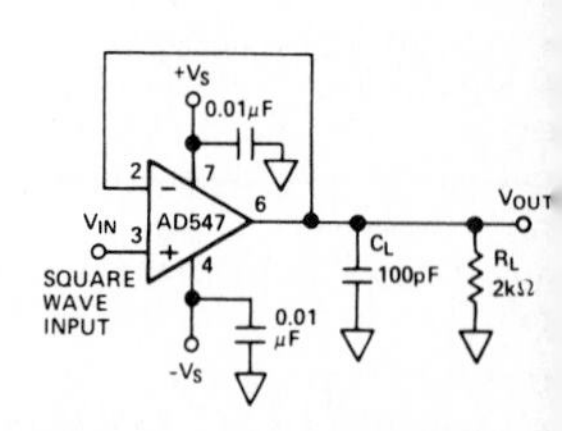

Figure 21c. Unity Gain Follower

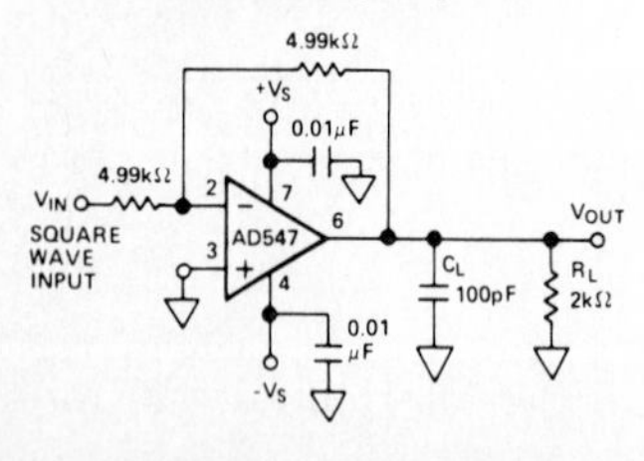

Figure 22a. Unity Gain Inverter

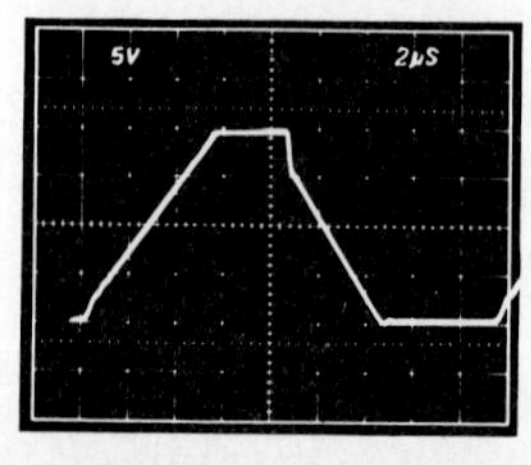

Figure 22b. Unity Gain Inverter Pulse Response (Large Signal)

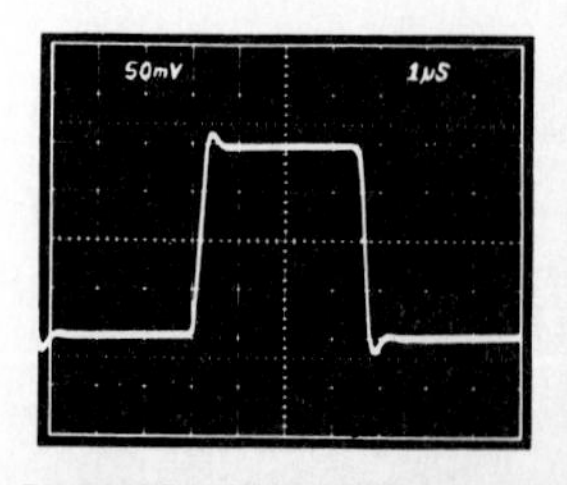

Figure 22c. Unity Gain Inverter Pulse Response (Small Signal)

·LICATION NOTES

AD547 was designed for high performance op-amp appli-ons that require true dc precision. To capitalize on all of performances available from the AD547 there are some :tical error sources that should be considered in using this :ision BIFET.

bias currents of JFET input amplifiers double with every C increase in chip temperature. Therefore, minimizing the :tion temperature of the chip will result in extending the formance limits of the device.

Heat dissipation due to power consumption is the main :ontributor to self-heating and can be minimized by reducing the power supplies to the lowest level allowed by the application.

The effects of output loading should be carefully con-sidered. Greater power dissipation increases bias currents and decreases open loop gain.

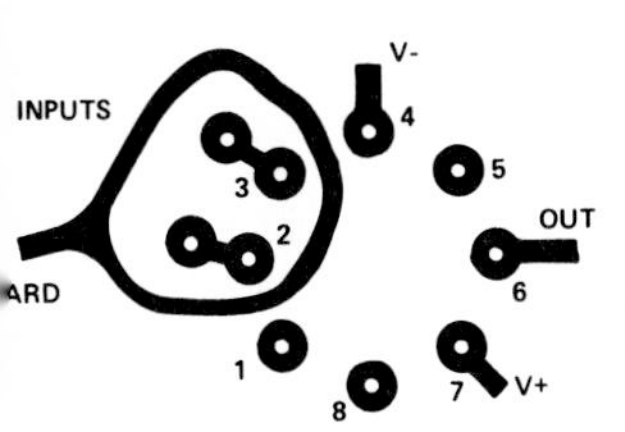

Figure 23. Board Layout for Guarding Inputs with TO-99 Package

'ARDING

e low input bias current (25pA) and low noise character-cs of the AD547 make it suitable for electrometer appli-ions such as photo diode preamplifiers and picoampere rent-to-voltage converters. The use of guarding techniques printed circuit board layout and construction is critical achieving the ultimate in low leakage performance avail-e from AD547. The input guarding scheme shown in ure 23 will minimize leakage as much as possible; the guard g is connected to a low impedance potential at the same el as the inputs. High impedance signal lines should not be :ended for any unnecessary length on a printed circuit; minimize noise and leakage, they must be carried in rigid elded cables.

'FSET NULLING

e AD547 has low initial offset voltage to compliment its :ellent drift performance. Even so, in some applications it necessary to null out even this low offset voltage. Precision polar amplifiers such as the AD510 or AD OP-07 ideally ve zero drift when their offset is nulled to zero, this is not : case for FET input amplifiers. A BIFET amplifier will pically exhibit a change of $3\mu V/^\circ C$ in drift for each mV of fset voltage nulled.

In view of this fact the circuit of Figure 24 is recommended for the most critical applications. The current in the AD590 is proportional to absolute temperature. This current through the 15Ω resistor generates a small drift proportional to the setting of the null potentiometer. This drift just cancels the drift induced by nulling. This circuit will typically remove all but $\pm 0.5\mu V/^\circ C$ per mV of nulled offset. For best results the 15Ω resistor should be connected directly to the V− pin of the AD547. This prevents any signal from coupling into the null terminals via changing currents in the supply rails.

INPUT PROTECTION

The AD547 is guaranteed for a maximum safe input potential equal to the power supply potential. The input stage design also allows for differential input voltages of up to ±0.5 volts while maintaining the full differential input resistance of $10^{12}\,\Omega$. This makes the AD547 suitable for voltage com-parators directly connected to a high impedance source.

Many instrumentation situations, such as flame detectors in gas chromatographs, involve measurement of low level cur-rents from high-voltage sources. In such applications, a sensor fault condition may apply a very high potential to the input of the current-to-voltage converting amplifier. This possibility necessitates some form of input protection. Many electro-meter type devices, especially CMOS designs, can require elaborate zener protection schemes which often compromise overall performance. The AD547 requires input protection only if the source is not current-limited and, as such, is similar to many JFET-input designs. (The failure would be due to overheating from excess current rather than voltage break-down.) If this is the case, a resistor in series with the affected input terminal is required so that the maximum overload cur-rent is 1.0mA (for example, 100kΩ for a 100 volt overload). This simple scheme will cause no significant reduction in performance and give complete overload protection. Figure 25 shows proper connections.

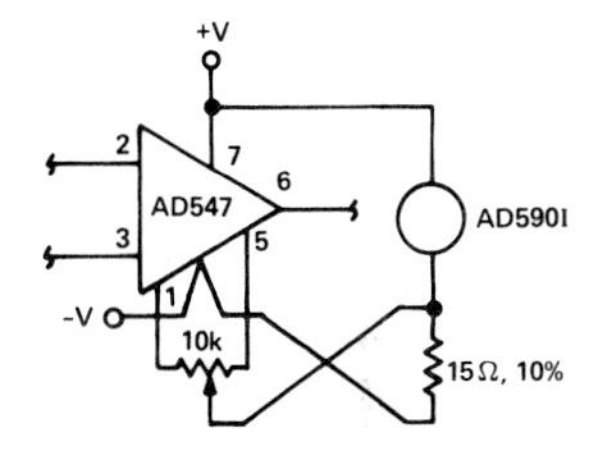

Figure 24. Offset Nulling Circuit

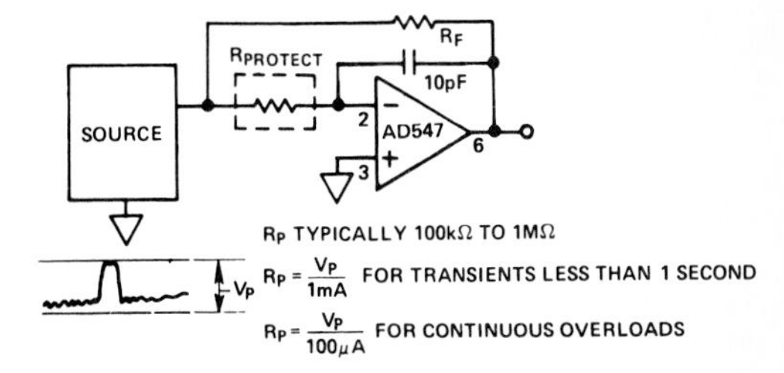

Figure 25. Input Protection

A PRECISION INSTRUMENTATION AMPLIFIER

The instrumentation amplifier of Figure 26 utilizes the outstanding characteristics of the AD547 to provide high performance at a reasonable cost.

The low drift, low bias currents and high open loop gain provide both high accuracy and linearity. The input amplifiers A1 and A2 are AD547Ls selected for their low offset characteristics (0.25mV of offset voltage and 1μV/°C drift) and low bias currents (25pA max). The use of the AD547Ls at the input guarantees a maximum input offset voltage drift of 2μV/°C with an input offset voltage of 0.5mV max untrimmed. A3 is an AD741JH and A4 is an AD547J. These serve two unrelated but critical purposes, A4 is the output amplifier and A3 is an active data guard.

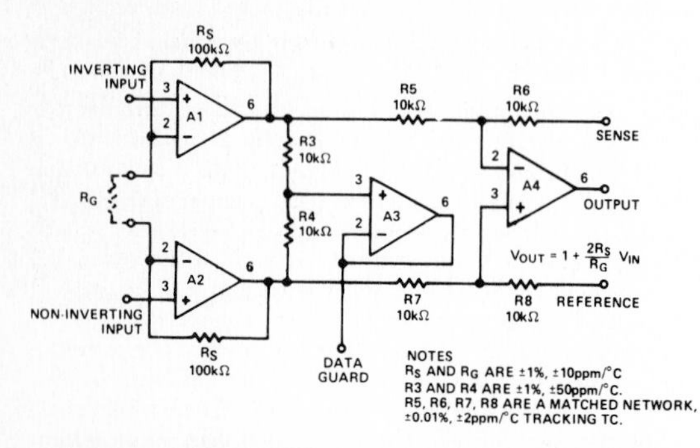

Figure 26. Precision FET Input Instrumentation Amplifier

USING THE AD547 IN LOG AMPLIFIER APPLICATIONS

Log amplifiers or log ratio amplifiers are useful in applications requiring compression of wide-range analog input data, linearization of transducers having exponential outputs, and analog computing, ranging from simple translation of natural relationships in log form (e.g., computing absorbance as the log-ratio of input currents), to the use of logarithms in facilitating analog computation of terms involving arbitrary exponents and multi-term products and ratios.

The picoamp level input current and low offset voltage of the AD547 make it suitable for wide dynamic range log amplifiers. Figure 27 is a schematic of a log ratio circuit employing the AD547 that can achieve less than 1% conformance error over 5 decades of current input, 1nA to 100μA. For voltage inputs, the dynamic range is typically 50mV to 10V for 1% error, limited on the low end by the amplifier's input offset voltage.

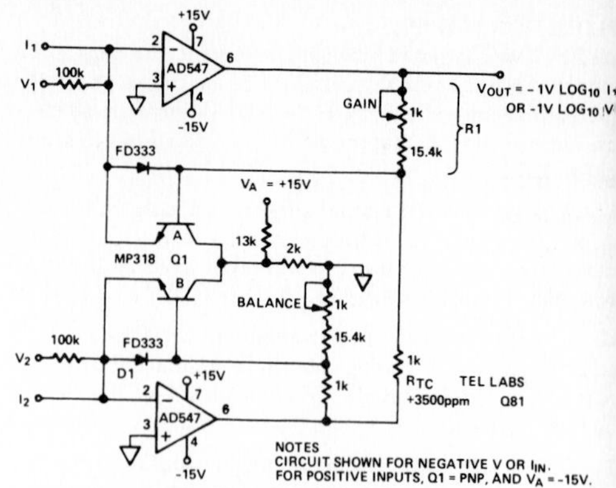

Figure 27. Log-Ratio Amplifier

The conversion between current (or voltage) input and log output is accomplished by the base emitter junctions of the dual transistor Q1. Assuming Q1 has $\beta > 100$, which is the c for the specified transistor, the base-emitter voltage on side is to a close approximation:

$$V_{BE\ A} = kT/q \ln I_1/I_{S1}$$

This circuit is arranged to take the difference of the V_{BE}'s o Q1A and Q1B, thus producing an output voltage proportion to the log of the *ratio* of the inputs:

$$V_{OUT} = -K\,(V_{BE\ A} - V_{BE\ B}) = -K\left(\frac{kT}{q}\right)(\ln I_1/I_{S1} - \ln I_2/$$

$$V_{OUT} = -K\,kT/q \ln I_1/I_2$$

The scaling constant, K is set by R1 and R_{TC} to about 16, to produce 1V change in output voltage per decade differen in input signals. R_{TC} is a special resistor with a +3500ppm/° temperature coefficient, which makes K inversely proportio to temperature, compensating for the "T" in kT/q. The log-ratio transfer characteristic is therefore independent of temperature.

This particular log ratio circuit is free from the dynamic pro lems that plague many other log circuits. The -3dB bandwid is 50kHz over the top 3 decades, 100nA to 100μA, and decreases smoothly at lower input levels. This circuit needs no additional frequency compensation for stable operation fro input current sources, such as photodiodes, that may have 100pF of shunt capacitance. For larger input capacitances a 20pF integration capacitor around each amplifier will provid a smoother frequency response.

This log ratio amplifier can be readily adjusted for optimum accuracy by following this simple procedure. First, apply V1 V2 = -10.00V and adjust "Balance" for V_{OUT} = 0.00V. Nex apply V1 = -10.00V, V2 = -1.00V and adjust gain for V_{OUT} +1.00V. Repeat this procedure until gain and balance readin are within 2mV of ideal values.

FD600

HIGH CONDUCTANCE, ULTRA-FAST PLANAR EPITAXIAL DIODE

ERAL DESCRIPTION - The FD600 is a silicon planar epitaxial diode that provides low capacitance, h conductance, and fast reverse recovery. With these features, the device is ideally suited for lications such as core devices, avalanche circuitry, logarithmic amplifiers for pulse applications for any critical circuit requiring high conductance and low internal power dissipation without rifice of speed capabilities.

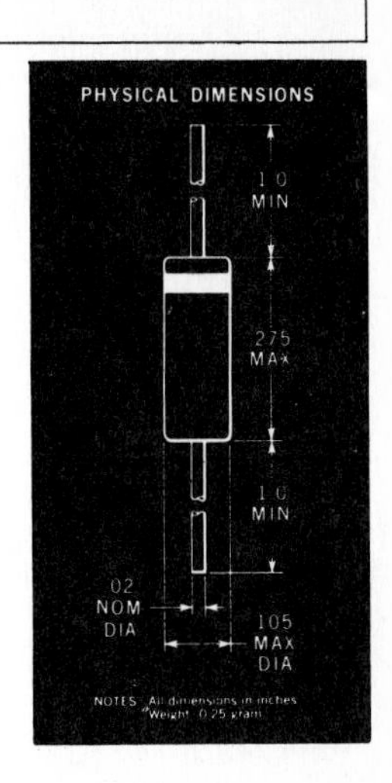

XIMUM RATINGS (25°C) (Note 1)

WIV	Working Inverse Voltage	50 Volts
I_o	Average Rectified Current	200 mA
i_f	Recurrent Peak Forward Current	900 mA
i_f(surge)	Peak Forward Surge Current Pulse Width of 1 second	1 Amp
i_f(surge)	Peak Forward Surge Current Pulse Width of 1 μsec	4 Amps
P	Power Dissipation	500 mW
P	Power Dissipation	170 mW at 125°C
T_A	Operating Temperature	-65°C to +150°C
T_{stg}	Storage Temperature, Ambient	-65°C to +175°C

CTRICAL CHARACTERISTICS (25°C Free Air Temperature unless otherwise noted)

Symbol	FACT Subgroup	Characteristic	Min.	Max.	Units	Test Conditions
$*V_F$	1a	Forward Voltage	0.87	1.00		I_F = 200 mA
V_F	1b	Forward Voltage	0.82	0.92		I_F = 100 mA
V_F	1b	Forward Voltage	0.76	0.86		I_F = 50 mA
V_F	1b	Forward Voltage	0.66	0.74		I_F = 10 mA
V_F	1b	Forward Voltage	0.54	0.62		I_F = 1 mA
I_R	1a	Reverse Current		0.1	μA	V_R = -50 V
I_R	1a	Reverse Current (150°C)		100	μA	V_R = -50 V
BV	1a	Breakdown Voltage	75			I_R = 5 μA
t_{rr}	1a	Reverse Recovery Time (Note 2)		4.0	mμsec	I_F = I_R = 10-200 mA, RL = 100 Ω
t_{rr}	1a	Reverse Recovery Time (Note 2)		6.0	mμsec	I_F = I_R = 200-400 mA, RL = 100 Ω
C_o	1a	Capacitance (Note 3)		2.5	$\mu\mu$f	V_R = 0 V, f = 1 Mc
ΔVF/°C		Change of Forward Voltage per Degree Change in Temperature	-1.8 mV/°C Typical			

TES:

The maximum ratings are limiting values above which life or satisfactory performance may be impaired.

Recovery to 0.1 I_F.

Capacitance as measured on Boonton Electronic Corporation Model No. 75-AS8 Capacitance Bridge or equivalent.

Leads are tinned. Gold plate with nickel strike may be obtained when specified.

4300 REDWOOD HIGHWAY • SAN RAFAEL, CALIFORNIA • (415) 479-8000 • TWX: 415-457-9100 • CABLE: FAIRSEMCO

MANUFACTURED UNDER ONE OR MORE OF THE FOLLOWING U. S. PATENTS: 2981877, 3025589, 3064167, 3108359, 3117260. OTHER PATENTS PENDING.

FAIRCHILD DIODE FD600

TYPICAL ELECTRICAL CHARACTERISTICS

FORWARD VOLTAGE VERSUS FORWARD CURRENT

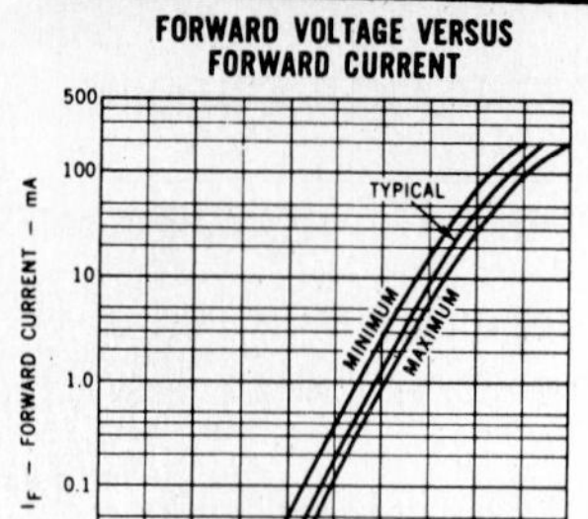

FORWARD CURRENT VERSUS TEMPERATURE COEFFICIENT

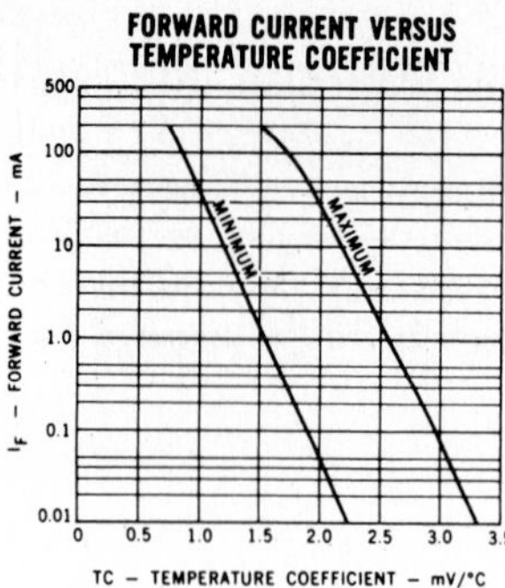

CAPACITANCE VERSUS REVERSE VOLTAGE

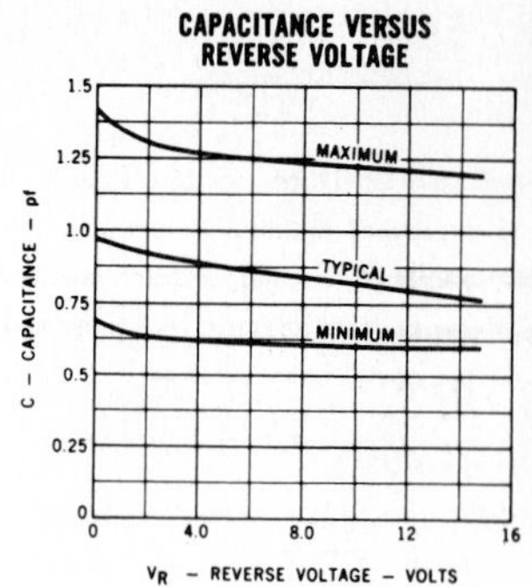

REVERSE VOLTAGE VERSUS REVERSE CURRENT

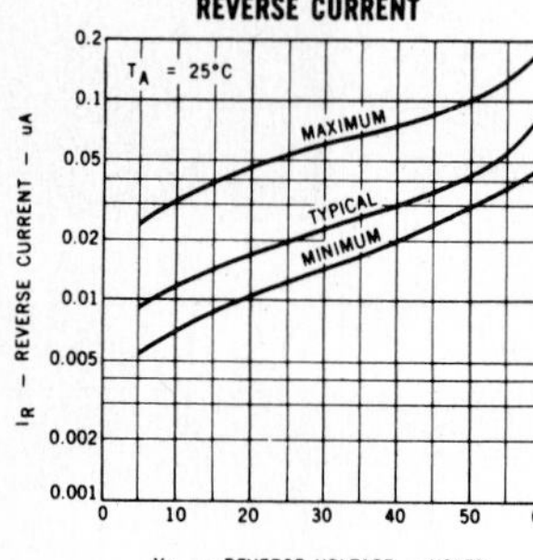

REVERSE CURRENT VERSUS AMBIENT TEMPERATURE

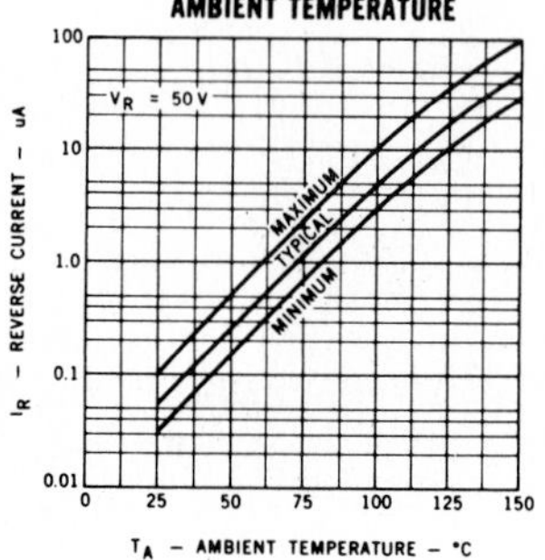

DYNAMIC IMPEDANCE VERSUS FORWARD CURRENT

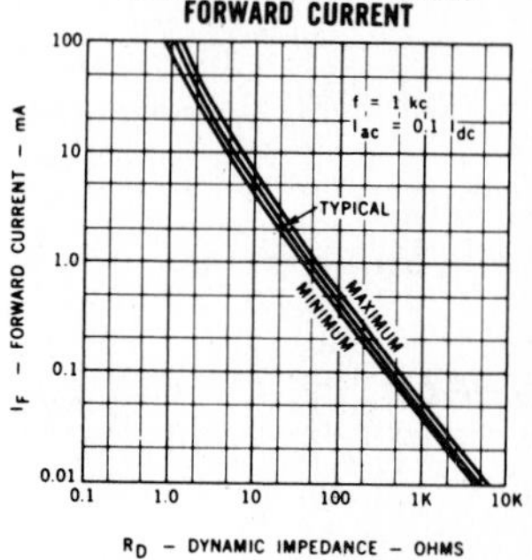

POWER DERATING CURVE

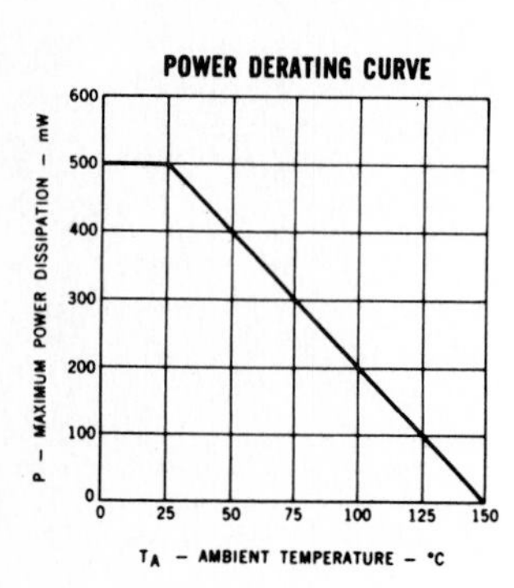

AVERAGE RECTIFIED CURRENT VERSUS AMBIENT TEMPERATURE

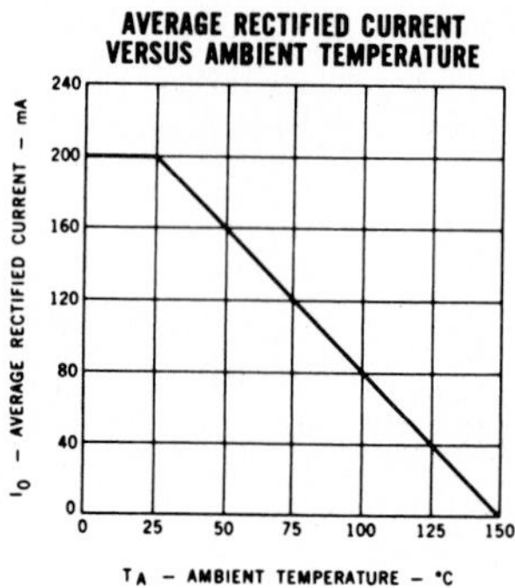

REVERSE RECOVERY TIME VERSUS FORWARD CURRENT ($I_F = I_R$)

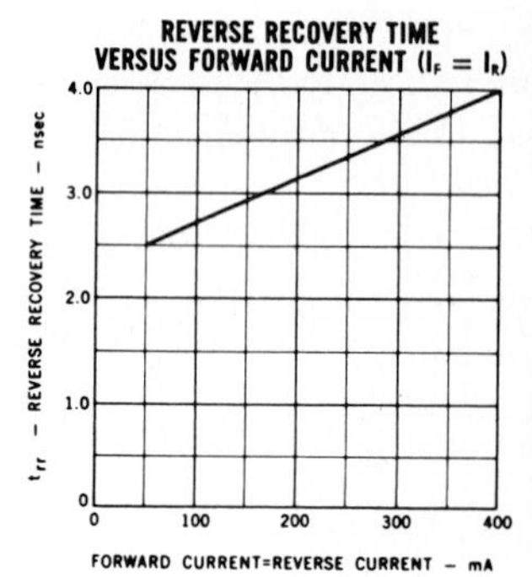

HARRIS SEMICONDUCTOR
N OF HARRIS-INTERTYPE CORPORATION

HA-2107/2207/2307

Operational Amplifiers

TURES

LOW OFFSET VOLTAGE OVER TEMPERATURE	3mV MAXIMUM
LOW BIAS CURRENT OVER TEMPERATURE	100nA MAXIMUM
LOW OFFSET CURRENT OVER TEMPERATURE	20nA MAXIMUM
LOW OFFSET VOLTAGE TEMPERATURE COEFFICIENT	$3\mu V/^{o}C$

NERAL DESCRIPTION

'he HA-2107, HA-2207 and HA-2307 are high performance eneral purpose operational amplifiers which are internally ompensated for unity gain stability.

'hese amplifiers have input and output overload protection. arge common mode signals do not cause latch-up. The IA-2107, HA-2207 and HA-2307 have guaranteed operating temperature ranges of $-55^{o}C$ to $+125^{o}C$, $-25^{o}C$ to $+85^{o}C$ nd $0^{o}C$ to $+70^{o}C$ respectively.

PACKAGES

CODE 2A

TO-99

170
500 MIN.
040
360
325
100 TYP.
200 TYP.
8 LEADS
.017 TYPICAL
SEATING PLANE
040
034
45°

ALL DIMENSIONS ARE IN INCHES.
ALL DIMENSIONS ± .010 UNLESS OTHERWISE SHOWN.

CODE 9W

TO-91

.010
100 TYP.
050 TYP
016 ± 002
250
250
750
005 ± 002
013 REF
SEATING PLANE
070 MAX

HEMATIC

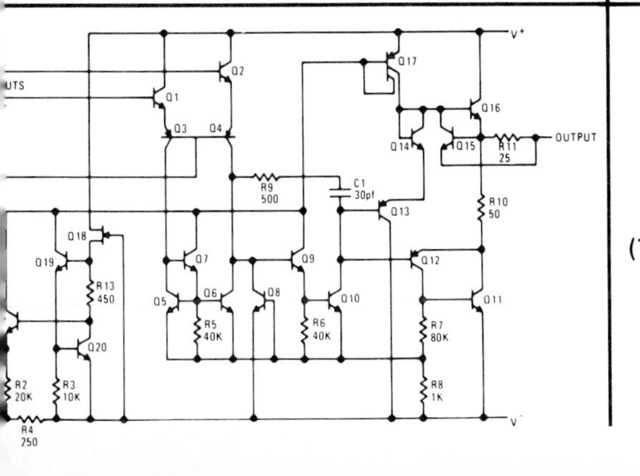

PIN OUTS

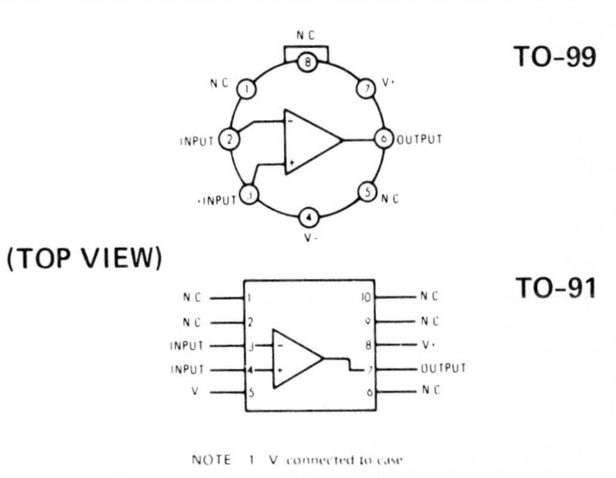

ABSOLUTE MAXIMUM RATINGS

SPECIFICATIONS

	HA-2107 HA-2207	HA-2307
Supply Voltage	±22.0V	±18.0V
Power Dissipation (Note 1)	500mW	
Differential Input Voltage	±30.0V	
Input Voltage (Note 2)	±15.0V	
Output Short Circuit Duration	Indefinite	

Operating Temperature Range	
HA-2107	-55°C to +125
HA-2207	-25°C to +85°
HA-2307	0°C to +70°
Storage Temperature Range	-65°C to +150°C
Lead Temperature (Soldering 60 seconds)	300°C

ELECTRICAL CHARACTERISTICS

PARAMETER	TEMPERATURE (Note 7)	HA-2107 HA-2207 MIN.	TYP.	MAX.	HA-2307 MIN.	TYP.	MAX.	UNITS
INPUT CHARACTERISTICS								
Offset Voltage	+25°C		0.7	2.0		2.0	7.5	mV
	Full			3.0			10	mV
Average Offset Voltage Temperature Coefficient	Full		3.0	15		6.0	30	μV/°C
Bias Current	+25°C		30	75		70	250	nA
	Full			100			300	nA
Offset Current	+25°C		1.5	10		3	50	nA
	Full			20			70	nA
Offset Current Temperature Coefficient	T_{Low} to +25°C		0.02	0.2		0.02	0.6	nA/°C
	+25°C to T_{High}		0.01	0.1		0.01	0.3	nA/°C
Input Resistance	+25°C	1.5	4		0.5	2		MΩ
Input Voltage Range (Note 4)	Full	±15.0			±12			V
TRANSFER CHARACTERISTICS								
Large Signal Voltage Gain (Notes 5 & 6)	+25°C	50K	160K		25K	160K		V/V
	Full	25K			15K			V/V
Common Mode Rejection Ratio	Full	80	96		70	90		dB
Output Voltage Swing								
R_L = 10KΩ	Full	±12.0	±14.0		±12.0	±14.0		V
R_L = 2KΩ (Note 6)	Full	±10.0	±13.0		±10.0	±13.0		V
POWER SUPPLY CHARACTERISTICS								
Supply Current (Note 4)	+25°C		1.8	3.0		1.8	3.0	mA
	125°C		1.2	2.5				mA
Supply Voltage Rejection Ratio	Full	80	96		70	96		dB

NOTES:

1. Derate TO-99 package at 6.8mW/°C for operation ambient temperature above 75°C and 4.9mW/°C above 50°C for the TO-91 package.
2. For supply voltages less than +15.0V, the absolute maximum input voltage is equal to supply voltage.
3. These specifications apply for ±5.0V ≤ V_S ≤ +20.0V unless otherwise specified.

(4) V_S = ±20.0V

(5) R_L = 2KΩ, V_{OUT} = ±10.0V

(6) V_S = ±15.0V

(7) Full = T_{Low} to T_{High} = -55°C to +125°C (HA-21
-25°C to +85°C (HA-22
0°C to +70°C (HA-23

ICAL

PERFORMANCE CURVES

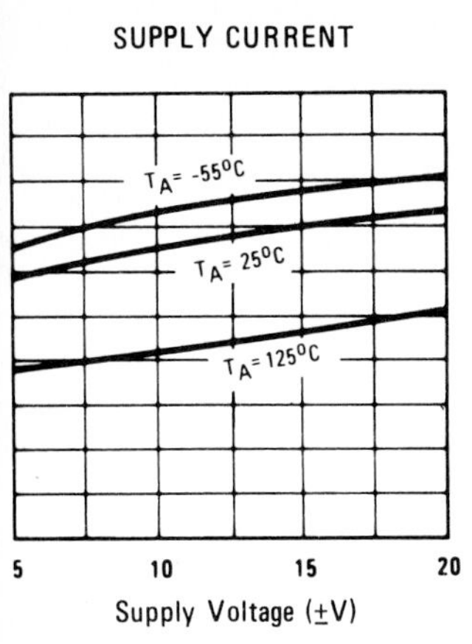

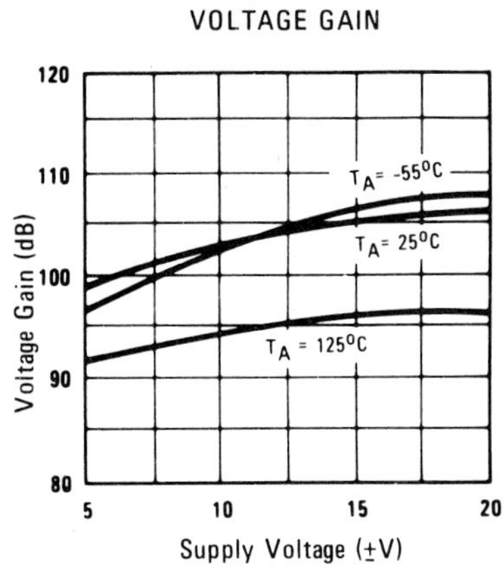

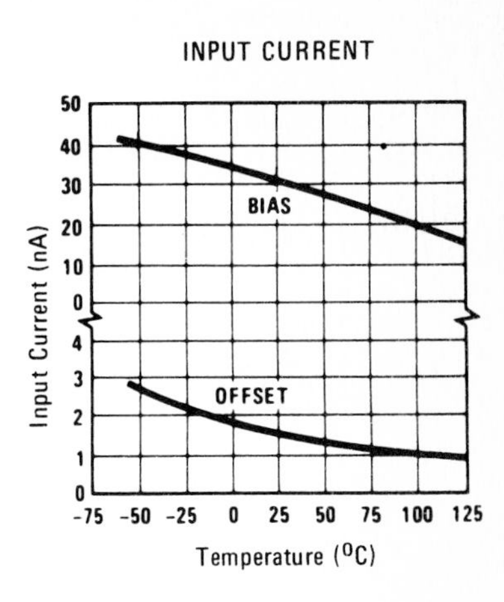

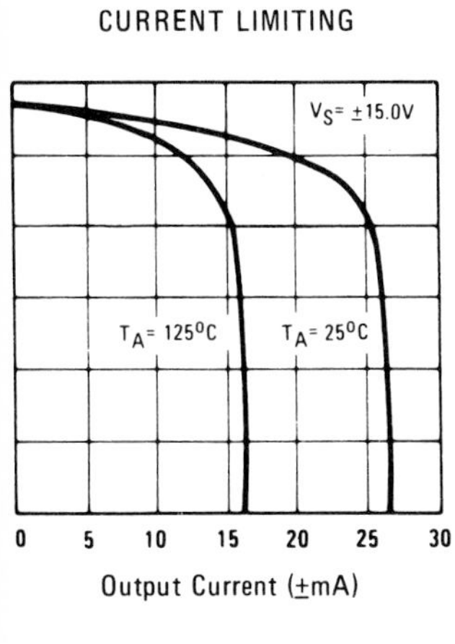

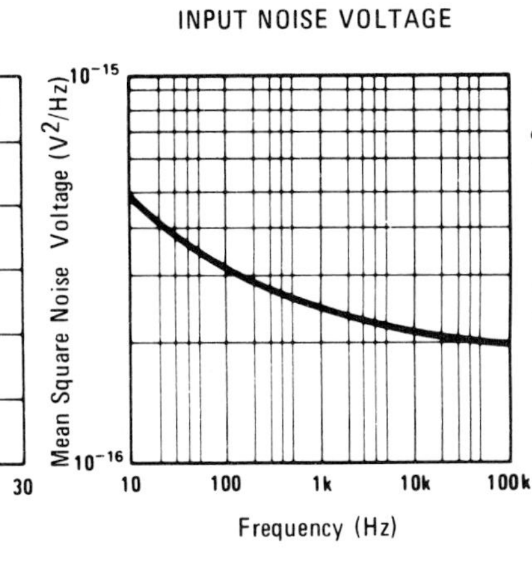

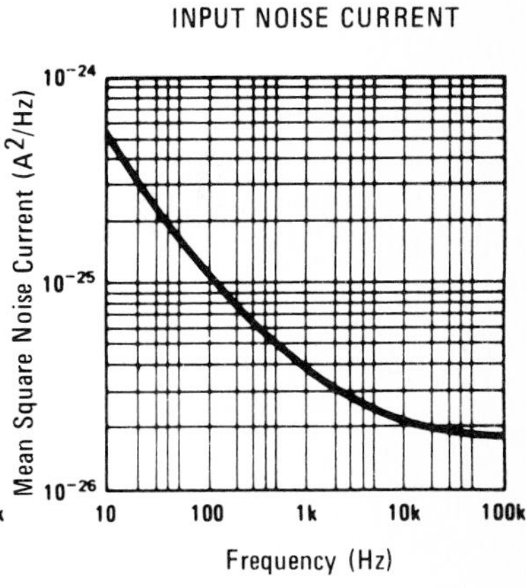

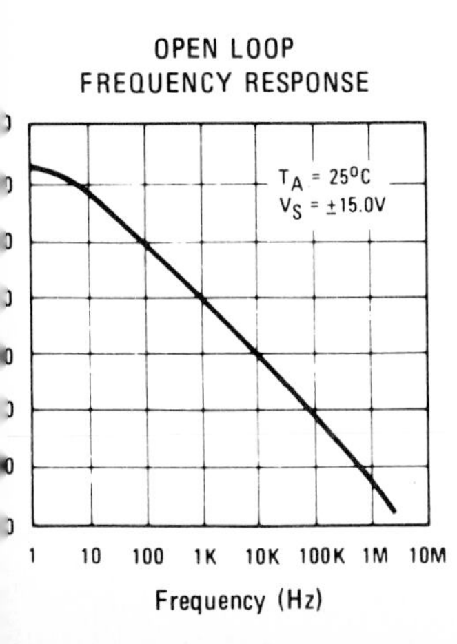

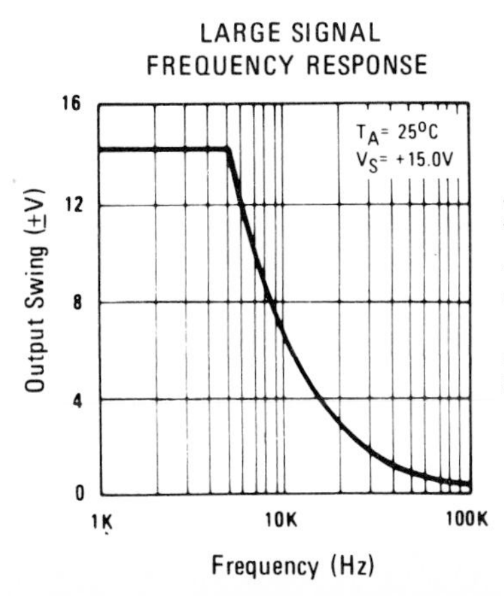

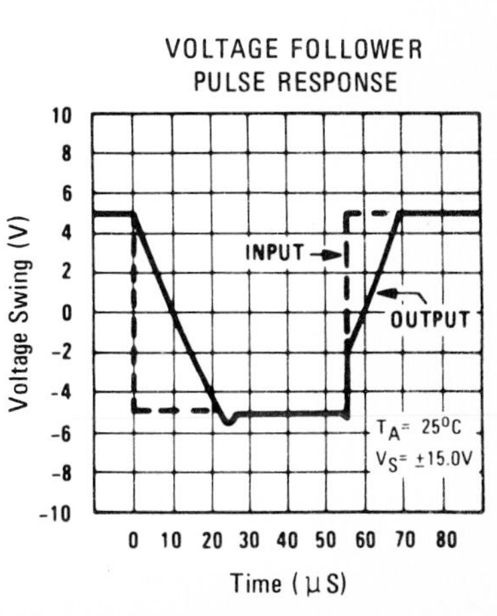

GUARANTEED **PERFORMANCE CURVES (continued)**

full operating temperature range (HA-2107: -55°C to +125°C; HA-2207: -25°C to +85°C; HA-2307: 0°C to +70°C)

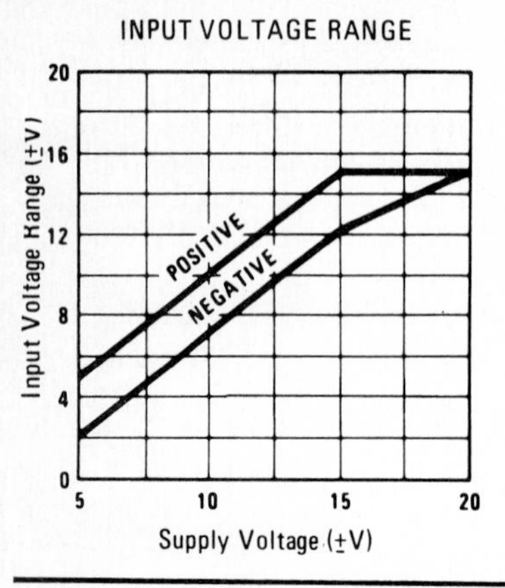

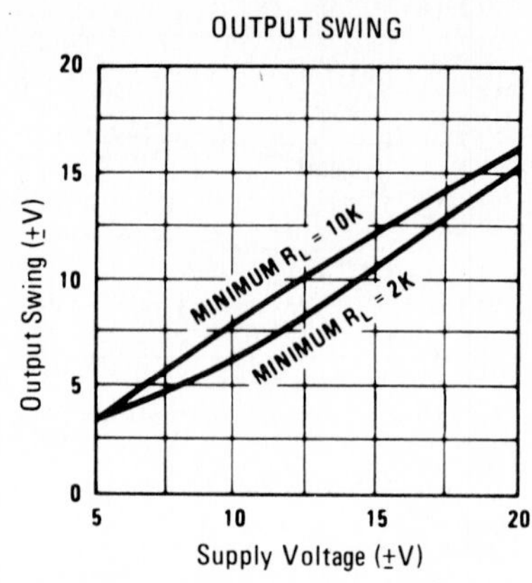

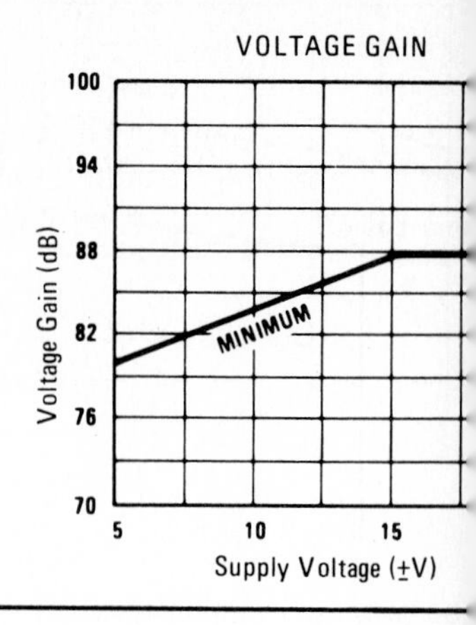

DEFINITIONS

INPUT OFFSET VOLTAGE—That voltage which must be applied between the input terminals through two equal resistances to force the output voltage to zero.

INPUT OFFSET CURRENT—The difference in the currents into the two input terminals when the output is at zero voltage.

INPUT BIAS CURRENT—The average of the currents flowing into the input terminals when the output is at zero voltage.

INPUT COMMON MODE VOLTAGE—The average referred to ground of the voltages at the two input terminals.

COMMON MODE RANGE—The range of voltages which is exceeded at either input terminal will cause the amplifier to cease operating.

COMMON MODE REJECTION RATIO—The ratio of a specified range of input common mode voltage to the peak-to-peak change in input offset voltage over this range.

OUTPUT VOLTAGE SWING—The peak symmetrical output voltage swing, referred to ground, that can be obtained without clipping.

INPUT RESISTANCE—The ratio of the change in input voltage to the change in input current.

OUTPUT RESISTANCE—The ratio of the change in output voltage to the change in output current.

VOLTAGE GAIN—The ratio of the change in output voltage to the change in input voltage producing it.

UNITY GAIN BANDWIDTH—The frequency at which the voltage gain of the amplifier is unity.

POWER SUPPLY REJECTION RATIO—The ratio of the change in input offset voltage to the change in power supply voltage producing it.

TRANSIENT RESPONSE—The closed loop step function response of the amplifier under small signal conditions.

GAIN BANDWIDTH PRODUCT—The product of the gain and the bandwidth at a given gain.

SLEW RATE (Rating Limiting)—The rate at which the output will move between full scale stops, measured in terms of volts per unit time. This limit to an ideal step function response is due to the non-linear behavior in an amplifier due to its limited ability to produce large, rapid changes in output voltage (slewing)...restricting it to rates of change of voltage lower than might be predicted by observing the small signal frequency response.

SETTLING TIME—Time required for output waveform to remain within 0.1 percent of final value.

SYMMETRICAL N-CHANNEL FIELD-EFFECT TRANSISTOR FOR VHF AMPLIFIER AND MIXER APPLICATIONS

TYPE 2N3823

- **Low Noise Figure: $\leq$ 2.5 db at 100 Mc**
- **Low C_{rss}: $\leq$ 2 pf**
- **High y_{fs}/C_{iss} Ratio (High-Frequency Figure-of-Merit)**
- **Cross Modulation Minimized by Square-Law Transfer Characteristic**

echanical data

N-CHANNEL EPITAXIAL PLANAR SILICON FIELD-EFFECT TRANSISTOR

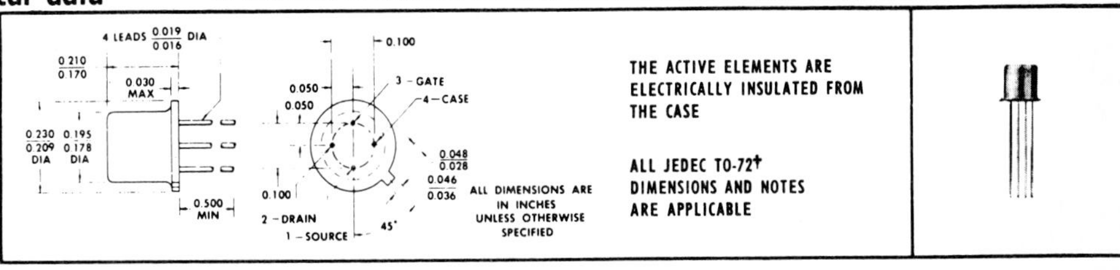

†TO-72 outline is same as TO-18 except for addition of a fourth lead.

bsolute maximum ratings at 25°C free-air temperature (unless otherwise noted)

Drain-Gate Voltage	30 v
Drain-Source Voltage	30 v
Reverse Gate-Source Voltage	–30 v
Gate Current	10 ma
Continuous Device Dissipation at (or below) 25°C Free-Air Temperature (See Note 1)	300 mw
Storage Temperature Range	–65°C to + 200°C
Lead Temperature 1/16 Inch from Case for 10 Seconds	300°C

lectrical characteristics at 25°C free-air temperature (unless otherwise noted)

	PARAMETER	TEST CONDITIONS‡	MIN	MAX	UNIT
$V_{(BR)GSS}$	Gate-Source Breakdown Voltage	$I_G = -1\ \mu a$, $V_{DS} = 0$	–30		v
I_{GSS}	Gate Cutoff Current	$V_{GS} = -20$ v, $V_{DS} = 0$		–0.5	na
		$V_{GS} = -20$ v, $V_{DS} = 0$, $T_A = 150°C$		–0.5	μa
I_{DSS}	Zero-Gate-Voltage Drain Current	$V_{DS} = 15$ v, $V_{GS} = 0$, See Note 2	4	20	ma
V_{GS}	Gate-Source Voltage	$V_{DS} = 15$ v, $I_D = 400\ \mu a$	–1	–7.5	v
$V_{GS(off)}$	Gate-Source Cutoff Voltage	$V_{DS} = 15$ v, $I_D = 0.5$ na		–8	v
$\|y_{fs}\|$	Small-Signal Common-Source Forward Transfer Admittance	$V_{DS} = 15$ v, $V_{GS} = 0$, f = 1 kc, See Note 2	3500	6500	μmho
$\|y_{os}\|$	Small-Signal Common-Source Output Admittance	$V_{DS} = 15$ v, $V_{GS} = 0$, f = 1 kc, See Note 2		35	μmho
C_{iss}	Common-Source Short-Circuit Input Capacitance	$V_{DS} = 15$ v, $V_{GS} = 0$, f = 1 Mc		6	pf
C_{rss}	Common-Source Short-Circuit Reverse Transfer Capacitance			2	pf
$\|y_{fs}\|$	Small-Signal Common-Source Forward Transfer Admittance	$V_{DS} = 15$ v, $V_{GS} = 0$, f = 200 Mc	3200		μmho
$Re(y_{is})$	Small-Signal Common-Source Input Conductance			800	μmho
$Re(y_{os})$	Small-Signal Common-Source Output Conductance			200	μmho

ES: 1. Derate linearly to 175°C free-air temperature at the rate of 2 mw/C°.
2. These parameters must be measured using pulse techniques. PW = 100 msec, Duty Cycle $\leq$ 10%.

icates JEDEC registered data.

fourth lead (case) is connected to the source for all measurements.

TEXAS INSTRUMENTS
INCORPORATED
SEMICONDUCTOR-COMPONENTS DIVISION
POST OFFICE BOX 5012 • DALLAS 22, TEXAS

TYPE 2N3823
N-CHANNEL EPITAXIAL PLANAR SILICON FIELD-EFFECT TRANSISTOR

* **operating characteristics at 25°C free-air temperature**

PARAMETER		TEST CONDITIONS‡	MAX	UN
NF	Common-Source Spot Noise Figure	$V_{DS} = 15$ v, $V_{GS} = 0$, f = 100 Mc, $R_G = 1$ kΩ	2.5	d

TYPICAL CHARACTERISTICS‡

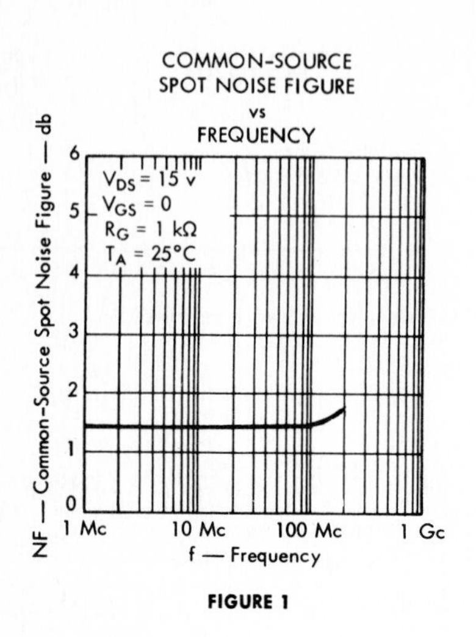

FIGURE 1

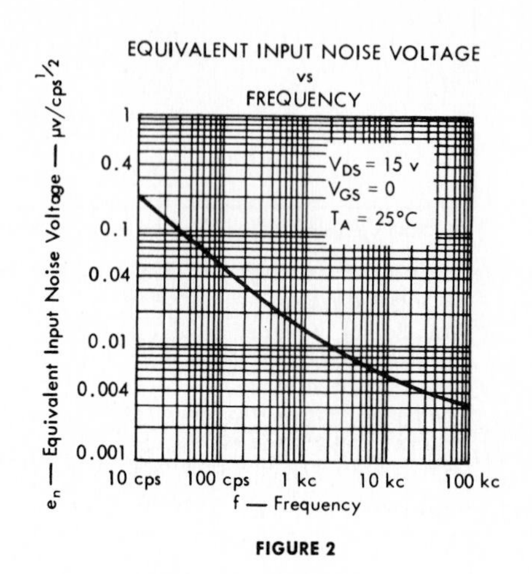

FIGURE 2

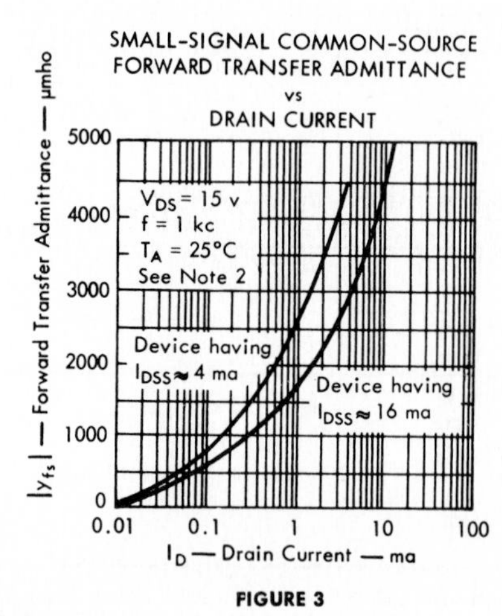

FIGURE 3

NOTE 2: These parameters must be measured using pulse techniques. PW = 100 msec, Duty Cycle ≤ 10%.

*Indicates JEDEC registered data.
‡The fourth lead (case) is connected to the source for all measurements.

TYPE 2N3823

N-CHANNEL EPITAXIAL PLANAR SILICON FIELD-EFFECT TRANSISTOR

TYPICAL CHARACTERISTICS‡

SMALL-SIGNAL COMMON-SOURCE FORWARD TRANSFER ADMITTANCE vs GATE-SOURCE VOLTAGE

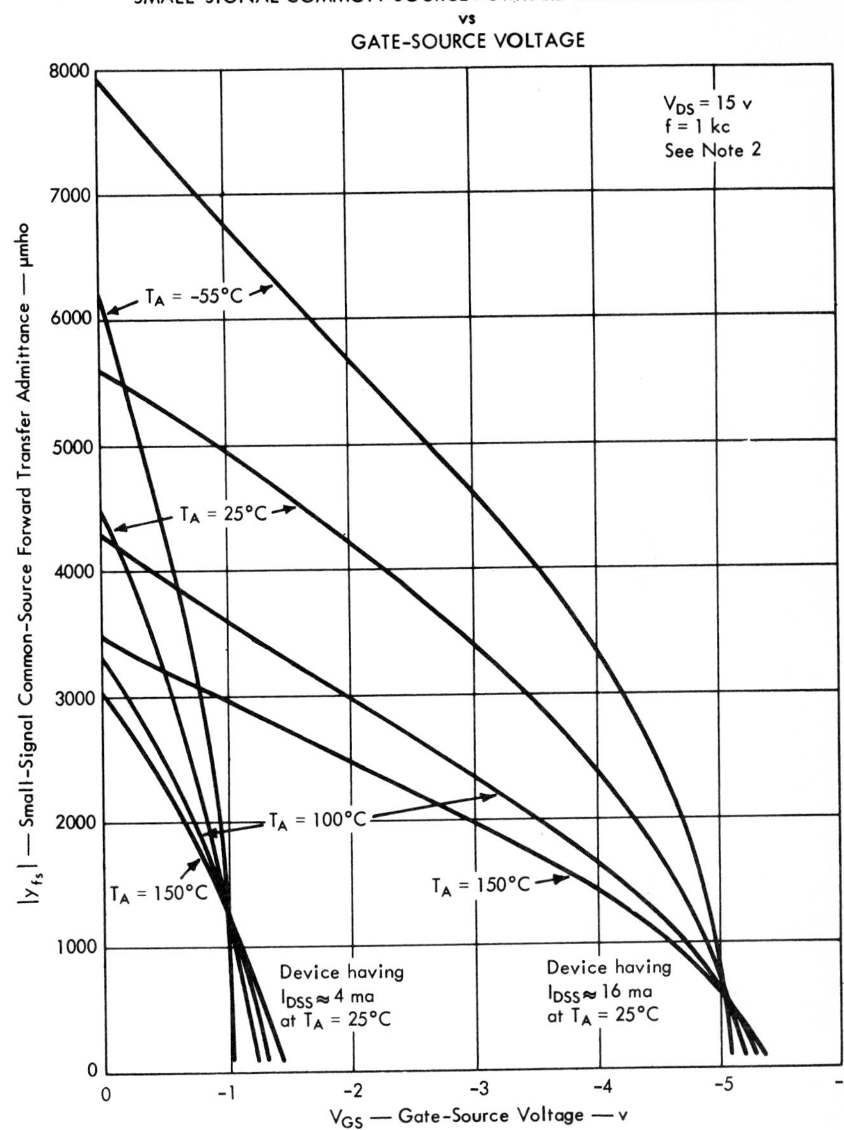

NOTE 2: These parameters must be measured using pulse techniques. PW = 100 msec, Duty Cycle ≤ 10%.

‡The fourth lead (case) is connected to the source for all measurements.

TYPE 2N3823 N-CHANNEL EPITAXIAL PLANAR SILICON FIELD-EFFECT TRANSISTOR

TYPICAL CHARACTERISTICS‡

GATE CUTOFF CURRENT vs FREE-AIR TEMPERATURE

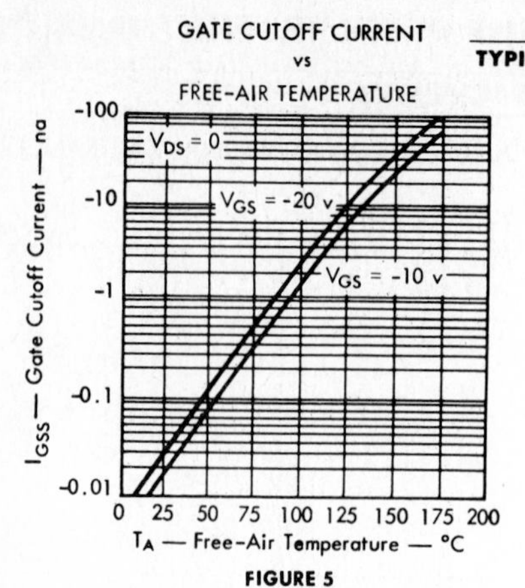

FIGURE 5

SMALL-SIGNAL COMMON-SOURCE INPUT ADMITTANCE vs FREQUENCY

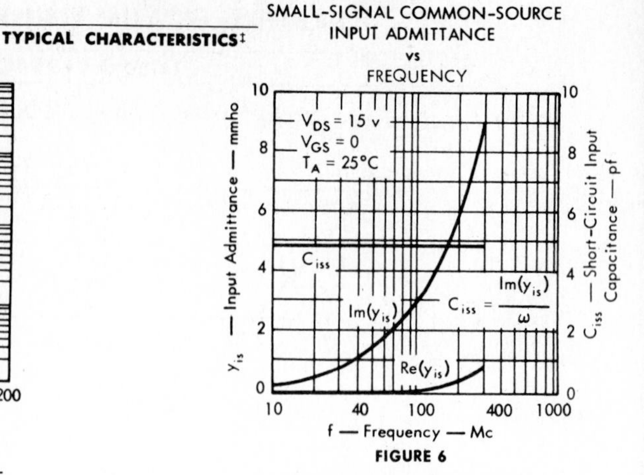

FIGURE 6

SMALL-SIGNAL COMMON SOURCE FORWARD TRANSFER ADMITTANCE vs FREQUENCY

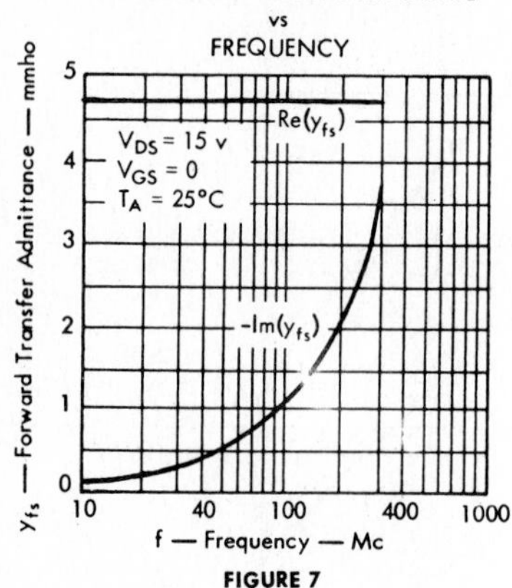

FIGURE 7

SMALL-SIGNAL COMMON-SOURCE REVERSE TRANSFER ADMITTANCE vs FREQUENCY

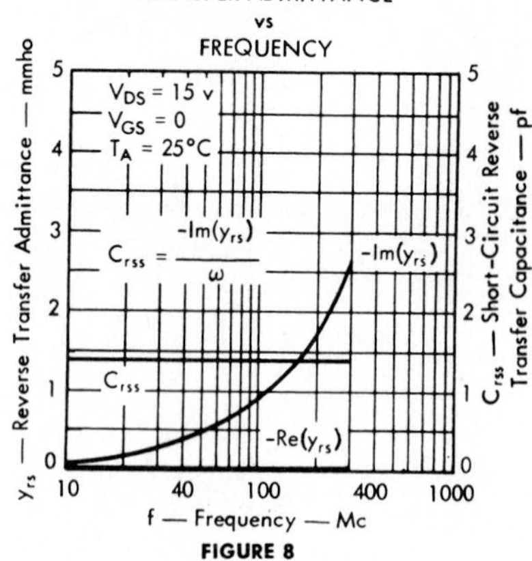

FIGURE 8

SMALL-SIGNAL COMMON-SOURCE OUTPUT ADMITTANCE vs FREQUENCY

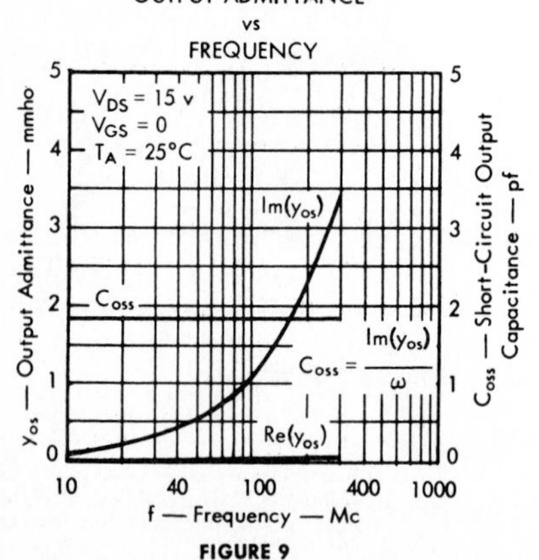

FIGURE 9

‡The fourth lead (case) is connected to the source for all measurements.

COMMON-SOURCE SHORT-CIRCUIT INPUT AND REVERSE-TRANSFER CAPACITANCES vs GATE-SOURCE VOLTAGE

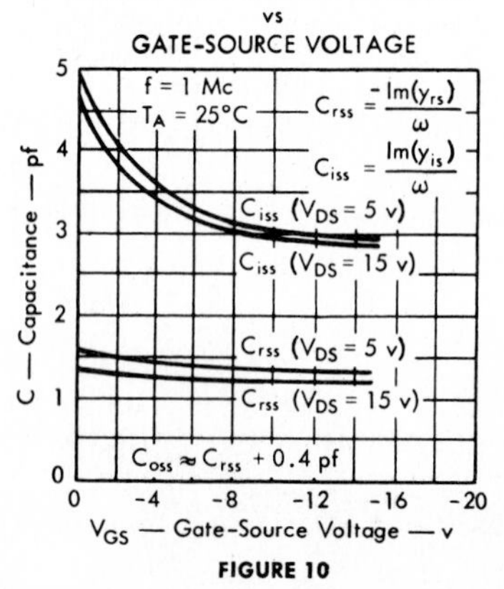

FIGURE 10

COMMERCIAL LIGHT EMITTING DIODES

5082-4487
5082-4488

Features

- LOW COST – BROAD APPLICATION
- LOW PROFILE – 0.18" LENS HEIGHT TYPICAL
- HIGH DENSITY PACKAGING
- LONG LIFE – SOLID STATE RELIABILITY
- LOW POWER REQUIREMENTS – 20mA @ 1.6V
- HIGH LIGHT OUTPUT – 0.8mcd TYPICAL

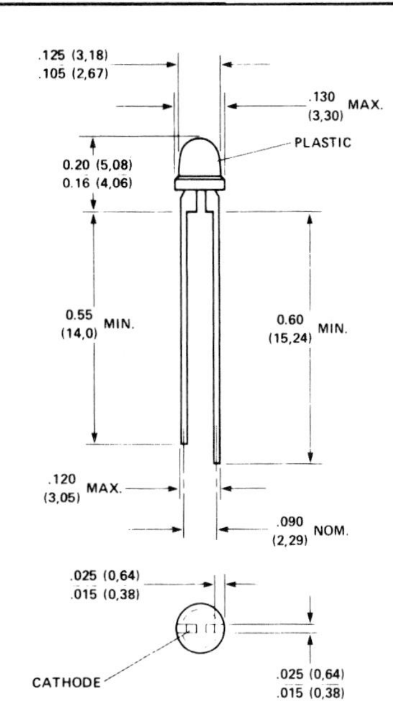

5082-4487/4488

Description

The 5082-4487 and 5082-4488 are Gallium Arsenide Phosphide Light Emitting Diodes for High Volume/Low Cost Applications such as indicators for calculators, cameras, appliances, automobile instrument panels, and many other commercial uses.

The 5082-4487 is a clear lens, low domed T-1 LED lamp, and has a typical light output of 0.8 mcd at 20 mA.

The 5082-4488 is a clear lens, low domed T-1 LED lamp, and has a guaranteed minimum light output of 0.3 mcd at 20 mA.

Maximum Ratings at T_A=25°C

DC Power Dissipation [Derate linearly from 50°C at 1.6 mW/°C] . . . . 100 mW
DC Forward Current . . . . 50 mW
Peak Forward Current [1 µsec pulse width, 300 pps] . . . . 1 Amp
Operating and Storage Temperature Range . . . . −55°C to +100°C
Lead Soldering Temperature . . . . 230°C for 7 sec.

Electrical/Optical Characteristics at $T_A = 25°C$

Symbol	Parameters	5082-4487			5082-4488			Units	Test Conditions
		Min.	Typ.	Max.	Min.	Typ.	Max.		
I_v	Luminous Intensity		0.8		0.3	0.8		mcd	$I_F = 20mA$
λ_{pk}	Wavelength		655			655		nm	Measurement at Peak
τ_s	Speed of Response		10			10		ns	
C	Capacitance		100			100		pF	$V_F = 0$, f = 1MHz
V_F	Forward Voltage		1.6	2.0		1.6	2.0	V	$I_F = 20mA$
BV_R	Reverse Breakdown Voltage	3	10		3	10		V	$I_R = 100\mu A$

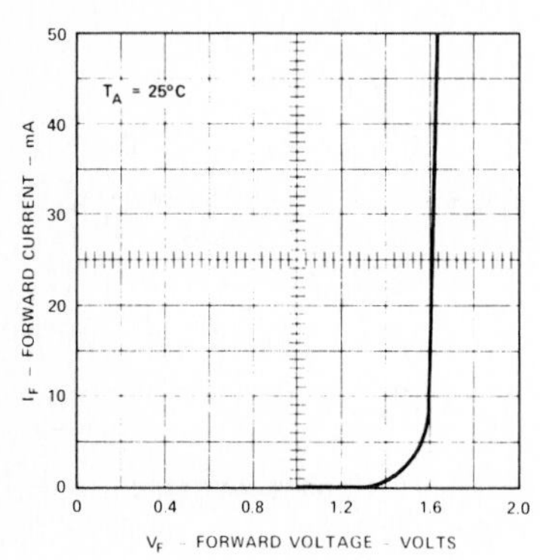

Figure 1. Typical Forward Current Versus Voltage Characteristic.

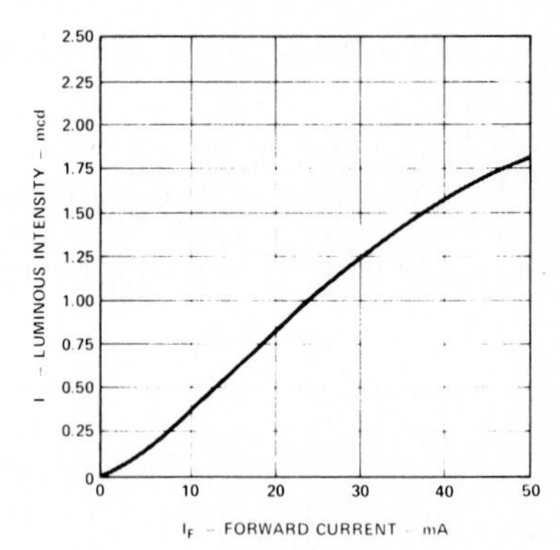

Figure 2. Typical Luminous Intensity Versus Forward Current.

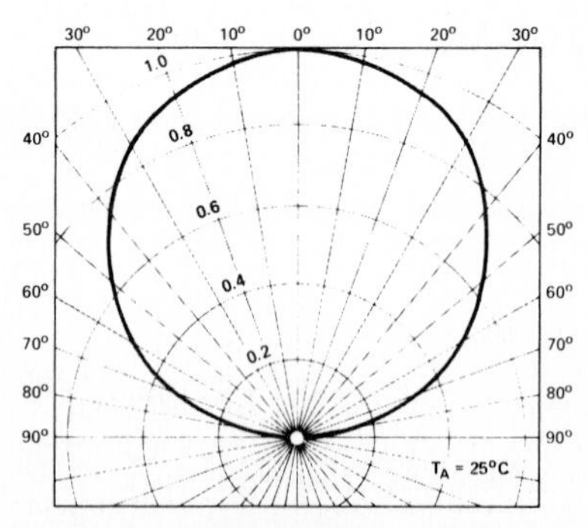

Figure 3. Typical Relative Luminous Intensity Versus Angular Displacement.

5088 (SILICON)
5089

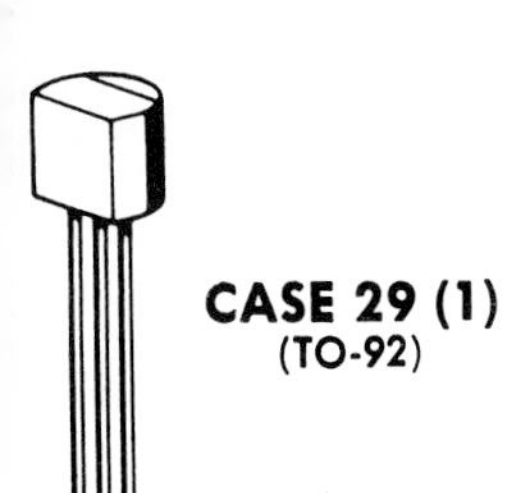

CASE 29 (1)
(TO-92)

NPN silicon annular transistors designed for low-level, low-noise amplifier applications.

MAXIMUM RATINGS

Rating	Symbol	2N5088	2N5089	Unit
Collector-Emitter Voltage	V_{CEO}	30	25	Vdc
Collector-Base Voltage	V_{CB}	35	30	Vdc
Emitter-Base Voltage	V_{EB}	4.5		Vdc
Collector Current	I_C	50		mAdc
Total Device Dissipation @ T_A = 25° C Derate above 25° C	P_D	310 2.81		mW mW/° C
Operating and Storage Junction Temperature Range	T_J, T_{stg}	-55 to +135		° C

THERMAL CHARACTERISTICS

Characteristic	Symbol	Max	Unit
Thermal Resistance, Junction to Ambient	θ_{JA}	0.357	° C/mW

2N5088, 2N5089 MOTOROLA Semiconductor Products Inc.

ELECTRICAL CHARACTERISTICS (T_A = 25°C unless otherwise noted)

Characteristic		Symbol	Min	Typ	Max	Unit
OFF CHARACTERISTICS						
Collector-Emitter Breakdown Voltage (I_C = 1.0 mAdc, I_B = 0)	2N5088	BV_{CEO}	30	-	-	Vdc
	2N5089		25	-	-	
Collector-Base Breakdown Voltage (I_C = 100 μAdc, I_E = 0)	2N5088	BV_{CBO}	35	-	-	Vdc
	2N5089		30	-	-	
Collector Cutoff Current (V_{CB} = 20 Vdc, I_E = 0)	2N5088	I_{CBO}	-	-	50	nAdc
(V_{CB} = 15 Vdc, I_E = 0)	2N5089		-	-	50	
Emitter Cutoff Current ($V_{EB(off)}$ = 3.0 Vdc, I_C = 0)		I_{EBO}	-	-	50	nAdc
($V_{EB(off)}$ = 4.5 Vdc, I_C = 0)			-	-	100	
ON CHARACTERISTICS						
DC Current Gain (I_C = 100 μAdc, V_{CE} = 5.0 Vdc)	2N5088	h_{FE}	300	-	900	
	2N5089		400	-	1200	
(I_C = 1.0 mAdc, V_{CE} = 5.0 Vdc)	2N5088		350	-	-	
	2N5089		450	-	-	
(I_C = 10 mAdc, V_{CE} = 5.0 Vdc)	2N5088		300	-	-	
	2N5089		400	-	-	
Collector-Emitter Saturation Voltage (I_C = 10 mAdc, I_B = 1.0 mAdc)		$V_{CE(sat)}$	-	-	0.5	Vdc
Base-Emitter On Voltage (I_C = 10 mAdc, V_{CE} = 5.0 Vdc)		$V_{BE(on)}$	-	-	0.8	Vdc
DYNAMIC CHARACTERISTICS						
Current-Gain – Bandwidth Product (I_C = 500 μAdc, V_{CE} = 5.0 Vdc, f = 20 MHz		f_T	50	175	-	MHz
Collector-Base Capacitance (V_{CB} = 5.0 Vdc, I_E = 0, f = 100 kHz, emitter guarded)		C_{cb}	-	1.8	4.0	pF
Emitter-Base Capacitance (V_{BE} = 0.5 Vdc, I_C = 0, f = 100 kHz, collector guarded)		C_{eb}	-	4.0	10	pF
Small-Signal Current Gain (I_C = 1.0 mAdc, V_{CE} = 5.0 Vdc, f = 1.0 kHz)	2N5088	h_{fe}	350	-	1400	-
	2N5089		450	-	1800	
Noise Figure (I_C = 100 μAdc, V_{CE} = 5.0 Vdc, R_S = 10 k ohms, f = 10 Hz to 15.7 kHz	2N5088	NF	-	-	3.0	dB
	2N5089		-	-	2.0	

NOISE FIGURE

V_{CE} = 5.0 Vdc, T_A = 25 °C

FIGURE 1 — FREQUENCY EFFECTS

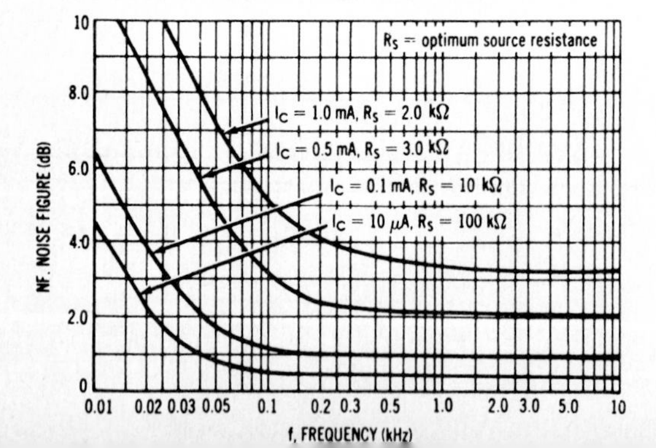

FIGURE 2 — SOURCE RESISTANCE EFFECTS

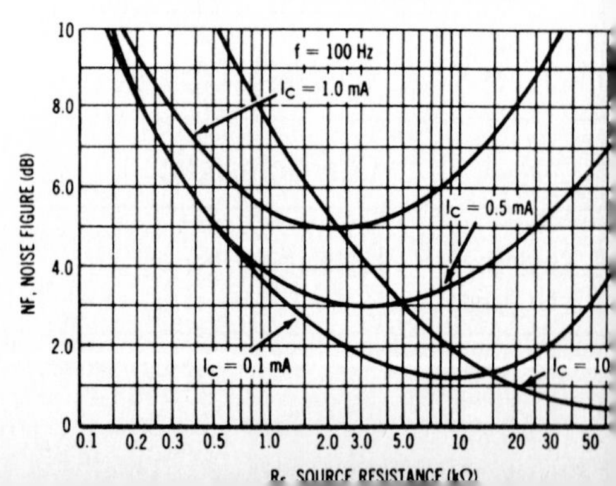

N5088, 2N5089 (continued)

h PARAMETERS

V_{CE} = 10 Vdc, f = 1.0 kHz, T_A = 25°C
(For Figures 3, 4, 5, 6, 8)

This group of graphs illustrates the relationship of the "h" parameters for this series of transistors. To obtain these curves, 4 units were selected and identified by number — the same units were used to develop curves on each graph.

FIGURE 3 — INPUT IMPEDANCE

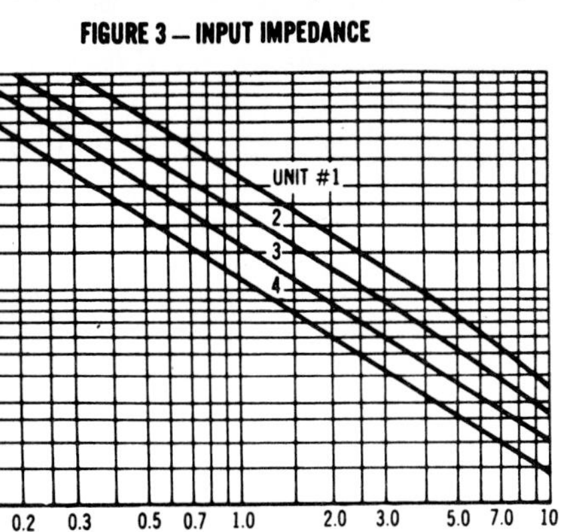

FIGURE 4 — VOLTAGE FEEDBACK RATIO

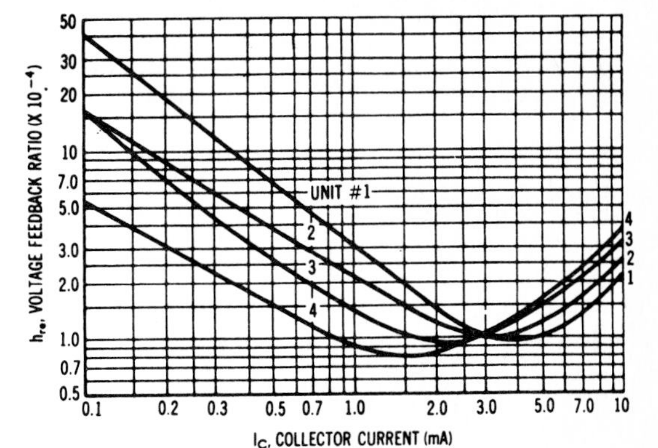

FIGURE 5 — CURRENT GAIN

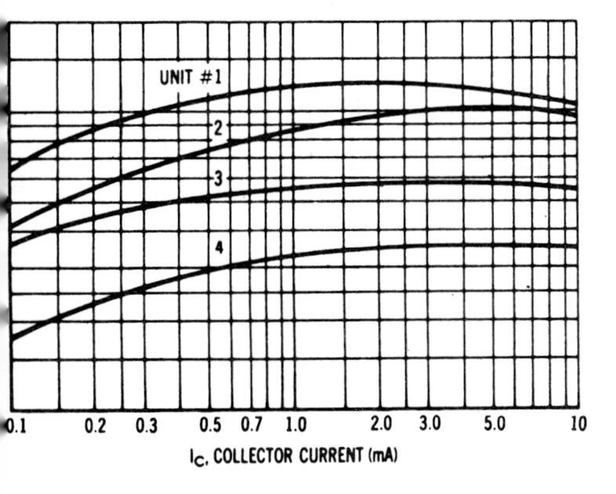

FIGURE 6 — OUTPUT ADMITTANCE

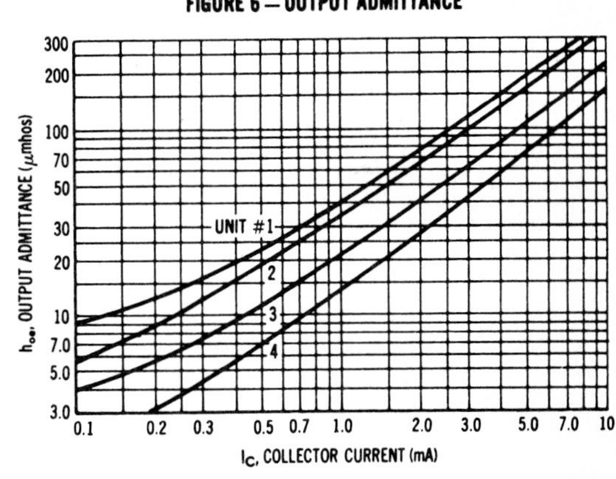

FIGURE 7 — EFFECT OF VOLTAGE

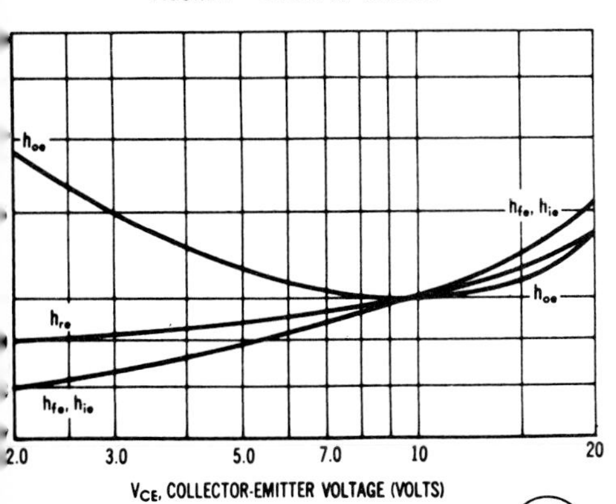

FIGURE 8 — DETERMINANT

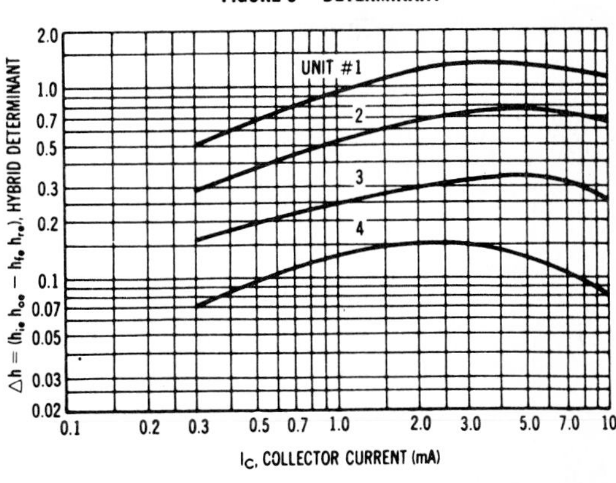

MOTOROLA Semiconductor Products In

2N5088, 2N5089 (continued)

FIGURE 9 — DC CURRENT GAIN

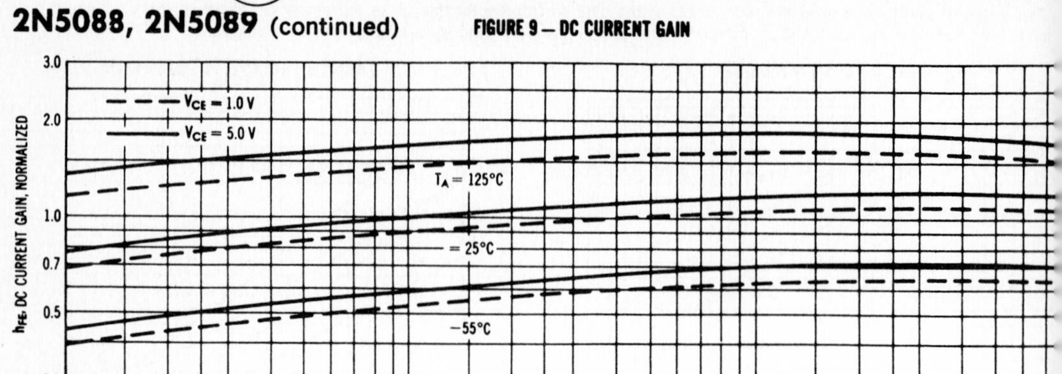

FIGURE 10 — COLLECTOR SATURATION REGION

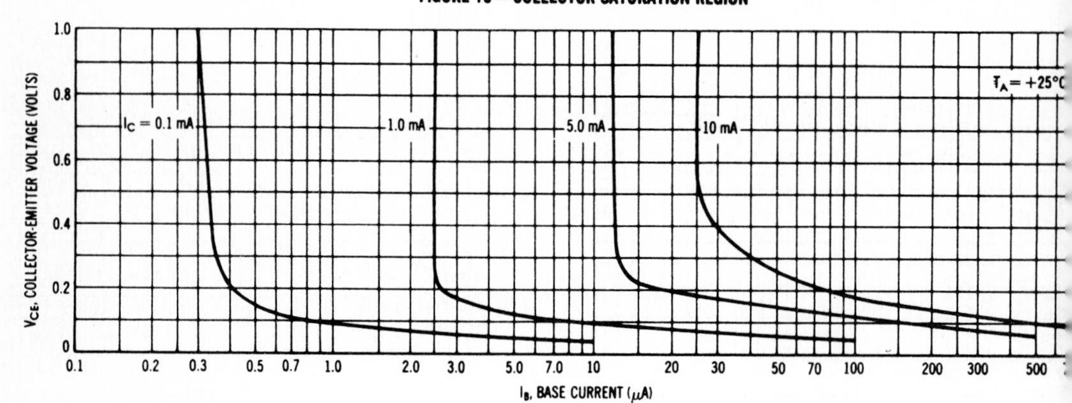

FIGURE 11 — CURRENT GAIN — BANDWIDTH PRODUCT

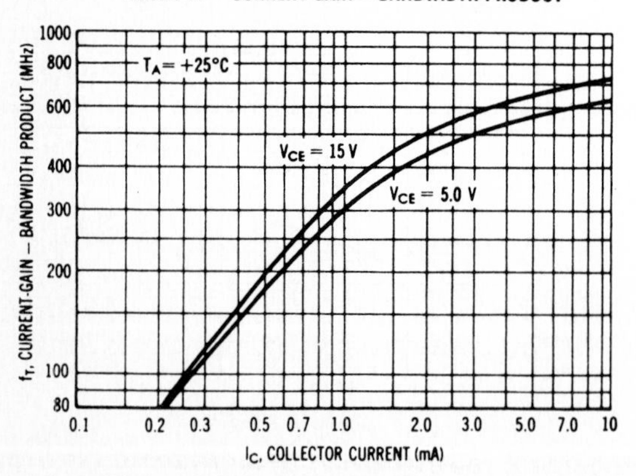

FIGURE 12 — CAPACITANCE

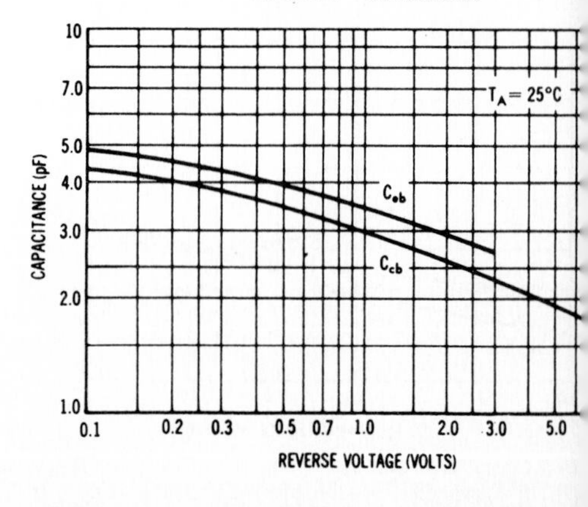

NPN Silicon Planar

TYPE 2N5376 AND 2N5377 PREMIUM PERFORMANCE ECONOLINE® TRANSISTORS

For Industrial Small-Signal, Low-Noise, Low-Power Audio Frequency Applications

ABSOLUTE MAXIMUM RATINGS at 25 C Free-Air Temperature (unless otherwise noted)

Collector-Base Voltage 60V
Collector-Emitter Voltage (See Note 1) 30V
Emitter-Base Voltage 5V
Total Dissipation at 25 C Free-Air Temperature (See Note 2) 360mW
Collector Current 500mA
Junction Temperature, Operating .... +150 C
Lead Temperature 1/16 Inch from Case for 10 Seconds .... 260 C
Storage Temperature Range .... −55 C to +150 C

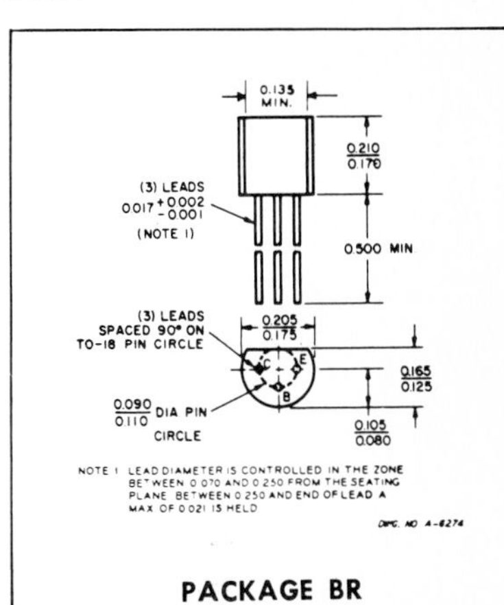

PACKAGE BR

ELECTRICAL CHARACTERISTICS: at T_A = 25 C (unless otherwise noted)

Parameter	Symbol	Test Conditions	2N5376 Min.	2N5376 Max.	2N5377 Min.	2N5377 Max.	Units
Collector-Base Breakdown Voltage	$V_{(BR)CBO}$	$I_C = 10\mu A$, $I_E = 0$	60	—	60	—	V
Collector-Emitter Breakdown Voltage	$V_{(BR)CEO}$	$I_C = 10mA$, $I_B = 0$ (See note 3)	30	—	30	—	V
Emitter-Base Breakdown Voltage	$V_{(BR)EBO}$	$I_E = 100\mu A$, $I_C = 0$	5	—	5	—	V
Collector Cut-off Current	I_{CBO}	$V_{CB} = 30V$, $I_E = 0$	—	10	—	10	nA
Static Forward Current Transfer Ratio	h_{FE}	$I_C = 10\mu A$, $V_{CE} = 5V$	100	500	40	200	—
		$I_C = 1mA$, $V_{CE} = 5V$	120	600	100	500	—
		$I_C = 10mA$, $V_{CE} = 5V$ (See Note 3)	150	—	120	—	—
Collector-Emitter Saturation Voltage	$V_{CE(SAT)}$	$I_C = 10mA$, $I_B = 1mA$	—	0.20	—	0.20	V
Base-Emitter Saturation Voltage	$V_{BE(SAT)}$	$I_C = 10mA$, $I_B = 1mA$	0.65	0.80	0.65	0.80	V

Notes:
1. This value applies when the base-emitter diode is open-circuited.
2. Derate linearly to 150 C free air temperature at the rate of 2.88 mW/°C.
3. Pulse test: Pulse width = 300 μsec, duty cycle ≤2%.

SEMICONDUCTOR DIVISION
SPRAGUE ELECTRIC COMPANY
Pembroke Road, CONCORD, N. H. 03301

ELECTRICAL CHARACTERISTICS - Cont. TYPE 2N5376 AND 2N5377

Parameter	Symbol	Test Conditions	2N5376		2N5377		Units
			Min.	Max.	Min.	Max.	
Small-Signal Short-Circuit Input Resistance	h_{ib}	$I_C = 1\text{mA}, V_{CE} = 5\text{V}$ $f = 1\text{Khz}$	20	32	20	32	ohms
Small-Signal Open-Circuit Output Conductance	h_{ob}	$I_C = 1\text{mA}, V_{CE} = 5\text{V}$ $f = 1\text{Khz}$	0.05	0.2	0.05	0.2	μmho
Small-Signal Common-Emitter Forward Current Transfer Ratio	h_{fe}	$I_C = 1\text{mA}, V_{CE} = 5\text{V}$ $f = 1\text{Khz}$	120	1000	100	900	—
Common-Base Open-Circuit Collector Capacitance	C_{cb}	$V_{CB} = 10\text{V}$ $f = 1\text{MHz}, I_E = 0$ (Note 1)	—	8	—	8	pF
Small-Signal Common-Emitter Forward Current transfer Ratio	$\lvert h_{fe}\rvert$	$V_{CE} = 5\text{V}, I_C = 500\mu\text{A}$ $f = 10\text{ MHz}$	3	15	3	15	—
Wideband Noise Figure	NF	$I_C = 10\mu\text{A}, V_{CE} = 5\text{V}$, $R_g = 10\text{ K}\Omega$, Bandwidth = 10Hz to 15.7 kHz	—	2.0	—	3.0	dB

Notes:

1. Measurement employs a three-terminal capacitance bridge incorporating a guard circuit. The emitter terminal should be connected to the guard terminal of the bridge.

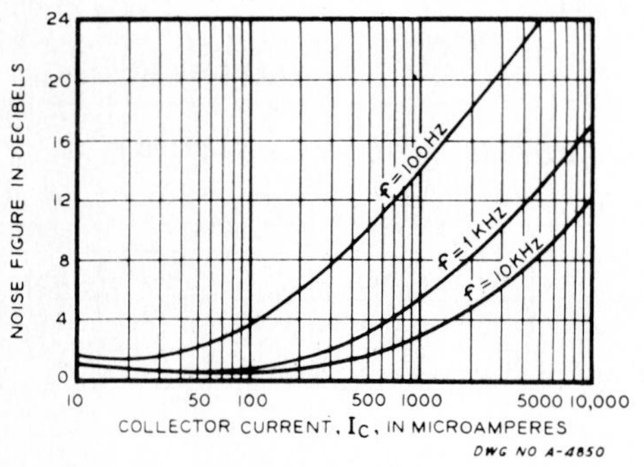

TYPE 2N5376
SPOT NOISE FIGURE AT 25 C WITH
$R_G = 10K$, $V_{CE} = 5$ V

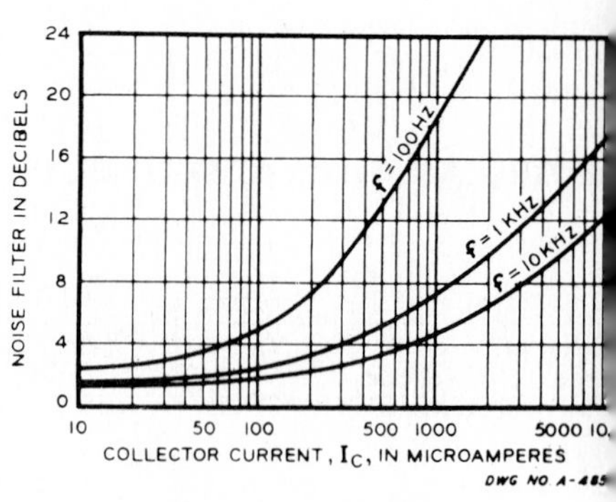

TYPE 2N5377
SPOT NOISE FIGURE AT 25 C WITH
$R_G = 10K$, $V_{CE} = 5$ V

ECONOLINE TRANSISTORS

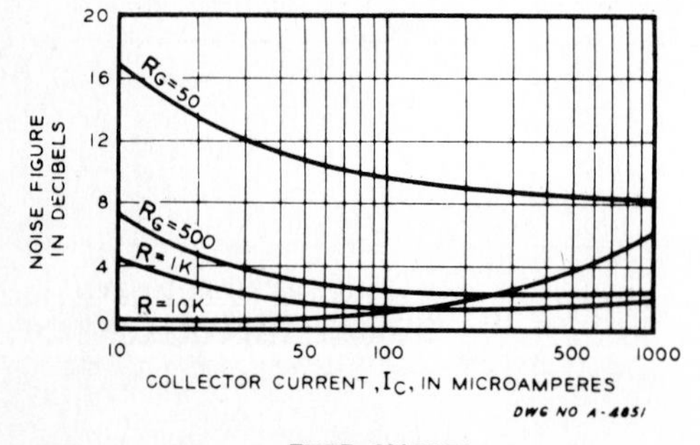

TYPE 2N5376
WIDEBAND (10Hz to 15.7kHz) NOISE
FIGURE AT 25 C WITH $V_{CE} = 5$ V

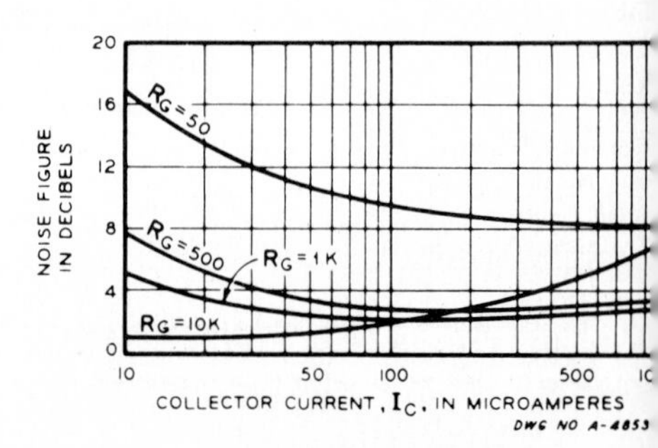

TYPE 2N5377
WIDEBAND (10Hz to 15.7kHz) NOISE
FIGURE AT 25 C WITH $V_{CE} = 5$ V

Chapter 1

1. (a) 1.050 (b) 5.00 fA (c) 0.926 mA
3. 766 pA, 23.7 mV, 1.53°C
5. (a) 3.23 mA/(K)^3 (b) 0.726 V and 0.647 V (c) (nk/q) [ln (I_D/K) $- 3(1 + \ln T)$] (d) −1.961 mV/°C
7. 1 point each curve: (0.919 V, 1 mA), (−0.081 V, 1 mA)
9. (a) 4.74 mA (b) 4.64 mA
11. (a) 0.305 V (b) 0.753 μA (c) 8.56 mA
13. (a) 73.2 mA (b) 25 mA (c) 1.667 V
15. (a) 0.708 V (b) 0.542 V (c) −4 V (d) −10.24 V (e) −11.18 V
17. (a) 0.437 V (b) 58.3 Ω (c) 4.44 Ω (d) $0.4370 \le V_D \le 0.4378$ V (e) $4.386 \le r \le 4.505$ Ω
19. 46.9 Ω
21. 25.6 cos $120\pi t$ μA; 379 + 25.6 cos $120\pi t$ μA
23. $i_D = 3.91 \times 10^{-3} + (v_s/26.4)$ A
25. (a) $V_S = 1$: $v_D = 0$, $v_R = 1$ (b) $V_S = 1$: $v_D = 0.7$, $v_R = 0.3$ (c) 34.3 Ω
27. (a) 1.552 Ω (b) 0.0525 Ω (c) 118.2 Ω
29. (a) −11.33 A (b) −8 V (c) Reverse (d) −2 A
31. $R = 40$ Ω, $14 \le I_S \le 21.4$ mA; or $R = 50$ Ω, $10.6 \le I_S \le 17$ mA
33. 1 point: (5 Mrad/s, $4.64\angle{-31.7^\circ}$ Ω)
35. (a) 2.14 pF (b) 68.7 kHz increase
37. (a) 194.6 pF (b) 212.4 pF (c) 155.5 pF (d) -6.53×10^{-13} F/°C
39. 1.904 pF, 0.218 μH
41. (a) 2 points: (0, 100 V), (5 ms, 0) (b) $v_{o(\mathrm{av})} = 39.3$ V, $v_{\sin(\mathrm{av})} = 31.8$ V (c) 2 points: (0, 50 V), (2.18 ms, 0)
43. (a) 57 V (b) 820 V (c) 3.33%
45. (a) 138.9 μF (b) 34.7 μF (c) 86.8 μF
47. 10:1 transformer, 1000 μF in parallel with 1.5 kΩ, 0.28%
49. 3.3 kΩ; $30.08 \le V_L \le 30.20$ V; $1.007 \le I_L \le 6.015$ mA; $3.05 \le |I_D| \le 8.02$ mA
51. (a) 5082-4487 in series with 632 Ω (b) reversed LEDs in parallel, 632 Ω in series

Chapter 2

1. 39.6 μA, 3.96 mA, 0.65 V, 8.85 V, 8.2 V, 35.1 mW
3. (a) 120 μA (b) 6 mA (c) − 6.12 mA (d) 0.7 V (e) 8 V (f) 8.7 V (g) 0.084 mW (h) 4.284 mW (i) 52.284 mW
5. (a) Both active (b) T1 saturated, T2 active (c) 4.802, 4.351, 73.650, 0.510, 0.639, 62.500, 0.108, 3.038 mW
7. (a) BE, F; CB, R; active (b) BE, R; CB, R: cut-off (c) BE, F; CB, F; saturated (d) BE, F; CB, R; active
9. (a) saturated (b) active
11. (Measure slopes of straight lines) (a) 90 Ω (b) 150 Ω (c) 210 Ω
13. (a) 64.00 mW (b) 4.74 mW
15. (a) *n* (b) 4.47, − 1.491 (c) 8.89 mA (d) − 0.879 V
17. (a) 2 mA, − 0.1 μA, − 1.9999 mA, − 0.5 V, − 2.5 V, 5 V, 5.5 V, 10.00005 mW (b) − 2 mA, 0.1 μA, 1.9999 mA, 0.5 V, 2.5 V, − 5 V, − 5.5 V, 10.00005 mW (c) *p*, A = S, B = D, C = G
19. (a) 5 mA, − 2 nA, ≐ − 5 mA, 6 V, − 1 V, − 3 V, 7 V, ≐ 30 mW (b) − 5 mA, 2 nA, ≐ 5 mA, − 6 V, 1 V, 3 V, − 7 V, ≐ 30 mW (c) A = S, B = D, C = G
21. (a) 1 mA, − 1 mA, − 2 V, 3 V, 3 mW (b) − 1 mA, 1 mA, 2 V, − 3 V, 3 mW (c) 1 mA, − 1 mA, 2 V, 3 V, 3 mW (d) − 1 mA, 1 mA, − 2 V, − 3 V, 3 mW
23. (a) T1 ohmic, T2 saturated, $I_D > 0$ (b) T1 ohmic, T2 saturated, $I_D > 0$ (c) T1 saturated, T2 saturated, $I_D > 0$ (d) T1 saturated, T2 ohmic, $I_D > 0$ (e) T1 cut-off, T2 ?, $I_D = 0$
25. (a) 556, 455 Ω (b) 1111, 909 Ω (c) 1667, 1364 Ω (d) 370, 303 Ω (e) 0.379, 0.290
27. (a) 800 Ω (b) 32 Ω (c) 0.949 and 11.86 mm
29. (a) 482 Ω (b) 1482 Ω (c) 1690 ppm/°C, 800 ppm/°C

Chapter 3

1. (a) 30.0 mA (b) 2.78 mA (c) 0.695 V
3. (a) 1.372 mW (b) 9.139 mW (c) 10.51 mW
5. (a) − 19.61 nA (b) − 0.834 mA (c) 1.701 mA
7. (a) 33.3 Ω, 1.2 kΩ, 0.56 V (b) 18.6 mA, − 18.9 mA, 0.62 V
9. (a) $I_C = I_{CEO} + (V_{CE}/R_0)$ (b) 1 point: (0.1 V, 3 μA)
11. 2.136 mA, 19.38 μA, 3.97 V
13. 1.923 mA, 0.603 V
15. 10.39 kΩ and 747 Ω
17. (a) − 140.0 and 144.0 mA, − 800 and 804 mA (b) − 1.429 mA and − 250 μA
19. (a) 1.984 mA, 3.03 V (b) 217 kΩ (c) 0.849 mA
21. R_C = 15 kΩ, R_B = 3.26 MΩ

23. 4.13 mA, − 1.240 V, 10.93 V
25. (a) 6 V (b) − 0.745 V (c) 11.21 mA
27. (a) 2.25 mA, 1.75 V (b) 250 kΩ (c) 424 kΩ
29. (a) − 4.5 V, 18 mA (b) − 1.146 V (c) 0
31. 3.66 V, 1 kΩ, 9 V
33. (a) 11.52 mA, 6.72 V (b) 4.5 mA, 16.35 V
35. (a) 3.63 V, 8.29 mA, 3.43 V (b) 28.41 + 137.30 + 0.92 + 1.38 = 201.16 − 33.14 mW (c) 8.43 m℧
37. 5.54 mA, 6.16 V
39. (a) − 8.5 mA, − 100 μA, − 3 V, − 0.8 V (b) − 11 mA, − 100 μA, − 9 V, − 0.8 V (c) − 4.5 mA, − 100 μA, − 2 V, − 0.8 V
41. 8 mA, 8 V
43. 7.1 mA, 11.3 V

Chapter 4

1. (a) 1.000 mA, 10.00 μA, 5.00 V and 1.031 mA, 2.06 μA, 4.39 V (b) 0.999 mA, 9.99 μA, 5.02 V and 1.128 mA, 2.26 μA, 2.45 V
3. (a) 1.878 mA, 3.27 V (b) 2.03 mA, 1.651 V (c) R_C = 5 kΩ, R_E = 4.64 kΩ, and R_2 = 27.0 kΩ OK
5. 116.1 kΩ, 35.7 kΩ, 15 kΩ, 5 kΩ, 25 V OK
7. (a) 0.815 mA, 5.24 V (b) R_1: 10.1%, R_2: 9.0%, R_E: 10.9%, R_C: 0% (c) R_1: 3.35 V, R_2: 3.55 V, R_E: 3.64 V, R_C: 3.77 V
9. 4.90, 14.71, 4.675, 2.825, 187.5 kΩ
11. 4.27 and 38 kΩ OK
13. 50 and 3.80 kΩ OK
15. (a) 500 mW at 25°C, 25°C, 150°C (b) 0.250°C/mW (c) 96.5°C
17. (a) − 95°C, − 15°C (b) − 120°C, − 40°C (c) 46.7°C, 82.2°C
19. (a) 12.5°C/W (b) 9.5°C/W (c) 2 W (d) 2.63 W
21. (a) R_C = 3.16 kΩ, R_E = 3.15 kΩ, R_1 = 22.2 kΩ, R_2 = 13.95 kΩ OK (b) − 50.9°C, 128.7°C
23. 52 kΩ, 103 kΩ, 16.8 kΩ, 6 V OK
25. (a) 12.24 V, 28.5 kΩ, 3.92 kΩ, 3.50 kΩ (b) − 1.097 mA
27. (a) 8.14 kΩ, 4 kΩ (b) 0.942 mA
29. (a) 3.16 mA (b) − 2.99 μA/°C
31. (a) 1.855 mA, 5.80 V (b) 2.22 mA
33. (a) 1.596 mA, 7.23 V (b) 7.2%
35. (a) 3.98 and 2.31 mA, 7.61 and 4.45 V (b) 12 V, 0.284 kΩ OK
37. (a) 2.26 mA (b) R_2 = 60.6 kΩ (c) R_2 = 36.6 kΩ
39. (a) R_D = 66.6 kΩ, R_{SS} = 16.74 kΩ, R_1 = 316 kΩ, R_2 = 100 kΩ OK (b) 99.1 kΩ
41. (a) 3 V (b) 5 V

43. $R_2 = \infty$, $R_{SS} = 1.083$ kΩ OK
45. (a) 1.458 mA, − 1.292 V, 4.96 V, 31.9 mW (b) 500 kΩ (c) 50 kΩ (d) 7.5 kΩ

Chapter 5

1. $A_V = -g_m(R_L \| r_d)(R_B \| r_\pi)/[R_s + (R_B \| r_\pi)]$
3. (a) 1.8 mA (b) − 510 cos 500t mV (c) 6 − 0.510 cos 500t V
5. 56.1 cos $10^4 t$ mV
7. 9.98 − j24.4 Ω
9. (a) − 29.3 (b) − 255
11. (a) 77.6 (b) 151.3
13. (a) 1.075 mA, 32.2 m℧, 2.80 kΩ, − 436 (b) 13.76 kΩ (c) 1979 kΩ
15. (a) 887 kΩ (b) 1.184 MΩ (c) 488 kΩ
17. (a) 9.8 pF (b) 15.6 pF (c) 27.7 pF
19. (a) 16.7 pF (b) 70.4 pF (c) 60.9 pF
21. 131.5 − j438 Ω
23. (a) 38.9 m℧, 5.14 kΩ, 74.8 pF, 2.5 Mrad/s (b) 4 pF, 200, 800 Mrad/s, 93.3 pF, 4 Mrad/s
25. (a) $r_d = 25$ MΩ, $g_m = 32.4$ μ℧, $r_\pi = 1.388$ MΩ (b) $r_d = 250$ kΩ, $g_m = 3.24$ m℧, $r_\pi = 19.73$ kΩ
27. (a) $I_C = 64.8/V_{CE}^2$ (mA) (b) $I_C = 9.62/V_{CE}^2$ (mA)
29. (a) 45.2 Ω, 2.99 kΩ, 65.0 pF, 3.98 pF, 44.9 m℧, 7.40 kΩ (b) 5 Mrad/s (c) Increase T, decrease I_C, increase V_{CE}
31. (a) − 60.8 (b) − 88.9
33. (a) $r_\pi = 8.8$ kΩ, $r_d = 24.4$ kΩ, $g_m = 65.9$ m℧, $n = 1.181$ (b) 387
35. 90.1 μΩ, 9.01 × 10^{-3}, − 0.910, 9.10 μ℧
37. 6.4 + 0.16 cos 1200t mA
39. 872 Ω
41. 483 Ω
43. (a) $C_{gs} = 3.4$ pF, $C_{gd} = 1.4$ pF, $g_m = 4.7$ m℧, $r_d = 20$ kΩ, $C_{ds} = 0.4$ pF (b) 2.5 pF, 1.3 pF, 2.35 m℧, 25 kΩ, 0.4 pF
45. 5.95 pF, 0.02 pF, 1.57 pF, 2.86 kΩ, 12.0$\angle -34.9°$ m℧, 1 kΩ
47. (a) − 1.096, 2265 (g_m in m℧) (b) 4.40 m℧, 3.44 m℧ (4.5, 3.3)
49. (a) All: 3.75 mA, 7.5 m℧ (b) JFET: 9.72 m℧, 17.84 pF, 3.54 pF; MOSFETs: 9.72 m℧, 10 pF, 2.5 pF

Chapter 6

1. (a) − 124.5, − 84.5, − 43.8, 5450, 2.11 kΩ, 6 kΩ (b) − 129.8, − 117.3, − 203, 26,400, 9.40 kΩ, 6 kΩ
3. − 54.6, − 47.0, − 167.9, 9170, 6.15 kΩ, 5 kΩ

5. −206, 15,310
7. $R_C = R_E = 6$ kΩ, $R_1 = 195$ kΩ, $R_2 = 137$ kΩ, $V_{CC} = 18$ V OK
9. 25 kΩ, 4 kΩ, 32.4 m℧ OK
11. 32, 20, 3, 2.7 kΩ
13. (a) −365 (b) −7.68
15. 124.6, 40.1, 0.492, 61.3, 23.7 Ω, 6 kΩ
17. $R_C = R_E = 5$ kΩ, $R_1 = 47.3$ kΩ, $R_2 = 30$ kΩ OK
19. 0.990, 0.944, 3.45, 3.41, 20.9 kΩ, 32.9 Ω
21. $R_E = 5$ kΩ, $R_1 = 234$ kΩ, $R_2 = 178$ kΩ OK
23. (a) −21.1 (b) 0.236 μA, 65.8 nA, 0.1974 μA, 0.1316 mA, 0.263 mA
25. (a) −19.75 (b) 1.635 kΩ (c) −21.6
27. (a) −16.43 (b) 147.0 kΩ (c) −17.25
29. 1 MΩ, 2.04 kΩ, 55.8 Ω OK
31. 1 MΩ, 2.02 kΩ, 39.6 Ω
33. 185.2 Ω, 8.25, 6.50, 0.306, 2.52, 3 kΩ
35. $R_D = 5$ kΩ, $R_{SS} = 500$ Ω, $R_1 = 1$ MΩ, $R_2 = 33.1$ kΩ, $V_{DD} = 15$ V OK
37. 255 kΩ, 0.786, 0.783, 40.1, 31.5, 185.2 Ω
39. $R_1 = 100$ kΩ, $R_{SS} = 4.96$ kΩ
41. (a) 1.961 kΩ, 2.61 kΩ, −48.2 (b) 1.891 kΩ, 2.70 kΩ, −49.4
43. 1 point: (2×10^{-3}, 1.942 kΩ)

Chapter 7

1. (a) 6.02 dB (b) 26.0 dB (c) 43.0 dB (d) 46.0 dB (e) sketch
3. Asymptotes: $|A_I|_{dB} = 20 \log [R_{P1}/(R_{P1} + R_{P2})]$, $|A_I|_{dB} = 20 \log [(R_{P1}) C_S \omega]$; corner at $\omega_L = 1/[(R_{P1} + R_{P2}) C_S]$
5. (a) 37.0 dB (b) 19.91 dB (c) −3.01 dB (d) −40.0 dB (e) $A = 100/[(1 + j\omega)(1 + j0.01\omega)]$; 40 dB, 20 dB, 0, −40 dB
7. (a) Proof (b) 6.20 Mrad/s
9. (a) ≐ 50.2 Mrad/s (b) −21.5 (c) ≐ 50.4 Mrad/s, −21.4
11. 1 point each: (a) (2 m℧, 198.6 Mrad/s) (b) (2 m℧, 3.31) (c) (2 m℧, 657 Mrad/s)
13. (a) j11 m℧ (b) −j9 m℧ (c) 3.54 + j5.31 m℧, −2.62 − j3.92 m℧
15. ≐ 361 Mrad/s
17. (a) 0.766 (b) ≐ 343 Mrad/s
19. 340 − j601 Ω
21. (a) −64.4, 5.19 Mrad/s, 334 Mrad/s (b) R_B: −64.6, 5.17, 334; g_m: −70.4, 4.81, 338; R_C: −68.1, 4.94, 336 (c) to increase $|A_{Vs(\text{mid})}|$, increase g_m; to increase ω_H, decrease g_m; to increase $|A_{Vs(\text{mid})}|\omega_H$, increase g_m
23. (a) Emitter 1 to ground: R_1, $g_m V_\pi$, $C_\pi(2)$, $r_\pi(2)$, $R_s + V_s$; Emitter 2 to ground: $g_m V_\pi$, R_C, R_L; E1 to E2: C_μ (b) $\tau_\pi = [R_1 \| (1/g_m) \| r_\pi \| R_s \| r_\pi] 2C_\pi$, $\tau_\mu = [R_L' + (R_1 \| 1/g_m \| r_\pi \| R_s \| r_\pi)(1 + g_m R_L')] C_\mu$ (c) −3.21, ≐ 91.8 Mrad/s

25. Proof
27. $\doteq$ 145 Mrad/s
29. (a) $\doteq$ 386 Mrad/s (b) $\doteq$ 361 Mrad/s
31. (a) 25, 25, 50, 10 rad/s (b) 0.577, 0.846
33. (a) 0.993 (b) $183 < \omega_L < 249$ rad/s (203.54 exact)
35. $195 < \omega_L < 234$ rad/s (202.49 exact)
37. (a) 26.1, 13.9, 577, 5 rad/s (b) $\doteq$ 577 rad/s
39. (a) 58.8, 6.25, 622 rad/s (b) $622 < \omega_L < 687$ rad/s (666 exact)
41. (a) 39.6, 6.7, 771, 6.7 rad/s (b) $771 < \omega_L < 817$ rad/s (801 exact)
43. 85, 5, 10 μF
45. (a) 80.1, 8.0, 7.8 rad/s (b) $80 < \omega_L < 96$ rad/s (84.96 exact)
47. (a) 5, 40 μF (b) 93.96 rad/s, exact

Chapter 8

1. (a) 11,310 (b) 6.87 kΩ (c) 5 kΩ
3. (a) 1879 (b) 2626
5. (a) 15,660 (b) 43.0 Ω
7. $\doteq$ 0.986 Mrad/s (1.005 Mrad/s, exact)
9. $\doteq$ 4.57 Mrad/s (4.635 Mrad/s, exact)
11. (a) −41.4 (b) $\doteq$ 20.9 Mrad/s (c) 21.31 Mrad/s
13. (a) $\doteq$ 0.310 Mrad/s (b) −5,820,000
15. $C_1 = 20$, $C_2 = 7.5$, $C_3 = 8$, $C_{E1} = C_{E2} = 400$ μF ($f_L = 17.43$ Hz, exact)
17. 35.0, 16.6, 30.8, 20, 437, 337, 687, $687 < \omega_L < 1564$ rad/s
19. (a) $506 < \omega_L < 760$ rad/s (581 exact) (b) $C_1 = 1$, $C_2 = 5$, $C_3 = 0.5$, $C_{E1} = 200$, $C = 0.02$ μF (246 rad/s exact)
21. $C_1 = 2.5$, $C_2 = 1.5$, $C_3 = 2$, $C_{E1} = 100$, $C_{E2} = 150$ μF ($\omega_L = 492$ rad/s exact)
23. $C_1 = 0.01$, $C_{SS} = 20$, $C = 0.01$, $C_3 = 0.1$ μF ($\omega_L = 1733$ rad/s, two zeros in numerator)
25. $R_{11} = R_{12} = 56.7$ kΩ, $R_{21} = R_{22} = 26.4$ kΩ, $R_{E1} = R_{E2} = 5$ kΩ, $R_{C1} = R_{C2} = 0.5$ kΩ ($A_{Vs(\text{mid})} = 302$, $f_H = 4.04$ MHz, exact)
27. Common-collector, 2N5089, $I_C = 1$ mA, $V_{CE} = 5$ V, $V_{CC} = 20$ V, $R_E = 1$ kΩ, $R_1 = 61.3$ kΩ, $R_2 = 5.51$ kΩ, $C_\mu = 1.5$ pF, $C_\pi = 18.6$ pF

Chapter 9

1. (a) 0.19968 mW (b) 1.5974 mW (c) 0.19984 mW (d) 1.5973 mW
3. (a) −1 (b) −0.990
5. (a) 10.880 V (b) −10.880 mV (c) 1 V (d) 0.989 V (e) −0.210 mA

7. $-45, \infty, 0$
9. $v_o = -6v_A - 4v_B + 2.75v_C + 5.5v_D$
11. (a) straight line, $v_o = 25v_2 - 5$ (b) 3.2 kΩ (c) 26 kΩ
13. 50 μA, 50 μA, 125 μA, 125 μA
15. (a) 25,000 (b) 15,000 (c) 300,000 (d) 1.41
17. (a) 6033 (b) 1408 (c) 21,480
19. (a) $A > 1748$ (b) $A > 2498$
21. $v_o = 99{,}989(i_{sA} + i_{sB})$
23. (a) 24 kΩ (b) 5 mV (c) 0.6 μV (d) 5.0006 mV (e) $|\Delta v_o| = 1.25$ mV

Chapter 10

1. (a) 1.018 298 V (b) 1.018 294 V (c) 1.018 300 V
3. (a) $R_2 = 717\ \Omega$, $R_3 = 5.7$ kΩ (b) 4 kΩ (c) 13 V
5. $0 < t < \pi/30$, $v_o = 15$ V; $\pi/30 < t < \pi/10$, $v_o = -15$ V; $\pi/10 < t < \pi/6$, $v_o = 15$ V; $\pi/6 < t < \pi/5$, $v_o = -15$ V
7. (a) $0 < t < \pi/3000$, $v_o = 12$ V; $\pi/3000 < t < \pi/600$, $v_o = -12$; $\pi/600 < t < 13\pi/3000$, $v_o = 12$ (b) $-\pi/3000 < t < 7\pi/3000$, $v_o = -12$; $7\pi/3000 < t < 11\pi/3000$, $v_o = 12$ (c) $0 < t < \pi/500$, $v_o = -12$; $\pi/500 < t < \pi/250$, $v_o = 12$ (d) $v_o = \pm 12$ V (constant)
9. Straight line through (−5.0054, −5) and (5.0046, 5) V
11. (a) 0, 20 μV, 20 μV (b) −0.4 V, 0, −0.4 V (c) −0.2 V, 10 μV, −0.19999 V (d) 0.2 V, 10 μV, 0.20001 V
13. (a) −0.15 V (b) −0.15075 V
15. (a) 6, 6, 0, 5 V (b) 6, 5.9970, −0.01499, 4.9975 V (c) 6, 6.0030, 0.01501, 5.0025 V
17. −32 dB per decade.
19. (a) $-8.15\angle{-5.19°}$ (b) 110 rad/s
21. (a) $70.7\angle{-45°}$ (b) $9.05\angle{-5.19°}$ (c) 110 rad/s
23. (a) 46.4 krad/s (b) 1732 rad/s (c) 182.6 krad/s
25. (a) 2 points: $\omega = 20$ (−3.91, 2.63), $\omega = 100$ (1.25, 1.25) (b) Yes (c) −8
27. 2 points: (2×10^4 rad/s, 42.55 dB), (2×10^5, 20.49 dB); Peak: 48.34 dB at 6.44×10^4 rad/s
29. (a) OK for $\omega < 10^6$ (b) 3.17 Mrad/s
31. (a) 1 point: (10^4 rad/s, 19.17 dB) (b) $30 < \omega < 3 \times 10^5$ rad/s
33. Proof

Index